HORSES: Their Breeding,
Care
and Training

OTHER BOOKS BY HEATHER SMITH THOMAS

A Horse in Your Life
Your Horse and You

HORSES: Their Breeding, Care and Training

Heather Smith Thomas

SOUTH BRUNSWICK AND NEW YORK: A. S. BARNES AND COMPANY
LONDON: THOMAS YOSELOFF LTD

A. S. Barnes and Co., Inc.
Cranbury, New Jersey 08512

Thomas Yoseloff Ltd
108 New Bond Street
London W1Y OQX, England

Library of Congress Cataloging in Publication Data

Thomas, Heather Smith, 1944–
Horses: their breeding, care, and training.

Bibliography: p.
1. Horses. 2. Horse breeding. 3. Horse-training.
I. Title.
SF285.3.T49 636.1'08 72-5176
ISBN 0-498-01072-4

PRINTED IN THE UNITED STATES OF AMERICA

To

JERRY RAVNDAL

who was perhaps the greatest horseman
I was ever privileged to know. His tact
and judgment, and above all his patience,
accomplished small miracles with horses.
It is to the ideals of good horse breeding
and good horsemanship which he fostered
within us—the budding young horsemen whom he
inspired—that this book is dedicated.

CONTENTS

PREFACE

My aim in this book is to provide useful information to the serious horseman in a form that is easily readable. My main emphasis is on horse care, or perhaps "intelligent horse-keeping"—trying to cover the breeding, raising, and using of horses, and all the various details that are involved. This is not a book on training or riding (though I've touched on a few points here and there), but rather a book on horse husbandry, which in my terminology is the art of keeping a horse fit and healthy in body and spirit. The keeping of horses involves a great complex of different things, and is certainly an art as well as a science.

This book covers some technical as well as some very untechnical subjects, but I've tried to write it in simple language. And, as in my first two books, I've used my own horses as examples and illustrations, for I think this makes a book more readable.

I owe special thanks to several people who helped to make this book possible. First of all, I should thank the many horsemen who have helped me and given me good advice over the years. This is one of the wonderful facts of being a part of the world of horses—other horsemen are always willing and eager to help you. I want to thank the editor of *The Cattleman* magazine, who gave me permission to reuse parts of my articles that have appeared in that magazine. And I especially want to thank Mr. Kenneth Stark, of American Horsemanship, Inc., who gave me permission to reuse portions of some of the chapters that I wrote for his horse course, *Enjoyment of Horses* in 1971. I want to thank my husband, Lynn, and my sister, Heidi, who helped me with many of the photographs.

Several books and periodicals have been a great help to me and I should list them here as references. Books I have used over the years in my own personal research have been:

Adams, O. R. *Lameness in Horses.* Philadelphia: Lea and Febiger, 1962.

Alston, J. M., and Broom, J. C. *Leptospirosis in Man and Animals.* Edinburgh: E. & S. Livingstone Ltd., 1958.

Blood, D. C., and Henderson, J. A. *Veterinary Medicine* (second edition). Baltimore: The Williams and Wilkins Co., 1963.

Bruner, D. W., and Gillespie, J. H. *Hagan's Infectious Diseases of Domestic Animals* (fifth edition). New York: Cornell University Press, 1966.

Self, Margaret Cabell. *Horsemastership.* South Brunswick and New York: A. S. Barnes & Co., 1952.

Wynmalen, Henry. *Dressage.* South Brunswick and New York: A. S. Barnes & Co., 1952.

Wynmalen, Henry. *Horse Breeding and Stud Management.* South Brunswick and New York: A. S. Barnes & Co., 1950.

I should also mention a few books and articles in passing which provided me with additional facts for this book:

Allen, A. D. "Horse Parasitism." *Horse World* July, 1969.

Eschle, James L. "Biting Flies of Horses." *Horse World* June 1961.

Harper, Frederick. *Top Form Book of Horse Care.* Merck & Co., Inc., 1961.

McChesney, A. E. "Adenoviral Infection in Arabian Foals." *Western Horseman* April, 1972.

Prickett, Milton E. "Utero-Ovarian Diseases." *The Washington Horse* March, 1969.

The Practical Stock Doctor (Farmer's Short Course in Livestock). F. B. Dickerson Company.

And I should give credit to the numerous articles by Robert M. Miller, D.V.M. that I have read and appreciated over the years in *Western Horseman* magazine. I should also list my two earlier books as references *A Horse in Your Life* (1966), and *Your Horse and You* (1970) for I have reused a few photos and several small portions of the subject matter.

And then I must thank my two children—Michael (born in 1968) and Andrea (born in 1970)—who are showing an early interest in horses, an interest that I want to nurture and encourage the very best I can; they are the horsemen of the future.

INTRODUCTION

Throughout the history of mankind, the horse has been a useful tool of man. He has been man's chief means of transportation until only recently, and one of his sources of power. From the earliest shadowy beginnings of man's civilized past, he has used the horse, prized him for his beauty, loved him for his companionship, and valued him for his speed and strength. The horse has been regarded as one of the most valuable and important of all domesticated animals.

The horse has been a central figure in art, sculpture, poetry, literature, and legend practically since time began. His pictures have adorned cave walls, appeared in ancient hieroglyphics and old Egyptian paintings. Far back in human history, where myth and fable mingle with the real, we even find the horse exalted as a semi-human being, the centaur—half horse and half man. The centaurs were evidently a tribe of expert horsemen that ignorance and superstition of the early time elevated into a superior race.

The horse is a symbol of man's achievement. In a very real way, man's success has been measured in horses down through the ages. In ancient times and primitive cultures, a man's or a tribe's wealth was often measured in horses. The peoples who had horses were much more successful in everything from achievements in agriculture to achievements in war than their neighbors who did not. The horse made a population or a tribe more mobile, increasing the possibilities for travel and trade, for attaining food, or for raiding their neighbors. Some of the most terrifying groups in history, including the Huns and the Mongols, were successful because they were on horseback. In war, the mounted fighter usually had great advantages over the foot soldier.

As civilization advanced, the use of the horse became more universal. Man's rise to civilization as we know it today is tied directly to his use of the horse. In our most recent history, before the coming of machines, the horse appeared in every phase of society. He pulled the plow and the harrow and the loaded wagons to market, and multiplied food production many times beyond what human strength could have done alone. The horse has been the steady, dependable slave of man, pulling heavy wagons, mine cars, stage coach, buggy, or harnessed to the more elegant and fashionable vehicles of transportation. He performed a thousand services that were indispensable to human comfort and advancement.

Many of the peoples whose whole livelihood depended upon the horse (including the desert Bedouin and the western American Indian after he acquired horses) had a very high regard for their animals and were excellent horsemen. They understood the horse, and this understanding was put to use in conquering him and winning his obedience. The same was not always true for the users of horses in Europe and America during the past few centuries. When a person reads old books on the subject of horses and horsemanship he is amazed by the high percentage of problem horses and various undesirable vices such as kicking, biting, striking, rearing, and running away. Almost every horse was a "spoiled horse" in one form or another. One reason for this was the fact that not everyone who owned horses was suited for working with horses. The horse was a necessity, like an automobile, and people owned horses whether they understood, appreciated, enjoyed them, or not. One book, written in the 1880s, made this statement: "Of the numerous faults ascribed to the horse, a very small portion are chargeable to his natural disposition, the remainder being the direct result of vicious training, or rather the absence of training and the substitution of something, which, under that name, first produces and then fosters the faults for which the animal is punished; while often the punishment is ineffectual, because the animal has no conception of why it is made to suffer."

After the Industrial Revolution the number of horses declined because they were no longer needed for transportation or on the farms. But in recent years we have experienced a tremendous upsurge in horse numbers again, a regular "boom" in the horse business, for horses are now being used for sport and pleasure. And the horse of today has one big advantage over his counterpart of the past. In yesteryear nearly every man owned a horse as a necessity, just as nearly everyone today owns a car. But now the horse is no longer needed as a means of transportation and the people who own him are the people who enjoy and appreciate him and own him because they want to. Today in our country, perhaps as no other time in history, the horse is owned by people who appreciate him, seek to understand him, and who will do their best to take good care of him. There are still problems with horses, of course, but most of these can be solved with a little time and patience and care, and seeking to understand the horse and why he reacts the way he does to what we do with him.

Now, the biggest curse in the horse world is perhaps a product of our machine age thinking—our wanting to accomplish everything overnight, whether it be building a highway or training a horse, traveling around the world or learning to ride. Our atomic age and our many wonderful modern conveniences and time-saving devices have spoiled us. Too many people want short cuts in everything they do. But there is still no shortcut to becoming a good horseman—no substitute for lots of time, patience, and the desire to learn all that one can, even if it takes a lifetime.

Dedication is what makes the horseman and separates the men from the boys, the horsemen from the dudes, the true enjoyers of horses from the "week-end cowboys." For only the person who seeks to *understand* the horse can truly enjoy him. The casual observer may admire the horse from a distance, but he is missing out on a great deal. The person who lives with horses, works with horses, sincerely tries to better his horsemanship and his understanding, who continually racks his brains over his horses, their problems and their idiosyncrasies, that person is going to get the most satisfaction from his horses.

Enjoyment of horses is tied directly to accomplishment and personal identification with them. Look at the child who has just won his first blue ribbon at a horse show. Or the old cowboy who has spent years in the saddle and whose means of getting a good job done depended on a good horse and upon his teamwork with that good horse. Or the farmer of yesteryear who worked his land with a good team, a team he was proud of for their strength and willingness and heart, their ability to work together and do any kind of job. Or the successful calf roper, or the competitors in a difficult dressage class. Look at the breeder who is finally succeeding in raising the kind of horses he has been shooting for for many years. They all have something in common. They are all proud of, and enjoy, their horses because they have worked with them for many hours and have accomplished something with them.

The horse is an athlete and part of man's admiration for him is due to his speed, strength, agility and gracefulness. There is life and beauty and power in the movements of a good horse, and something very appealing to the eye in the way a good horse is put together. But there is more to enjoyment of horses than admiring one on canvas or in stone or watching one gallop across a pasture. The key to real enjoyment of horses is *doing*, whether it is feeding them, grooming them, breeding them, showing them, punching cows with them, hunting, racing, jumping, or riding for pleasure. The ultimate enjoyment of horses comes from achievement, from making a horse good, or a good horse even better. It comes from having done something of value—having a good horse and a personal ability that one can be proud of. It gives a person a sense of accomplishment and personal worth. We all need a sense of identity, of belonging, a cause or a realm in which we feel at home and with which we can identify, and for those of us who enjoy horses, working with horses and accomplishing something with them gives us this.

Understanding, experience, achievement. These are some of the necessary ingredients that make up the good horseman. Another ingredient is a "feel" for horses—a feel for proper care, for what a certain horse needs at a certain time. Part of enjoying a horse is caring for him as best you know how, and being able to tell when he is fit and healthy and when he isn't. And if he isn't, being able to tell what he needs, or whether or not he needs professional veterinary attention. You want a horse you can be proud of, and that kind of horse needs proper care—the right amount of the right kind of feed, enough of the right kind of exercise to keep him fit and in condition for the kind of work he is to do or the use to which he will be put.

In other words, the good horseman who truly enjoys his horses has a sort of feel for his horses, knowing what they need and when they need it, and cares for his horses as regularly and responsibly as possible—as he would for his own child or anything else he loves. The dedicated horseman is somewhat of a perfectionist as far as his horses are concerned, for he will do anything and everything in his power to take proper care of his animals, and to learn exactly what proper care entails. There are many fads and gimmicks in the horse world today and it is up to the individual horseman to learn how to sort and choose among them, to know what is valid or essential and what is superfluous or harmful in caring for his horse, whether in feeding him, fitting him, training him or riding him.

Horses are individuals and each one is different; the horseman has to know and understand each horse to know what is best for each in a given situation. There is no substitute for the "eye of the master," whether it is in feeding, doctoring, or training a horse. The man who understands his horse will give him the proper amount of feed, the proper amount of work and rest, the proper amount of reward and the proper amount of discipline. He knows what the horse can take and what he can't; he knows where the breaking point is and doesn't ask the horse to go beyond it. He knows how much a certain horse can take without becoming bored or soured or upset, whether in practicing barrel runs, schooling him to jump, work cattle, change leads, or open a gate. Too much practice or nagging at a horse is usually worse than too little.

Horses are good for people. There is an old saying that the outside of a horse is good for the inside of a man, and it is very true. There's something about a horse that is good for the soul. The farmer of ages past, the man who depended on good teams of horses to get his farming done, knew it. Machinery may have sped up production but it didn't do much for man's peace of mind. The pace of a horse is just the right speed for contemplation and reflection, for getting a little bit better in tune with oneself and the world around one. The speed and noise of a machine hardly gives a person time or opportunity to think. The farmer, trudging along behind or beside his horse, or driving his wagon loads, was a free man. The horse did the work and knew where to go and could function almost on "automatic pilot" if need be, while his driver's imagination could soar.

There is something soothing and peaceful in the rhythmic gaits of the horse. As our modern life gets more hectic, as our schedules become more complex, we seek means of relief, for we all need moments of peace just to get back into proper focus again. And what better way is there to find this tranquility than on the back of a favorite horse, riding for pleasure along a woodland trail or practicing figure eights in the horse pasture. For the nature lover who enjoys the beauty of out-of-the-way places, the horse is a good means of getting you there and back again. Scenery and wildlife seen from horseback are thoroughly enjoyed. Motor travel is usually too fast and too noisy. The speed of the horseback traveler is just right for enjoying nature's handiwork or for enjoying the peace and quiet from the hustle and bustle of man. For most of us, the change of pace is welcome and healthful.

And, speaking of health, horseback riding is good exercise. It is not so strenuous that it limits its participants to athletes, as some sports do; horseback riding can be accomplished by the very young and the very old and even the handicapped. Yet it is valuable exercise. Like swimming, riding properly uses *all* of your muscles. It is a good way to keep physically fit without expending tremendous exertion. It is especially good for the waistline. And if you have ever ridden much at a good trot, you'll know that trotting can build up your wind as well as that of your horse.

Today, enjoyment of horses can take many forms; there are practically limitless opportunities for adventure with horses. Down through the ages the horse has been man's faithful friend and companion, and men have esteemed and enjoyed the horse for all that he provided and for all that he meant to them. But in one sense there are even more opportunities today to find enjoyment with horses because today there are so many things a person can do with a horse. Many of the shows and competitions for horses that we now have were not even thought of a century ago. And riding is often a leisure-time activity now; people have more leisure time than ever before, so they have more time to ride just for enjoyment.

The world of horses is as varied and colorful as it ever has been. There are horseback sports to suit every horseman's fancy. The age we live in has its problems, but perhaps horses and horsemen can be thankful that they live in this time and not in another. Scientific advances have conquered many of the horse's health problems. And today most horses are owned by people who enjoy them and care about them. Horses as a whole in this country are better off than they ever have been for they are enjoyed. And the horseman who truly enjoys his mount will care for him properly, feeding him, and training him the very best he knows how.

The horse is a fascinating and complex animal. Perhaps this is one reason that he attracts us, one reason that so many of us are bitten by the horse "bug" and get "horsitis." The horse is a challenge; there is much to learn about him. The serious student of horsemanship is fortunate in that there are many books on the market today which can aid him in his quest for knowledge about the horse. This book is only one of many; it cannot possibly hope to cover all aspects of horses and horsemanship, for no one person can know and understand all there is to be known and understood. This particular book will put its main emphasis on horse care with the serious horseman in mind, touching on the raising of horses, and their health and soundness. It is hoped that this book might fill a gap between the often too-simple books written for the novice, and the all-too-technical book written as a veterinary textbook. It will touch on some aspects of training and conditioning, but will leave much of the basics of simple training untouched, for the serious student of horsemanship will have already learned these elsewhere. This particular book does not claim to be a complete text for the advanced horseman, but it is my hope that it will tend to cover some of the gaps and fill in a few more corners.

HORSES: Their Breeding, Care and Training

1

RANCHING WITH HORSES

CATTLE AND HORSES HAVE ALWAYS BEEN LINKED TOGETHER HERE in the American West, ever since the days of the first *vaquero.* The Spaniards brought cattle to the southwest more than 300 years ago, and from that time on horses have been a necessity to the cattleman for moving, sorting, branding and all the other chores that go with taking care of a herd of cattle.

The horse has long been a companion and tool of man; he has been man's chief means of transportation for centuries. But now in the last few decades most horses have been replaced by machines—so completely replaced on the highways and on the farms that many people today consider him useful only as a pleasure animal, for leisure time riding, and for sport (racing, showing, competitive events, and so on).

Yet here in the western United States there is still one field in which he holds his own and proves to be as indispensable as ever—working cattle. In steep and rough country the cow horse is still the best worker for the job, and I bet it will be an awfully long time before anyone invents a machine that can do the job as well, even though some people are trying to use jeeps, hondas, and airplanes.

In some of the flatter areas of the country, farmers and livestock men run a few cattle with a minimum of horses, and some perhaps with no horses. But how frustrating it must be to try to work the cattle with no horses. Horses are so much more efficient than men on foot, and herding cattle with motor vehicles is the height of hilarity and ridiculousness. My first experience with this type of "cowboying" occurred the first summer Lynn and I were married. We were living on a dairy farm in southern Idaho, trying to get a start and get some cattle paid for so we could plunge into what we really wanted to do—to start buying a ranch and raise beef cattle. Our dairy farm was small and flat, sandwiched in among other similar farms, and we had no place to keep horses and no room to ride them in if we did. What little pasture we had was for our dairy heifers. I had a young filly with me, confined to a corral, but my two mares were home on Dad's ranch, waiting for the time we could move to a larger place.

So I'll never forget the time the heifers got out. Some of them wandered over a plowed field, and some of them had gone down the highway by the time we discovered they were out. My husband leaped on his diesel tractor and proceeded to try to round them up. It was frustrating, indeed, to try and maneuver that roaring, smoke-belching, 40-horsepower beast across a rough plowed field after a bunch of sassy, agile Holstein heifers, when a horse could have done the job so easily and quickly. Those heifers practically ran circles around those "forty horses." A couple of hours, lots of black smoke, and many cuss-words later, he finally ran them all down. It would have been funny, if it hadn't been so sad!

Lynn and I were both raised on cattle ranches, and that one summer on the dairy farm was enough for us. As soon as we were able, we "headed for the hills" and went back to the land of our childhood—the mountainous country of eastern Idaho. Our present ranch is near Salmon, and though it is just a small place (900 acres) it is a very beautiful and satisfying one. It is a rough place, strung out over two miles of creek bottom, with an adjacent half-section and quarter-section of dry-land grass. And now our horses are indispensble; we couldn't begin to run this ranch without them.

The summer range (6000 acres) for our cattle is right out our back door and it is steep and rugged country. Some parts of it we can reach by jeep, but other places are most easily accessible on horseback. We pack out most of our salt and maintain our water troughs and fences with horses. We often take materials part way in the jeep, but to many of the areas we must go the rest of the way on horseback.

When we built the present fence around our half-section in the mountains, there were places we couldn't get to in the jeep; we ended up packing rolls of barbed wire on a horse, dragging bundles of posts with a lariat dallied around the saddle horn, and stringing the wire from horseback. To illustrate just how steep some of this country is, I should mention that when my younger brother and I unloaded one of the rolls of barbed wire from the horse we lost it; it got away from us on the steep hillside and rolled and bounced about a quarter of a mile down the slope. We had to take the horse back down to get it.

This land is steep and dry (our average yearly rainfall is only 10 inches) yet it grows some of the best native grasses in the country, including western blue bunch wheat grass, Idaho fescue, and Sanborg's blue grass. These native bunch grasses are very palatable; the horses and cattle like them and do very well on them. Our home ranch in the creek bottom (on Withington Creek) is rough; there is a lot more marginal land than cultivated land. But this marginal land (the steep hillsides surrounding the meadows) provides much of our fall pasture.

Let's look at the horse end of our operation. We've had many different horses over the years, but our present work string consists of four mares—Khamette, Nikki, Bambi, and Fahleen—a young gelding, Ahmahl, that my younger sister Heidi is training, and Surrabu, Heidi's Arabian stallion. We also have an old mare, Nellace, who is now in "retirement."

Nell is a grand old mare of 20 who had still put in a good day's work in the

Views of parts of our cattle range

Nikki

Heidi and Khamette

hills until a year ago when some old injuries made her front feet unsound for hard work. She is 15/16ths Thoroughbred and has had four foals for us; she is the dam of three of our present horses—Nikki, Fahleen, and Ahmahl.

Khamette is a 14-year-old half-Arabian mare. Her dam, Scrappy, was a grade Thoroughbred mare that Lynn raised when he was a boy. I raised Khamette as a 4-H project when I was in high school. She has appeared as an example and illustration in many of my articles and was the subject for many of the photos that illustrated my first two books, *A Horse in Your Life*, and *Your Horse and You*. She is a willing traveler with lots of stamina, and a real pleasure to ride all day in rough country. My sister, Heidi, rode her on a competitive trail ride in 1971 and did very well. Khamette is one of our better cow horses. Yet she is gentle and patient enough that anyone can ride her. She is our "guest horse" and "kid horse" when she's not busy out on the range.

Nikki (whose registered name is Nokhomis) is an 11-year-old half-Arabian mare out of Nell, by the Arabian stallion El Khamis. Nikki is a small horse (14.3 hands) with a big heart and is the best cow horse we have. She is very willing and agile and will go anywhere to head a cow—over logs and brush, through a bog, down a rocky mountainside. She's sure-footed over rocks and sagebrush and steep terrain even at breakneck speed. She's never stumbled or fallen down with me, and we've done some crazy things; she's done more than any other horse to make me grateful. And, she can travel all day in the mountains or work cattle all day in rough country. This ability to travel long hours over steep terrain is one of the

Heather on Nikki, and our dog Pepper, moving cattle up over a ridge

Heidi and Khamette checking cattle on summer pasture

qualities we look for in our horses because we have many rough miles to travel and we need a horse that still has plenty of zip to head a cow even after miles of steep country.

Bambi is Lynn's 11-year-old registered Quarter Horse mare. She is speedy at short distances and very gentle. We're hoping to raise several half-Arabian foals from her. Fahleen is a six-year-old half-Arabian mare out of Nell; Fahleen is a full sister to Nikki. This young mare is learning to work cattle (I didn't start riding her steadily until she was five) and she is showing a lot of promise. She's a better traveler than Nikki and may make a good competitive trail horse. Ahmahl is a full brother to Fahleen and Nikki. He belongs to my younger sister, Heidi, who spends the summers with us helping ride range and look after the cattle while Lynn is busy haying. Lynn seldom gets time to ride during the summer months, so Heidi and I take care of the range riding.

Surrabu is a seven-year-old Arabian stallion owned by Heidi. He has a sweet disposition and wonderful manners. Heidi has had him for two years now and he is becoming a good cow horse. She also uses him for competitive trail riding. During the school year, while she is in Boise, Idaho, she rides him English and stands him at stud. She shows him English and Western. Heidi also has two young purebred mares and is raising some foals.

Last year (1972) we had plenty of "stallion power" on the ranch, for we kept another Arabian stallion, Sur Amir, who was raised by Jerry and Velma Ravndal. We bred three of our mares to him and are hoping for some nice foals in 1973. We also boarded a few horses, including a small herd of purebred

Arabian mares and foals owned by Jim Hopkins, of Twentynine Palms, California.

You may think we have a lot of ranch horses for an operation as small as ours, with so few riders. It's true that we do own horses partly just because we like them and enjoy them, but on our strung-out ranch they also do a lot of work; they more than pay their way. Most summers we go enough miles and have enough work to keep four or five horses in good shape. Heidi rides Surrabu, Ahmahl and Khamette and I ride Fahleen and Nikki, and we use Bambi as an alternate. Lynn rides either Bambi or Khamette whenever he gets a chance to ride or help us move cattle.

A competitive trail ride goes through our cattle range every summer (the 35-mile Salmon River Competitive Trail Ride). Three years ago Heidi was riding Khamette in that area of the range, checking cattle, and joined the trail riders at their noon stop. One of the ride officials (Jerry Ravndal) took one look at the mare she was riding and asked why she hadn't entered the ride; Khamette was as hard and fit as many of the trail ride horses. So two summers ago (1971) Heidi entered Khamette in the ride—after conditioning her for only a month, riding bareback no less, in an effort to heal up an old saddle sore. The bareback riding paid off; the sore was healed sufficiently by trail ride time to use the mare in the ride (horses must be ridden with saddles in the ride). She conditioned the mare just by riding range, checking and moving cattle. It was no easy task, at first, chasing cattle bareback up and down the steep mountainsides, but it made a better rider of her. For this first attempt at a

Heidi and Khamette at the noon stop on the 1971 Salmon River Competitive Trail Ride. Khamette's temperature, pulse, and respiration rates are being taken

Heidi and Khamette starting up a ridge at the half-way point along the trail

trail ride, Khamette did quite well, placing second in the junior division, and winning the trophies for best Lemhi County horse and best Idaho horse.

Some people may wonder why most of our horses are of Arabian and Thoroughbred breeding; many people feel that cattle ranching and Quarter Horses go together. The answer here is stamina. We ride our horses very hard and we like the natural toughness and endurance of the little Arab. For us, the Arab-Thoroughbred cross has worked out very well. There are good horses in every breed, and the Quarter Horse has certainly made a name for himself as a stock horse. On many ranches he is perhaps the best horse for the job. He is quick—good for working cattle and unexcelled for roping because he is stout. But for our particular ranching operation, where we have so many hard miles to travel, we prefer some Arabian blood—for the same reasons that you'll find when looking at the type of horse that wins endurance rides and competitive trail rides; the highest percentage of winners are Arabians or part-Arabians. I don't think there is really any question here; it is just a fact that the Arabian is usually the best endurance animal, and toughest for his size. And for working cattle in rough terrain, a small horse is usually more agile and sure-footed than a large one. Bambi is the type of Quarter Horse we like—fairly small and trim and not muscle bound. We've found that those big "strong" horses with lots of extra muscle to pack around don't last as well all day in the mountains as the wiry little Anglo-Arabs. Some people may say we're prejudiced, but over the years we've had both kinds. And it's interesting to note that when any of the three of us has to make a long, tough ride, we always choose one of the part-Arabians.

Foreground—Lynn on Bambi, sorting cattle on winter pasture

This isn't meant to be a discussion of the merits of various breeds. I just want to say that for our particular circumstances and situation, the part-Arabian has worked out better than the Quarter Horse. Someone else's situation may be entirely different.

Some ranchers turn their cattle out on summer range and practically forget about them, but in our small herd (180 cows) every cow and calf counts and we like to check on them often, keeping them moved to good feed and water, checking for sick ones, and so forth. Sometimes we need to bring home a sick one to doctor for footrot, pink-eye, lump jaw, etc. Checking the cattle frequently pays off in many ways. Sickness in an animal can be detected in time to do something about it—before it gets really bad or kills the animal. We can keep the bulls properly distributed among the cows, and we can get breeding dates (and therefore calving dates on the cows for next spring). We can keep a close check on the grass and move the cows to higher range before the lower grass is all gone, before the cows get restless or start to lose weight. We also keep track of the springs and water troughs; occasionally a trough or pipeline will need fixing, and we can unplug a trough, clean out a spring box or pipe, or repair leaks, and we can do it as soon as the problem develops—before the cattle have to go without water. And we can keep the cattle out of some of the areas where larkspur is bad, keeping them out until this poisonous weed is past its bloom stage and no longer such a deadly threat.

We also use our horses to pack salt to the cattle and to check and repair fences. In the winter we don't ride as much and we only keep one or two

Heidi and Nell moving cattle on the range

horses down at the main ranch for checking and moving cattle on winter pasture. Occasionally we use a horse to pack pellets to the cattle when they are in areas that are too steep or slick for the jeeps to get to.

During most of the year we use our horses often here on our home ranch (as well as on the range during the summer), moving cattle from pasture to pasture and checking through them. We raise Hereford cattle and a few crossbreds (we have quite a few Angus cows which we breed to Hereford bulls), and all our cows are individually numbered with brisket tags; we keep extensive records on every cow—breeding dates, calving dates, medical records, data on each calf, etc.—so it is essential that we check them often. And this checking is best done on horseback.

There are just too many miles to cover to try to check the cattle on foot, or even in a jeep. The fact many people don't realize is that a horse travels much faster in steep and rough country than a jeep can. A horse can go over and through obstacles and rough places that a jeep has to go around, and a horse can trot through terrain where a jeep has to maneuver slowly and cautiously.

Trying to check cows on foot has several disadvantages. There is always the danger of snakebite in the summertime (each summer we kill from 10 to 20 "rattlers"). But sometimes even more dangerous than snakes to a person on foot are aggressive mama cows. Those same cows always respect a person on horseback. And, when you're mounted on a horse you are taller and can see a lot more territory, especially in brushy areas.

Heidi and Khamette sorting cattle

Fahleen in training as a five year old

Out on the range it is very impractical, verging on the impossible, to try to check cattle from a jeep, for the jeep just can't go to many of the places that the cattle do. And then, if you did happen to find a cow that should be brought home, how would you ever get her home? If you have ever in your life tried to cut out a cow or round up a calf with a jeep, you know that you just can't beat a horse. Hondas fall into the same category as jeeps in our opinion, except that they are even less useful because they can't carry as much. A jeep is handy for lots of things on our ranch; we use it rather like a "super" pack horse, though it is a clumsy one. But when it comes to handling livestock, we rely on our horses.

We used to have a neighbor who occasionally tried to herd cattle on his Honda. It worked all right when he was just following them down the road, but his "cow horse" became somewhat limited as soon as the cows got off the beaten path. One time he tried to get in front of a herd to head them off in a certain direction, and after bouncing bravely over the sagebrush and rocks he got near the crucial point at the front of the herd, and his motor beast sputtered and died. Never yet have I seen a *horse* kill its engine.

Horses may be expensive to keep and to feed, but on a ranch a horse is still the best (AND CHEAPEST!) for the job. A horse may eat hay, but a jeep burns a lot of gas. Next time some motor cowboy tells you how expensive it is to keep a horse, remind him that a jeep gets very low mileage. When a jeep is in four-wheel drive and traveling over rough country, it will do well to get four miles to the gallon. At the price of gas, this can add up rather fast. The horse has several advantages over the machine. How many horses ever need new parts or a major overhaul? The horse never needs a new battery, points, or plugs. Horseshoes cost a lot less than tires. And many horses can produce a new model when you need one!

The initial investment is usually greater for a jeep than for a horse. A jeep will cost anywhere from just a few hundred dollars for an old, beat-up model to $4500 for a new one. A good horse might cost $300 to buy, give or take a little depending on the circumstances. It will probably cost you as much or more to raise a horse than to buy one when you consider breeding fee, pasture, hay and grain, plus medicines and vaccine, to say nothing of equipment you might need to buy.

But even though it might cost as much or more to raise a horse instead of buying one, there are many advantages. We prefer to raise our own for use here on our ranch, because when we raise our own we know what we are getting, and when we train them ourselves they are accustomed to us and our handling, and are adapted naturally to our conditions (feed, climate, terrain, etc.) and the way we'll be working them. There's no retraining or unspoiling to do. And a horse that grows up with you and trusts and respects you is truly "your" horse and has a little more chance of becoming a top cow horse for you than a horse that has had several trainers and riders.

One more thing about our young horses—we halter break and gentle them as foals, do some ground work with them as long-yearlings or two year olds

Heather and Nikki working cattle on the range

(longeing, driving, and so on) but we don't get on them until they are nearly three and we don't make long rides on the range or work cattle with them until they are four. Even then we lay off on the long, hard rides (working cattle all day in steep country, etc.) until they are five. We think very highly of our horses because they do a lot of hard work for us, and when we have a good one we want him to last 20 or 30 years; we don't want to burn him out or make him unsound by too much strenuous riding or cattle work when he is young and immature.

On a ranch like ours, a person might say at first glance that since the cattle come first, the horses are secondary—just a hobby. It is true that on this ranch the cattle are our cash crop, our source of income. But though the cattle are the main thing, taking care of them properly determines the outcome of that "cash crop"; over the years it determines whether or not we stay in businesss. In other words, we can't afford *not* to take care of the cattle as very best we can. And in our particular operation, we can't even begin to take proper care of them without horses. Horses on our ranch are indispensable and here to stay. And if we can breed and train good ones that can do the job well and pleasurably—horses we are proud to own—so much the better.

Even on a ranch more flat and well laid out than ours, horses have their place. The only kind of ranch I can think of that might get by without horses is a flat ranch with excellent fences, very small pastures, with corrals and cattle-handling facilities easily accessible from every pasture. It would be a tired cowboy's dream, but the cost of such a layout would be such that only a rich man could afford it.

Heidi and Ahmahl

Heidi and Surrabu setting out to ride range and check cattle

Not many of us have a perfect setup for working and handling cattle. Even though we improve our pastures and fences and corral set-up year after year, we'll still always need our horses, especially on a ranch as strung-out and rough as ours. Besides, we who enjoy a good horse and find working cattle a challenge and a lot of fun instead of work, would really rather hate to see that "ideal" set-up anyway.

Another point that might be mentioned in this defense (horse versus machine as "ranch hand") is the fact that a horse can give you friendship. What kind of jeep or Honda can be good company on a long hot day and share your experiences in the mountains working cattle? The hours can at times seem long when you are on the range alone all day; people would think me strange if I talked to a machine, but it's all right to talk to my horse. And what kind of machine can give you its all on a hard cattle drive, carry you safely home after dark, appreciate the grain and the final pat of farewell when you get home, and give you a friendly nicker in return?

Horses are not only neccessary to us in our business. They also provide part of the reason we enjoy our business so very much. They are a lively, sparkling ingredient that helps to make ranch work a joy rather than a chore. And because we think so highly of our horses, both as necessary tools for the job and also as interesting and sensitive individuals in their own right, we try our best to take excellent care of them. We want them to stay healthy and sound for a long life of usefulness and we want them to enjoy their work. A horse that enjoys his work is much more pleasant to ride than one who is tired, sour or resentful. Therefore, we must also strive to be good horsemen as well as good ranchers, trying to be tactful and patient, understanding and respectful, learning all we can about good horsemanship and proper horse care. One thing we have found about horses—there is always more to learn. And as we learn, we try to pass on to others what we have found, or what worked well for us.

That, really, is what this book is all about; it is our attempt to share some knowledge and experience with other horsemen who are eager to learn. This book is written for the person who enjoys working with horses and wants to learn all he can about them, the person who owns horses and perhaps raises a few himself. The first part of this book is geared most specifically to the horse breeder, but I think it will also be of interest to anyone who ever raises a horse, whether he raises many or just one from a favorite mare for his own enjoyment. The second part of the book deals more generally with horse care and problems that any horseman may face, and covers facts of horsemanship that may be of interest to the serious horseman. It is my hope that this book may be of help somewhere along the way. But remember that experience is the best teacher. Books can be a great help in many areas, but they are really no substitute for actually getting out there and doing it.

In many of the following chapters, our own horses will be used as illustrations and examples. I think that this helps to make a book more lively and less like a textbook. Many books on horse care and horsemanship are written

Heather and Nikki and the dog Pepper on a ridge at the upper end of the "320" (half section)

either too simply (geared for the novice) or too technically (written for the veterinary profession, and therefore most difficult for anyone else to decipher unless he has a medical dictionary). It is my hope that this book will be somewhere in between, easy to read, but containing worthwhile and "meaty" information for the horseman who is seriously interested in learning.

2

SELECTING A HORSE

WHEN YOU SELECT A HORSE YOU MUST FIRST DECIDE WHAT TYPE of horse you want. What the horse will be used for will make a difference in your decision. There may be a vast difference in the horse chosen for use as a child's first horse, a competitive calf roping horse, a potential show horse for driving in harness, a race horse, and a broodmare prospect. What you plan to use the horse for will narrow your field of selection somewhat as you consider types and ages of horses, breeding, and the training or handling they have had.

There are good horses and poor horses in every breed; each breed has its outstanding and versatile individuals. If you have a favorite breed, you will of course choose your horse from within its ranks, especially if you are breeding horses. If partiality for a certain breed does not enter the picture, you will probably be looking for a horse of the breed or type which as a general rule is best suited for the things you want to do with your horse. If you are looking for a horse for your child, you might choose a pony, or a dependable grade horse, or a gentle horse with the proper temperament within the ranks of any major breed. If you are interested in competitive timed events such as team roping, calf roping, barrel racing or games on horseback, you would probably choose a Quarter Horse. If you are looking for a horse with lots of stamina—for endurance riding, trail riding, or many miles of ranch riding—you might choose an Arabian. If you are looking for a hunter or a jumping prospect, you might choose a Thoroughbred or part-Thoroughbred. There are good horses in every breed, and if you look long enough within a certain breed for the kind of horse you want (even if you want a Quarter Horse jumper, an Arabian barrel horse or an American Saddler for endurance riding) you'll eventually find him.

There are four main points to consider when buying a riding horse: disposition, training, and more important, conformation and soundness. When buying a breeding animal there are a few other considerations as well, such as bloodlines, but we'll touch on these later in the chapter.

First of all you'll have to decide what kind of horse you want and what kind of price you want to pay. You'll have to decide whether you want a gelding, a stallion, a mare, or a young horse, and whether you want a purebred or a grade. There are many good grade horses, that in most cases are less expensive than purebreds. A good grade horse can be quite satisfactory if you plan to use the horse for pleasure riding, ranch work, trail riding, and most general purposes. But if you plan to eventually breed the horse or do a lot of showing, you'll probably want a purebred.

If you are looking for a horse for pleasure riding, or a horse for someone to learn to ride upon, you'll want to choose a horse that has had some good general training, and not a horse that has been trained and used specifically for some special event (such as barrel racing, roping, other timed events). A horse that has only done one thing for most of his life, especially if it has been a speed-timed event, can give the beginner or the pleasure rider many problems in ordinary use. Even though the horse may have been a champion in his particular field, he may be worthless for ordinary riding unless he can be retrained to the point where he can forget about competition and be able to settle down and relax in the presence of other horses. Sometimes a "jazzed up" game horse can be retrained, especially if he hasn't been spoiled by improper handling, but it is usually better just to start fresh and select a horse that hasn't had one-sided training, so that you can develop him in the directions you wish him to go, with not so much unspoiling or retraining to do.

To get a horse that is steady and dependable, yet full of life, you must find a horse that combines a good disposition with good training. Disposition and training are very closely tied together in determining whether or not the horse will make a good, safe mount. The horse may be friendly and have the nicest disposition, but if he is untrained or improperly trained he may still be dangerous for the novice. On the other hand, all the training in the world won't make a good beginner's horse out of one that is naturally cranky and resentful. You'll want to find a horse that has been trained and handled properly, and that has a good disposition and basic trust and confidence in his handlers. When it comes down to choosing between two young or inexperienced horses that are equally good-natured and nice as far as disposition goes, I would always prefer to take the one that had not been handled much, if what little handling he'd had was not detrimental. It's always easier to start out fresh with a horse if you plan to train him yourself; there's less retraining to do and and less wondering about what happened to the horse in the past to make him react the way he does. I hate to follow someone else's mistakes. It can take years to undo some of the things that improper handling does to a horse.

Disposition is very important and should never be neglected when choosing a horse. When choosing a horse for his athletic ability (as a barrel prospect, cow horse, endurance horse, and so on) people tend to look at his conformation, speed, etc., and forget about disposition. But a horse's basic nature can make that little bit of difference; it can determine whether that horse will be a champion or not. If a horse respects you and wants to work for you, he'll put a lot more heart into whatever he's doing. Some horses are too easy going to ever get very far; they couldn't care less. They do what is asked of them but they never put any fire into it. Other horses are too resentful and cranky; you don't make a good team when you are always fighting over something. Horses respond differently to different riders. Sometimes one person can get more from a horse than another person can, even though they might both be good horsemen. Teamwork is what counts, and you need to choose a horse that you think will be compatible with your way of training and riding. A strong-willed horse is not necessarily one to be avoided, for with the proper kind of tact and handling, that strong will can be bent to do your bidding and you will find you have a horse with a lot of guts and heart. But a strong-willed horse and a strong-willed untactful rider will get nowhere, for if the horse must be "broken" before it will do the rider's bidding, it will be worthless, or even vicious, resenting the rider or reacting only out of fear.

Some horses are timid, some are bold, some are standoffish and independent, some are sweetly friendly, others are aggressively friendly. Some horses are just naturally crankier than others. You can usually tell in your first meeting with a horse whether or not he is friendly. When you approach the horse, watch his ears. They will show whether he is alert and inquisitive, attentive to you, drowsy, or indifferent. If he puts his ears back or plasters them back against his neck, it's usually a sign that he is cranky and doesn't want to be bothered. Some horses put their ears back and don't mean anything by it, but most horses mean business and generally follow this warning with a bite or a kick. Be alert and careful with a strange horse that puts its ears back.

Every horse is different. Some are eager and friendly, others are more withdrawn, while some are downright standoffish and a few are actually ill-tempered and mean. How a horse has been treated makes some difference in his attitude toward people, helping to determine whether he is trusting or fearful, respectful or spoiled, or aggressively resentful. Sometimes if a horse has a basically good disposition to start with, damage from improper treatment can be undone, but it usually takes much time and effort to overcome the bad memories. Some things can never be fully erased.

Some types of horse personality are desirable over others. I would sooner choose a timid, fearful horse than a spoiled one with no respect. The timid horse will overcome his fears with time and patience and kindness, learning that you will not hurt him. The only danger here is that the horse has had such a bad experience with some phase of his experience with people (call it "training" if you will) that he retains a lifelong phobia about it. For instance, a horse might continually panic at noises behind him, or, he might fear a whip so much that he startles at any extra movement, such as the wave of your arm, the swinging of a rope, or the waving of your reins or hat. The spoiled horse, on the other hand, is potentially dangerous because he has absolutely no fear or respect. He may be headstrong, or playful to the point of being obnoxious, or so stubborn that he only wants to do his wishes and not yours. In many ways the spoiled horse is more difficult to work with and harder to train than the timid horse; it takes a lot of work to undo the spoiling, and to make the horse respect you.

The friendly, easygoing, eager-to-please horse is a joy and a jewel, and a pleasure to work with. The more reserved, standoffish horse may be potentially a great animal, but he needs more work, more time, more tender loving care and understanding. It's hard to win the confidence of some of these independent horses, but once you do you are truly a team. Some of the very best horses I've owned were really not outwardly friendly. They tolerated me and that was about all, but as time went on and we worked many hours and many miles together, these horses blossomed for me and did a great job. Most of them were sensitive horses, and had a lot of heart once they decided to work with me and for me instead of against me. So don't overlook or underestimate the indifferent horse if in all other respects he looks like a good one. He may be just reserving judgment on you until he gets to know you better.

Be wary of the cranky horse, the resentful one that offers to bite or kick you. Some horses, either naturally or through mistreatment, have bad dispositions and can cause you a lot of problems or even hurt you. I wouldn't want one of this kind even if he was the most outstanding looking animal in the country. A person can form a kind of truce with a mean horse and get along with him if he is careful and knows how to handle him, but there is always the danger of the horse hurting someone else. It's not worth taking a chance on. There are many good horses to choose from, so I feel that a bad disposition is justification for being disqualified from the field of selection.

One last point about disposition. When judging a horse's disposition, be sure you know what motivated the horse's actions. If the horse is startled, he may kick. Don't chalk him up as being mean or bad tempered if he is merely exhibiting a natural reflex action because he was frightened. Another example of what I'm trying to say: a novice might throw a saddle on a horse (if the stirrup bangs the horse's elbow on the off side the horse may jump) and then cinch up tightly and swiftly, pinching the horse's skin. If the pained animal reaches around and nips at his tormenter, it's only in self-defense and that novice should not blame the horse or judge him as a nasty-tempered beast. If a horse is handled in a calm, intelligent manner, his true disposition should soon be readily apparent.

Be observant while the owner is getting the horse ready for your inspection. Is the horse easy to catch? Does he bridle easily? Does he stand quietly for saddling? Will he stand still while you mount him? Will he walk away from the barn or from other horses without hesitation? Handle easily at all gaits? Change gaits smoothly when you want him to? Stop easily? Turn smoothly? Does he fight the bit? Will he stand quietly after you stop him? How does he behave around other horses? Manners are important. A well-trained horse is a quiet-mannered horse. Some novices think it is "smart" to have a nervous, prancing horse; they think it shows what fine riders they are. Actually it shows what fine riders they *aren't*—betraying the horse's lack of training or his rider's abuse of him.

Before buying a horse, check on his bad habits if you can. If you know some of his personal history you can be better prepared to cope with him, or to decide not to buy him if he is a real lemon. You don't want a confirmed kicker or nipper, a rearer, or a halter puller (by the way, you'd best check on the horses tying habits; not all horses are trained to tie up with bridle reins). A conscientious owner will usually warn you if the horse has any really bad habits, but it is always a good idea to ask.

So far we have overlooked the most important factors in choosing a horse—conformation and soundness. The way a horse is put together has an important part to play in his ability to hold up under strenuous work. A sound horse is a healthy horse, one that is without weakness or injury or other undesirable factors that might impair his abilities or make him go lame.

Some unsoundnesses can be inherited (such as crooked legs or weak pasterns) and others are caused by injury (such as certain types of splints, or bowed tendons). These will be discussed more fully later. There is a difference between an unsoundness and a blemish. An unsoundness impairs the horse's action and usefulness. A blemish is merely unsightly, like a swelling or a scar caused from injury, but does not interfere with the horse's action and ability to move or his ability to perform ordinary work without tiring or going lame.

When selecting a horse you should know what to look for and what to look out for. You don't want a horse that might go lame with ordinary riding, or a horse that is built in such a way that he jars the rider at every step or is dangerous to ride at fast gaits. You don't want a horse that might trip himself with his own feeet, or a horse that nicks one leg with another, or constantly stumbles.

Never let color overcome your good judgment in selecting a horse. Some people never see beyond color or a pretty head. A horse may be unsound or may have poor conformation or a poor way of traveling but if he is their favorite color they do not see anything else. They forget that they can't ride color, and they can't ride the head; a horse has to have a lot more going for him than his color and the set of his head.

You'll want a well-built, well-balanced animal. He should have a smooth, pleasing overall appearance. He should not be gawky and clumsy or rough looking. Size is an important factor to consider also. If you are a large or heavy person, you will want a fairly large or tall horse, for you would look out of place on a small animal; a small horse might find your weight difficult to carry. If you are short or small, you probably won't want a huge tall horse. Aside from the difficulty of getting on and off, you would look a little funny perched on top of him. Choose a horse that is compatible with your size and weight.

The horse is an athlete and his main purpose is motion. If he is clumsy instead of agile, or stumbling instead of surefooted, he won't be a very good horse, and he might be a dangerous one as well. How well a horse moves determines how useful or how valuable he is to his owner, whether he is a race horse, show horse, ranch horse, endurance horse, jumper or pleasure horse. Because our primary concern in considering a horse's good or bad points of conformation is to determine how good an athlete he is (how well he can move and how well he will hold up under normal use), our main interest will be in his feet and legs, and in the upper structures of his body that play a part in the movement of his feet and legs. The proportions of his body can also be a factor in determining whether or not there will be limb interference when the horse is in motion.

When you are judging a horse's conformation, remember that he should be studied while he is moving as well as while he is standing still. He should have a good way of going and should travel "straight and clean" (this means that his legs move in straight lines and do not interfere with one another). A horse with good conformation has straight legs and feet that move forward in straight lines. Stand directly in front of the horse as he is being led toward you, at both the walk and the trot. Then have him led straight away from you at the walk and trot and stand directly behind him. Lameness, and certain unsoundnesses and points of bad conformation will show up more sharply when he is moving. Watch his feet and legs as he travels so that you can see any possible faults or deviations from straight forward leg action. If his front feet paddle outward or wing inward, or if his hind feet come dangerously close to one another when in motion, or if his hocks roll outward, the horse does not have good conformation. We will discuss these various conformation faults in more detail in a later chapter. Poor conformation of feet and legs can contribute to some types of lamenesses (making the horse more likely to go lame) and in some cases may actually be the cause of lameness. A

A horse should be judged while he is moving, as well as while he is standing still

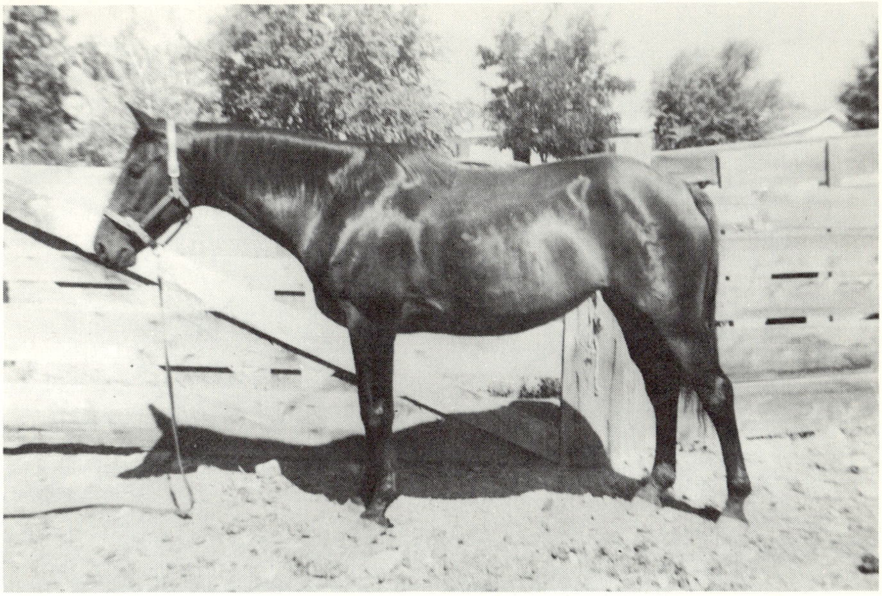

The horse in these photos is Nell, our 15/16 Thoroughbred mare

good judge of horses starts from the ground up when selecting a horse, for any horse is only as good as his feet and legs. If they are weak or poorly formed, or crooked, they are likely to go bad with steady use, and then the horse will be lame and worthless. Two later chapters, one on conformation and one on unsoundnesses, will cover the horse's legs more fully.

To continue a quick look at the horse's conformation when selecting a horse: good withers are important. A horse with good high withers holds a saddle well and you don't need a painfully tight cinch, double rigging, breast collar and breeching to hold the saddle in place. A round-backed (mutton-backed) horse with low withers must be cinched very tightly and still the saddle may turn with you when you mount. When traveling up or down steep mountainsides, or changing directions swiftly (as when cutting cattle) it is difficult to keep the saddle in the proper place on a round-backed, low-withered horse. Choose a horse with good withers and you'll have a minumum of saddle problems if you use a saddle that fits him (don't use a saddle with a Quarter Horse tree on a horse with good high withers, or you'll rub his withers raw; the Quarter Horse tree was designed for mutton-backed horses with low withers and a wide front end).

A horse with good high withers usually has a long shoulder that slopes and is well laid back. A shoulder that is too straight (up and down instead of sloping back) limits the horse's freedom of action and gives a shorter, choppier stride. A horse with short, upright shoulders usually has a short neck and short pasterns, that creates more jarring and concussion, hence more

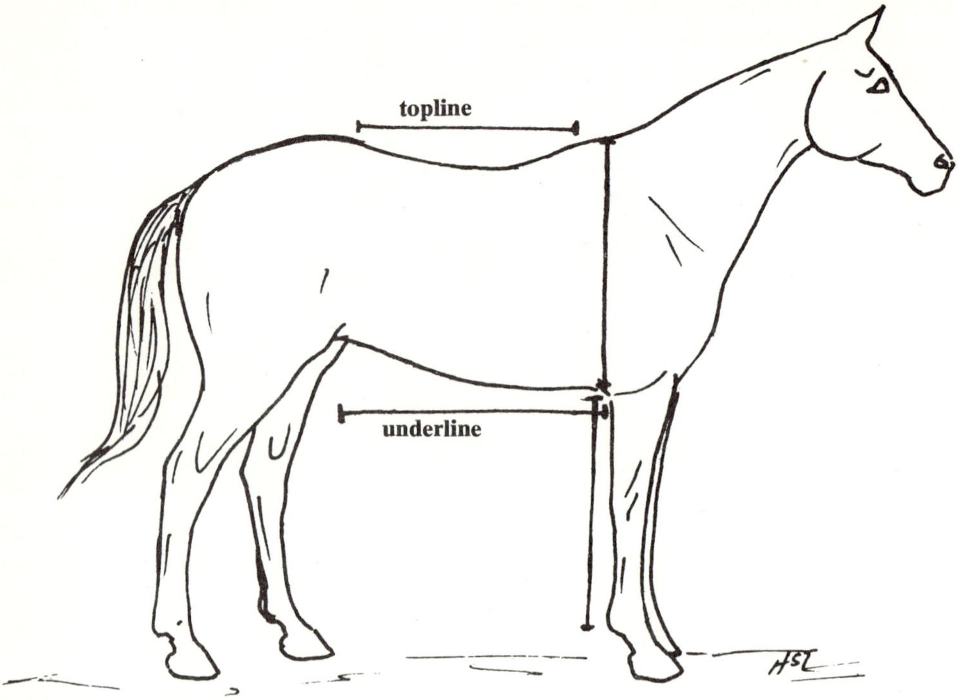

A good horse has a long underline, a short topline, and the same height at the withers and hips. The distance from the withers to the underline is the same as the distance from the underline to the fetlocks

leg problems.

Ideally, a horse should be the same height at the withers as at the highest point of the hips. A well-balanced horse has about the same distance from fetlocks to his underline as from his underline to the top of the withers. In other words, his front legs are about the same length as the depth of his body from withers to underline.

The angle of the horse's neck will vary a little with his breed, but as a general rule you will want a fairly long and slender neck. It will be more flexible, making it easier to set the horse's head properly for best collection and maneuverability, and the horse's balance will be easier and more graceful with a long neck than with a short, stubby one. The short, thick "pony neck", the upside-down "ewe neck" and the thick, heavy crest are all undesirable types in a riding horse. A horse with this type of neck is less able to flex properly; he is less able to respond readily through the bit and it is harder for him to keep in best balance performing difficult maneuvers with a rider on his back.

You don't ride the head, but you should never overlook the head when judging a horse. Most breeds have certain desirable specifications for heads, and you should be aware of these when selecting a purebred. As a general rule, short, well-formed ears are desirable, and so is a short head. The head, ears, and eyes can give a hint as to the character, intelligence, and disposition of the individual horse. Most horsemen like good width of forehead between the eyes, and a calm but alert eye.

When selecting a horse, you may want to look at his teeth. The age of a horse can usually be fairly accurately determined by careful examination of the teeth (see chp. 10). Also, teeth are important when considering a horse's overall conformation. His digestive processes depend on how well his teeth perform their job. Some horses that are "hard keepers" are unthrifty because of poor teeth. They stay thin because their food is not chewed well and much is wasted through improper digestion. These horses are often helped by good dental care. A horse may have rough or irregular teeth that make it hard for him to chew his food. A horse with a parrot mouth (overbite) is usually a poor keeper. Parrot mouth is considered to be an unsoundness.

Undesirable necks in a riding animal

a short, thick neck

a ewe neck

a balanced neck

An unsoundness in a horse is any defect that interferes with the usefulness of the horse—any physical problem that makes him unsafe to ride, lame, or in any other way unable to perform or hold up as he should. Tendencies toward certain unsoundnesses may be inherited through poor conformation, making a horse more likely to have such problems as ringbone, sidebone, splints, curb, etc., or may be due to injury (dislocated stifle, blindness from a blow on the eye, sidebone caused by a wire cut, and so on) or from strain and overwork (such as bowed tendons, certain types of splints, and so forth). If you are selecting a horse for breeding purposes you definitely do not want to choose a horse that has an unsoundness due to poor conformation, for this may also be a problem in his foals.

The severity of an unsoundness depends partly on what the horse will be used for. There are some defects that would hinder and handicap a horse used for racing or for some other types of strenuous work, but which might not bother a horse that was being used strictly as a pleasure horse or as a child's mount. There can also be a difference in certain cases between working soundness and breeding soundness. A horse may not be sound for work (he may have a bowed tendon, a bruised sole, a badly cracked hoof, foundered feet, or some injury that impairs his action) yet still be sound for breeding purposes so long as his problems are not heritable or resulting from poor conformation which might be heritable.

Good judgment must also be used to differentiate between certain unsoundnesses and blemishes. For instance, a splint can be a minor problem, merely a blemish, if it is not accompanied by poor conformation (such as bench knees, which are actually off-set cannon bones) or lameness, and if it is well away from any moving part such as the knee or back tendon. The same is true for thoroughpins and wind puffs. If they are caused by strain or injury and are now healed and causing no pain or lameness, they can be considered blemishes provided the horse has good conformation. If the horse has poor conformation, putting extra strain on certain parts of his leg and making him more likely to have these problems, they are considered to be unsoundnesses for they are likely to recur or be reinjured.

In checking a horse for soundnesss, don't forget to check his vision. Blindness in one eye may not be noticed by a casual observation of the horse, especially if the eye does not appear abnormal. Blindness can be caused by injury, or by certain diseases. If you are going to use the horse for breeding purposes you may not care if he is blind in one eye. But if you are going to use him out in the woods or in some types of competition, a blind eye may be a serious or even a dangerous handicap. Pass your cupped hand carefully in front of each eye (be careful not to create an air current or touch his "eye whiskers") and watch him for a response. If he can see your hand, he should blink. Occasionally a blind horse (one that is actually blind in both eyes or partially blind in both eyes) will adjust well to his handicap and be passed off as a sound horse, especially if the owner can show him to you under ideal conditions. Stumbling can be due to impaired vision as well as due to poor conformation. So check that horse's vision before you buy. If you are unsure about any physical aspect of the horse you are considering, have a vet check him.

Now let's take a look at selecting breeding stock. There are several things to consider here that have not been covered under the selection of a riding animal.

When selecting the foundation stock for a breeding program, there are several routes that you can take. You may already have a few horses and your wish might be to incorporate them into your breeding program if they are mares, buy a stallion and a few additional mares and go on from there. If your mares are purebred, you are already pretty well on your way. If they grade, you'll want to eventually get some purebreds, for the grade horse usually won't command the kind of prices that purebreds will bring.

Your purposes in raising horses will have a great deal to do with the kind of foundation stock you select for your breeding program. If you are primarily interested in breeding purebred horses that will be sold as breeding stock to other people in the purebred business, then you'll want to get the best horses you can buy in that particular breed—horses with excellent conformation *and* bloodlines. If your stock is good enough, people will come to you when they want a stallion or a filly that might help upgrade or expand their own herd.

Or, you might want to raise horses with a particular use in mind—racing, showing, stock horses, roping horses, endurance horses, or something else. A well-trained grade horse may command a good price if he is a champion in his field. Good grade horses can compete very well in fields such as roping, bulldogging, barrel racing and so on. Yet many of the bigger horse shows are sponsored by various breed associations and a horse must be purebred to compete. A good, well-trained grade horse will compete well in certain "open" events that are not restricted to any certain type of horse. But you will spend more time and effort raising a horse to maturity and selling it as a well-trained, specialized animal (for the amount of money you will get for the horse) than if you raise purebreds to sell as breeding stock. However, if you raise purebreds and train a few of them to show what their (and their offspring's) potential might be in certain fields, they will command a greater price. People want horses that can do things.

Our purposes in raising Arabians will be to raise part-breds, and perhaps someday some purebreds, that excel as ranch and endurance horses. The part-Arab has been the best horse that we have been able to find for the job in our own ranching situation and we want to raise more of these horses with agility and ability—to show people what the Arabian can do. Too many stock horse people pooh-pooh the Arabian as a pretty picture-book horse. This image is starting to change, with Arabians winning most of the endurance rides, but more people need to use Arabians as cow and ranch horses just to show the public what this tough, quick little horse can do. The few Arabians that have been trained as cutting horses have done very well in open competition with other breeds; it's time this happened. The Arabian is not necessarily better. It's just time to recognize his versatility and acknowledge the fact that he can do a lot more than just look pretty. But now, back to the selection of breeding stock.

When buying fillies or mares, you'll find a great variation in prices, partly depending on the breed. A fairly good weanling or yearling filly may cost anywhere from $200 to $2500. A purebred Arabian filly will probably cost you at least $1500 to $2500 and on up (my younger sister just bought two purebred fillies—a weanling for $1500 and a two year old for $2500—and felt she got a bargain) whereas a Quarter Horse filly might cost you $300 to $500 for a good one (we bought a registered Quarter Horse three-year-old filly just a few years ago for $300). This difference in price is due to the fact that there is still a somewhat limited quantity of Arabian breeding stock in this country, while there is practically an unlimited supply of Quarter Horses, the Quarter Horse being the most numerous breed of horse in the world today by a large margin. It may be harder and take longer to get into a higher-priced breed if you haven't much money, but your foals will be worth more when you sell them.

It will cost you more to buy a national champion or a son or daughter of a national champion. Popular bloodlines will also be more expensive than those that aren't. But here again, the foals you will sell from championship bloodlines may be worth more. But don't spend more than a horse is actually worth. Never sacrifice conformation, action, quality, disposition, and so on for name alone. A few of the "top" horses got to the top partly because their owners had a lot of money to publicize them. Some of these "top" horses don't always produce "top" offspring. There are many other horses running around that are equally as good but won't cost you as much. When buying breeding stock, look at quite a few horses before deciding, buy the best you can afford, and then try to improve upon it by careful and selective breeding.

The highest priced female stock will usually be the young stock with good parentage and the producing young and middle-aged broodmares that have already produced good foals. The most economical age groups are the very young (unproven) and the older matrons. You can sometimes buy a good quality older mare fairly reasonably and with luck still raise a few more fillies from her to replace her and add to your breeding herd. When you buy an older mare you are gambling on the fact that she will keep on producing for a few more years, but if she is in good health and is sound and has had a record of being a good producer, it is a gamble worth taking. But beware of buying an older mare if she hasn't produced a foal for several years. Older mares can become infertile or sterile just from old age, and a mare that has consistently produced a foal every year for the last few years will be more likely to keep producing than an old mare that hasn't had one lately.

Above all, in selecting mares, try to get good conformation. Conformation is so important in a horse, whether he is a show horse or pleasure horse, ranch horse or race horse, because everything he does is linked with how well (or how poorly) he is put together. After all, we are breeding *riding* animals, and this should never be forgotten. The breeder who never rides his animals or who concentrates primarily on "halter propects" sometimes tends to overlook the facets of conformation that determine how well the horse can *move* as an athlete.

The horse with good conformation will move better, be more pleasant to ride, have nicer action, stay sounder longer and be a safer mount than the horse with poor conformation. You are trying to breed the best, so don't settle for something less; you'll have more trouble selling the offspring. Besides, we want to improve and upgrade our particular breed, whatever it is, rather than tear it down. Breeding horses with poor conformation hurts the breed.

One other thing should be mentioned that is often overlooked when talking about broodmares and breeding stock. Try to select mares (or fillies from mares) that are good mothers. A good broodmare should be feminine and maternal just as a good stallion should be masculine. You'll have less problems. Too many mares are selected for their masculine characteristics—speed, powerfulness and strength, etc.—and as a result some of them are not very feminine, not very fertile and not very good mothers when they do produce. Keep this fact in mind when looking at prospective broodmares.

When buying a stallion you'll want to buy the best one you can afford, for it is the stallion you'll be counting on to upgrade your stock. He will produce many foals, whereas a mare will only produce a dozen or so during her lifetime. If your stallion is very good you may be lucky enough to raise foals that are better than their mothers. Every good breeder tries to improve his breeding stock and the easiest way to do it is with a good stallion. It is very foolish to breed a good mare to a stallion that isn't as good as she is; if you can breed her to one that is as good or better, you can begin to upgrade your herd. And if you are fortunate enough to breed horses that complement and improve upon one another you may actually produce a foal that is better than either of his parents. This is the big challenge and part of what the selective breeding game is all about—trying always to improve upon what you already have.

You may want to buy a weanling colt that looks as if he will be a good stallion prospect and then raise and train him yourself. In many ways and many cases this is the cheapest way to buy a stallion, but it is also uncertain, for you can't always tell from the weanling or yearling how he will turn out as a mature stallion. Judge him pretty closely. Then if you really like what you see, scrutinize his parents (and half siblings, and full brothers and sisters if possible) to get a glimpse of what he might be when he is mature. Whenever you buy breeding stock try to see as many of the horse's relatives as possible; if there are some poor ones in the batch you'll have to realize that your horse may carry some genetic possibilities for poor offspring, too. But if the relatives of your stallion prospect are of consistently good quality, go ahead and gamble. If there are inconsistencies and characteristics you don't like in some of them, you'd better look a little longer somewhere else.

A young stallion (two to four years old) will probably cost you more than the weanling or yearling. The young stallion is starting to mature and you can get a glimpse of what he may look like (and he has probably already had some training and time and effort put into him). When judging a young stallion, don't expect yearlings and two year olds to look too mature. If they are—if they look almost like mature horses at that tender age—they may be a little too large and coarse or heavy when they are mature. A two-year-old stallion should "still have something to look forward to," to put it in the words of one good horse judge. Also keep in mind that some breeds mature more slowly than others. The Quarter Horse is probably the fastest maturing, as far as reaching his size and growth.

The mature stallion of age five or six to about twelve or fifteen will probably be the most expensive age in which to try to buy. He is right at his peak, he is easily judged, and a person can see what kind of offspring he produces to see what kind of a sire he is. It's always an advantage to be able to see a horse's offspring, for then you have a better guess as to what he can do for you with your mares. Not all good-looking horses are the best sires, and sometimes some of the plainer ones will surprise you with exceptional foals. Through usually "like begets like," it's best to see for yourself what a stallion produces; his offspring are his proof. If a stallion is of breeding age but has never sired a foal, be cautious. You want to be sure he is fertile.

The older stallion may be somewhat cheaper than the stallion still in his prime, although an older stallion with a good show record or famous offspring may still command a high price. But if you can get a good older stallion reasonably, it might be worth a try, in the hope of getting several more good years of use from him. Or, you may not feel you can afford a stallion of good quality at all at first. In this case you might breed your best mare to a stallion of your choice (with a stud fee you can still afford) and hope to raise a colt from her that might suit your purposes for your first stallion. Or, there is also the possibility of leasing a stallion rather than buying one. In that event you might hope to get a good colt or two from your own mares that might make future stallion prospects.

So, before you buy a horse, whether he is to be for riding or for breeding, be sure to take a good look at all the angles, all the possibilities, and consider all of the problems and all the desirables. Check on the horse's age, soundness, conformation, disposition, training, and bloodlines, and shop around before you make that final decision. When you consider that there are nearly eight million horses in the United States, you don't have to be too eager to buy the first horse that you find for sale. Be sure you get a *good* horse, for there are still a lot of them around.

3

KEEPING A STALLION

IF YOU ARE A BREEDER OF HORSES, YOUR STALLION OR STALLIONS are a primary concern, being perhaps the most important part of your whole operation in one sense. It is your stallion or stallions that come most before the public eye, giving your operation its reputation. The stallion—his looks, ability, condition and fitness, and of course, the foals he sires—is what people are most interested in when they come to see your breeding program, for they want to see him and his offspring. If they are favorably impressed they may decide to buy a son or daughter or breed some of their own mares to him. The stallion or stallions at a breeding farm are what give a breeder his good name. Good mares are equally important of course, but since it is the stallion that many people come to see, he must carry more than his share of the responsibility for a good impression.

Because a stallion must be in good condition to breed a large number of mares, and because you want him fit and sleek and at his best anyway because he is always on show, stallion management is indeed a very important concern if you are in the business of raising horses. A stallion should be properly fed and exercised to be able to best perform his duties and to look his best. And he should have adequate housing or pasture. His surroundings should be clean, hazard-free, and well built. A shoddy, run-down enclosure detracts from your stallion's good impression and can be dangerous to him besides.

There are two ways to go about keeping a stallion. You can keep him in a small, safely fenced enclosure where you will want to exercise him periodically, and hand breed him to your mares and perhaps some visiting mares as well. Or, you can let him run at pasture during the breeding season with his own group of mares. There are several factors to take into consideration when doing the latter, some of which I'll touch on briefly, but other than these few precautions, pasture breeding is in many ways the simplest method and has several advantages. Pasture breeding is more natural, and the conception rate with this type of breeding is usually higher than with hand breeding, because the stallion always breeds the mare at the proper time. Also, pasture breeding is usually cleaner. Mares that breed and foal at pasture have *less* infection problems than those that are kept in confined quarters. And if the pasture is large, the stallion and mares get enough exercise on their own, eliminating the extra daily chore of exercising the stallion.

But if you do pasture breed there are a few things you'll want to consider. Not every stallion will pasture breed *and* hand breed. You are fortunate if you have an easygoing individual who will run with his mares, let you come catch him and take him away from them to hand breed a visiting mare every now and then. Some stallions that have been used only for pasture breeding do not cooperate very well when you decide to hand breed a mare with them; they are used to courting and breeding a mare at their own leisure and pace, as nature intended, and may be hesitant to just up and breed a mare that is suddenly presented for breeding. Perhaps being handled and confined at breeding time rather than being able to do it his own way is distasteful to some stallions that are used to breeding naturally out at pasture.

If you pasture breed, you'll want a very large, clean, and grassy pasture, and one that is adequately fenced with strong and smooth fencing. And definitely *no* other horses right through the fence in an adjoining pasture unless they are very young stock or perhaps some pregnant mares. Another stallion through the fence is just inviting trouble, and mares that may come in heat will cause problems too if they are adjacent. Geldings are also taboo, for a stallion hates a gelding. But, if you have an adequate, well-fenced pasture with no adjacent "trouble-makers" through the fence, and can pick out a group of mares for your stallion and let him run with them throughout the breeding season, pasture breeding is the ideal way to settle your mares. Many of the larger breeding ranches in the West do it this way to a large extent, giving each stallion twenty to forty mares to run with during the breeding season. It saves time and labor, and usually insures getting the mares with foal.

But now let's look at the more common method practiced because many breeders do not have the acreage or room to let each stallion run at pasture with a group of mares, and also partly because they are also breeding many outside mares. In this method the stallion is confined to his own special place (a stall, or perhaps a stall and paddock, or a small stallion pasture) and mares are brought to him to be bred. A mare is brought when the person in charge of the breeding operation feels the mare is in the proper stage of her heat period to accept the stallion and to conceive a foal. The breeding may take place in a small pen or enclosure, in a breeding chute, or whatever arrangement the stallion owner has devised for the hand breeding of mares to his stallion.

The young stallion is considered a colt until he is four years old. At that age he is considered to be mature. A few people start using a stallion for breeding purposes in a very limited capacity as young as two or three years of age, but usually it is better to wait until he is older. Unless a colt is very mature for his age, he should not be used for breeding at the tender age of two or three. Even at four, he should have a limited number of mares. A young stallion can be used to breed eight or ten or maybe twelve mares during his first breeding season.

A mature stallion can breed many more, depending of course on the individual horse, his physical fitness, and his situation. Forty is usually considered to be an adequate number of mares for any one stallion to breed in a season, but with careful hand breeding (trying to breed each mare at the proper time in her heat period and therefore not having to breed her more than once, or twice at the most) a stallion can successfully breed more than that, sometimes up to a hundred mares. But for best conception rates, he should not have to breed mares very close together in time. One a day is plenty. If he must service two mares in a day, the breedings should be spaced out, breeding one in the morning and the other in the evening. Some stallions are used more often than this, but usually you cannot count on them settling every mare. A stallion will be more fertile if he is used sparingly. You are being more fair to the mare's owner if you do not overwork your stallion; he will have a better chance of settling the mares he breeds. An old stallion should not be used hard. Like the young stallion, his number of mares should be somewhat limited. He will last longer if you do not overwork him.

The enclosure where you keep your stallion should be stout, well-built and smooth, and high. A box stall for your stallion should be large and roomy. It is convenient if you have the space to make an adjoining paddock or small pasture where he can be turned out to exercise himself. If you want him to have enough room to graze, the pasture will have to be large enough that he won't tramp out most of the grass by his exercising and by his "walking the fence." Most stallions are very exuberant and use up a lot of energy worrying about what goes on around them, and spend hours walking or cavorting up and down the borders of their enclosure, looking out and letting the world know their opinions about things. This type of activity soon beats a deep path along the stallion's favorite routes, and if the paddock is small to begin with there soon won't be much grass left. If you expect him to have any good grass intact for grazing, you'd better give him a large area.

A stallion enclosure should be well fenced, preferably with a fence high enough so that the stallion cannot get his head over it. It can be made of poles, board, or pipe, or anything that is safe, smooth, and very durable. The spacing of the poles or boards should be such that he cannot get his head between them. If his enclosure is made of wood he may chew at it, and you will be wise to put a "hot" (electric) wire around it, perhaps one along the top and one along the side to keep him clear away from the fence. This will keep him from chewing up the fence and will also prevent his kicking or pawing at it or trying to get over or through it. Horses thoroughly respect a "hot" wire and will not touch it if they can possibly help it.

A stallion is very sensitive about his surroundings. He will consider all the mares on the breeding farm as "his" and will be quite concerned about them. He will notice more things than a mare or gelding; all the goings-on about the place will interest him for he feels they are his personal concern. If you bring a mare in from the pasture he will notice it; if you take one of the horses to ride, he will be watching. And he may be the first to become aware of your return.

Temporary "stallion quarters" for Sur Amir when he first came to our ranch—our old round corral with a "hot" wire around the top of it (about a foot or so above the top pole)

He knows where each horse on the place belongs. Changes upset him, such as when strange horses are brought in, or if his daily routine is changed for any reason. Some stallions are more easygoing about changes than others. Most are a little fussy. If he is in a paddock, stall, or corral next to another one, make sure that the other place is always occupied by the same horse. A strange horse next to him may upset him. He can be stabled or corraled near other horses without problems, so long as he cannot get close enough to the other horse to actually sniff noses with him. If he is in a stall next to another horse there should be a solid wall between them. If he is in a paddock next to another horse the fence should be such that he cannot get close enough to smell the other horse over the fence, unless he is used to the other horse and gets along well with him. It is not ideal to totally isolate a stallion from other horses. He will be mentally and emotionally healthier if he has company even if he is just observing other horses in the next field or from his stable door.

Stallions must have some special care in their handling, for they are very different from mares and geldings. The stallion is proud and strong, of high courage and exuberant. He has more drive to excel than do some mares and most geldings and therefore is a challenge and a pleasure to work with. As a general rule stallions have a little more heart and guts when it comes to endurance. This is why some horsemen like to ride a stallion as an endurance horse or in competitive events. Because of his highly exuberant and sensitive nature, the stallion

Sur in the round corral

The stallion is proud, high couraged and energetic

Sur and Lynn going back to the corral after breeding a mare

should always be treated with firmness and respect but also with much sympathy and intelligence. To abuse a stallion is to invite trouble, and potentially dangerous trouble at that.

If a stallion has a good disposition to start with and is handled properly (with tact, understanding, and respect as well as firmness) from the time he is young, he will not be a problem to handle. His handler should be the kind of person who understands him and who is able to keep a good rapport with that horse, a person who has the ability to communicate with him and to demand (and get) the horse's obedience with a mere word or touch of a hand. A good stallion handler is tactful and kind, but he also constantly remembers that the horse is a stallion. The stallion handler should be able to command respect and instant obedience wherever necessary. Thus, the man or woman who can combine intelligent understanding with quiet discipline will be a good stallion handler. There has been much controversy over whether or not women should handle stallions, but this is foolish; it is entirely an individual matter, depending upon the woman and upon the stallion. There are many men who should never handle stallions, just as there are also many women who should not. It is not a matter of generalizing about one sex's ability or inability; it is a matter of personal aptitude. Some women are better than men when it comes to handling a horse because a woman often has more patience. However, no generalizations should be made about women and stallions. It all depends upon the person and upon the horse. Some horses need a strong and steady hand, while others need a lot more tact and patience and sympathetic understanding.

Rough handling is bad experience for any horse, but it is especially bad for the stallion because he is such a proud and courageous animal. If treated with roughness, he will usually fight back. Rough handling will make the stallion irritable and bad tempered, and a bad-tempered stallion can be a dangerous animal. It is better to be gentle but firm, communicating to the stallion that you have some respect for and confidence in him. The stallion usually enjoys human companionship and will try to please his master, but if his handler is nervous or afraid, the stallion can immediately sense this lack of confidence and become nervous also, reacting adversely if the handler resorts to force. No one who is afraid of a stallion should ever handle one, for fear is usually covered up by forceful and rough handling and a loud voice. And any self-respecting stallion will soon resent that sort of nonsense and retaliate. The use of force just leads to more force. The more you can treat your stallion just as you would any other horse (using good judgment, of course, remembering his male characteristics), the easier it will be to keep him gentle and good tempered, trustworthy and easy to communicate with.

To be in the best condition for breeding, a stallion should have plenty of good feed and exercise. Good feed is essential to fertility, but so is good health and physical fitness; a horse that is too fat can be infertile. Proper exercise is very important to a horse's virility. The best way to exercise a stallion is just to ride him a little every day, using good judgment when in the company of other horses. My sister uses her stallion to ride range and move cattle. If you are not able to ride him daily he can be led. I know of one stallion owner who leads his stallion out the pickup window—in a smooth pasture—jogging the horse several miles each day. Each person can work out his own method of exercising his horse, but that horse does need exercise, whatever form it may come in. Boredom and lack of exercise, especially if a stallion is confined, tends to make him sluggish or bad tempered, depending on his personality. A high-strung horse that is well fed (to keep in "breeding shape") but confined can just about go nutty because he has no way to burn up that extra energy. Taking a high-lifed horse and feeding him well but not exercising him is a good way to ruin his disposition. Good feed and good exercise should go hand in hand toward getting a stallion in shape and keeping him that way—fit, healthy, and happy.

On cold or hot days the stallion should have access to shelter. On hot days he should be able to have shade or to go into the cool of his stall. Standing out in the heat of the sun all day can take a lot out of any animal. In some of the hotter climates, some stallion owners like to keep a fan in the horse's stall during the heat of the day, or give him a bath after his exercise, to refresh him. It all helps to keep him feeling tip-top. His surroundings should always be kept as clean as possible.

When the breeding season begins, the stallion should not be expected to be perfectly fit and eager for his duties; his sexual instincts can go somewhat dormant during the months in which he is not used for breeding, and it will take a little time for him to become fully up to his job. It is wise to have the stallion's semen

Heidi Smith and Surrabu 38018, six-year-old Arabian stallion

The best way to exercise a stallion is to ride him

checked by a vet before the breeding season starts. If he is a little infertile, or sterile, it's better to find out before the breeding season begins than after a number of mares have been bred fruitlessly.

A few last remarks about your breeding operation before we end this chapter: If you are seriously going into the horse business as an occupation, try to select a location that is good for raising horses and yet easy to find—a place where people can easily come to see your horses. We haven't done this ourselves because we raise horses only as a sideline (raising them mainly for our own use) and because we like isolation. But I can see where it would be handier for someone who was seriously in the horse business (and wanting to deal with the public) to be located a little nearer. Keep this in mind.

Another plus factor in any breeding setup is to really have a good stallion or stallions. As mentioned earlier, the stallion helps to make a name for you and helps to sell your horses. If you have a really good stallion people will want to breed mares to him or come to buy his offspring. Keep your registrations up to date. Join your breed association and register your foals according to its rules. Foals will always be worth more if they are properly registered.

If a mare owner "books" his mare to your stallion, a breeding contract should be drawn up and signed by both parties (you and the mare owner). A breeding contract protects both the mare owner and the stallion owner and should be fair to both parties. The mare owner knows what the breeding fee will be and what the board bill will be on his mare per day. The mare owner should know whether or not his mare will have a return privilege to be bred again if she does not settle or produce a foal, or whether or not the breeding fee guarantees a live foal. If you are standing your stallion at stud, you will be wise to insist on a veterinary check of each mare before she is bred to your stallion. A check for breeding soundness is beneficial to both the mare and stallion owner, for if any problems exist they can then be cleared up (hopefully) before the mare is bred, and you will be more sure of having only healthy mares bred to your stallion.

A stallion's breeding fee should be reasonable, especially if he is a young stallion or one that is not very well known. You can start your stallion's fee rather low and always raise it later if he becomes well known and sought after as a good sire, but if you start out with the fee too high and later want to lower it, it can be an embarrassing matter. If the fee is reasonable, there are more people who will be able to afford to use your stallion; if the fee is too high you might get better mares coming to your door (if the stallion is really that good), but there won't be very many of them and the stallion will have less offspring to represent him and to show people what kind of foals he can produce. His offspring are the final proof, showing whether he is a good sire or not. If he can consistently produce good quality foals out of a variety of mares, people will want to breed their mares to him. If you are out to really sell your stallion's services and make him well known, start him at a reasonable fee and then as his foals start to make a name for him, you can go on up from there.

4

RAISING HORSES

HORSES ARE BIG BUSINESS TODAY. THEY HAVE MADE A FANTASTIC comeback as a pleasure and recreation animal; the pleasure horse actually abounds in greater numbers today than his working counterpart a century ago on the farms across the nation. There are many facets of the horse industry—everything from selling horse equipment and western wear, horse trailers and feed products, to the raising and selling of the horses themselves. There are nearly eight million horses in the United States today and the number is steadily growing.

All kinds of people raise horses. Ranchers, businessmen, movie stars, kids with 4-H projects, urban dwellers with a horse in the back yard, and professional horsemen. The range is very wide—from the one-horse family to the large stud farm breeder whose only occupation is raising horses to sell, and all the various breeders in between.

This chapter will deal with the horse breeder and his problems; it is written for the person who wants to go into the horse business in a large or small way, as a profession or as a hobby. Our own situation here on the ranch includes both aspects, for we breed horses out of necessity, for use on our ranch, and for our own pleasure. And someday we may also breed a few horses for sale. Some of the aspects of raising horses will be covered in other chapters—selecting breeding stock, keeping a stallion, breeding the mare, infertility in mares, care of the broodmare, and caring for the abnormal foal—so this chapter will try to hit the rest of the high spots along the way.

Selecting breeding stock is covered in chapter 2, but a few more words should be mentioned here. When selecting a mare or stallion, you'll want to check closely on conformation, disposition, training, *and* pedigree. If you are going into the horse business bloodlines are very important because in a very real sense the ancestors of a horse are what produce the horse. Try to see as many of the individuals in the horse's pedigree as you can (sire, dam, grandparents, and so on), in pictures if they are not available to look at in the flesh. The first fourteen ancestors (sire and dam, grandparents and great grandparents) are probably the

The object of successful breeding is to produce a live and healthy foal. Nell and her first foal, Amahl

Quarter Horse mare and foal owned by Jake Frank, Park City, Montana

most important ancestors to consider when looking at a horse's pedigree. Ancestors farther back than that are not likely to have nearly as much influence in the characteristics of the horse, though occasionally it can happen. If the individuals close to your horse's pedgree are consistently good, that horse will very likely produce good quality foals when mated to an equally good horse.

Perhaps a word about inbreeding and linebreeding should be mentioned. If a horse is bred to another that is closely related to him (immediate family: sister to brother or half-brother, mother or father to son or daughter) it is called inbreeding. This is sometimes deliberately done with outstanding individuals. Inbreeding is always risky because it doubles up many characteristics and recessive traits can crop out that were hidden in both parents. But if a horse is outstanding and has no faulty genes, if genetically he carries mainly good characteristics, and is bred to a close relative that carries the same, the offspring has a chance of inheriting good qualities and being truly outstanding, particularly in the genetics he will then pass on. With luck, a horse that is the product of inbreeding from

example of inbreeding: A son of Skowronek was bred to a daughter of Skowronek to produce Sureyn, an Arabian stallion found in the pedigree of some Arabians today.

Sureyn 1886	*Raseyn 597	Skowronek GSB	Ibrahim db
			Jaskolka PASB
		Rayya GSB	Rustem GSB
			Riada GSB
	*Crabbet Sura 1199	Skowronek GSB	Ibrahim db
			Jaskolka PASB
		Sardhana GSB	*Nureddin II 974
			Selima GSB

good-quality parents will carry only these doubled-up good qualities in his genetic make-up (he will be genetically "pure" for these characteristics, carrying no others along with them) and he may be outstanding as a sire (or dam), passing on consistent characteristics to his foals. But inbreeding should always be done with care—with horses that are not known to carry undesirable traits—for if the inbreeding produces a doubling-up of undesirable characteristics it can be very harmful rather than beneficial.

Linebreeding is a form of inbreeding, but in a more cautious manner. Linebreeding is being done when you breed related horses (uncle to niece, grandsire to granddaughter or great-granddaughter, and so on), just not quite as close a relationship as inbreeding. Careful linebreeding is done by many breeders of

Example of linebreeding: A son of Sur Valiant was bred to a granddaughter of Sur Valiant to produce "Jirre."

"Jirre" (chestnut Arabian filly)	Sur Amir 33189	Sur Valiant 13894	Sureyn 1886	*Raseyn 597
				*Crabbet Sura 1199
			Gharisita 6484	Gharis 623
				La Platita 2831
		Miriam 7184	Bazal 3388	Baarouf 2163
				Fazala 1247
			Saera 670	El Sabok 276
				Kalurah 481
	Lyttl Miss 43762	Lyttl-Fix 14801	*Serafix 8955	Raktha GSB
				Serafina GSB
			Amirta 6101	Tamarlane 3274
				Amiri 2111
		Demah 24826	Sur Valiant 13894	Sureyn 1886
				Gharisita 6484
			Il Dalala 11908	El Khamis 9145
				Tavaletta 8290

horses and other livestock to improve the breed. It can produce outstanding animals by doubling up good qualities and concentrating on certain desirable characteristics in families of horses, but isn't quite as risky as inbreeding.

An outcross is when you breed a horse of one family strain to a horse of another family strain that is not a bit related. In other words, if the two sides of a horse's pedigree are not the same (none of the horses on his sire's side are the same as any of the horses on his dam's side) he is the result of an outcross. A horse that is the product of an outcross may be an outstanding individual, but his genetic makeup will be more varied than the inbred or linebred horse and there may be more variance among his offspring; no two may be quite the same.

Crossbreeding between two different breeds gives the same results. The individual produced may be exceptional but there will likely be a variety of traits within the horse's genetic makeup. For instance, we have raised four good foals from Nellace (15/16ths Thoroughbred) by an Arabian stallion (El Khamis 9145). Even though the four foals (two colts, two fillies) are all full brothers and sisters, none of them are alike. There was great variety in the combinations of genes to draw upon, mating two horses of totally different breeds. Of the three offspring we still have, Nikki is a small bay mare, timid, and exhibiting rather modified Arabian characteristics. Fahleen is a flashy, hot-tempered, aggressive chestnut whose moods are like her Thoroughbred dam, her friendliness and conformation (short back, long graceful neck, elegant action) are like her Arabian sire. Her color came from her two grandsires, carried as a recessive gene in both her sire and dam (Nell is brown, Khamis is bay). Ahmahl is a dark, dark bay like his brother Amahl (who died) but has a totally different disposition and build, tall and leggy like a thoroughbred. It will be interesting to see what we get when we breed the two half-bred mares, Nikki and Fahleen, back to an Arabian stallion.

As long as we are touching on genetics, maybe we ought to mention a few facts about color. Some colors are recessive and others are dominant. Chestnut is always recessive. In other words, a horse has to get two chestnut genes (one from each parent) in order to be chestnut. A chestnut horse has no other color genes, so all he can pass on to his offspring is chestnut (even though it may be overshadowed in the offspring by the other parent if the other parent passes on a dominant gene such as bay or gray). So a chestnut bred to a chestnut always will produce a chestnut and no other color. Bay is dominant over chestnut. If two bay horses are bred, they may or may not produce a bay foal, depending on whether the bay parents were pure bay genetically. A bay horse may carry a chestnut gene (recessive) as well as bay, but because bay is dominant the horse himself will be bay. However, if he is mated with another bay that carries a recessive chestnut gene, there is a chance that the foal may inherit a chestnut gene from each parent and be chestnut rather than bay (like our Fahleen). But a genetically pure bay (carrying only bay genes) will produce a bay regardless of what he is mated with unless he is mated with a gray.

In the first cross, the "pure" dominant parent will decide the color of the foal when mated to a recessive. The foal will then carry a recessive color gene that can be passed on to its offspring

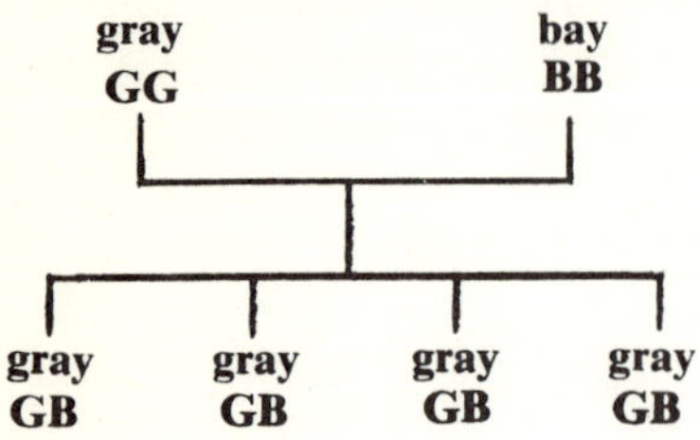

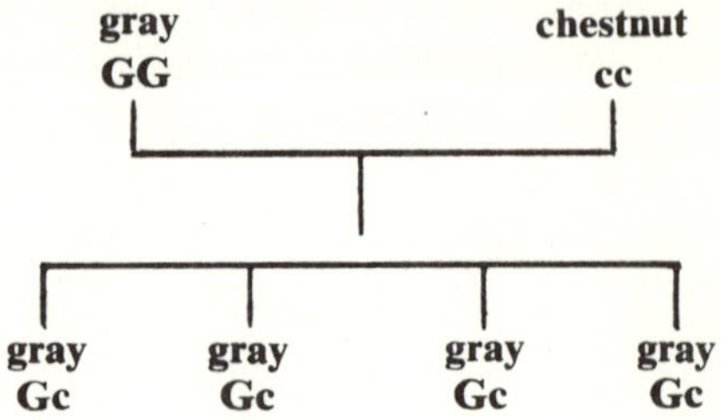

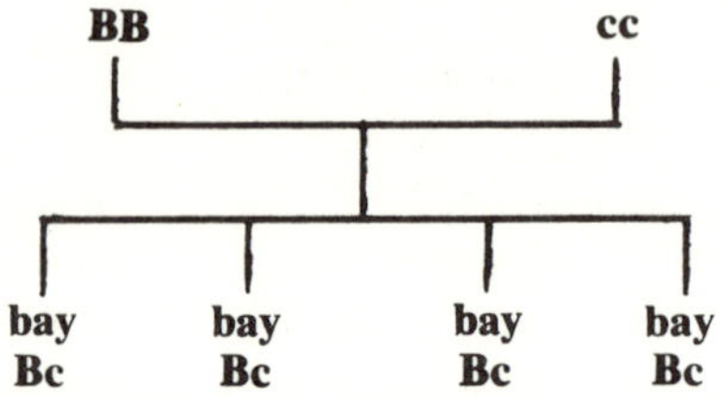

A gray horse that carries a bay gene will produce mostly gray foals but still can produce a bay foal. When mated with another gray horse that carries a chestnut gene the foal has these four possibilities: pure gray, gray but carrying a chestnut gene, gray but carrying a bay gene, and bay carrying a chestnut gene.

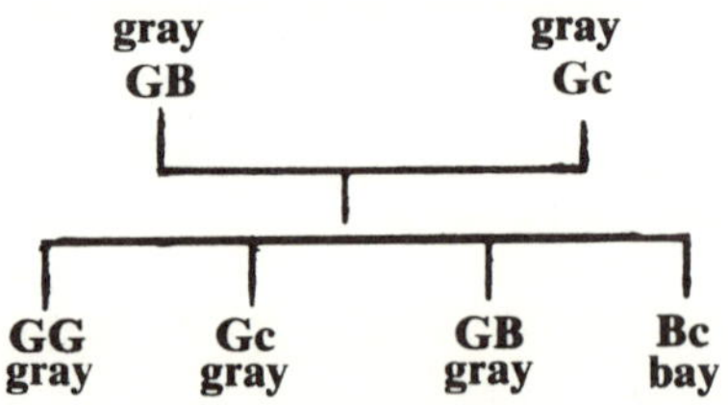

If a gray horse that carries a recessive chestnut gene is mated to a bay horse that carries one also, the foal has these four color possibilities: gray carrying a bay gene, gray carrying a chestnut gene, bay carrying a chestnut gene, and chestnut—if he inherits both recessive genes

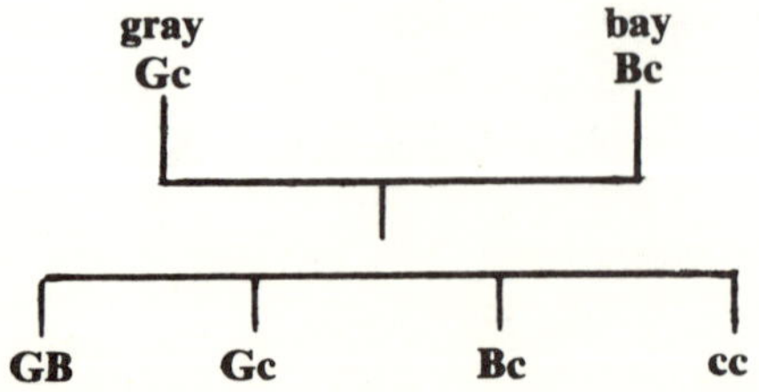

gray = dominant over all colors

bay = dominant over chestnut

chestnut = recessive

A pure gray mated to an impure gray (carrying a bay gene) will produce nothing but gray foals, but some of them may carry a bay gene and pass the bay color to their offspring when mated with another horse that carries a bay gene. A pure gray mated to a bay carrying a chestnut gene will produce nothing but gray foals, but some will carry bay genes and some will carry chestnut genes. Two gray horses carrying bay genes can produce a bay foal, but the chances are one out of four

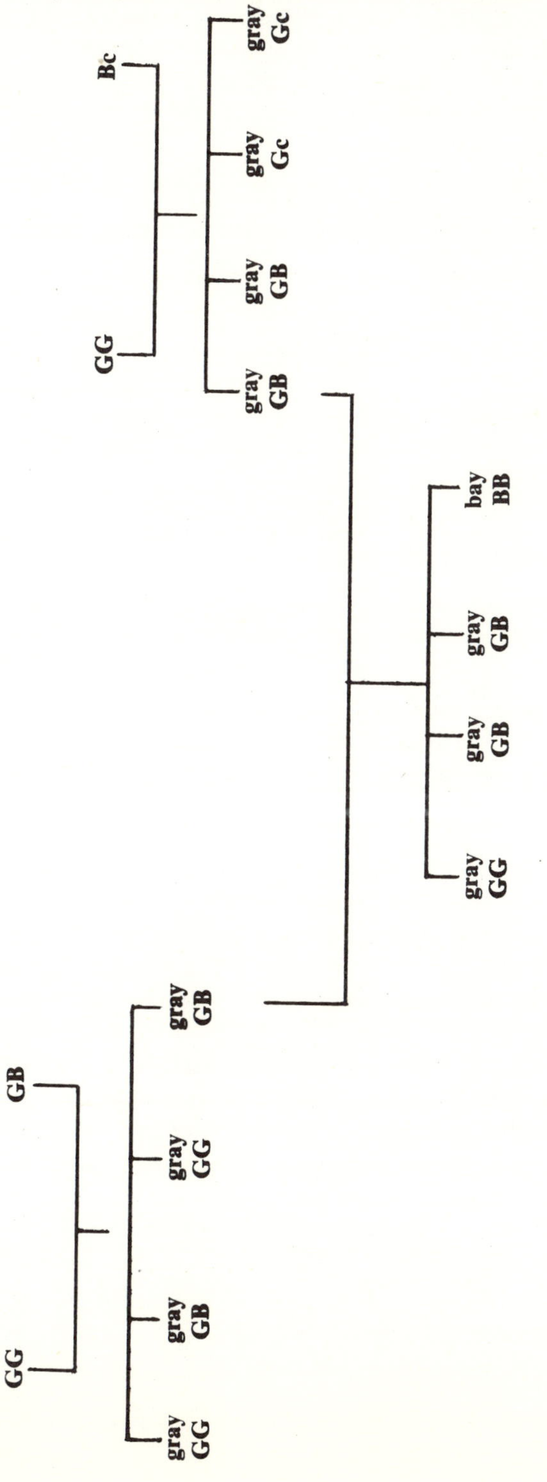

A pure bay mated with a bay that carries a chestnut gene will produce bay foals, but some of the foals may carry the recessive chestnut gene. Two bay horses carrying a chestnut gene can produce a chestnut foal

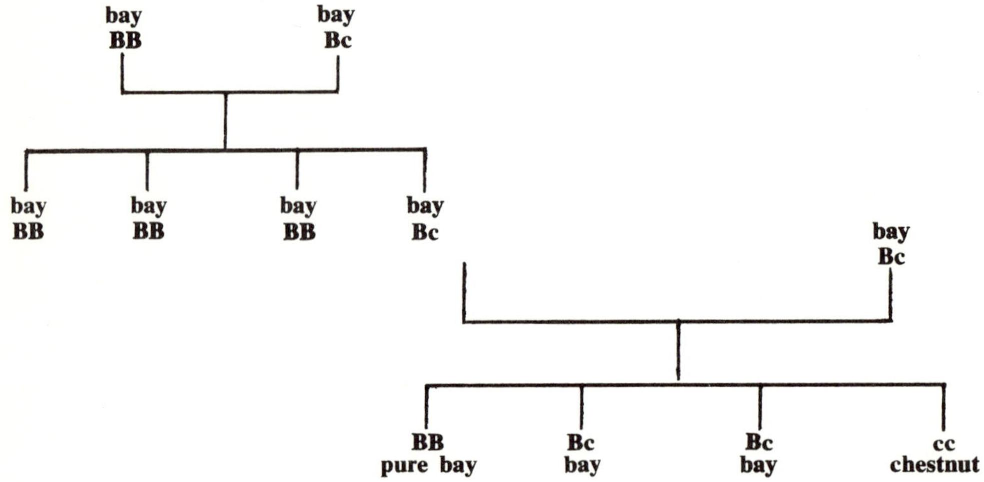

The graying gene seems to be dominant over all the others. A bay mated with a gray will produce a gray unless the gray parent carries a recessive gene. A pure gray (no other genes) will always produce a gray foal no matter what he is mated with. To be gray, a horse must have a gray parent, for gray cannot skip a generation as chestnut can. If a horse receives a gray gene from a parent he will always be gray, even though he might also receive a recessive gene along with it that he might be able to pass on to his own offspring. Gray is always dominant, so if a horse inherits a gray gene he will be gray though he may be born another color and turn gray later. But he can receive a recessive gene from a gray parent and be another color. Then he will have no gray gene to pass on to his offspring.

Brown and black can be considered as related to bay color in horse color genetics. They are merely darker varieties. It is not yet fully known what factors are involved in producing brown or black except that they usually originate from bay parents or from a bay-chestnut cross. They never originate from two chestnut parents. Bays can produce blacks and blacks can produce bays, so it is not a simple dominant-recessive question; there are several unknown factors involved.

White markings are usually dominant. If a horse has white markings (star, blaze etc., on his face, and white feet or legs) he will generally pass them on to his offspring.

Gray is dominant and chestnut is recessive, but two gray parents can produce a chestnut foal if each carries a recessive chestnut gene

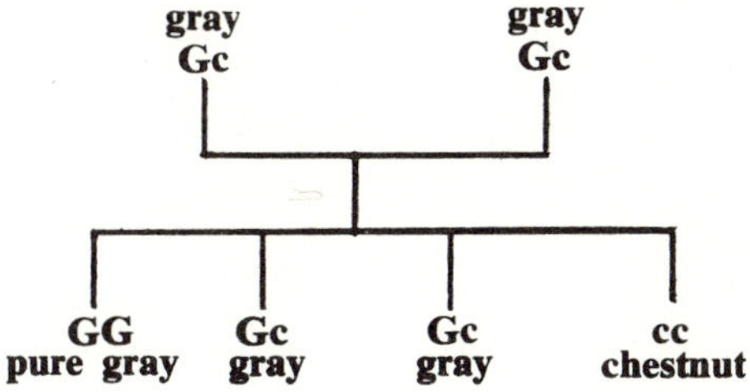

Chestnut to chestnut *always* produces chestnut

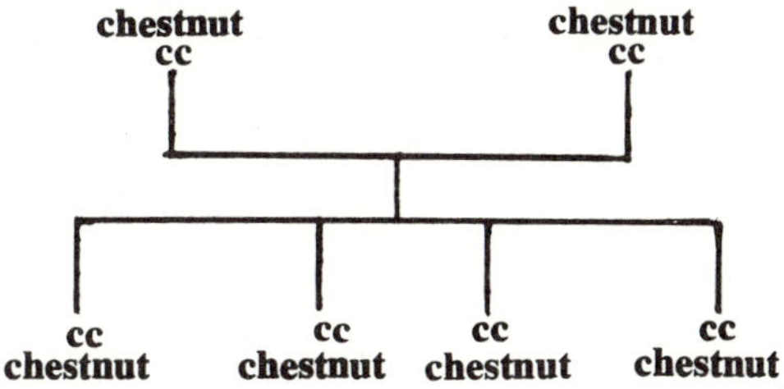

Chestnut to any other color (provided the other color doesn't carry a chestnut gene) will always produce the other color, or its genetic possibilities. In this case, a chestnut was bred to a gray that carried a bay gene. The foal will be either gray or bay, and will carry a chestnut gene

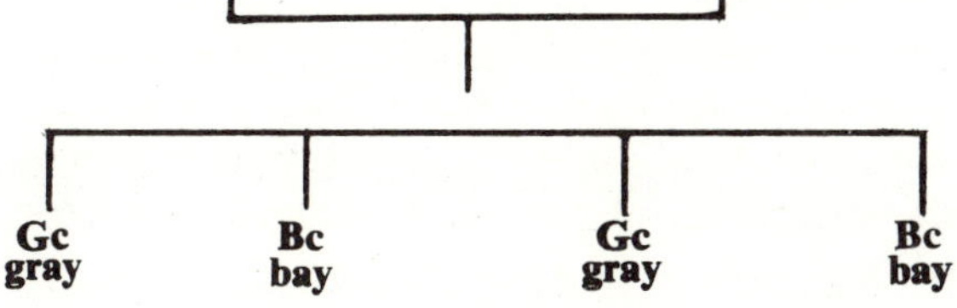

If you are going to breed horses, you will need adequate facilities. You'll need a good place to keep your stallion, and adequate pastures or paddocks (with proper fencing) for the mares and foals and young stock, and hopefully some kind of barn or stalls for bad weather (for foaling mares, sick horses, and so on).

Horse pastures should be on good land that is well drained, clean, and fertile. Swampy, acid soils do not make good horse pasture. Pastures should have good palatable types of grass for horses, and should never be overgrazed. Grass is most palatable and nutritious when it is growing; when it is mature it is rank and coarse. Pastures should be grazed when still green and growing, and should be rotated so that they are not eaten right down into the ground. The grass will grow back much faster and won't dry out so badly if it is still two or three inches long when the horses are taken out and put into a different field. Varieties of grasses make the best horse pasture. Alfalfa and clover are not so good because they can cause colic and digestive problems. Too much clover in a pregnant mare's diet can also cause her to lose her foal if she is in advanced stages of pregnancy.

Horses are very selective grazers and will leave some areas of grass untouched to grow rank while grazing others right down to the ground. It is best if you can rotate cattle and horses, for their grazing habits are different and seem to complement each other. Cattle do not graze as close to the ground as horses for they break the grass off with a twist of their head while horses crop it with their

Don't feed hay on the ground. Horses may pick up worm eggs from manure, or may pick up some dirt or sand

Horses tend to use some areas for their droppings—and manure that accumulates like this makes ideal breeding spots for flies and also harbors worm eggs. Piles of manure should be broken up and spread over the field

teeth. Rotating horses and cattle is also a good practice because it helps to break the parasite cycle. Cattle and horses both pick up parasites from their own droppings, but the parasites of cattle will not continue their life cycle in a horse and vice versa. So after cattle have grazed a horse pasture for a time, the horse parasites will tend to die off and be eliminated.

Another good practice is to harrow your horse pastures periodically, spreading and breaking up the droppings. If droppings accumulate in an area (and horses seem to use some areas just for this purpose), they will be good breeding spots for flies and parasites. Also the surrounding grasses will be sour and the horses won't eat them. But if the droppings are broken up and spread over the pasture they will serve as good fertilizer and will also be opened to the sunshine, killing many of the worms and worm eggs.

Another good management procedure is to clip your meadows and pastures periodically if some patches get tall and coarse. If these patches are just left to get taller and ranker, the horses will never eat them. But if they are clipped, the horses will eat some of the new growth which is more nutritious and palatable.

Horses should never be pastured together in great numbers, especially in small confined areas. Not only are they hard on the pasture, but they will be restless and irritable and there will be a great deal of quarreling amongst themselves. If you have large pastures they can be ideal for six or eight horses, but if your pastures are small and you don't have a lot of room you should divide your horses up carefully in the small pastures (according to age groups and personalities) and then manage the small pastures well with proper rotation and care to assure healthy grasses and to prevent overgrazing. On our ranch we never like to put more than four or five mature horses together in the same pasture.

Don't keep mares and geldings together in the same pasture. One gelding with a group of barren mares that he gets along with may be fine. But two or more geldings in a group of mares will create jealousies and there may be too much rough playing around. And never put geldings with pregnant mares. The teasing and playing around that may result can make a mare abort her foal. Also, do not mix young stock with older horses. Horses are bossy and selfish by nature and the younger, weaker ones will be picked on and run around rather badly. It's best to put horses together of like kind, age, and size. Young colts can run together, and with geldings, up to about three years of age, but should never run with fillies or mares—from the time they are weanlings on. Some colts are mature enough at one year of age to be a real nuisance around fillies and mares and may actually do some successful breeding. Mature stallions should not run with any other horses unless they are with a small group of mares to pasture breed. A stallion may be kept with one or two mares that he has already settled. This situation is often better for the stallion than keeping him totally isolated, for the mares keep him company and he will be happier. Stallions usually hate geldings and should never have to be near them, unless they are through the

Mares and foals at pasture, relaxing in the sunshine

Pastures should be harrowed periodically to spread out the droppings and break them up, exposing parasite eggs to the sunlight

Young stock should be wormed regularly or they may become weedy, stunted, rough, or pot bellied from heavy infestation of internal parasites. This yearling filly is an example

fence from one that they get along with. But if the gelding is with mares, the stallion will be inclined to be jealous.

Pastures and pens must have good, smooth, safe fencing so that horses will not injure themselves. Barbed-wire is the worst kind of fencing for horses, for invariably a horse will at some time get entangled in it or put a foot through it and cut himself severely. On our ranch we are still stuck with a few barbed-wire fences that we haven't had the time and materials to replace yet, so we use a "hot" wire along the top (and an additional one sticking well out along the side if the fence is really a poor one) to keep the horses out of it. Pole fence is very good if it is kept up properly and is always stout and smooth. If you must have wire, use smooth or woven wire and then use an electric wire in conjunction with it so that under no circumstances will the horse ever come in close contact with the wire fence. Horses will respect a "hot" wire very much and it can be used to great advantage to keep horses out and away from fences and also to keep them from chewing on posts and poles. But an electric fence must be regularly checked to make sure it is working properly and not shorting out on grass or bushes or bad insulators. Some horses get rather wise about "hot" wires, and though most will never ever question it, a few (like our redhead Fahleen) will know when it is not working and will then not respect it.

Make sure there are no hazards in your pasture—no sharp edges on water

Mares and foals in a large mountain pasture. During the breeding season a stallion runs with them. These are Quarter Horse mares and foals on Jake Frank's ranch near Park City, Montana

Beware of hazards in your pastures

tanks (and make sure the water is not situated in a corner where horses may get into trouble trying to get away from a bossy companion), no obstacles such as fallen trees, branches, machinery and so on. There should be nothing in the pasture upon which a horse could possibly hurt himself when galloping in play or attempting to dodge a bite or kick from the boss horse. Gates should be wide, and situated in places where there will not be much chance of horses crowding through. Each pasture should have an ample supply of fresh drinking water. If you feed the horses hay during part of the year, feed it in feed bunks rather than on the ground. Horses that eat their hay from the ground may pick up worm eggs from manure, and they may breathe in dust that will make them cough. Too much dust can lead to heaves. And if a horse is fed on sandy ground or gravel, he may eat bits of sand which will collect in his caecum and later might lead to fatal colic.

Buildings, barns, box stalls, and so on, should be adequate in size and sturdily constructed. Box stalls should have proper drainage and should be at least twelve by sixteen feet in size, with plenty of head room. Doors should be eight feet high and at least four feet wide, for narrow doorways are dangerous; a horse may knock a hip or stifle or scrape himself badly in a narrow doorway.

To be a good horse breeder you must be a good judge of horses. When breeding two horses, you must consider that you are mating not only two individuals, but two separate ancestries as well. You must in some way be able to predict what the product of that mating will bring, trying to select matings that will complement both the sire's and dam's characteristics, creating a foal that will

Horses should have shade during the heat of summer

inherit good qualities from both. There are no perfect horses. So you have to use good judgment in trying to mate horses that will improve upon themselves and produce offspring at least as good (and hopefully better) as themselves. If your horse has faults, mate it with a horse that is strong where yours is weak. Never mate two horses with the same faults or you will likely produce a foal that is even worse than its parents. It is never a good idea to double up bad points of conformation.

Many horsemen are very capable of seeing the faults in other people's horses but are unable to see the faults in their own. This can be very detrimental to their breeding program. To be a successful horse breeder you must be very able to critically judge your own horses, especially their weaknesses and shortcomings, so that you will be able to mate them in the best possible way to improve upon them.

When breeding horses we must remember that disposition, conformation, action, and soundness are all greatly inheritable and must be taken into consideration. If a mare is fairly good but hasn't quite got perfect straightness of action, she should be mated with a stallion that is perfectly straight in his way of going. Lameness or unsoundness that is due to injury or accident can be overlooked to a great extent because that kind of defect will not be hereditary. But if the problem is partly due to faulty conformation the horse will most likely pass it on.

Breeding good horses is a time-consuming job that involves much work and expense. It really costs no more to keep a good broodmare than it does to keep and feed a poor one, so every effort should be made to try to buy or raise the best stock you can possibly afford. And a good stallion will give your breeding farm and your horses a good reputation. The more you can constantly upgrade and improve your horses, the better. If you consistently sell good foals, people will come back for more, but if you sell a few poor ones, you'll find yourself with a more limited market. Word gets around about the quality of your stock. Remember that when you are in the business of breeding horses, you have several responsibilities—to your stallion (his good name and reputation as a sire), to the general public, to the mare owners that come to your door to breed to your stallion, and to the breed you are trying to improve and promote. If you are truly conscientious about your role as a breeder of good horses, you won't let money or fads overrule your very best judgment in what is best for the breed and for the horse industry in general. If you are consistent in your goals and your goals are set high, there will always be a market for your horses even if you aren't following the lastest fad. Best of luck.

5

BREEDING THE MARE

LET'S TAKE A QUICK LOOK AT THE MARE YOU ARE ABOUT TO breed, whether she is part of a broodmare band or the favorite mare of a pleasure horse owner who wants to raise a foal "just like her."

The time to give the broodmare proper care for conceiving, carrying and raising a foal is before she is ever bred. The importance of adequate care and proper feeding cannot be overemphasized for the horse has the lowest conception rate and the highest mortality rate among newborns of any domestic animal, partly because horses are bred and foaled in such unnatural conditions and partly because of improper care, feeding, and sanitation. Proper care can do a lot toward insuring a foal from your mare.

The mare should be in good physical condition before she is bred. She should not be too thin or too fat. Mares that are thin and undernourished may not come in heat. Mares that are too fat are hard to settle. A thin mare in a gaining condition will actually settle faster than a fat one. Fertility is highest in the mare when she is healthy, carrying good flesh but not fat, and getting regular exercise.

She should be fed a high protein hay or be on good pasture. Some mares will need some supplementary grain even with good hay or pasture (especially if they are still nursing last years foal), and some mares won't. Know your individual mare and feed her accordingly. If your pasture is poor or you are feeding poor hay, you'll need to feed a larger grain ration. Additional vitamin and mineral supplementation may be necessary too unless your pasture or hay is top quality and not grown in deficient or depleted soils.

You will want to worm the mare before she is bred. You'll want everything in her favor when she is carrying her foal so that it will be a healthy one; you won't want internal parasites to rob her of proper nutrition and vitality. Also, you should have her checked thoroughly by a vet before you have her bred. If you are breeding her to someone else's stallion a vet's certificate on her breeding soundness may be required anyway.

A mare should not be bred until she is mature and fully grown. Breeding a filly too young is hard on her and may stunt her growth. If she is still growing herself, she won't do justice to herself and the foal she is carrying, and she may not give as much milk to her foal as an older mare would. Some horses mature faster than others and can be bred at three years of age, to foal at four. Most mares however are better off if you wait until they are four years of age to be bred, to foal at five. I would never recommend breeding a two-year-old filly.

Breeding good horses, if that is going to be your business or hobby, is a time consuming job requiring considerable work and thought, and some expense. If you are raising horses or even just a foal or two, remember that it costs no more to keep and feed a good mare than it does to keep and feed a poor one, and the costs involved in raising a foal will be nearly the same whether the foal is a good one or a poor one. If you are raising horses, it is foolish to try to use anything but the best stock available that you can afford. But if you only have one mare or two and want to raise a foal or two for the enjoyment of it, I wouldn't discourage you even if your mares may not be the best. A mare may be of uncertain quality or old, but if you have enjoyed her and want to raise a foal from her, it's worth a try. Such a mare probably won't raise a top-quality foal, but if you breed her to a good stallion the foal will have chances of being better than its mother. Even though your chances of getting a really good foal from a mare like that is rather poor, since you are planning to keep the foal for yourself rather than sell it, merely breeding the mare and raising the foal for your own enjoyment, the foal will be worth your time. If the mare has given you satisfaction, you will probably get some of the same qualities in the foal that you enjoyed in the mare.

If a mare is old she may be hard to settle, but it is not impossible to raise a foal from her unless she has a definite sterility problem. If a mare is over ten years of age and has never raised a foal she may be more difficult to settle than a younger mare or a mare of the same age that has been producing foals consistently. But though the chances of getting a mare in foal decrease with each year of age she does not produce one, it may still be worth while to try, if you are fond of the mare. There have been cases of mares as old as 20 raising foals successfully when bred for the first time, but this is rather rare. If you decide you want to breed an older mare, have the vet check her first.

A mare can be bred practically any time of year if she is healthy and her estrus cycle is regular, but there are many advantages to breeding in the spring. Many mares do not cycle all year round; they remain somewhat dormant in the winter months and begin cycling again in the spring. The first few heat periods in the early spring may be long and irregular; she is more apt to conceive if bred in later spring when her cycle is normal and regular again. Ovulation (egg release, during her heat period) is not completely automatic; the mare may not be entirely regular. Outside factors (season of the year, the mare's health, social stimulation) seem to affect it.

A healthy normal mare usually comes in heat every 21 days, if she is not pregnant. This figure is an average, however. A mare's individual cycle may be more days than this, or less. The heat period in mares is relatively long, lasting three to seven days. When a mare comes in heat she may be restless, or she may play up to the other horses in the pasture or across the fence, or be more cranky or quarrelsome than usual. Sometimes she will urinate frequently, or there may be a discharge from the vagina. Some mares are always violently obvious about their heat periods while others are not noticeable at all and are difficult to diagnose.

If you are going to have the mare bred somewhere else it is a good idea to leave her at the stud farm for awhile so that when she does come in heat she can be bred at the proper time. She'll be more apt to come into heat properly and conceive if she has a chance to get acquainted with the strange surroundings and the stallion before she comes in heat.

Leaving your mare at a stud farm will probably cost you about $1.50 to $2.00 a day for her board and care. This is really quite reasonable when you consider the value of the feed she will be eating, and the time involved in caring for her and getting her bred. When you send a mare to be bred, have her in good breeding condition, in good shape but not too fat. If you have been keeping track of her heat periods and have an idea when she might come in heat next, this fact may be taken into consideration when you send her and you'll also want to advise the stallion owner of the time she may possibly come in heat.

If she is a maiden mare (one that has never been bred before) or a barren mare (one that has had foals before but does not have a foal at side this year) you should send her to the stud farm about a week before she is due to come in heat. This will give her a few days to get adjusted and acquainted and she will be more likely to come in heat on schedule. If she is sent too close to the time she is to come in heat she may be upset by strange surroundings and may not come in heat on schedule. If the mare is pregnant and you are sending her to be foaled out there and bred shortly after foaling, send her at least two months before she is due to foal; sending her too close to her foaling time can result in accidents if she is bumped or upset during traveling. If you are sending her to be bred with foal at side, you can send her a few days after foaling. Some mares are most easy to settle at foal heat, which occurs at about the ninth day after foaling, but breeding at foal heat is risky if there is any trace of infection or if the mare is not yet properly healed from foaling.

When sending her to be bred, have her well groomed and clean, with her feet properly trimmed, and with no shoes on her hind feet. This latter consideration is for the safety of the stallion. You should also send her with a good halter.

For successful breeding, timing is important. Ovulation usually takes place about 24 to 48 hours before the end of heat. Because mares vary in their length of heat, they are often bred the second or third day of heat and every other day thereafter for as long as the heat period lasts—even though one good breeding is enough, if given at the proper time. Usually when a mare is ready to conceive, she will be much more cooperative about being bred. If your mare is fairly regular in her heat periods and always seems to be in heat a certain length of time, it would help the stallion owner to know these details.

Let's take a closer look at the actual breeding of a mare to your own stallion. When breeding mares by hand (not pasture breeding) you must try to determine when the mare is in heat, and then try to determine when she is most likely to accept the stallion and conceive. A mare may be in heat for as long as five to seven days but there will be a time during that heat period that she is ready to breed. Breeding her too early will not conceive a foal.

There are several ways in which you can try to determine when she is ready to be bred. Some mares are obvious. Others are not, and may have to be taken near the stallion every other day or so to discover their reactions. This introducing of a mare to the stallion to see whether or not she is ready to breed is called teasing. Sometimes teasing can be done with a gelding if there are many mares to check and you want to "save" your stallion: some mares will react to any male horse.

There are several ways you can introduce a mare to the stallion. You can put the stallion in a box stall with the top door open so that he can see the mare as she is led up toward him, but he will be somewhat protected if she suddenly decided to strike or kick. It is always a good idea to introduce the mare to the stallion a time or two before she is actually ready to be bred, for then she will be somewhat acquainted with him and more likely to accept his services when the time comes. If she has never seen the stallion before, she may become a bit frightened (especially if she is a maiden mare) and refuse to accept him. The first time our Bambi was introduced to a stallion, she was terrified and would not go near him.

A good way to check your mare is just to walk the stallion daily past the mares' paddocks or pastures. This will reveal a great deal as to which mares are likely to be about ready to breed. A mare that is not in heat at all will practically ignore the stallion and go on about her business after a glance or two, whereas a mare that is coming into heat will be more curious and come over to the fence and maybe flirt a little. A mare in heat that is about ready to breed will come up to the fence and show her inclinations. But if a mare is shy or is not one that shows obvious and easy-to-see signs of heat, you may have to bring her to the stallion every day or so and try her at closer quarters.

Some breeders use a "teasing chute" to determine whether or not a mare is in heat. The mare is confined in the chute and the stallion is led up to her in a manner so that she can see him coming. The stallion should always be under complete control, with excellent manners. The stallion approaches the mare and makes her acquaintance, nibbling and talking to her but is never allowed to bite or scream. If the mare is in heat and ready to breed she will show it. If she isn't, she will reject the stallion rather obviously and firmly.

When a mare is ready to breed she will submit to the stallion willingly and stand for him. A mare that kicks or fusses a great deal is probably not quite ready. Some breeders use kicking hobbles and other restraints on a mare as a matter of course, but this is not always a good idea. If a mare neeeds hobbles to keep her from kicking it probably means she is not ready to be bred. And

Sometimes a mare can be teased effectively with a gelding to determine whether or not she is in heat

Khamette and Sur Amir. A good way to check a mare is to bring her to the stallion and let her make his acquaintance. She will soon make her feelings obvious. If she is in heat, she will express willingness to be bred. If she is not in heat, she will refuse the stallion loudly and perhaps strike at him

breeding her before she is really ready will not likely produce a foal. It is better to use a little patience and work with nature. If a mare is not ready to be bred, she will show her disinterest quite firmly, and may kick or strike. It's no use wasting time trying to force her. And as to the use of hobbles—they are quite unnecessary on a well-mannered mare, and very hard to put on a flighty, nervous mare. It's better not to use them.

When a mare is in heat and ready to breed, the proper time will come when she will accept the stallion willingly, without any inclination to kick. However, maiden mares that have never been bred might be a little nervous. They may need quite a bit of steadying to keep them from moving around too much.

When a mare has been introduced to the stallion and you have decided that this is the proper time to breed her (she shows signs of being willing to stand for the stallion), have someone lead her to the place where she will be bred. It's best to have at least two people—one in charge of the stallion and one to lead and hold the mare. The mare should have her tail wrapped in a clean tail bandage or enclosed in a clean nylon stocking. Her genital area and buttocks should be washed and thoroughly rinsed. The stallion's penis and sheath should be washed with a very mild soap and then thoroughly rinsed, with water at body temperature. The mare can be briefly teased again so that she and the stallion will both be ready.

The stallion should then be led up behind the mare, about three horse lengths behind her and a little to one side so that she can see him. The area where the mare stands should be level, with good footing. The mare may be nervous and (especially if she has never been bred before) she may try to turn around and face the approaching stallion. The person holding her must try to keep her calm and steady. If the mare is very nervous and upset you may want to lead the stallion around her at a safe distance a time or two so that she can see him better and perhaps lose some of her fear. The stallion must, of course, be well-mannered and very manageable.

But if the mare is still nervous and won't be still, yet you are fairly sure that this is the right time to breed her, you can use a twitch to make her stand quietly. The twitch should be used only temporarily and should be removed immediately after the stallion has begun to mount her. It won't be needed any longer than that. We've never had to use a twitch or restraint of any kind on any of our own mares, even as maiden mares. When they are in heat they are very willing to be bred and stand nicely for the stallion.

If the mare to be bred has a foal at side, you can confine the foal in a safe enclosure some distance away with someone to stay with the foal and keep it quiet, or you can put the foal in a little pen right in front of the mare. If she can see the foal she will not be so worried about it.

The stallion should be led up to the mare a little to one side, just in case she might try to kick. He should not be led up until he is fully ready to breed her, and then when he is ready he should be led up with no delay. But do not let him rush or he may frighten the mare. The stallion handler should make sure of a proper entry as the stallion mounts. The mare will step forward a step

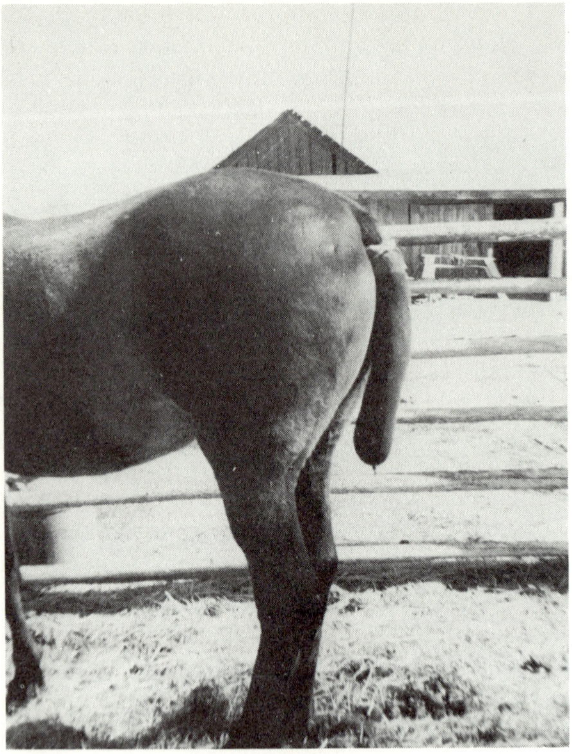

A nylon stocking makes a good tail wrapping for a mare about to be bred. It fits easily over the tail and can be tied securely. And it can be washed and reused several times

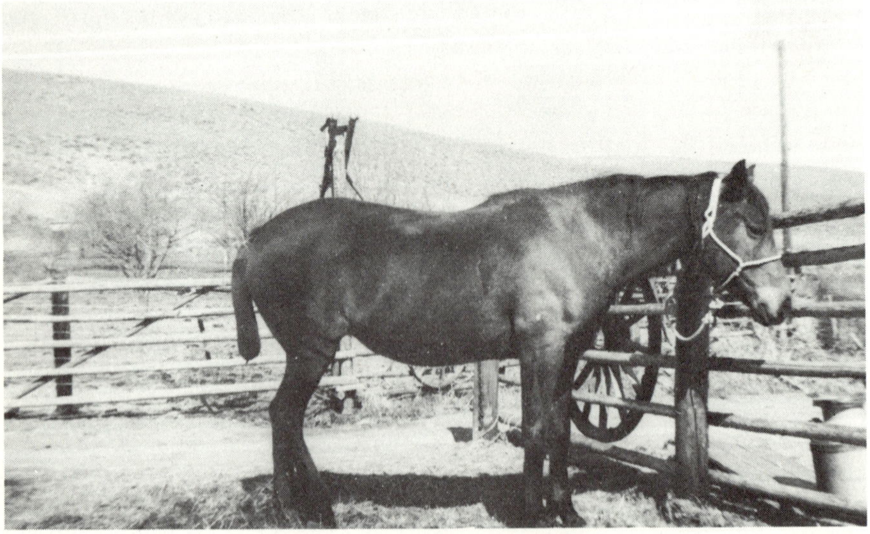

Khamette

Sur Amir and Khamette, discussing the matter

or two as the stallion mounts, but this is all right and probably best, to prevent tearing the mare inside and to allow the stallion to find his proper position.

The person holding the mare should try to hold her head rather high as the stallion comes up to her, as this will tend to inhibit her kicking if she feels so inclined. After the stallion has finished his service, the person holding the mare can move her a step or two to the side, which will help the stallion dismount. Then the mare should be led for a few minutes to keep her from standing and straining as some mares do, discharging the semen. The mare can then be put back in her pasture and the stallion taken to his stall or paddock and his sheath washed with a mild, nonirritating disinfectant. If ever any irritation develops from repeated washings, vaseline can be liberally applied.

If a mare does not conceive, she will come in heat again at her next period, about 21 days later. The mare should be checked at her twenty-first day after service and then again about the forty-second day, but if all indications are that she is safely in foal, these trials should be as short as possible. Because mares can be very irregular in their heat periods it is good to watch every bred mare closely for several weeks after service to determine whether or not she settled. A mare can be considered safely in foal if she shows no signs of heat for the next two possible heat periods.

Pasture breeding is much more simple than hand breeding and some breeders like to do it this way. The stallion runs with his group of mares during the breeding season. The pasture should be large enough to accommodate them, with no other stallions, geldings or mares that might come in heat in adjacent pastures. Pasture breeding is the cheapest and easiest way to breed horses, if a person is set up to do it that way. Mares out in a large pasture don't have near the infection problem that mares can have in a more confined and contaminated area. And mares that run with a stallion are more likely to come in heat and conceive for they are bred at just the right time. And there is no problem with "shy" mares that escape human detection when in heat. The stallion knows more about mares and breeding them than his human handlers do.

If you pasture breed, you'll want to pick out the mares to breed to a certain stallion and have him run with the same mares all season. You can't change mares on him. And, if you take a mare out of his group for some reason, you may not be able to put her back later.

The pasture breeding stallion knows just when to breed a mare. Conception rate in pasture breeding is much higher than that of hand breeding, but a stallion carefully hand bred can service more total mares during the breeding season. In a pasture breeding situation there may be two or three mares in heat at once but the stallion will breed the one that is most ready. He won't ever breed a mare until she is ready. He knows how to handle his mares, how to tease a mare and how to avoid kicking heels and striking forefeet. He may tease several mares at once and then breed the one that is most ready. Running with the stallion, and the lengthy teasing that results when a mare is in heat, seems to help a mare be fully ready to breed when the time comes.

6

INFERTILITY IN MARES

INFERTILITY IS ONE OF THE MAJOR PROBLEMS IN THE HORSE industry today. The average conception rate of mares runs as low as fifty percent in many breeding herds. A figure this low would put any other livestock producer out of business (for instance a cattleman tries for at least a 95% calf crop and feels quite a disappointment and loss if his cattle do any more poorly than that). A fifty percent foal crop can almost put a horse breeder out of business too, for he cannot really afford to keep mares that do not produce foals.

Some of the reasons for this terribly low figure in the horse industry are poor management on the part of mare owners, lack of understanding about the normal breeding functions of the mare, and physical problems of individual mares. A physical problem making the mare infertile may be due to any number of things. Sometimes the infertility is inherited, sometimes it is man-made and sometimes just the result of improper management.

Some mares are not properly fed and conditioned for best breeding soundness. Some mares are too fat and therefore hard to settle; too many layers of fat inhibit the ovaries. A fat mare should be dieted (on a well-balanced diet of course) and exercised. When she has lost a great deal of weight and actually looks thin (remember that the last fat to leave the body is that internal fat around the various organs), you can breed her as she starts to pick up weight again and she will probably settle.

If a mare lives in a small enclosure and does not get enough exercise to keep herself in good physical condition with proper muscle tone, she may be infertile. If your mare is a problem breeder and you think it might be due to lack of exercise, ride her. A mare that is being ridden and exercised will have less internal fat and her ovaries will be stimulated by the increased circulation of her blood and by the "massage" of being ridden and worked, performing various gaits and movements.

Other mares may be too thin, or have heavy infestations of worms, bad teeth, or some other problem that interferes with proper nutrition. A thin mare that isn't getting proper feed will not even come into heat, for this is nature's way of protecting her; she could not support a growing fetus.

Season of the year and psychological factors are often disregarded by the horse breeder when breeding a mare but they can play a very important role in the success or failure of getting a mare bred and settled. Nature intended the mare to be bred in late spring or early summer; she is usually gaining weight then if she was thin towards the end of winter, and she is shedding her winter coat. Her health is blooming. The increasing light of the long summer days has an effect on her endocrine glands and the new spring grass helps ready her for breeding. Under normal and natural conditions this is the most ideal time for a mare to be bred. Her foal will be born the next spring during good weather, and he will have the summer ahead of him during which the feed will be abundant and nutritious—advantageous to his early growth and to his mother's milk supply.

But many breeders try to get foals as early in the year as possible, even though mares are harder to settle then. Breeders want their foals in January and February because of the fact that a horse's birthday is January first, no matter what month he was actually foaled. Thus if a young horse is to be raced or shown, he will have an advantage if he is a few months older. Or rather, he will be at a disadvantage if he is a July foal competing against January foals. If he is born early in the year he will be larger and stronger. Horse breeders who are in the racing business or who try to show or sell their young stock at an early age want as much age and growth on those youngsters as they can get. This fact has actually hurt the horse industry in several ways.

First of all, horses are raced, trained and ridden much too young. The competition is fast and furious to try to have a "finished" horse at age two or three. This is ridiculous because it only results in unsound horses. A horse isn't mature enough to take this kind of strenuous activity until he is four or five years old.

Secondly, the demand to breed mares early and produce foals in January and February (still in the dead of winter) has been one of the factors contributing to the low conception rate of the horse. Breeding a mare in the winter goes against nature. In winter and very early spring the mare's reproductive cycle is usually irregular (it usually goes somewhat dormant in late fall and winter) and her early spring heat periods may be long and irregular. Many mares do not cycle at all until later in the spring. Those that do cycle in winter and early spring are usually irregular and may fail to conceive if bred at this time. If you want to insure a good conception rate among your mares it is better to work with nature rather than against her. Breed a mare at the most advantageous time for the mare and her foal even though it might not be the most advantageous time for the people who want the foal to be half grown at "age one" and ready for the track or the show ring.

When breeding a mare, psychological factors are important and should never be ignored. A mare's reproductive cycle is not entirely automatic. Loneliness can hinder normal ovulation. A mare kept in a pasture all by herself may not come in heat as regularly as a mare that lives with other horses or is where she

In early spring some mares have long, irregular heat cycles. Khamette (shown here) was in heat for more than four weeks during March and April of 1972, but she had more normal cycles later that spring

can see a stallion from time to time. Social stimulation is more important to ovulation than many pleasure horse owners who want to breed their mares actually realize.

Improper handling of a sensitive mare, or strange surroundings (a long trailer ride, strange place, strange feed and water, strange horses) may throw her off schedule and upset her so much that she will not breed successfully. When a visiting mare goes to be bred at a stud farm, give her time to get acquainted and to adjust to her new surroundings. Don't be surprised if she doesn't come in heat for several weeks; she may have been in heat soon after she got there but was too nervous or tense to show it.

A sensitive mare should always be handled with care and sympathetic understanding during teasing and breeding. Shy and sensitive mares can easily become discouraged by improper handling. The stallion also should be properly handled; he should have excellent manners. If the stallion is too aggressive, the mare may become frightened. A shy mare may have become shy because at some time or another she may have been thoroughly frightened by a stallion that was allowed to be too aggressive and abusive during teasing. Or, she may have been forcefully bred sometime when she was not really ready, and she may hesitate to ever show signs of heat again. Mares are very sensitive creatures and the psychological aspects of breeding have more to do with successful breeding than many people know.

Inheritance of fertility traits is another factor that many horsemen rarely think about when breeding horses. There is always much more emphasis placed upon speed, agility, gaits, athletic ability and powerfulness, color and so forth, than upon fertility. None of the traits we usually look for in a horse and breed for actually contribute to fertility and there is some evidence that great athletic ability in a mare may actually be conducive to poor reproductive performance and poor milking ability as a dam. Most livestock breeders take into consideration the fact that males should be masculine and females should be feminine and maternal, for best fertility and production of offspring. A cow that doesn't produce a calf, or one that doesn't feed hers well is culled, and within a few generations the infertile lines of cattle are pretty well bred out and eliminated. But this hasn't been true in the horse industry. Many powerful, masculine-type mares are selected for breeding in the hopes that they will pass on their athletic ability to their foals. But when they do reproduce (often after many unsuccessful tries to get them settled) their offspring are likely to inherit this tendency toward infertility and the problem is perpetuated. This is one reason that there are so many good performance and racing mares that are very hard to settle. After much effort this type of mare may be finally gotten with foal, only to produce a filly with her same problems, or a colt that grows up to be semisterile.

Poor breeding techniques can also contribute to a mare's inability to conceive a foal and carry it to term. Breeding must be done at just the right time. And then some breeders simply rebreed a mare if she comes in heat again, without first checking her for pregnancy. This is not always wise because occasionally a mare will show one or more heat periods after becoming pregnant, and if she is rebred she will usually lose her fetus. It helps to know the individual mare. Some mares exhibit signs of heat during pregnancy, but will not actually accept a stallion. If forcefully bred (as some breeders do as a matter of course, using hobbles, etc.) the mare would probably abort. Be sure the mare is exhibiting a normal heat before you rebreed her.

Sanitation is another very important factor and should be of utmost importance on breeding farms, but sometimes it is lacking. This will be discussed a little more fully when we talk about infection in mares, later on in this chapter.

Man-made fertility problems, in which a mare's hormones have been tampered with, are usually the saddest cases for sometimes they are unsalvageable. Giving a mare a shot of hormone can have its place occasionally when dealing with problems, but hormones should never be used indiscriminately. The hormones in an animal's body are part of an intricate and delicately balanced interlacing system and should never be treated lightly. Giving a mare a shot of hormone to assure she doesn't come in heat (so she won't be nervous and unpredictable at a major show or something) is inexcusable; there may come a time later that you'll wish to goodness that she *would* come in heat.

Another factor that can contribute to infertility in mares is just plain old age. Cattle and other livestock are usually culled out and sold when they get old, and senility doesn't become a problem in reproduction. But mares are often kept until they are very old because of their individual value, and are bred as long as

Nell at age 19. She produced four foals between the age of 8 and 15—taking time out from the work as a ranch horse. She ceased showing signs of heat periods at age 17. At age 19 she showed signs of heat again, perhaps because there was now a stallion on the ranch, but did not conceive after several breedings. Later in the summer she showed evidence of a genital tract infection

they will conceive. A good mare is often still valuable as a broodmare even after she is no longer able to perform very well physically (on the race track, in competitive events, and so on). As a result, old age is often a cause of infertility and it should be recognized as such. The ovaries may simply become inactive, even in a mare that has regularly been producing foals. One condition sometimes found in older mares is fibrosis; as a result of scarring, ovulation does not take place. Another change often associated with aging is the presence of cysts in the ovulation fossa, which may result in delayed ovulation or none at all. As a mare ages there may also be changes taking place in her uterus which make it difficult or impossible for her to conceive.

Hormone imbalance sometimes causes infertility. Some of the conditions caused by abnormalities of the endocrine glands can be easily detected, but others can be determined only after many internal examinations to discover what is happening in the ovulation cycle. Some hormone problems respond to treatment with the proper hormone, but others do not respond to treatment at all. A complete and honest history of the mare may help your vet diagnose and treat her problem. If she has had injections at some time or another to keep her from coming into heat, tell your vet and give him the name of the drug that was used. If this is the case, your mare may get back into proper balance again (if you are lucky) if you turn her out to pasture for a year or so to "let down" and shed all the accumulated artificial hormones in her body and then reestablish her own cycle. Fresh air, sunshine, and green grass, plus the exercise of moving about and grazing, plus *time* can often do wonders. When spring comes again, the mare may begin to cycle again.

Cystic ovaries can cause hormone problems in a mare. The mare does not ovulate, her body acts as though it were pregnant, and she does not come in heat. Occasionally a mare with just one cystic ovary will go ahead and cycle normally on the other one. Cystic ovaries can sometimes be treated successfully with hormone shots, internal massage, or increased exercise.

Our mare Bambi had never shown any signs of heat from the time we purchased her as a 3 year old. When she was 8 we had our vet check her because we wanted to breed her. He found that she had a cystic ovary and was not cycling at all. We tried the recommended hormone shots, but they did not help. We do not know exactly why she began to function normally two years later. Perhaps exercise and *time*, a new season of the year (spring, when mares usually start ovulating again after a dormant winter) and a stallion nearby were what helped. The psychological aspect may be more important than we know; that spring there was a stallion on the ranch, and the other mares in Bambi's pasture were being teased and bred. The next thing we knew, Bambi came in heat also—for the very first time. And she was then bred successfully.

The largest single cause of infertility in mares is infection of the genital

Bambi. Eight years old when this photo was taken, this mare had a cystic ovary and had never been observed to be in heat from the time she was three years old (when purchased). But as a 10 year old, this mare came in heat and was bred

tract. Once a uterine infection becomes well established, it is very very difficult to clear up, even with extensive treatment. Many infections go undetected until too late and then the future usefulness of the mare as a broodmare is greatly endangered.

If an infection is detected early enough and treated properly (within the first few months, or even within the first breeding season in which it occurred) the chances of recovery may be as high as 75 percent. But if treatment is delayed until the second year from the time the infection started, chances of complete recovery drop to forty or fifty percent. By the third year of infection, the figure drops as low as 10 to 20 percent. After this much time has gone by, it is highly improbable that the mare will ever produce another foal.

Uterine infection is caused by many different organisms, and the infection can be acute or chronic. Infections are most often caused by the bacteria that are normally found in the surrounding environment; they become a problem in the reproductive tract only through contamination—at foaling time, during breeding if unsanitary methods are used, and so on. Some mares are naturally more susceptible to infection than others; it can tend to be a hereditary problem.

The uterus is an ideal place for bacteria to grow because it is warm, dark, moist, and nourishing. Nearly all mares pick up a genital tract infection at foaling time, for the open genital tract comes in contact with a certain amount of contamination. When the foal is expelled, the void it leaves in the uterus creates a vacuum (negative pressure) and the stall air rushes in, perhaps carrying infectious agents with it. And if a mare doesn't expell her placenta as promply as she should (perhaps because of poor health or poor nutrition or because she is too fat) the retained afterbirth may act as a route for infection to enter; it is hanging out, contaminated by the outside environment, and leads directly into the uterus, which at this point is somewhat weakened and susceptible to bacteria because it has just been under great stress.

Most mares throw off this infection that they pick up at foaling time and are able to throw it off in a relatively short time (one or two weeks) and without treatment. But if the mare is in poor condition or for some other reason has a lowered resistance at this time, she may not throw it off so easily. If there has not been extensive damage to the tissues during foaling, the mare's natural resistance is usually able to combat this minor infection and ward off serious problems.

But, the fact that most mares *do* have minor infections for a short time after foaling is one good reason for not rebreeding at the foal heat (which occurs about the ninth day after foaling). Some mares are just not yet completely recovered at this time and the breeding may not be successful, or may create further problems if the mare does conceive. Breeding the mare before her infection is cleared up just seals in the infection if she conceives, and will probably result in an abortion later.

Many different organisms can cause uterine infection, but the most common is *Streptococcus zooepidemicus.* All strains of this organism seem susceptible to penicillin, so many infections in the mare could be cleared up if treated properly in time, before extensive or permanent damage is done to the genital tract. Chronic or repeated infection results in the loss of the glands in the mucus coat of the uterus and the replacement of these glands with scar tissue. The scarring can become extensive enough to result in permanent sterility.

Retained placenta is one of the most common causes of deep-seated uterine infections. Retention of the placenta is often due to a low-grade infection that is present even before the mare foals. The foal may or may not be diseased at birth from the infection. If a placenta weighs more than fifteen pounds when it is freshly passed, it is usually infected.

One type of infection that can occur in association with retained afterbirth is caused by the bacteria *Escherichia coli.* When this happens, the uterus thickens and fills with a foul-smelling bloody fluid. The entire mucous membrane of the uterus lining and its glands are destroyed and the mare dies within 24 hours. Deaths following complications of a retained placenta are not uncommon.

Sometimes infertility in a mare is caused by an abnormality in her anatomy. Some abnormalities are obvious to the layman but others can only be discovered by a vet's internal examination. Some can be corrected surgically but others cannot. A few of the problems in anatomy that can cause infertility are imperforate hymen, closed tubes, hermaphroditism (the presence of both male and female organs) and so on.

"Windsucking" in mares can be a cause of infection and infertility. This problem is anatomical in that it is usually due to a defect in the relationship of the anus to the vulva—a sinking of the anus and a tipping forward of the vulva. The anus sinks forward into the body and the vulva is drawn upward and forward so that its upper part is horizontal instead of vertical. Thus, when the mare has a bowel movement, some of the feces may fall through the vulva into the vagina. This can be a conformation problem or may be merely due to poor condition. Even mild cases can cause trouble, for when the mare gallops, air is sucked into the vagina, carrying with it bacteria that are present on the vulva from defecation. The constant irritation and contamination of bacteria can finally overcome the natural resistance of the tissues and an infection gets started. This irritation can also create an unfavorable chemical reaction with the fluids of the genital tract, shortening the life of the spermatazoa when the mare is bred, making it almost impossible for her to conceive. Windsucking can also drive the infection further on into the uterus. A mare with this problem can be settled (provided the infection is merely local and hasn't gotten into the uterus) by "capsuling," bypassing the local irritation and chemical upset. Capsuling is done by collecting some of the semen from the stallion and placing it inside the mare in a small capsule. Sometimes in mild cases a douche can be given to the mare 24 hours before breeding (a pint of warm water containing disinfectant) to neutralize the bacteria temporarily. A surgical operation (the simple Caslicks repair) in which the upper edges of the vulva are sewn together eliminating the easy access of feces into the vagina can usually prevent windsucking and its problems.

Infection can be spread from mare to mare by the stallion. Obviously cleanliness and good breeding hygiene are very necessary to prevent this. Mares that are known to carry any infection at all should not be bred. It is a poor risk to breed a mare with any kind of infection anyway because even if she did settle the chances are good that she would abort later. It is much better to "waste" a few weeks treating the infection and clearing it up than to breed the mare and have her abort the fetus at three to six months and then have to start all over again.

For a stallion to spread infection from one mare to another, it is usually necessary that he breed the second mare within 24 hours or less from the time he bred the infected mare. Sometimes a stallion will have an active infection himself but he rarely passes it on to a mare unless the mare's tissues have low resistance (such as in a mare not fully healed from foaling). Mature stallions seem to be quite resistant to becoming infected from breeding mares with infections, but young stallions are rather susceptible to picking up infection.

Other common causes of mechanical contamination that can lead to infection of the genital tract in mares are unsanitary capsuling techniques when collecting semen and depositing it in the mare, poor breeding hygiene, the use of unsterile instruments, and the outmoded procedure of "opening up" a mare. Many old-time horsemen thought that every mare had to be opened up before she could conceive. It was a common practice to insert dilators into a mare's cervix, and many infections were started this way. Once in a great while a mare does have an abnormal cervix that needs to be opened up, but these cases are rare.

There are several diseases that can invade the uterus and make a mare slip or abort her foal. One of the most common of these is leptospirosis (see chapter 9). This is usually a systemic infection, affecting the entire body. The horse will usually recover from the original infection, but then the spirochete that causes the disease may localize in some part of the body and cause problems. It often localizes in the eyes, causing periodic ophthalmia ("moonblindness"—the most common cause of blindness in horses), or in the kidney, or in the uterus. If the mare is pregnant she aborts. Lepto is a very common cause of abortions and can also be a factor in the production of weak or dead foals carried full term. Lepto is a common disease in many animals, some of which become carriers. Horses can be vaccinated against leptospirosis quite cheaply and all broodmares should be vaccinated.

There is still much to be learned about the various causes of abortions in mares. Some abortions are caused by disease, some are caused by nutritional deficiencies or other factors. Abortion can occur at any time in the gestation period. It is more common in the early weeks of pregnancy than many people realize, because at this early date the aborted fetus is so small that it goes unnoticed, and the abortion has little effect on the mare. Abortions can be caused by disease, uterine infection, lethal factors, and injuries. Sex-linked lethal factors occasionally occur, in which some mares might, for example, be able to produce only fillies if the male chromosome carries with it a lethal factor. And sometimes if the dividing cells that make up the embryo take a wrong turn somewhere along the way the embryo may die. The mare's body automatically sheds the dead embryo. If she conceives twins, one may die for some reason (perhaps insufficient circulation or lack of room for proper development). If for some reason the mare is aborting, it is usually better to let her go ahead and abort rather than trying to hang onto a precarious pregnancy and make her carry it full term. If the foal is dead for some reason, she must get rid of it. And if it is not dead, it may be abnormal. Occasionally an abortion can be forestalled, only to have the mare give birth to an abnormal foal when foaling time comes. Nature usually has reason for terminating a pregnancy in abortion.

Once in a great while you might have a mare that habitually aborts at a certain stage in her pregnancy, not from any disease or infection, but rather from some deficiency in her chemical makeup. If you work closely with your veterinarian on a case like this and give the mare prescribed medication, you may be able to prevent the problem and have the mare bring a foal to full term.

A mare aborting later on in pregnancy (particularly if the abortion is due to an injury such as a kick or a fall) may have a hard time of it—much worse than if she were actually giving birth normally at full term. Her body is not prepared for this sudden state of affairs and the birth may be a difficult one. A normal mare approaching a normal birth is gradually prepared for it by her changing hormones. If a mare suffers a traumatic abortion she should then be handled with care and not rebred until she has fully healed and her uterus is back to normal. If your mare is aborting and having trouble you may likely need veterinary assistance to help her extract the dead fetus.

A few last words about infections in the genital tract, since infections are the most usual cause of fertility problems. When a mare has an infection it is not always obvious. This is one reason that uterine infections sometimes go unnoticed and go too long, becoming a cause of infertility or even sterility by the time they are discovered. A mare with an infection may or may not have a discharge. She will usually have normal heat periods, as regular as ever, but she won't become pregnant even after many breedings. And if she does conceive, she may abort early in pregnancy (sometimes between two and four months) and suddenly come into heat again.

Infection usually prevents conception, but there are cases in which the mare does become pregnant and carries the foal full term only to have it born dead or seriously diseased. Occasionally a mare with a low-grade uterine infection carries a live foal full term but the foal will be infected. The foal may have swollen joints and a severe septicemia, and sometimes only immediate treatment with antibiotics will save its life. This happened to me with my very first foal, when I was trying to raise a foal as a 4-H project. Our mare, Scrappy, a grade Thoroughbred, had a uterine infection and we didn't know it. When I was 13 years old my father sent her to be bred to a stallion owned by a rancher. The stallion owner was of the old school of thought that believed in opening up a mare, and when Scrappy came back to us he told us that he had had to open her up. We didn't think much of this at the time for we didn't know all that it implied.

Scrappy didn't settle. A year later I took her as a 4-H project and had her bred to the Arabian stallion, El Khamis 9145, for a mare and foal project. This time she settled and we didn't realize how lucky we were. A nice filly (Khamette) was born the next May. The foal was born early in the morning and within a few hours her fetlock joints seemed somewhat large. I didn't realize the meaning of this and went gaily off to school. But when I came home that afternoon

Penicillin saved Khamette's life and she grew up to be a useful animal, but she was the only foal Scrappy was ever able to have

and saw the filly's legs swollen clear up to the knees and hock joints, I knew something was really wrong. Khamette had navel ill, apparently born with the disease already in progress. Massive doses of penicillin saved her life and caught the disease in time to prevent the usual crippling of the joints.

We tried unsuccessfully for two more years to settle Scrappy again for another foal, not knowing that she had a uterine infection. She had a hormone problem as well, and we thought that it was the cause of her infertility. Later we sold her and her new owners tried unsuccessfully to raise a foal from her. Twice she did conceive but the fetus died during the pregnancy. They discovered that Scrappy had a long-established low-grade uterine infection and were unable to get rid of it even after extensive treatment. This infection was very likely started the first time Scrappy was bred—when she was opened up, probably with unsterile instruments. Only years later did I realize what Scrappy's problem had been, why we were unable to raise more foals from her, why her one filly was born with navel ill and how very very lucky we were to even have Khamette at all.

Never breed a mare if infection of any kind is known to exist in the genital tract. The mare should be checked for infection before breeding, and if infection exists she should be treated to clear it up before she is bred. A cervical culture must be taken in order to diagnose a low-grade uterine infection; a sterile sample can be taken from the cervix by a vet when the mare is in heat. If bacteria are present, the diagnosis can be confirmed after they are cultured, and a sensitivity test run to see which antibiotics would be most effective against them.

Most stallion owners won't accept a mare for breeding unless she has been examined by a vet for breeding soundness, and if you have a stallion, you shouldn't either. The examination should include a check on the uterus and ovaries (this is done by rectal palpation) and examination of the vagina and cervix through a speculum. The examination should be made while the mare is in heat. A breeding examination is important to both the mare owner and stallion owner. It protects the stallion owner by allowing only healthy mares to be bred to his stallion, and it is beneficial to the mare owner by determining whether or not his mare is sound for breeding, before he goes to all the time and expense to get her bred. A breeding exam can locate potential trouble, and if treatment is necessary, the mare can be treated and the trouble corrected before it interferes with conception and pregnancy.

Much valuable time can be lost if a mare has a problem that interferes with fertility and is never checked before breeding so that the problem can be corrected. And if by chance the mare has a definite uncorrectable problem that would prevent her from conceiving or carrying a foal, it is good to find out the facts, to save the time and trouble of trying unsuccessfully to get her bred.

7

THE BROODMARE—PREGNANCY AND FOALING

WHETHER YOU OWN ONE HORSE OR MANY, THERE MAY COME a time when you decide you want to raise a foal. You might breed and raise horses as a business, or just raise a foal or two from a favorite mare. The situation and facilities of the professional horse breeder and the pleasure horse owner raising just one foal may differ a great deal, but many of the problems and factors to be considered are still the same.

The birth of a foal is always an exciting event, and eagerly awaited. But it is also an event that should be very carefully planned for; you'll want to make sure that everything goes well. You have waited approximately 11 months for the new arrival and you want to be sure that it is a healthy foal, one that you will be glad to own and raise.

Once a mare is bred, she is considered a broodmare. You will hope that she has settled, and if she doesn't show signs of coming into heat again at the time you would expect her to, you can be fairly sure that she is with foal. Many mares take a long time to begin to show definite signs of pregnancy. Other mares become fairly obvious as soon as six months—the belly will begin to drop and widen towards the lower part of her flanks. This drooping belly will look different from a nice round belly that is merely fat. When you look at the mare from the rear as she is standing still or traveling, if she is pregnant one side of her belly toward the flank may appear a little larger than the other side, and as the mare moves there may be a slight swinging motion. but some mares, especially with their first foals, do not look obviously pregnant and may be difficult to determine merely from observation, right up to foaling time. And some fat mares may look deceptively pregnant even though they are not. After about the eighth month there may be signs of movement back toward the mare's right flank as the unborn foal kicks and squirms. But these signs are not always observed. If you see movement, you will know that the mare is definitely carrying a foal, but not being able to see it does not necessarily mean that she isn't.

During the time the mare is carrying her foal she should have the best care

The pregnant mare has a drooping belly in later stages of pregnancy. Nell, carrying her fourth foal

possible. She should be kept in good condition, but not overfed. A pregnant mare's feed requirements increase significantly only during the last third of her pregnancy. Many horsemen start feeding a mare extra as soon as she become pregnant and as a result they have a mare that is too fat when her time comes to foal. Excessive fatness should always be avoided in broodmares. The proper amount of feed is important, but perhaps even more important than the quantity is the quality of the feed. Huge amounts of feed won't do the mare much good if it is poor quality and lacks the nutrition she needs for the growing foal inside her. A pregnant mare should be fed a well-balanced diet so that she will have the proper elements needed for the growth of the foal.

Exercise is also very important. The mare's muscles should be in good condition, not soft and flabby from nonuse. Muscles are required to expel the foal in the act of labor, so good muscle tone is important. Don't make the mistake of giving the broodmare little or no exercise for fear of hurting her or the unborn foal. On the contrary, she is not as fragile as some horsemen seem to think, and she should have exercise.

A large pasture where she can move about freely can provide needed exercise. If you have no pasture for her and she must be kept in a stall or corral, ride her or lead her for an hour or so each day. If you were riding her before she was bred, keep riding her after she is pregnant, but just use good judgment in how and where you ride her.

When a mare is pregnant, keep her with just a few other horses she gets along with, preferably other mares. Several pregnant mares in a pasture by themselves will get along the best. Never keep a pregnant mare in the same field with young horses or geldings. Young horses, geldings, or large numbers of any type of horses always increase the chances of injury to the mare and the unborn foal from being kicked or from slipping and falling while running. Geldings should never be in with her, or even across the fence from her. Geldings are apt to tease and fuss with a mare and have been known to make mares abort their foals, especially in early stages of pregnancy.

You can keep the mare by herself if she has some kind of company across the fence, other mares or even a goat or a cow. If she is out on good pasture and requires no supplemental feeding, a pregnant mare will need very little care as long as she has good feed and water and shelter from the flies and hot sun. Remove any obstacles or hazards in her pasture. Avoid narrow gateways and doorways when moving her out of the pasture or into a barn or stall; you won't want her to be bumped. If she is with a group of other horses, don't let them crowd through a gate. In her stall, remove all projecting edges that she might run into.

The quality and development of a foal depends somewhat on the feed the mare eats before foaling. If she is undernourished, run down, or has a deficiency of vitamins or minerals, the foal may be underdeveloped or may not even survive. If you live in an iodine-deficient area you should give the mare supplementary iodine—with iodized salt. Lack of iodine in the pregnant mare's diet can cause contracted tendons in the foal. You should also ask your vet about giving the mare supplementary vitamin A and perhaps a little calcium.

During the winter months if your climate is severe you may want to let the mare have access to shelter in her pasture or keep her in a stall at night. Horses do not mind cold weather, but if it is windy and wet as well as cold, shelter is a good thing. If a mare is kept in a box stall part of the time, she should be turned out or exercised every day. Also, her box stall should be thoroughly cleaned daily and fresh bedding put down.

There are some diseases that should be kept in mind when preparing for the arrival of the new foal. One of the worst of these is called navel ill or joint ill. The bacteria that cause it (usually *Streptococci*) are often present around farmyards and wherever livestock are housed or confined in close quarters. Once the bacteria are present they seem to live in the soil almost indefinitely. These bacteria (from uterine discharges from infected dams or from discharging abscesses in other animals, etc.) can enter the newborn foal through the navel after the umbilical cord has ruptured following birth. If the foal gets navel ill, the first sign is usually a painful swelling of the navel and surrounding abdominal wall. His joints swell and become painful (particularly the hock, stifle and knee joints), there may be a flat swelling around the navel area, 6 to 8 inches in diameter, and he has a fever. The painful joints make him lame and he may spend much of his time lying down. This is a very serious disease because it goes through the bloodstream to all parts of the body and can do internal damage to organs such as the heart and liver as well as affecting joints. There may be an accumulation of pus in the front part of one or both eyes. Symptoms usually develop about two weeks after the foal contracts the disease. It is hard to save a foal with navel ill once the symptoms develop, and if you do save him, he may be a cripple (with chronically arthritic joints) for the rest of his life. It is better to try to prevent the disease in the first place.

The best protection against navel ill and other diseases of the newborn foal is to keep the mare's environment as clean and sanitary as possible, prevent contamination, and to vaccinate the foal against navel ill immediately after birth, as well as saturating his navel stump with iodine. Even if you think your stalls and pastures are free of navel ill bacteria, it is always a good idea to vaccinate against it. You should have the vaccine on hand well ahead of the time your mare is due to foal, in case the foal comes early. If you wish, a mixed equine vaccine can be used, as it protects the foal from several other diseases at the same time. Check with your vet.

Some horsemen like to vaccinate the mare before she foals, to start building up immunity in the unborn foal. In rare cases the mare herself may be the source of the disease and her foal will be born with the infection already in progress. This can happen when the mare has a low-grade uterine infection that was not properly cleared up when she was bred (see chapter 6). Usually an infection will prevent conception or kill the foal before it is full term, but sometimes a low-grade infection may merely cause the foal to be born diseased. In this case, antibiotics must be given to the foal immediately to try to save its life and prevent the permanent crippling of the joints. If you decide to vaccinate the mare before the foal is born, do it about three weeks before she is due to foal.

After the foal arrives, it should be vaccinated immediately whether or not the mare was vaccinated. And the foal should then be given a second vaccination a week after the first one. Use a small horse needle on the foal and vaccinate him in the large muscle below the buttocks rather than in the neck (a vaccination in the neck muscle might make his neck sore and interfere with his trying to nurse). When giving shots, make sure your equipment is sterilized, and give the shots in thick muscle tissue; avoid hitting a bone.

Well before the mare's foaling time approaches, you should give some thought as to where she will foal. If she must foal in a small enclosure such as a shed, corral or box stall, it must first be thoroughly cleaned and then provided with clean bedding. If your mare is foaling in early spring before the weather is good, you may want her to foal in a sheltered place like a box stall, or have her out in a clean pasture during the day when the weather is nice and in the stall at night and when the weather is bad.

The best place for a healthy mare to foal (especially if she has had a foal or two already and has a history of easy deliveries) is in a level grassy pasture where she can be by herself. It should be a clean, dry place with safe, smooth fencing and no rough underbrush, rocky ground, gullies or other hazards—no loose nails sticking out, machinery and the like, that might injure her or the foal. And there should be no water puddles or deep ditches. Foals have been known to drown in just a tiny pool of water before they had a chance to get to their feet. Even a dry ditch can cause trouble if a foal happens to roll or squirm into it in his efforts to get up and about. A foal cannot stay on his back very long; abdominal pressure begins to cramp the heart and lungs and in a short time he will suffocate. Foals have also died because they got down on their backs under a fence.

But if the pasture is a safe one and the season of the year permits, there are many advantages to having a mare foal out in pasture rather than in a confined box stall. In the first place there is the factor of sanitation. A good clean pasture is the healthiest place for the mare to foal; there is less chance of infection out in a pasture than in a stall or corral. The problem of infection is always present when livestock are kept in confined areas. Another advantage to pasture foaling is that it is natural for the mare to seek her own privacy; she will be less upset and nervous. Also, a mare out on pasture usually has an easier time foaling. It is a much more natural environment. If she has been out on pasture she will have had some exercise traveling around and grazing and her muscle tone will be better than that of the confined mare unless the confined mare has been regularly exercised. And the mare on pasture will have looser, more natural bowel movements than the confined mare that is fed hay and grain. The mare on pasture will therefore have a looser, more flexible digestive tract and she'll have an easier time foaling than the hay-fed mare that is more constipated.

But foaling out on pasture has its disadvantages too. It is harder to keep a constant watch on the mare. Most horsemen like to have the mare close at hand so that they can check on her often and be able to help her in case of trouble, especially if it is the mare's first foal. And of course, pasture foaling is not advisable in cold or wet weather. You may want to let the mare out at pasture during the daytime and keep her in a stall at night where she can be more readily observed. Most mares do not need help when foaling and are better off without it, but it doesn't hurt to be on hand so that you can help or get help for the mare if by some chance she needs it.

If you are going to have the mare in a stall when she foals, or have a stall available for her at night or in bad weather, you must first thoroughly clean and disinfect the stall. It should be large enough to give the mare plenty of room to move around, for when she goes into labor she will be very restless. A box stall 16 by 16 feet is a good size. The stall should be built or arranged so that you will be able to check on the mare's progress without her seeing you.

Clean out any old bedding from the stall and then carefully scrub the whole stall with disinfectant—floor, walls and all. Sprinkle the clean floor with slaked lime before putting in new bedding. Clean, bright straw makes good bedding; don't use straw that might be old or dusty and moldy. Dusty straw may make the mare cough. Moldy straw is never good for bedding because if a horse picks and nibbles at the bedding some of the mold may be eaten. Mold is very dangerous for a pregnant mare to eat because it may make her abort. Cover the stall floor with clean, deep straw, deep enough to be comfortable to the mare but not so deep that it would make it hard for the new foal to get up and about. Sawdust or shavings should not be used as bedding for a foaling mare because these can sometimes be sucked into the newborn foal's nostrils and lungs, and will also stick to the new foal and create a problem when the mare tries to lick him dry. Remove all obstacles from the stall—feed tubs, buckets, anything else that might get in the mare's way during her labor.

About 336-340 days from the time the mare was bred, the foal should arrive. Gestation time for a mare is approximately 11 months. But this length of time is just an average; not many mares foal on precisely the day they are supposed to. The period of gestation may actually vary greatly; mares have been known to foal as early as the 287th day and as late as the 419th day, with normal foals. Thus you should begin to watch the mare more closely for signs of foaling from about the tenth month on. The light breeds (saddle horses) tend to have a longer gestation period than draft breeds. Another factor to take into consideration is that gestation periods ending in summer and fall are usually shorter than those ending in winter and spring. Why mares carry their foals longer in winter and early spring is not clear, but it may be due to the greater intensity of the sun's rays in summer and fall, longer daylight hours, or better nutrition during certain phases of the gestation period, bringing a summer or fall foal to full term faster. This has been borne out by our own mares. Nell, for instance, had three summer foals (two colts and one filly) that arrived about two weeks early. Her one late winter foal (Fahleen, a filly, born March 7) came several days late.

Watch the mare carefully as her foaling time approaches. Some mares will be early, others late, depending on the individual and also upon the season of the year as just mentioned. Two or three weeks early is not uncommon, nor is two or three weeks late. But if your mare goes more than two weeks past her foaling date you should have a vet check her to make sure nothing is wrong.

About a month before the mare is to foal, her udder begins to spring a little, looking fuller and larger, especially at night when the mare is at rest. But at first this enlargement will tend to recede again during the daytime when the mare is moving about. But as her foaling time approaches—about a couple weeks before foaling—the udder will remain larger, filling with milk, and will look shiny. There may be a secretion from the teats. This "waxing" from the teats is one of the most common signs that foaling is near. The wax is formed by the congealing of secretions from the udder that are forced out the end of the teats. Most mares will wax profusely within 24 to 36 hours of the beginning of labor, but there are always exceptions and individual differences. Some mares don't wax at all. Others will wax for as long as 10 days before foaling. Still others may make up a large udder and leak streams of milk before they foal, the milk dripping down their hind legs. The mare may leak milk for just a few hours before foaling, or sometimes for several days. At about this same time (just a few days before foaling) the muscles at each side of her tail around her pelvic bones will become relaxed and droop away from the root of her tail. Her vulva will appear somewhat relaxed and swollen.

When these signs are evident you can expect that her foaling time is near and you should now keep her under close observation. If she is at pasture, let her be by herself. She should not be with a group of other horses, for some of them may bother or pester her when she starts to foal. And sometimes another mare will steal the new foal, chasing off the mother (especially if the mother is a younger or more timid mare) and lick up the foal herself. The foaling mare will want to get away by herself to have her foal in privacy and safety. Ideally, the mare should be let out at pasture during the daytime in a place where she is close and handy and can be checked on at frequent intervals. Then she can be put in a stall at night, unless you want to go out at frequent intervals during the night to check on her in the pasture. If the mare seems content and is grazing, she has probably not yet begun labor. But if she is standing off in some secluded corner, or appears restless, or grabs a bite to eat and then paces a bit before grabbing another, you can suspect that things are about to happen, and start checking on her more often.

As the mare prepares for actual labor she may become quite agitated and nervous. She may walk restlessly, fuss with her feed, sweat a little, or just stand and tremble. She may switch her tail, kick at her belly, nose at her side, or show other signs of discomfort.

Mares often foal at night (more foals are born at night in the quiet seclusion of darkness than in the daylight), so if you suspect that her foaling time is here you should check on her periodically throughout the night. This may mean several nights of checking if she doesn't foal on schedule or shows signs of foaling before she actually goes into labor. Check on her every few hours but do not alarm her or make her nervous. She will be less upset if there are no distractions so if all is going well, stay out of sight.

The mare should be kept clean and well groomed. The cleaner the mare, the less chance of infection when the genital tract is exposed during foaling, and the fewer bacteria will be picked up by the foal as he noses about the mare to begin to nurse. It's a good idea to wrap the mare's tail in a clean tail bandage when you clean her up for foaling; this can keep a lot of filth from coming in contact with the open genital tract as she foals, and also keep the mare's tail clean and free from blood and mucus.

If she is on hay, she should have her feed cut down for a few days before she foals and should be kept on low rations for several days after foaling. The birth of the foal can be accomplished much easier if the mare is as empty as possible at that time. Another reason for cutting down her feed at foaling time is to decrease the milk flow; too heavy or too rich a diet at foaling time will usually cause the mare to produce too much milk. The foal won't be able to take all of her milk at first. If the milk is too rich or if the foal is getting too much, he may get diarrhea. When the mare begins to show signs of actual labor, do not feed her at all. Let her have a little water if she wants it, but not too much.

If you suspect that the mare is in the early stages of labor, she should be checked at least every 15 minutes, but do not disturb her. There are three stages of labor. The first stage is the longest and will take from 4 to 24 hours. It will take longer for a mare with her first foal than it will for one that has had foals before. In this first stage there are contractions of the uterus as the foal is turned and moved into the position for birth, turned so that his head and front legs are aimed toward the birth canal. Once in a great while a mare will have trouble; the foal will not be turned to the correct position and she will have to be helped in order to give birth.

The inexperienced horseman should make arrangements to have a vet or an experienced horse breeder present. Usually everything goes according to nature's plan and the mare has her foal safely with no trouble at all. But if problems arise, the inexperienced person should never try to help the mare for he may only make things worse. He may endanger the life of the foal or the mare and foal both. The time to contact the vet, or whoever you have decided upon to help you, is when the mare is in the first stage of labor, when she is merely restless and periodically uncomfortable as the foal is shifting inside her. Keep him posted on the mare's progress so that he can be there when she goes into the second, or active stage of labor.

During the first and earliest stage of labor the mare is usually restless, short tempered, and nervous. She is preparing to give birth to her foal; her pelvic muscles relax more fully, her tail rises and her sides appear caved in just ahead of her hip bones. The foal has "dropped"; it is shifting position and heading into the birth canal. As contractions occur the mare will show distress, switching her tail, nervously pacing, pawing, or perhaps nosing at her flank and grunting. When you observe these signs of labor, it is time to bandage her tail, supply her with fresh clean bedding if she is in a stall, and then leave her in privacy.

The second or active stage of labor is when the foal is expelled from the uterus in a series of hard contractions. This stage does not take very long. During the first stage the mare moves around quite a bit and is fairly comfortable between contractions, but after she starts the second stage of labor she is more constantly uncomfortable and will move around less; the contractions are coming much faster now. When the mare shows more constant signs of distress and perhaps lies down and gets up you know she is starting the second stage and the foal will come very soon now. She may get down only once and start to strain, producing the foal in a short time. Be sure she has soft bedding, for she may throw herself down rather violently as the pains hit her. I know of one mare that died during labor because she threw herself down so hard that she ruptured internally.

During the second stage of labor her bag of water will rupture; you'll see some fluid rushing out. When this happens, you'll know that the second stage is in progress and the foal should then come within 5 to 30 minutes. Cows can safely be in this active stage of labor for several hours and still produce a live calf, but not a mare. If the mare takes longer than 20 minutes, she needs help and she needs it right away. A first foal will often take a little longer than other foals, but if nothing has happened within 30 minutes after the bag of water is broken, the mare is in trouble. Either the foal is being presented wrong and cannot be born, or it is exceptionally large and is not moving readily through the birth canal.

The normal position for a foal being born is head first with front feet extended. A membrane filled with fluid will be first to appear and then in a moment the feet should be visible within it. As the feet protrude still farther, the foal's nose should appear, tucked between them. Sometimes the feet will not be even; one

Correct and normal presentations: *top,* **correct front presentation (most common);** *bottom,* **correct rear presentation**

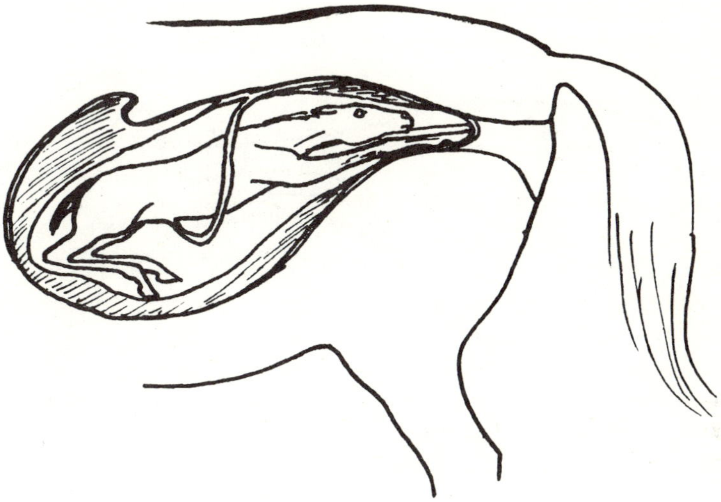

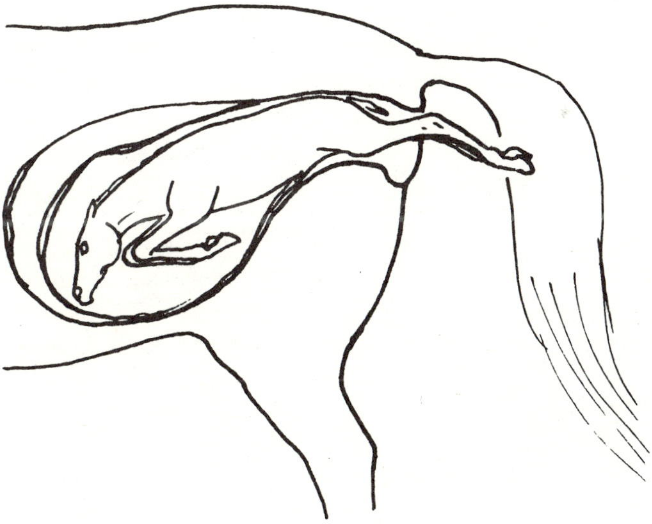

front foot will appear a little in advance of the other as the foal edges through the birth canal one shoulder at a time. Any deviation from this front-feet-first position (head turned back, one or both front feet turned back, and so on) is abnormal and professional help will be needed at once, the sooner the better.

Abnormal presentations: *top,* **head turned back (fairly common)—the head must be brought into the birth canal;** *bottom,* **head turned back—it too must be brought into the birth canal**

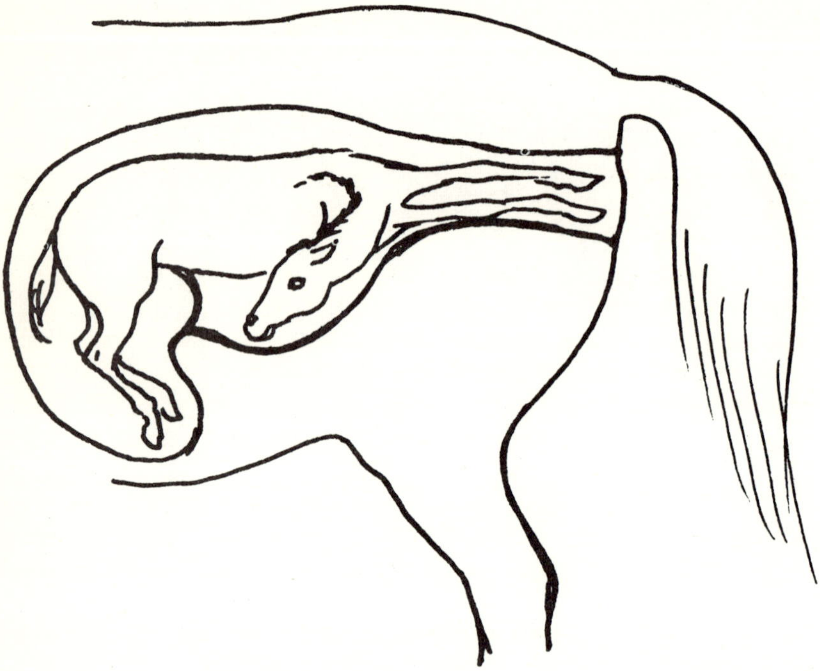

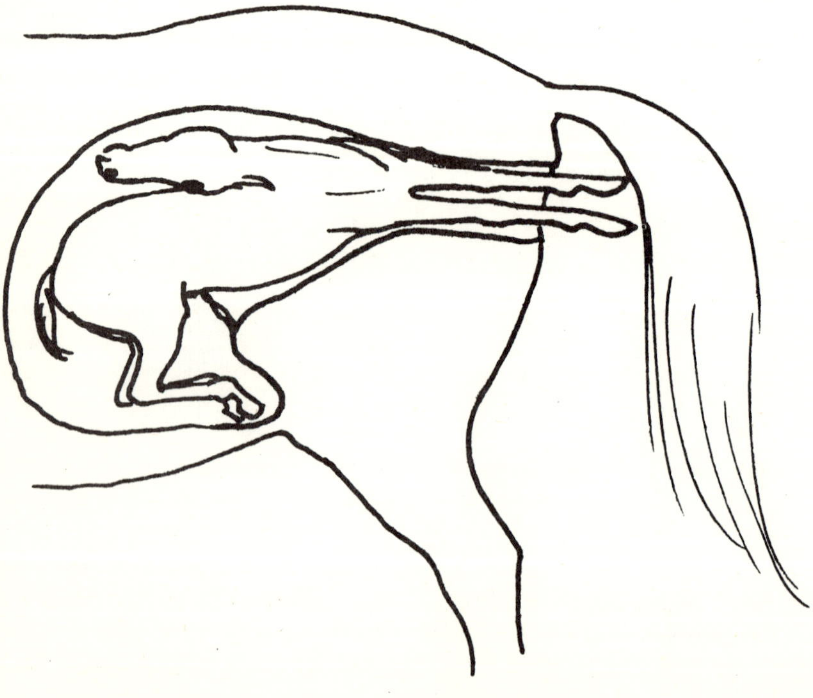

A foal in an abnormal position cannot be born and it and the mare may both die. If the hind feet are presented first, the foal can be born, but the birth must be accomplished quickly so that the foal will not suffocate or breathe in some of the surrounding fluid before he is completely out of the birth canal. You will know that they are hind feet if the bottoms of the feet are up, showing the little frogs. Front feet will be bottoms down, frogs and fetlocks down. If a foal is being presented hind feet first, grasp the feet and assist the mare as she strains, pulling only when she strains, to help assure a quick delivery so that the foal can begin to breathe. You may have to help the backward foal start breathing; clean the mucus from his nose and mouth, press on his ribs and breathe into his nostrils if necessary.

If the front feet appear but there is no head between them, you can suspect that the head is turned back. If only one front foot appears you'll know there are grave problems. If any other portion of the foal's anatomy appears first (tail, three or four feet presented all together) you must get help at once. In an abnormal presentation the vet must push the foal back into the uterus and turn it to the proper position. This can be accomplished much easier if the problem is detected early and the vet is there to correct it before the mare strains and strains in vain—before the foal is pushed fruitlessly against the pelvis and part of the foal is already well along in the birth canal. To push the foal back into the uterus to where it can be turned and straightened out properly will be pushing against the mare's labor contractions and is very hard to do if part of the foal is already well along in the birth canal. If the foal is being presented wrong, do not attempt to correct the situation yourself if you are inexperienced, but get skilled help immediately.

It won't hurt to have a few items on hand when the mare goes into labor—a clean bucket filled with lukewarm water and disinfectant, clean towels, a bottle of iodine. If you assist the mare yourself, have very clean hands, plenty of "liquid soap" (from your vet) to serve as disinfectant and lubricant, and some obstetrical gloves to cover your hand and arm. You can get disposable OB gloves from your vet. If you do have to help the mare, you must take every precaution to prevent infection. Everything that comes in contact with the mare's genital tract must be sterile and clean. Be ready in case the mare needs assistance, but do not disturb her if all seems to be going well. She may get up and down a few times, or she may just go down once and stay down. Mares usually foal lying down, but occasionally a mare will foal standing up. If she is foaling in a stall make sure she is in the center with her hindquarters well away from the wall.

The foal's shoulders are its widest part to come through the narrow birth canal, and they usually must come through one at a time. This is why one front foot may be extended farther than the other. If both front feet are protruding exactly the same and the mare seems to be having a little difficulty moving him out, you can pull a little on one front leg as she strains, to advance that shoulder so the foal can move along more easily. But in most cases you should not even try to assist in the birth of the foal because the mare will manage fine on her own, and too much pulling and tugging at the foal may injure the mare. The mare will alternately strain and rest, moving the foal gradually but steadily through the birth

Abnormal presentations: *top,* **hind feet must be brought into the birth canal;** *bottom,* **hind feet must be pushed back (fairly common). There are many other abnormal presentations—one or more front feet turned back, foal sideways or upside down in uterus, etc.—but they are not as common**

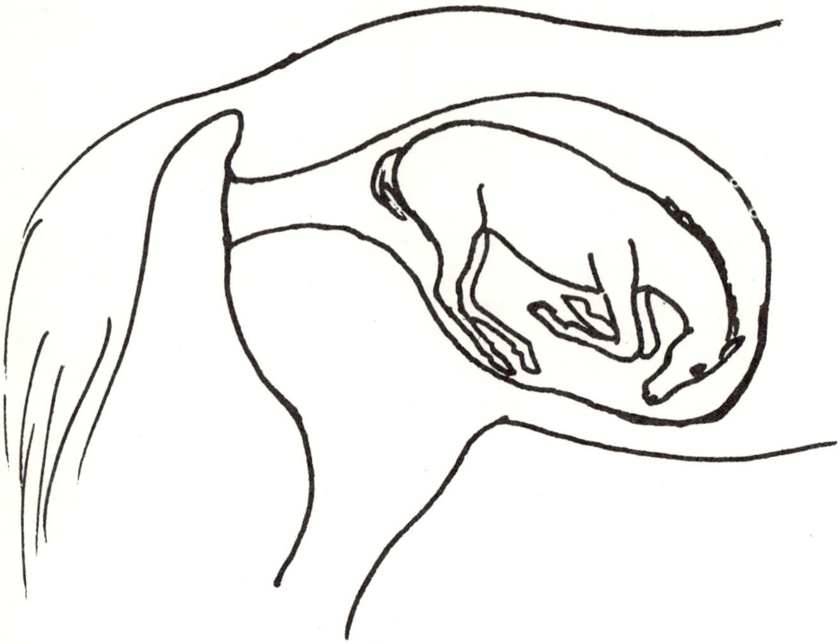

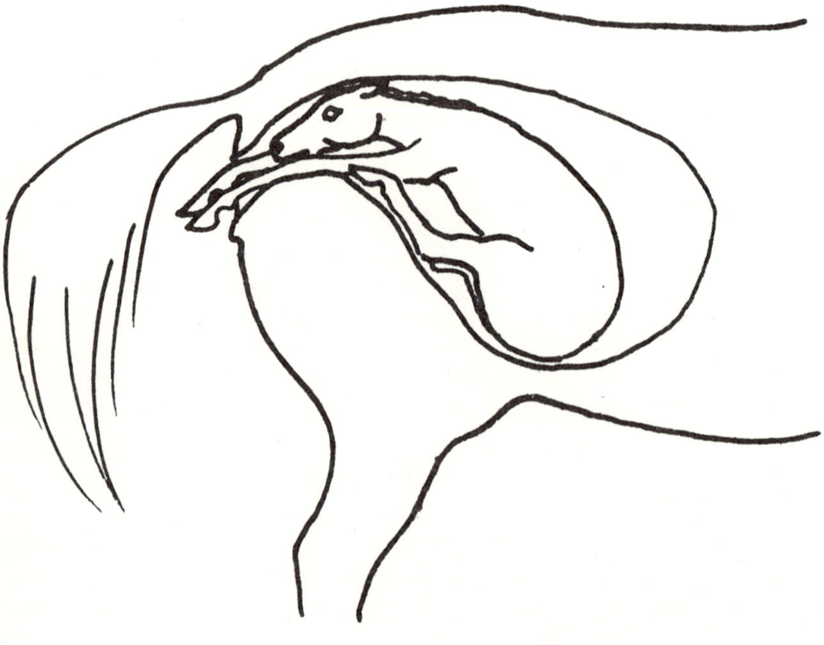

canal. After his shoulders pass through he should slide out easily and quickly.

But occasionally the birth may be a hard one even if the foal is being presented normally. This is unusual, but once in a great while a mare may take longer than normal having her foal and may need some help. Once in active labor, the mare must deliver the foal quickly or the blood supply to the foal will be hindered and it will die. If the foal is part way along the birth canal and the umbilical cord is being pinched, or if there is great pressure on the foal's ribs and body, the foal will not be able to survive this position for any great length of time. If the foal is unusually big, or if for some other reason the mare is taking too long, grasp the foal's forelegs and help the mare in the delivery. A small towel around the foreleg will give you a better grip; the legs will be quite slippery. If it is a difficult pull you may need one person on each foreleg. Exert a downward pull toward the mare's hocks, directly in line with the birth canal. Pull strongly with each contraction, but *do not* pull when the mare is not straining. If you pull too hard when she is not pushing, you may injure the membranes of the birth canal. If you tear the mare too badly she will hemorrhage.

Be careful how you assist her, but assist her if she needs it, for a long and difficult birth will exhaust both the mare and the foal. The more quickly the mare accomplishes labor, the more quickly will she recover and return to normal, and the better start the foal will have. If the foal is too exhausted from being under great pressure in the birth canal too long, it will take him longer to get up the strength to get to his feet and nurse.

When the foal is born, he will usually be still wrapped in membrane, but he should quickly free himself by his struggles. If he does not immediately free himself, you should break the sac (any touching of the mare or newborn foal during the act of birth or shortly thereafter should be done with very clean hands) and wipe the mucus from his nose and mouth so that he can begin breathing. Make sure he is breathing normally. If a foal is slow in taking his first breath (and sometimes he will be, if he has undergone a difficult birth), you may have to massage him and press his ribcage in and out; usually with a little stimulation he will start to breath.

The umbilical cord usually breaks when the mare gets up. If she doesn't get up, you can pull the foal away from her until the cord breaks. Never cut or tie the cord, and do not touch it with your hands unless they are very clean and disinfected. The blood vessels of the cord have a greater elasticity than the outer covering and if the cord is broken rather than cut, these blood vessels will draw back into the stump of the navel and help seal off the opening to infection. Wait at least ten minutes before breaking the cord. Much of the foal's blood supply is still in the placenta, within the mare. If the cord is broken too soon the foal will be weak.

The stump of the navel should immediately be disinfected with tincture of iodine. This is done by saturating a piece of sterile cotton with iodine and holding it firmly over the stump of the navel for a moment, or can be done very easily with the foal in a standing position by placing a small jar or glass of iodine up to the navel stump and making sure the entire stump is thoroughly dipped. Be careful not to spill any iodine on the foal. Disinfecting the navel with iodine is a very important step to take immediately after the foal is born; it helps to close the navel stump to infection and reduces the chance of navel ill and other problems. Do not touch the navel with your hands, and do not tie it. There will be a certain amount of natural drainage; tying would prevent it and cause swelling and trouble.

Move the foal to the front of the mare if she has not yet gotten up, so that she can see him and nose him. Place him with his back toward her so that she can lick and nose at him but won't lick at the freshly iodined navel. Some mares will get up immediately after the delivery is complete, but others will lie for a short while afterward, recovering their strength. Most mares will be up within 20 minutes. If the weather is cold and the mare is not yet up to begin licking and stimulating the foal (the mare's licking stimulates his circulation and encourages him to move and try to get up and about) you can dry the foal a bit with a rough towel, but otherwise the mare herself can manage nicely. Remove any wet or soiled bedding from the stall, then leave the mare alone. You can thoroughly change the bedding a little later; at this point it is best to give her and the new baby a little privacy. She will do a better job of mothering the foal (especially if it is her first one) if there are few distractions to upset her and make her nervous. If she is in a box stall, keep the doors closed if the weather is cool or breezy. The foal is damp and the mare may be wet with sweat; you don't want them to become chilled in a draft.

Nell's fourth foal, Ahmahl. A good way to restrain and hold a new foal for iodining the navel or for giving his shots is to use one arm around his buttocks and the other around his chest

The third and final stage of labor is when the mare expels the afterbirth. This should occur soon after foaling, and the afterbirth can usually be seen hanging down soon after the mare gets up. If it takes her longer than two hours to "clean" (to shed the afterbirth) this is abnormal. If it takes longer than two hours you can suspect problems, and if she hasn't cleaned within eight hours consult your vet, for the mare will be in grave danger if she goes much longer than this. She is likely to get a very bad infection, and a serious infection can sometimes prove fatal.

If the afterbirth is hanging down and doesn't seem to be coming readily, you might tie it up to the level of her hocks so that neither she nor the foal will step on it and pull it out forcibly. But use very clean hands. *Never* try to remove a mare's afterbirth; it must come away by itself or you may tear it, leaving part of it within the mare, which is always very serious.

When the afterbirth is shed, it should be checked for completeness, for if any part of it is retained within the mare it will cause serious infection. When the membranes are spread out on the ground to be checked, they will look like a pair of trousers with the ends of the legs closed. There should only be one tear or rip in the membrane and it will be at the "waist" of the "trousers." If any part of the membrane is missing (still within the mare) it will usually be the tip of one of the closed trouser legs.

The foal should be on his feet and nursing within an hour or possibly two from the time he is born. Make sure the foal gets that first feeding. Do not rush or force him; you probably won't have to help him at all, for nature usually takes care of her children very well. Just stay out of sight and be patient. Eventually he will find the right place to nurse, even though to an observer it may seem like an eternity as he noses about here and there. But once in awhile you may need to assist a foal in nursing, if the foal is exceptionally tired (from a long or hard birth) or weak, or if the mare kicks at him.

The mare's udder is usually a little swollen and tender at first and she may be a little nervous about this first nursing, because it hurts her. Sometimes a mare may move around quite a bit or squeal or kick at the foal, but her kicks are usually only half-hearted and she won't really hurt the foal (she usually isn't really trying to kick at the foal; she's just stamping around because her udder hurts). Rarely will a mare kick viciously at a foal, but if this ever happens, you will have to intervene. A mare with her first foal may take a while before the foal succeeds in nursing; this nursing business is new to her and she may move around because her udder is sore and tender.

If the foal has not succeeded in nursing by the time he is two hours old, you should help him before he gets tired and discouraged. If the foal is reluctant to nurse, or fails to nurse properly (even with assistance) within eight hours, you should seek the advice of a veterinarian.

If the mare won't let her foal nurse, you'll have to assist. Catch the mare and hold her up against the wall of the stall, and have someone guide the foal to the mare's udder. If the mare still insists on kicking at the foal, hold up a front leg on that side (or tie it up if you have to) so that she won't be quite so able to pick up

Nell with her second foal, Nikki—first meal

Nell and Nikki—eight hours old

that hind foot to kick. Talk to her and soothe her and try to keep her calm and still. In some cases it may be necessary to use a twitch on the mare at first to make her stand quietly. Use as little force as possible, however, to try to keep the mare calm and quiet; you don't want her fighting against you as well as against the foal. Usually by the second attempt at nursing (which should be tried about 20 minutes or half an hour later) the mare will be a little easier. In most cases by the third or fourth nursing the mare will accept the foal. You may merely need to hold her a little until the foal starts nursing and then you can release her gently and she will let the foal continue to nurse. Occasionally a mare will be stubborn and try to refuse her foal for 24 or even 48 hours, but usually if you are patient and persistent, aiding the foal at each nursing, she will in time come to accept it. Nursing seems to stimulate the maternal instinct; after a foal has nursed several times the mare will usually more readily accept it even if she was hesitant or reluctant at first. Also, nursing helps to shrink the uterus back to normal and gain its muscle tone again.

The mare's first milk (colostrum) acts as a laxative for the foal and cleans out his digestive tract. The colostrum also stimulates gastric and intestinal secretions in the foal. These secretions are important in warding off bacteria and intestinal disturbances in the newborn foal. Unlike the human baby who picks up much of his mother's immunity while still in the womb, the foal is born with very little resistance to infection. Also unlike the human baby, the foal has very little time in which to absorb immunities through the mother's milk; after that the foal's intestinal membranes are too thick. This is why it is very important to get the foal nursing as soon as possible, for the foal has only 24 to 36 hours after birth to absorb resistance to disease through the colostrum. After a few weeks the foal will be able to produce his own antibodies.

If the foal is too weak to get up during the first hours after birth and too weak to stand and nurse even with help, you can milk out the mare a little into a sterile container, put the milk in a sterile bottle, and feed the foal from a nipple. Be careful not to gag or choke the foal. After the foal gains a little strength you can assist him in nursing from his dam.

The foal's first bowel movements will be dark, and sometimes very hard. This is the material that was in the foal's intestines before he was born. While in the uterus, the foal continually swallows some of the fluid in which he floats, but never evacuates his intestinal tract until after he is born, so a great number of hard pellets accumulate in the intestinal tract and are sometimes packed so hard that the foal has a hard time getting rid of them. Then, after he has had a meal or two and the colostrum (which has a laxative effect) begins to work, the bowel movements will become soft and yellow, but they should not become runny.

A foal may die if you neglect to make sure that his innards are working properly. If the foal doesn't have a bowel movement within a few hours, catch him and have someone hold him (an arm around his chest and one around his hindquarters) while you remove a few pellets with your finger (well greased with vaseline) very carefully. Then you can give him an enema to soften up some of the hard pellets that may have been out of reach of your finger. The enema can be half a pint of warm soapy water, at body temperature. Watch him or check the stall for droppings periodically to make sure he has that first bowel movement. Constipation is one of the most common causes of death in newborn foals.

If the foal's bowel movements become runny, he is probably getting too much milk. This problem will usually correct itself if the mare is not fed excessively and is not fed grain; the foal will adjust to the mare's milk in a few days. You may want to keep his hindquarters well greased, however, to prevent irritation and loss of hair. But if scouring (diarrhea) persists, and especially if the bowel movements become watery or of unnatural color or odor, consult your vet. The foal may have an infection and will become weak from loss of fluid if scouring persists for any length of time. If you suspect infection, you can give the foal a small shot of antibiotics and talk to the vet about medication to help combat the infection in his digestive tract. He may also suggest giving the foal additional fluids if dehydration is severe. After the mare has foaled, if all seems to be going well, you can give her clean, dry bedding. Remove the afterbirth from the stall and check it, then burn or bury it. It should always be immediately removed from the stall because sometimes a mare will try to eat it. Cows often eat their afterbirth; this instinct to get rid of the afterbirth may be due to the fact that out in the wilds, the newborn animal would be safer from predators if there were no decaying placenta nearby to attract prowling wolves or coyotes with its odor. I imagine that deer and other wild animals eat theirs, for I have never ever seen evidence of any where they have their fawns.

If you are leaving the mare and foal in their stall during the first night, you can leave a light on in the stall if you wish. Mares generally will not step on a foal, but a light will be a good idea if the mare is nervous or if this is her first foal.

Nell's first foal, Amahl—six hours old

Nell's third foal, Fahleen—two days old

Nell's fourth foal, Ahmahl—six hours old—with his young owner, Heidi

If the birth has been a normal one and both mare and foal are doing nicely, they can be turned out into a clean pasture if the weather is nice. If the weather is cold or wet and sloppy you may want to wait another day or so. And in this case you might also wish to turn them out only during the day and put them back in a warm dry stall at night until the weather is better. If the ground is cold, the pair should be turned out only briefly; as soon as the foal becomes tired and starts to lie down, they should be brought in again. A foal can chill quickly lying on the cold ground. But they should be turned out, for both of them need exercise. Exercise will help the mare gain back her strength and muscle tone and she will recover from foaling more quickly. Exercise helps her uterus get back into the proper position. After a few days they can be put into a pasture with other mares and foals. If you were using the mare before she foaled, you can begin riding her again moderately about 10 or 12 days after foaling, if you have a good place to put the foal (a box stall or similar confinement) while you are riding the mare. Don't take the mare away for long periods of time, for the foal nurses every two or three hours.

It is very important that the mare and foal be kept in a good clean place for several days after foaling; both of them are susceptible to infection for a few days. It doesn't do a bit of good to foal the mare in a good clean box stall and then move her to a place that is not clean, or to neglect to keep her stall clean after she has foaled. And if you have several mares to foal, be careful about using the same boxstall for one mare right after another. If you must use the same boxstall take extra care to make sure it is thoroughly disinfected and cleaned between occupants, and that there is no chance of infection being spread from one mare or foal to another. If a case of navel ill ever occurs, be doubly sure that the stall or area in which it occurred or in which the foal was kept is thoroughly cleaned and disinfected, and then do not use it again for foaling or for a young foal for quite awhile. Navel ill is highly contagous to young foals (if the navel is still moist or open or draining) and very hard to get rid of in a place if it once occurs.

Nell and Nikki. Foals nurse every two or three hours and grow very rapidly

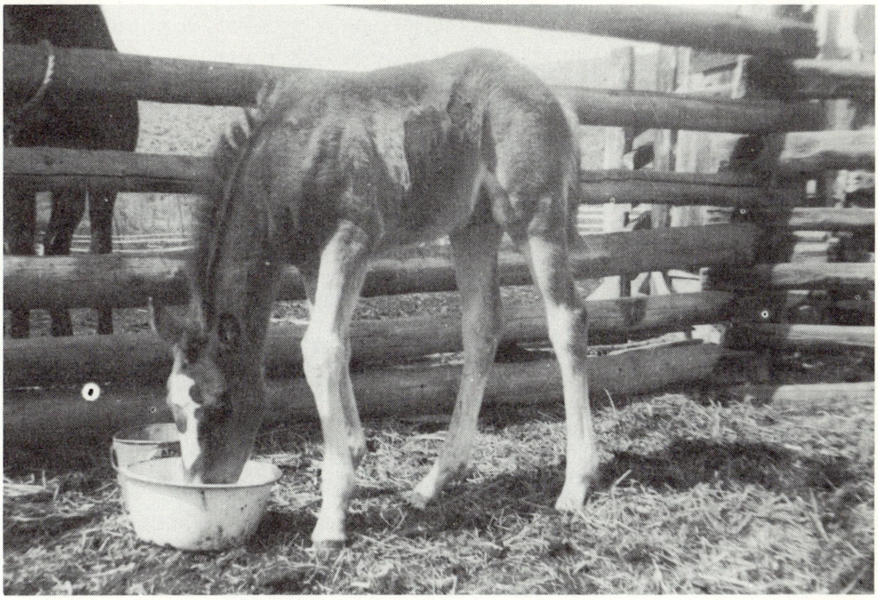

Fahleen. The foal can be started on grain at two or three weeks of age

There are a few uncommon occurrences that sometimes happen in conjunction with foaling, and we shall touch on a few here just briefly. Once in a great while a mare will begin to exhibit all the obvious signs that foaling is imminent—she will have a full udder, relaxed vulva and pelvic muscles, and so on—but she will hold off for days and days. If she goes on and on like this and still does not foal, you can begin to suspect that something might be wrong. This sometimes happens when the mare has twin foals, with one of them dead already inside her. If you ever suspect that something is not quite normal with your mare, consult your vet.

During the actual foaling process if the foal is being presented normally but the mare gets tired after the front feet and head have come, it may mean that she is having trouble passing the shoulders through the birth canal, since the shoulders are the widest part of the foal. If this is the case, the mare's efforts may become weaker and the periods of rest between straining may become longer. If the mare is tiring, you will want to assist her, in the manner already described, pulling on the foal's forelegs as the mare strains. It is always good to give the mare assistance (the foal being presented normally and correctly) if the mare does not make any visible progress for four or five minutes when trying to push the foal on through the birth canal.

If your mare goes into labor and the water bag appears but there is no further progress, she is in serious trouble. She may strain, with no results, or she may give up straining. The appearance of the water bag with no further developments usually means that the foal is being presented wrong and cannot be born.

If all is going well, the mare and foal should be turned out into a clean field, for they both need exercise

It is important that the mare and foal be kept in a good clean place for several days after foaling, for they both are highly susceptible to infection for a while

One other serious problem, which is fortunately quite rare, is the prolapsing of the uterus after the foal is born. The uterus comes on out after the foal and hangs down from the vulva, down to the mare's hocks. This is a very serious accident and will be fatal to the mare unless you get prompt help from your vet. While you are waiting for the vet to come, you can saturate a clean blanket with a warm disinfectant solution and use it to support the uterus. Tie the mare so she won't move around, and try to keep the inverted uterus supported so that it won't come out any further and put great strain on its attachments. It may take two persons to support it, holding it on the blanket between them. If possible, it should be held up nearly as high as the vulva. Prolapses are rare, and are usually due to the mare's continued straining after the foal is born; it is most likely to happen if the mare has had a difficult time foaling, but can happen even under good conditions. If a mare seems to be straining after the foal has come, or just to be on the safe side after a difficult birth, you can give her some medicine to keep her from straining. Talk to your vet about this in advance, for you may want to have some medicine of this sort on hand before the mare foals.

After foaling, it is best to give the mare some laxative types of feed for awhile. This will make it easier for her; you'll want to avoid constipation in both the mare and the foal right after foaling. A bran mash, or some green feed, will be ideal for the mare. If there is green grass available, it is the best and most natural feed for the mare at this time. After she has passed her first heat period at about nine days (during which the foal may have a tendency toward loose bowels) you can begin to add a little grain to her diet.

If you plan to breed the mare back again, it can be done at the foal heat (approximately nine days after foaling) or at the next heat period. If she has had a very normal and easy foaling, she can be bred back at the foal heat, for many mares are fairly easy to settle at this early heat period (the hormone level is still at a high because of the recent pregnancy). As summer progresses and the days begin to get shorter again, a nursing mare may not continue to come in heat. Some mares are preoccupied with their foals, and some, especially those that are giving a great deal of milk, do not seem to come in heat as readily and may wait until the foal is weaned in the fall or may not even begin cycling until the next spring. But if your mare has had any trouble at all during foaling, or if she was torn at all, or has any hint of an infection, you should wait until she is fully healed and not try to breed at foal heat. On our ranch we never breed a mare at foal heat, just as a matter of principle. We've always been able to breed the mares at the second or third heat period. Besides, we are not in any great hurry to have the foals born exceptionally early in the spring; May, June, or even July is fine for us.

Breeding a mare while there is still any chance of infection means that she may not settle, or if she does settle, the infection present in the genital tract will be sealed in and may cause problems later with the developing fetus. Almost all mares pick up a genital tract infection at foaling time, but a healthy mare (particularly if she was in a clean environment and she did not injure the reproductive tract during foaling) will throw off this infection very soon. It isn't

The pregnant mare who is still nursing a foal has a tremendous job to do. This mare (Lyttl Miss, in foal to Sur Amir) and filly (by Sur Amir) are registered Arabians owned by James Hopkins of Twentynine Palms, Calif.

a good practice to breed a mare at foal heat after her very first foal. If it was her first one she may have stretched and injured herself a little more than a veteran broodmare would, and it may take her a little longer to get back into proper breeding shape again. And besides, if it is her first foal, you'll want to give her a little more time between foals; it won't be so hard on her.

And while we're on this subject of time between foals, it should be mentioned that if you breed a mare back while she is nursing a foal, you will want to wean her present foal at a fairly early date so that she will not endanger her unborn foal by trying to feed her foal at side. In other words, a mare cannot really do justice to both the foal at side and the foal she is carrying toward the latter half of her pregnancy. So the nursing foal should not be left on too long. When she reaches the later stages of pregnancy the demands on her system increase greatly and if she is nursing a foal as well, she may become anemic. So if she is pregnant again, and especially if she was bred back quite soon after she foaled, wean her foal at about 5½ to 6 months of age. Never leave the foal on her longer than 6½ months if she is carrying another one. The weaned foal will do fine at this age, especially if he is being fed a daily ration of grain.

The pregnant mare should be adequately fed. The nursing mare, even more, should be given special consideration feed-wise because the demands on her are even greater. Thus if a mare is both nursing and pregnant, feed her a well-balanced and adequate ration; she has a tremendous job to do.

8

THE ABNORMAL FOAL

NOW LET'S TAKE A QUICK LOOK AT THE FOAL, FOR HE IS THE end product, the hoped-for result of your breeding program. Hopefully, he will be born normally, easily and swiftly, and will be healthy and vigorous. But just in case he isn't—in case he may be an exception—it will be a good idea for you to become somewhat familiar with the difficulties and diseases that can beset the new foal, so you can be prepared to cope with them.

You have probably heard of Rh factor foals. This problem is also called hemolytic disease, neonatal isoerythrolysis, isohemolytic disease, and iso-immune hemolytic anemia of the newborn. Foals with this problem are sometimes called icteric foals, jaundiced foals, or Rh foals. A foal born with this difficulty has an allergic reaction to the antibodies in the mare's colostrum, caused by incompatability of the mare's and foal's blood. This reaction affects the foal's red blood cells (erythrocytes) and is usually fatal if not treated. The foal inherits erythrocyte antigens from the sire and these pass through the placenta into the mare's blood stream. If these antigens are not also part of the mare's normal blood makeup, antibodies are produced in her blood stream—antibodies against the foal's oxygen-carrying blood cells. These antibodies are usually produced in large amounts by the eight to tenth month of pregnancy but they usually don't affect the unborn foal because they are too large to pass through the placental barrier from the mare to the foal. So no reaction occurs in the foal until he is born and drinks the mare's colostrum which contains a high concentration of these antibodies.

The antibodies in the mare's milk are absorbed into the foal's bloodstream after he nurses, where they attack his red blood cells and destroy them, inhibbiting the flow of oxygen to his body tissues. The results are anemia, bloody urine (caused by break up of the red blood cells) and jaundice. The jaundice (yellow color of the foal's tissues such as the whites of his eyes, skin, and mucous membranes) is caused by the hemoglobin pigments that are scattered throughout the blood stream from the destroyed cells. These pigments are then picked up by the body tissues.

But in acute cases, the first noticeable symptom (even before the foal has time to look jaundiced) is weakness. Very soon after birth—after the foal has nursed a few times—the foal will show signs of weakness and listlessness, even though he was probably lively and vigorous earlier. He is dull and lethargic and doesn't seem interested in nursing. There is no fever, but his pulse rate may go up to 120 beats per minute. His respiration rate remains normal until the anemia begins to take its toll; then he has difficulty in breathing and starts yawning or showing other signs of respiratory inadequacy and his respiration rate may go up to 80. To save the foal, he must be taken away from the mare immediately, preferably before the symptoms get this bad. He must not be allowed to nurse the mare.

Some cases do not become obvious until the foal is two to four days old. These cases usually show jaundice, but only moderate anemia and just a little blood in the urine. Very mild cases may not show symptoms until the fourth or fifth day after birth and these mild cases often recover untreated. But if a foal shows symptoms within a day or two after birth, it is serious and very prompt action should be taken to save him. Have your vet run tests on the foal's blood and upon the mare's colostrum to make a definite diagnosis.

If the foal is not taken away from the mare and treated for this problem, he will usually die within four days. Some foals die within the very first day of life (and jaundice does not have time to develop this soon) so you must act quickly if you suspect this kind of problem. If you remove him from the mare soon enough and put him on a formula, milking out the mare, you may save his life. The mare should be milked out often (every two hours) to keep up her milk flow, hasten the coming of regular milk, and to milk out all the last traces of colostrum which contain the antibodies that are so deadly to the foal. The vet should take a blood test of the foal to see whether or not he will need a blood transfusion. Sometimes a transfusion can save him if he is already very weak and anemic. The blood used must be from a horse with a compatible bloodtype, to avoid servious reactions in the foal.

To prevent this kind of problem, a blood test can be taken from the mare and the stallion. If the mare is already pregnant and you are suspicious that this condition might exist (for instance if she has produced Rh foals before, or if the stallion has produced Rh foals when bred to other mares that are related to your mare, or some other reason that makes you wonder) you can test the mare's blood serum against the sire's blood cells a week or two before the mare is due to foal. If you get a positive reaction, meaning that the foal will be allergic to his mother's milk, you must not let the foal nurse his mother when he is born, until he is fully three days old. Usually after the second day of life the foal's intestinal membranes are thickened and he no longer absorbs antibodies from his mother's milk and will not be harmed by the ones he is allergic to. Be present when the foal is born and make sure he doesn't nurse. The foal can be muzzled to prevent nursing, the mare's milk can be tested, and if it is harmful to the foal you can feed the foal a formula for the first three days. The mare should be milked out regularly in the meantime. Once a mare has produced an Rh factor foal you'll want to try to prevent its happening again, breeding her to a stallion with a more compatible blood type. A formula which can be fed to the Rh foal if you ever have one will be discussed in a moment when we talk about caring for orphan foals. Vitamins A and C should be added to the formula to give the foal a better "start" since he will not be getting the benefit of his mother's colostrum.

Diseases are sometimes a problem with a newborn foal because the new foal has very little resistance to infection. This is a major reason for keeping the foal's environment as clean and sanitary as possible. Foals born out in a large clean pasture usually have less chance of contamination.

One of the worst foalhood diseases is navel ill, sometimes called joint ill. This disease was discussed in the chapter on the broodmare. Another foalhood disease is called actinobacillosis (sleepy foal disease). This is a serious bacterial infection in young foals and used to be called viscosum infection, or shigellosis. This is an acute and highly fatal disease of newborn foals. The foal is sometimes called a "sleeper" or a "dummy" because he becomes groggy and inattentive and may wander aimlessly about or lie around. Affected foals may be born with the disease, having become infected *in utero*, or they may pick up the disease from unsanitary surroundings, via the navel after birth. They get swollen joints, diarrhea, high fever and become very weak. They usually are debilitated and refuse to nurse their mothers. Any time a foal refuses to nurse you can suspect that he is sick. You can tell that the mare hasn't been nursed if she has a full and uncomfortable udder, and if she seems worried about her foal, nickering and trying to encourage him to nurse.

If you become aware of the problem soon enough, you may be able to save the foal affected by actinobacillosis—with intensive antibiotic treatment and good supportive treatment (usually additional fluids, I. V.'s and so on given every few hours) which your vet can help you with. This disease is not the same as European "dummy" foals. In Europe this name is applied to a convulsive syndrome that is not due to the bacteria that cause actinobacillosis, and is usually a milder type of problem. Actinobacillosis is caused by an organism that is often found in the intestines and tissues of normal horses, occasionally in the uterus. A mare that carries a foal which shows symptoms of the disease at birth does not seem affected by it herself. In foals the disease causes a severe septicemia (bacterial toxins are spread throughout the blood stream) often causing death within just a few hours. Foals that survive more than 24 hours develop inflammation and pus in the kidney, joints and intestines. The foal may be sick at birth, or begin to show symptoms within a few hours or days—up to 3 days of age. There is sudden fever, weakness, rapid respiration, and diarrhea. Death usually occurs within 24 hours if the foal is not treated, or between the 2nd and 7th day if he survives the early fever stage, showing signs of lamenesss and abdominal pain.

Diarrhea in a foal can be caused by many things—too much milk, the laxative effect of the mare's milk during her first heat period, the foal eating strange things such as dirt or manure, and so on. But diarrhea should never be neglected or considered lightly, for it may also be a sign of infection. It may be an intestinal infection (mild or serious) or actinobacillosis or some other problem that might easily prove to be fatal. The cause of the diarrhea should be determined and prompt measures taken to correct it if it is not merely due to too much milk or the mare's heat period. A foal will lose body fluid rapidly if he has diarrhea for any length of time, and with his body defenses lowered by this dehydration he may become susceptible to other serious problems as well.

Constipation is a common problem in foals and is covered in more detail in the chapter on care of the foaling mare. Constipation is a major cause of death in newborn foals; neglected cases usually die from bowel obstruction. If the foal strains and strains to have a bowel movement but doesn't succeed, or switches his tail and stamps his feet in discomfort or shows colicy signs, you may suspect he is constipated and should take immediate measures to try to correct the problem with an enema.

At some time during your career with horses, you may be confronted with a problem of a different sort—an orphaned foal. Such a foal may just be one that the mare will not accept, or a foal whose mother doesn't produce enough milk, or a foal whose mother dies. In any case, the foal needs milk, and since that milk isn't going to come from his mother, it will have to come from you.

One of my first horses was an orphaned filly named Ginger, that I got when she was a yearling. She was a spoiled and lovable pet. Her mother was an old Thoroughbred mare that died out on the range. Her owner knew she was with foal and spent several days looking for her and found her, dead, with a new little half-starved foal. The old mare had apprently died during or soon after foaling. He brought the little filly home on his saddle horse and his wife nursed it day and night, feeding it with a medicine dropper until it gained enough strength to suck a bottle. Ginger seemed to thrive on ordinary cow's milk and tender loving care and grew into a nice, healthy animal, but she was lucky. Not many foals would have survived that long after birth with nothing to eat from mama. Ginger may have nursed the dead or dying mare and gotten some colostrum, we don't know. Foals sometimes will nurse the mare before she gets up, and this may have been the case with Ginger.

Because the mare's first milk colostrum contains vitamins and disease-fighting antibodies, plus a laxative effect on the bowels to help clean out the intestinal tract, a newborn foal that is deprived of his mother's milk is therefore more susceptible to infection, constipation, and other problems. To make up for this lack, you should give the newborn foal several boosts—an injection of vitamins, some antibiotics, and probably several enemas. You may also want to talk to your vet about giving the foal some protective serum to make up for its lack of antibodies from the mother's milk.

The orphan will need a warm, dry place, and proper food. At first the foal will have to be fed every hour and fed about half a pint at once. After the foal is a few days old it will need feeding only about every two hours and the amount of formula slightly increased. By the time the foal is about ten days old (and you are going crazy from lack of sleep) you can feed it every four hours. Then when the foal is at least a month old and doing well, you can cut down to just four or five feedings a day (that's every five or six hours). The ideal situation is when

Ginger at two years of age—a filly who was orphaned at birth and raised on cow's milk and tender loving care

you can locate a nurse mare and have her adopt the foal. Occasionally a mare will lose a foal and still have milk by the time you are lucky enough to locate her and you can graft the orphan onto her, or you may be lucky enough to have a mare of your own that is giving plenty of milk and able to feed an extra mouth occasionally. But if no nurse mare is available, you will have to feed the foal a formula from a nipple bottle.

Mare's milk is not the same as cow's milk, but cow's milk can be altered to make it more nearly approximate mare's milk. Mare's milk is lower in fat and protein than cow's milk, but much richer in carbohydrates. So in making a formula for the foal, using cow's milk, you can modify the cow's milk by lowering the fat and protein content and increasing the sugar. This can be done in many different ways. There are many different formulas that can be used, but they all have these same objectives in mind. Some formulas contain eggs, while others do not. You may want to talk to your own vet about what he would recommend in a formula for your foal. Here are a few sample formulas, none of which contain eggs, for unless you have chickens, buying eggs to make foal formula will run into quite a bit of money. There are many different formulas; these are only samples. If one doesn't seem to agree with your foal, try another, and don't hesitate to get help or advice from your vet if you have questions about how your foal is doing.

A. 1 can evaporated cow's milk mixed with equal part water
   4 tablespoons limewater
   1 tablespoon sugar or corn syrup

B. 1 pint low fat cow's milk (butterfat content about 1% or 2%)
4 ounces limewater
1½ teaspoons sugar (white sugar is harder to digest than corn syrup or lactose or some of the other types of sugar; you may want to use another form of sugar)

C. 1 pint low fat cow's milk
2½ ounces water
2½ ounces limewater
2 teaspoons lactose

Goat's milk, even though it is not too similar to mare's milk (though it is high in carbohydrates) is sometimes used successfully for raising an orphan foal. A good goat might be able to take care of your feeding problems if you build a platform for the goat to stand on while the foal nurses. The foal will grow much taller than the goat and will be able to nurse more comfortably if the goat is standing on something. A goat can become quite accustomed to this arrangment for feeding her adopted child and can be happily eating grain on a platform while the foal (who may soon become twice her size) has dinner.

If you are going to use a formula and feed the foal with a bottle, get a good bottle with a nipple that doesn't run too fast; you don't want the foal to drink too fast and choke or get digestive upsets. Be sure the bottle is very clean, and warm the formula to body temperature (about 100 degrees) before you feed the foal. If the foal ever shows signs of sickness—refusal to eat, weakness, fever, diarrhea or constipation—call the vet immediately for many kinds of problems can soon prove fatal to the young foal. Modern drugs, blood transfusions and intravenous feedings can often save a foal, but never let sickness get the upper hand; treatment should be started just as soon as you suspect the foal is sick, or it may become too much of an uphill battle.

Very soon you should try to get the foal interested in grazing, or in eating some good alfalfa hay if there is no grass for it. Alfalfa hay should be very tender, green and leafy—palatable and nutritious for the young foal. You can start him eating grain at about two weeks of age, and after he is accustomed to grain and eating it readily you can give him about a half a pound daily (split up into two or even three portions, to do him the most good) and then increase his ration gradually as he grows.

When the orphan is very young, don't leave him all by himself for very long. Psychological factors are just as important as physical factors in determining whether the orphan will live or die. The will to live is very important in animals as well as in people. Sometimes it is the deciding factor. You don't want that foal to be too lonely; let him have a companion of some kind. Usually any sort of friendly animal will do—a calf, a dog, cat, duck, goat or burro, just anything to keep him company.

If by chance you ever have a mare produce twins or a premature foal, you'll find that these will take as much care (or more) than the orphaned foal, for a while. You'll want to consult your vet, for he can look into the situation and perhaps give you some good individual instruction for your particular case. For a premature foal or a set of twins, you'll want a very good place for the mare and her fragile offspring—a place that is clean, warm, and dry. It should be a roomy place, and well away from other horses. These tiny babies will have to be in isolation for a time until they gain strength and resistance to disease. Be very careful that you don't bring them any disease germs.

If the weather is cold, you'll want a heat lamp for the foal (or foals) in a dry, warm corner, in a place where the foal can snuggle in but the mare cannot. A pole or bar placed across a corner about chest high on the mare will keep her out, but she can still see and smell the foal and won't be nearly so upset as she would be if you took him clear away from her. Make sure the heat lamp is high enough that the foal can't bump it. I know of one case in which the whole barn and all its occupants were burned because a calf knocked its heat lamp down into the straw.

If the mare has twins and they both live, she may not have enough milk for both of them. You may have to give some supplementary bottles at first, until the foals get big enough to start eating more feed of other types. Keep the stall absolutely clean, changing the bedding often. With good luck and determination you can often save a premature foal if not *too* premature, and sometimes you can also be fortunate enough to save twins. Twins are, thank goodness, quite rare in horses, for it is not common to save them both. When twins occur sometimes one will be born dead, or very weak. If you save them both you are quite lucky.

Before ending this chapter, let's take a quick look at a few more problems that occasionally beset the foal. We'll briefly discuss wobbles, rickets, and abnormally crooked legs. These problems are not common, but because they *do* occur, you should be somewhat familar with them in case they ever should happen to you.

Wobbles (also called enzootic incoordination, spinal ataxia, foal ataxia, and equine sensory ataxia) is a disease of young horses that affects the spinal cord and vertebrae in the neck area. It is characterized by incoordination, defects in gait and posture, and loss of condition. The affected foal may drag his toes when he walks, and show a weakness in his rear quarters. The problem may start suddenly, with symptoms so severe that the horse goes down and in some cases never gets up again, or the symptoms may appear gradually, or come and go. The young horse's temperature, pulse, respiration, and appetite remain normal. The incoordination is often evident if the young horse is made to change direction suddenly, or if a hind foot is picked up.

The cause of the disease is unknown although there is some evidence that it may be inherited. The problem seems to be a congenital defect of the cervical vertebrae, resulting in a narrowing so that the spinal cord is compressed. The condition may be inherited, or the susceptibility to this disease (perhaps brought on by nutritional deficiencies) may be inherited, but this has not been actually proven. The spinal cord is damaged by pressure in the neck region—pressure exerted by vertebrae which are too mobile because of disease in the intervertebral discs and ligaments. Most cases appear when the foal is a suckling or weanling, but occasionally a case is not obvious until the horse is nearly two years old.

There is no treatment that will help this problem, and affected horses are usually useless for riding. Horses with mild cases have been known to live for years, but it is not unusual for these horses to suddenly get worse, or to die if the problem is aggravated by an accident such as falling or being pulled on the neck or performing any violent activity which may cause the neck vertebrae to cut into the spinal cord.

Rickets is a disease of young growing animals, characterized by improper calcification of growing bones. Joints are enlarged, and the poorly mineralized bones are usually bent out of shape because they are not strong enough to bear the weight of the horse. Rickets is very rare in foals under natural conditions, for in a natural environment the mare and foal receive adequate levels of calcium, phosphorus and vitamin D from natural feeds and sunshine. But rickets can occa-

Rickets. Fetlock joints are cocked outward on the front legs of this filly, her pasterns and toes turning inward

sionally occur if the pregnant mare (and then the newborn foal and nursing mare) are confined and fed inadequately, or fed a diet of very poor quality feeds.

Rickets is evidenced by a stiffness of gait, enlargement of the limb joints (especially front legs), and abnormal curvature of the long bones—usually forward and outward and outward at the knee. The one foal I have seen with rickets had a curvature of the cannon bones, accompanied by abnormal knee and fetlock joints, with pasterns and hoofs drawn inward from the fetlock joints.

Rickets can be treated by giving the foal (and the nursing mare) an adequate diet. Mild cases may recover completely, but severe cases usually retain the deformities that occured when the bones were too soft.

Sometimes a foal is born with crooked legs, or weak bones or tendons. These abnormalities can be caused by a great number of different things including heredity, nutrition, or even just lack of room in the womb. Sometimes a very large foal that has been severely cramped in the womb will take several days or even weeks or months to straighten his legs, having been bent so long. Lack of iodine in the mare's diet during pregnancy can cause contracted tendons in the foal. Lack of vitamin A or vitamin D can also cause problems. The list goes on and on. It is not ususual for foals to be born with weak, deformed or crooked legs, but most cases tend to straighten up by themselves within a few days or weeks, especially if the mare and foal are given good care and adequate nutrition. Some foals are born knock-kneed, others bowlegged. Some have contracted tendons which make the legs knuckle forward at the fetlock joint. Other foals are born with weak pasterns, toes turning up and fetlock joints pressing down to the ground. Some of these do not straighten up by themselves, and these require veterinary attention. Prompt veterinary assistance is needed, of course, if the foal is so weak or deformed that it cannot stand up. In serious cases of crooked or deformed legs, surgery, braces, casts, splints, or stretching manipulations will be needed. With these devices, a crooked-legged foal can often be salvaged to grow into a more normal animal that can be used for riding or for breeding (if the defect was not an inherited problem and will not be passed on to the horse's offspring).

Stainless steel staples are sometimes used to help correct a crooked foal. This procedure is called epiphyseal stapling. At the end of the long bones of the leg there is an area called the epiphysis; this is the growing area of the young bone. After the horse is mature and has stopped growing, this area calcifies.

If the long bone is growing crooked in a young horse, a stainless steel staple can be inserted on one side of the bone in such a way that it bridges the epiphysis, inhibiting the bone growth on that side, making it grow slower or not at all. The other side of the bone grows normally, gradually straightening the leg. The staple is then removed at the proper time. This procedure has been quite successful in correcting several cases in which a crooked-legged foal failed to correct itself or grew progressively worse. But the stapling must be done early, while the foal is growing, to accomplish the straightening.

A new type of foalhood disease has been discovered recently, and should be mentioned here. It is caused by a virus, called *adenovirus*, and seems to affect Arabian foals more often than any other kinds. Adenoviral infections in foals had been recognized in many areas by the early 1970s—the Midwest, the West Coast, Pennsylvania, and perhaps England and Australia.

Antibodies for this disease have been found in many horses of various breeds, showing that these horses have come in contact with the disease at one time or another but have recovered spontaneously, without obvious signs of illness. The fatal form seems to hit Arabian foals predominantly, and these are foals under three months of age. It does not seem to be highly contagious, because in a herd of 20 mares with foals, only one or two of the foals will develop the disease symptoms even when exposed to an infected foal.

If a foal develops symptoms, the disease progresses relentlessly and is fatal; treatment has very little effect. The first symptoms are pussy discharges around the eyes and nostrils. Then the foal's breathing beomes difficult and he develops a cough. His temperature may range from normal (normal can range from 99 to 100.5 degrees, depending on the individual horse, the time of day, etc.) up to 106 degrees. Respiration and pulse rates are higher than normal, and the foal's eyes are glazed and dull. It is rather common for pneumonia to infect the foal as a secondary invader. About one-third of the foals affected with this disease will have diarrhea, intermittently or persistently. As the disease progresses the foal will become thin, his hair coat will be rough, he will tire easily, and his appetite may be somewhat reduced. The foal will usually seem worse after exercise.

In a foal with adenoviral infection, white blood counts go progressively lower, and his lymphocytes (the white blood cells that produce antibodies) gradually disappear. Antibodies are needed to combat the disease, and the few foals that survived this disease were ones that did not lose their lymphocytes and were able to produce enough antibodies to eventually overcome the infection.

Treatment is usually to no avail. Antibiotics can help to keep secondary bacteria and fungi from taking over, but they do not touch the virus. If the foal is dehydrated or does not eat or drink well, additional fluids can be given into the stomach (by nose tube) or intravenously. Even temperature (about 72 degrees) in a dust-free stall with low humidity can help the foal if he has trouble breathing.

There is no vaccine yet available to help prevent this disease, but passive immunity can be given to a sick foal by giving him blood or serum transfusions from his dam or from the dam of a foal that has already had the disease. Experiments are showing that this may be helpful in treating the disease.

If a young foal dies and you are suspicious that adenoviral infection might be the cause, a postmortem examination and microscopic examination of tissues can give you the answer. If you have a sick foal with a respiratory problem, have your vet check the foal to determine whether it might have this disease. He can examine in his laboratory smears made from the inner linings of the nose or eyelids and give you a diagnosis. More research is being done to find out additional facts on this disease and to try to find better ways to prevent and treat it.

There are other problems occasionally encountered with young foals, but in this chapter I've tried to touch on the most common ones. If you are ever in doubt as to the diagnosis or seriousness of a problem, consult your veterinarian; foalhood diseases can often prove fatal or permanently damaging if not treated early.

9

DISEASES OF HORSES

YOUR HORSE'S HEALTH IS A VERY IMPORTANT MATTER, WHETHER you breed and raise horses, use your horses in ranch work, showing, competitive events or pleasure riding. You can't afford sickness; treating a sick horse can be time-consuming and costly, to say nothing of the fact that the horse might be useless to you during the time you might need him. In this chapter we'll take a look at some of the most common communicable diseases. Many of these you can now vaccinate your horse against, and it is certainly well worth the time and trouble, especially if it saves a horse's life.

SLEEPING SICKNESS

Equine encephalomyelitis, also called sleeping sickness and brain fever is a viral disease of horses which has occurred in the United States for many years. There are three distinct types of encephalomyelitis virus found in North America, and other types that appear in other areas of the world. Three known strains of this particular virus exist—the eastern, western, and Venezuelan strains. In 1971 the Venezuelan strain (VEE) crossed the United States border from Mexico after moving northward from South America and created panic and problems because North American horses had no immunity against this strain. Many horses in Texas were lost and great efforts were taken to try to stop the progress of the disease northward into the rest of the United States, and to create a suitable vaccine for horses that would immunize against this strain. During the emergency, horses were vaccinated with live virus vaccine which had been manufactured for use in humans.

The eastern and western strains of the disease have been in the United States for many years, and before the days of vaccine and immunization, the disease took a heavy toll among horses and mules. In the summer and fall of 1912 it was estimated that more than 35,000 horses were lost in the Midwest. In 1930 the disease had spread to California and appeared in epidemic proportions in the San Joaquin Valley. The virus was finally isolated and studied that year. It continued to spread until by 1937 it had been recognized in every state west of the Mississippi River and in several states east of it. In 1938 more than 184,000 horses died from this disease (estimated by the U.S. Bureau of Animal Husbandry).

Meanwhile the disease had appeared along the East Coast. Symptoms were similar to those of horses suffering from the western disease, but the mortality rate was higher, approaching 90% (mortality rate in the western disease is about 30%). It was found to be caused by a different strain of the virus. Both viruses produced sleeping sickness in horses and man but the eastern type is more deadly. Also it was discovered that horses immunized against the eastern type are not immune to the western disease, and vice versa.

The virus that causes equine encephalomyelitis was named for the horse because it was first discovered in horses, but now we know that the horse is not the natural reservoir for the disease. Birds are the primary source of the disease, and some snakes may also be involved. The virus is transmitted from birds (and snakes) to horses or man by bloodsucking insects, especially mosquitoes.

In the U.S. and Canada the disease is seasonal, occurring during the warm summer months when mosquitoes are numerous. In the warmer climate of the southern states sporadic cases may occur throughout the winter. But in most states a few cases occur during early summer, with the highest number of cases appearing during the hotter months of August and September. Then the number diminishes as mosquitoes are killed by frost in the fall.

Horses of all ages are able to contract the disease but young horses are the most susceptible. A horse that has suffered from the disease and recovered will have a natural immunity for about two years.

Several types of mosquitoes, including *Aedes* and *Culex* are the chief vectors, and the virus can live within the mosquito for several generations, passing through all stages of the mosquito's life cycle.

After a horse comes in contact with the disease (after being bitten by a virus-carrying insect) the incubation time is one to three weeks. The first signs of infection are fever and depression, but at this time the symptoms are so mild that they may go unnoticed. A very brief viremia occurs at the height of the fever in the eastern and western type of disease and this is the only time that the virus can be passed on to other animals via bloodsucking insects that bite the infected horse. For all practical purposes in the western and eastern disease, the horse can be considered a dead end host; there is not much chance of spread from an infected horse. But in the Venezuelan strain of the disease the viremia persists throughout the course of the illness and the blood thus provides a source of infection for biting insects. Also, the virus is present in the saliva and nasal discharges in a horse affected with VEE and the disease can be spread by direct contact. This is one reason that VEE can be spread so rapidly.

The next state after the fever, in all types of this disease, is invasion of the nervous system. This does not occur in all horses affected by the disease, however. If it doesn't occur, the horse recovers (and may not have been observed to be ill) and then possesses neutralizing antibodies in his blood serum for about two years thereafter.

If the virus attacks the nervous system, the horse usually shows signs of deranged consciousness by the time the fever reaches its peak. Then by the time the nervous symptoms have been evident for about 24 hours, the horse's temperature is back down to normal again. By this time the horse's blood is no longer infective (unless the horse has the Venezuelan strain); the disease can't be passed on by bloodsucking insects.

Soon after the nervous system becomes involved, the horse may be restless, or show mild excitement, or may seem blind. He may be overly sensitive to touch and sound. He may walk in circles or crash through fences or wander blindly into obstacles of any kind. Involuntary muscle movements may occur; there may be tremors of shoulder and facial muscles. Severe mental depression appears, a sleepy attitude develops and the horse may stand with his head hanging low, perhaps resting it on the manger or on the fence. The horse can be aroused, but he soon lapses back into the sleepy posture when not being prodded into activity.

A stage of paralysis follows. The horse may sit on his hindquarters or stand with his front legs crossed, or assume some other unnatural position. He is unable to hold up his head. His lower lip hangs down and his tongue hangs out. He may press his head against something solid or hang back on his halter if he is tied. He has difficulty in walking because he lacks full control of his legs. Defecation and urination are suppressed and the horse is unable to swallow.

The final stage is complete paralysis. The horse goes down and is unable to get up, and this happens within two to four days from the time the first signs of illness appeared. A few affected horses do not develop paralysis and they survive, but they are often mentally deficient afterward because of permanent brain damage.

The course of the disease can vary. In some cases horses die within just a few hours from the time the first symptoms are noticed. Most deaths occur within two to four days. Horses that survive the effects of the sleeping sickness virus may develop terminal pneumonia and die after a week or more. Others may recover completely. Others that survive may show various paralytic effects for many weeks, or even permanently. The death rate with the western type may be as high as 50% but usually averages from 20% to 30%. The Venezuelan type is more similar to the eastern type in its symptoms and severity.

Prevention is the best cure for sleeping sickness. Once a horse has contracted the disease there isn't much that can be done except to keep him comfortable and try to help him survive the danger period. Hyperimmune serum is the only specific treatment and is sometimes beneficial if given in large doses in the earliest stages of the disease, but this serum is not available commercially. If the horse is able to bear some weight, he should be supported in a sling and fed a nutritious, laxative diet, through a nose tube if necessary. If the horse is down, he should be given ample bedding and he should be turned frequently. He should be shaded and protected from flies and heat.

After experimenting for a while with live viruses and inactivated viruses, researchers finally created a very successful vaccine in 1938, made from infected chick embryo tisssues. This vaccine is still used today, administered intradermally (between the layers of the skin), the dose repeated within a week or ten days. Good immunity is established in about 15 days from the time of the first dose, and lasts one season. Annual vaccination should be given in the spring before the disease is prevalent. In areas where the possibility of both eastern and western types of virus exist, vaccine should be used which gives immunity against both types. In 1971, intramuscular vaccine was made available for the first time, making sleeping sickness vaccinations much easier to administer.

Now the annual death rate in the U.S. from eastern and western types of sleeping sickness numbers low in the hundreds instead of in the thousands, and even these deaths could be prevented if more people vaccinated their horses. In recent years there has been an upswing in cases, probably because some horsemen consider sleeping sickness a disease of the past (until the VEE scare hit the U.S in 1971) and neglect vaccinating. They forget that the local bird population may be the prime source of the virus via bloodsucking insects. Even if no other horses in your area contract the disease, you are still taking a chance if you leave your horses unvaccinated; as long as there are mosquitoes, there will be possibility of infection. Vaccine is quite cheap when weighed against the possibility of losing a good horse. There is now a vaccine available for VEE as well (it must be administered by a veterinarian), for this new strain may become a permanent threat to U.S. horse owners if efforts are unable to halt its northward spread during the warm summer months.

TETANUS

Tetanus is a highly fatal disease of all species of animals, but horses and man are especially susceptible. This disease is caused by the toxin produced by *Clostridium tetani,* a spore-bearing type of bacteria that is commonly present in the digestive tract of many animals and in soils that are rich in humus. Tetanus occurs in all parts of the world and is most common in heavily populated areas that have been under intensive cultivation for many years. The spores formed by tetanus bacteria are quite resistant to heat and can last in the soil for years if somewhat protected from air and light.

The organism was first isolated in 1884, from white mice that had been inoculated with garden soil. The first pure culture was obtained in 1889 by heating cultures taken from infected wounds; the heat destroyed the ordinary bacteria, leaving the heat-resistant tetanus spores.

Tetanus bacteria are commonly present in the feces of animals, particularly horses, and are present in the soil wherever there has been horse manure or other forms of organic contamination that contained tetanus bacteria.

Infections in animals and man occur as the result of wound contamination, deep puncture wounds being the most common portal of entry for tetanus. Clean wounds rarely result in tetanus; it is the dirty wound containing foreign matter, especially soil (particularly in the enclosed puncture wound) that is most dangerous. Puncture wounds in the feet are a common site of entry in horses.

Spores may lie dormant in the body tissues for some time and produce the disease only when conditions turn favorable, producing toxin only if for some reason there is a lowering of the local oxygen level in the tissues. This can happen immediately after the bacilli enter the wound if there has been enough tissue damage, or may be delayed for weeks or even months until another injury at the site produces tissue damage. The original wound might be completely healed by that time.

The study of tetanus became more intensive during World War I because so many soldiers developed the disease while fighting in the cultivated fields of Flanders. It was learned that tetanus spores would not grow and germinate in living tissues because of the presence of too much oxygen, and that germination happened only when the tetanus spores were introduced into destroyed or damaged tissue. When dirt or foreign matter accompany the tetanus spores into a wound, creating local irritation, the conditions become ideal for tetanus to germinate and produce its deadly toxins. Tetanus toxin is one of the most poisonous substances known, being 100 times as toxic as strychnine.

The toxin reaches the central nervous system of the animal by traveling along the nerve trunks. When the central nervous system becomes affected, muscle spasms begin to occur and many types of normally mild stimulations to the animal's nerve endings cause exaggerated responses by the animal. Death from this disease is usually due to suffocation because the muscles with which the animal breathes become paralyzed and do not function properly.

Susceptibility to tetanus varies greatly among different kinds of animals. The horse is the most susceptible domestic animal, and birds are the least susceptible. For instance, the amount of tetanus toxin needed per gram of body weight to kill a chicken is about 350,000 times as great as that needed to kill a horse. The amount that will kill a dog is about 600 times as great as for a horse.

The incubation period for this disease is usually one to three weeks for a horse, with occasional cases occurring even later than that. The first symptom is muscle stiffness and a beginning of muscle tremor, then chronic muscle spasms. Usually the facial muscle and chewing muscles are affected, making it difficult for the horse to open his mouth. Hence the common name "lockjaw."

The third eyelid membrane protrudes, and the horse's hind legs become quite stiff, making his gait unsteady. He holds his tail out stiffly. The protrusion of the third eyelid is one of the first signs of tetanus. Other symptoms to look for are an anxious and alert expression, with ears up, eyes wide, nostrils dilated, and the horse exhibits exaggerated responses to ordinary stimulations. He may eat and drink normally in the early stages, but as the disease progresses his chewing becomes difficult and he cannot swallow. Saliva may drool from his mouth.

The horse becomes constipated because of lack of muscle control, and is unable to urinate, partly because of the difficulty he has in assuming the natural position for urination. His temperature and pulse rates are normal at first and then rise as the disease progresses (temperature sometimes getting as high as 107 degrees) and muscle spasms increase. The horse assumes a "saw-horse" posture, with his legs stiff—hind legs stuck out stiffly behind and front legs forward. His muscle contractions may be erratic and uneven, making his spine curve and his tail stick out to one side. He has great difficulty in walking, and if he is suddenly startled he may fall down in his scramble. If he falls he may injure himself because of his lack of coordination; his limbs do not work properly. If he falls or gets down on the ground it may be impossible for him to get up again. As the disease worsens, he may have occasional convulsions, especially if he tries to respond to stimuli (especially sound or touch), then later the convulsions may arise spontaneously.

Tetanus is almost always fatal, the horse dying within five to ten days after symptoms appear. In fatal cases there is often a brief period of improvement that may last for several hours, only to be followed by a final spasm that is very severe, affecting the horse's respiration so that he cannot breathe. Occasionally, however, a mild case will be followed by recovery. If symptoms are mild (and especially if they don't appear for some time after the wound introduced the tetanus bacteria) and the disease runs a long course, the horse may recover. In the few cases where a horse *does* survive, it takes him a long time to recover completely, the muscle stiffness gradually disappearing over a period of several weeks or even months.

Treatment of a tetanus case involves several things including elimination of the tetanus bacteria, the neutralization of the remaining toxin, relaxation of the horse's muscles to avoid the spasms which can result in suffocation, and keeping the muscles relaxed until all the toxin has been eliminated. The main thing is to try to keep the horse alive during the critical stages of the disease. Antitoxin is given, the earlier the better, and penicillin is administered in large doses. Muscle-relaxing drugs can be given. The horse should be supported in a sling during the time he is unable to navigate on his own, and he should be fed intravenously or with a stomach tube during the critical stages when he cannot eat or drink. He should be kept as quiet as possible, preferably in dark surroundings. Good bedding and plenty of room will help avoid injury if he has convulsions. He may need enemas and catheterization to relieve his discomfort from his inability to defecate and urinate.

The best protection against tetanus is vaccination, in which the horse is actively immunized with "toxoid." The usual procedure is to give two shots several weeks apart, then a yearly booster. It takes a few weeks after the shots to build up or increase this immunity, so if a wound occurs shortly after the first vaccinations, or towards the end of the year before you give the booster, you should give the horse a shot of antitoxin. Antitoxin develops the horse's resistance much more quickly, it gives quick protection, but doesn't last long. Yearly vaccination is the safest protection, but if the horse suffers a wound when his vaccinations are not up to date, antitoxin may temporarily (for that particular wound) prevent tetanus.

STRANGLES

Strangles, also called distemper, is a disease of horses caused by the bacteria *Streptococcus equi.* Distemper is a misleading term for this disease because distemper of cats and dogs is an entirely different type of disease, caused by a virus.

Strangles in horses is characterized by inflammation of the upper respiratory tract and swollen lymph nodes that often rupture and become abscesses. Young horses seem to be more susceptible than adult horses. The disease begins with a respiratory infection (cough and runny nose), then produces swelling and abscesses in the lymph glands of the head and neck. Deaths occur when internal lymph glands rupture and abscess.

Strangles is a world-wide disease and was a serious problem in horse and buggy days when there were many horses in cities and working on farms; there were often major epidemics of strangles. It is still a problem when strange horses come together (especially at horse shows, race tracks, etc.), unless they have been vaccinated. It is spread by direct contact or by contaminated feed and water. If adequate treatment is given in the early stages of the disease, the mortality rate is usually very low. But deaths can occur from strangles if the infection spreads to other organs of the body such as the kidney, liver, spleen, brain, etc. Infection can also spread to the tendon sheaths and joints, causing lameness.

The bacteria are present in the nasal discharge of affected horses, and this nasal discharge can contaminate feed, water and pasture. An infected horse can continue to spread the disease for at least four weeks after he contracts strangles. The organism that produces strangles is relatively tough and may live for several weeks on the ground or bedding and may infect other horses even a month or longer after the sick horse has been removed from that place.

Strangles can strike any horse of any age, but horses in the one to five year old group are the most susceptible. Outbreaks of strangles can occur any time during the year, but are most likely to happen when the weather is wet and cold and when horses do not have adequate shelter from bad weather. But the most important factor in the spread of the disease is contact with a horse that has a case of strangles. Bad weather merely makes exposed horses more susceptible. If a horse with strangles is introduced to the group, and if the weather is wet and cold or if most of the horses in the herd are young and susceptible, with no immunities, the whole group may contract the disease. After a horse recovers from strangles, he will have good immunity for a while, but if the infection is a persistent and virulent one, horses in the herd may have repeated attacks of strangles about every six months unless the disease is totally eradicated within the group.

After coming in contact with disease, and after an incubation period of about 3 to 8 days, the horse will begin to show sudden symptoms. The horse then has a sudden high fever (103 to 105 degrees), complete loss of appetite, a clear nasal discharge that soon becomes very thick and profuse, and respiratory irritation. Infection and irritation in the throat area may become so severe that the horse cannot swallow, and if he tries to eat or drink he may regurgitate through his nose. Another typical symptom is a painful cough that is soft and moist.

The fever usually goes down in a few days, but then rises again as abscesses develop in the lymph nodes in the horse's throat. These swollen nodes are hot and painful and may begin to drain fluid about ten days after the horse's first symptoms appeared. The nodes may then rupture, giving forth a thick, creamy pus.

If the infection is severe, many other lymph nodes may become swollen and abscess at this same time. Abscesses may appear anywhere on the horse's body, particularly on the face and legs, and there may be a great deal of swelling. The horse's lower legs may swell to several times their normal size.

If abscesses form in other body organs, there may be serious complications. If the infection spreads to the lungs, the horse may develop an acute pneumonia. An abscess in the brain usually causes a meningitis and the horse then shows signs of excitement, extra sensitivity to stimulations, and then becomes paralyzed and dies. Abscesses may also develop in the liver or spleen and cause death if they rupture, and this can happen several weeks or even months after the horse seems to be recovered from the actual case of strangles.

If a horse contracts strangles, he should be isolated and treated immediately, with proper types and doses of antibiotics. Penicillin is usually effective if given soon enough in the early stages of the disease, but if treatment is undertaken after the disease is already well in progress, the tetracyclines are more effective. Usually proper treatment can prevent the spread of the disease through the horse's system and avoid dangerous complications, but if the lymph nodes are already swollen they may continue to enlarge and rupture. The horse being treated for strangles should have adequate shelter that is warm and dry (he should also be blanketed if necessary), and he should be fed a soft and palatable diet. If swelling in the throat area makes it hard for him to breathe, the lymph nodes can be treated surgically.

After the horse is removed from the stall or barn, the area in which he was kept should be thoroughly cleaned and disinfected. The old bedding should be burned. Any other items that were used for that horse (brushes, blankets, drinking tubs, and so on) should be carefully disinfected. If you have several horses and you fear that some of them may have been exposed to the disease, talk with your veterinarian. Vaccination is usually an effective means of preventing strangles, but some vaccines for strangles should not be given to a horse that has already been exposed and has the disease in its incubation stage. Your veterinarian can help you control an outbreak of strangles if you have a group of susceptible horses. Vaccination is the cheapest and easiest route, but should be done before a horse has a chance to come in contact with the disease, preferably in early summer before bad weather, and before the horse comes in contact with strange horses (at shows, riding stables, trail rides and so on) that might bring him the disease.

INFLUENZA

Equine influenza is an infectious disease of the respiratory tract in horses, caused by a virus. In past years there has been some confusion concerning the viruses that affect the respiratory tract in horses, for there are several virus diseases that produce similar symptoms. Some of these same viruses also cause abortions in pregnant mares.

Equine viral rhinopneumonitis (EVR) is a mild disease of the upper respiratory tract caused by a specific virus that can also cause abortions. Equine viral arteritis (EVA) also causes abortions, and sometimes causes acute upper respiratory tract infection. Equine parainfluenza virus and equine rhinovirus are also associated with respiratory infection.

Influenza is a term that has sometimes been applied to several respiratory diseases; in the past it was almost a general term for any respiratory tract problems with "cold symptoms." But now the term influenza, or more specifically equine influenza, is narrowed to apply to disease that can be caused by two of the viruses in the influenza A group (termed equine influenza A-1 and A-2). Both of these viruses produce respiratory illness and can also cause abortion, but abortion is not a very common symptom. Other names for this disease in past years have been shipping fever, stable pneumonia, pinkeye, and epizootic cellulitis of horses.

Influenza is a very contagious disease, and one that is hard to treat, for the virus does not respond to antibiotics. Antibiotics given to a horse with influenza are for the purpose of preventing or controlling the secondary bacterial invasions that often follow the initial virus attack. When a horse is weakened by a virus such as influenza, bacterial complications often follow, causing permanent damage in some cases, or even death.

Equine influenza spreads rapidly among susceptible horses. This disease can affect horses of any age, but occurs most commonly in young horses that have not yet acquired any immunity. Young horses that have been moved to new places, especially when they come in contact with strange horses, are very susceptible. This is why influenza has always been such a problem where large numbers of horses are brought together, particularly in places like sales yards, horse dealers' herds, army remount stations, riding schools, and so on. In stables where new horses arrive from time to time, the newly arrived and highly susceptible horses contract the disease and provide a constant source of infection, helping to keep the disease going in very virulent form. Under certain conditions the death rate among infected horses may become very high.

In the past there have been periods in which the disease reached epidemic proportions across most of the country. One of the worst epidemics of influenza in the United States happened during the winter of 1872–1873 in which almost every horse in certain areas was affected. Traffic in many of the large cities across the nation nearly came to a halt because there was such a shortage of horses healthy enough to work. During World War I, thousands of army horses died of influenza. A more recent epidemic in 1956, struck the horses of eastern Europe and Sweden. And in 1963 a serious epidemic spread across the U.S. after first appearing in Florida—brought in by horses flown to Florida from South America for racing. That epidemic hit many race tracks and horse shows before the year was over. The cause of this epidemic was a new strain of the flu virus, named A-2 equine influenza virus.

When the number of horses in our cities and farms declined as they were replaced by machines, many of the prevalent horse diseases, including influenza, declined also. But now some of these diseases are on the upswing again. In our modern age when horse numbers are rapidly increasing again, because the horse is assuming a new role as a pleasure horse and contest horse, diseases like influenza are increasing too, and become a further problem because horses do so much traveling—to horse shows, races, rodeos, and trail rides all over the country (and often to other countries), and horses today are often sent clear across the country just to be bred. In this day of trailers and vans and jet-age horse transportation, there are many opportunities for diseases to spread. Vaccination for the most prevalent contagious diseases, like influenza, is a must today, especially if you ride with other horsemen, keep your horse at a stable, take him to shows, or have a breeding farm or stable where strange horses come to board or be bred.

Because equine influenza is highly contagious, practically all the young stock and many of the older horses will contract the disease if it appears in a specific area such as at a stable or on a farm. The mortality rate is usually not very high though, unless there are complications. The usual loss is mainly in time—time during which the horse is too sick to be worked or ridden or in training or showing. The horse may be sick for one to three weeks, or even longer. The length of the disease depends greatly upon whether or not complications develop. Uncomplicated cases may be fully recovered after one week.

Incubation period after contact is about five to ten days. The symptoms of influenza appear suddenly. The horse shows a depressed attitude, stands with his head down, and doesn't eat. He has a fever of 103 to 106 degrees and this usually lasts about three days. In many cases the horse is sensitive to light and his eyelids and conjunctivae seem swollen; the conjunctivae may stick out between the horse's closed eyelids. The eyes water (sometimes the cornea becomes cloudy) and the nose runs. Lymph nodes around the head may become swollen, but not always. Sometimes pneumonia occurs as a secondary complication, in which case the horse usually dies.

In some cases of influenza, swellings appear on the belly and the legs; the tendon sheaths are inflamed (this form of the disease was often called epizootic cellulitis in the past). Kidney damage can sometimes occur with a case of influenza also. A thick nasal discharge often follows the acute stage of the disease, and this usually means that there is some secondary bacterial infection.

The usual means of transmission is direct contact. A healthy horse that comes in contact with one that has influenza is likely to get the virus from the sick horse's discharge. The virus can also be passed in the semen from a stallion that has had the disease, sometimes for several months after recovery. The disease usually leaves a recovered horse with some immunity; this is one reason that older horses that have had previous exposure to influenza are not as susceptible as young horses.

No specific treatment will affect the action of the virus if a horse contracts the disease, but antibiotics should be given to prevent or treat secondary bacterial invasion. The horse should be isolated, kept in a warm, dry place free from drafts, fed a laxative food, and should be given a constant supply of fresh drinking water.

There is now a vaccine for equine influenza. The usual procedure is to give two doses of vaccine several weeks apart, and then one booster shot each year thereafter, or whenever the horse may be exposed or an epidemic seems likely (if he hasn't yet had his booster for that year). Epidemics of this sort, however, would soon become a thing of the past if more horsemen protect their horses against influenza by using a conscientious yearly program of vaccination.

LEPTOSPIROSIS

Leptospirosis, or lepto as it is called in our part of the country by cattlemen, is a disease of animals and man caused by a bacteria, a spiral shaped germ called a leptospirochete. There are many types of leptospirochetes; over 50 have been found. Some cause leptospirosis in dogs, others cause the disease in cattle, pigs, and horses, still others cause disease in rodents and humans, wild animals, and so forth. In humans leptospirosis can be very serious (Weil's disease) or mild (seven-day fever, canicola fever), depending upon the type of leptospirochete that causes it. It is time more horsemen were aware that this disease exists and that it can affect horses.

There are two main types of leptospirochetes, the nonpathogenic (harmless) type which live and multiply in ponds, lakes, and other bodies of standing fresh water, and the pathogenic types which cause disease, living and multiplying in the tissues of animals. Unlike the harmless kinds, the pathogenic leptopirochetes do not live very long outside the body, depending on environmental conditions (especially temperature and moisture) which I'll discuss in a moment.

Leptospirosis is a disease of great economic importance to the livestock raiser, for it causes abortions. It can affect an entire herd of cattle or horses under certain circumstances, and losses from abortions can be tremendous.

Leptospirosis is often spread by a carrier animal. A carrier is an animal that contracts the disease and recovers, but still harbors leptospires in its body that can be spread to other animals. The most common carriers of lepto are rats, mice, and other rodents, although some lepto serotypes are most commonly found in dogs, pigs, or cattle. These pathogenic leptospires are parasites, and each serotype seems to have a favorite host. Rodents carry several types. Pigs are the main carriers of *Leptospira pomona,* the species that is the most common problem in livestock, including cattle and horses. There is a vaccine that will immunize livestock, including horses, against *L. pomona.* Dogs are the main carriers of *L. canicola.* Cattle are occasional carriers of *L. grippotyphosa.* Rats seem to be able to harbor leptospires in their kidneys throughout their lives with no apparent damage, whereas dogs that are carriers of lepto suffer extensive kidney damage, and may eventually die.

Cattle that contract *L. pomona* may shed leptospires in their urine for ninety days afterward, and then they cease to be carriers. But a cow that contracts the serotype *L. grippotyphosa* may remain a carrier for a longer time. Rats are carriers for the rest of their lives after contracting the disease. Dogs have been known to be carriers of *L. canicola* for more than three years after recovery. Each lepto serotype tends to be associated with a particular species of carrier host (*L. icterohaemorrhagiae*—which causes the serious Weil's disease in humans —with the brown Norway rat; *L. canicola* with dogs; *L. pomona* with pigs, and so on) but there are many exceptions because one serotype may be carried by several different hosts, and one species of animal may act as host to several serotypes. As far as we know now, the horse is not a carrier after complete recovery from the disease, but he is susceptible to infection from several serotypes, including *L. icterohaemorrhagiae, L. canicola, L. pomona* (the most common source of leptospirosis in horses), *L. grippotyphosa,* and *L. sejroe.*

Lepto can be transmitted to man, usually indirectly, from rodents or other animals which harbor leptospires and excrete them in urine. Man picks up the leptospires when his skin comes in contact with wet contaminated ground, or when he swims or bathes in contaminated water. In humans, leptospirosis can be very mild or very severe, depending on the serotype involved.

In horses, the disease itself is usually not serious. When a horse contracts lepto he may or may not appear sick. Different horses react differently to this disease. A horse may have many symptoms, or he may have only one or two. Lepto is a very versatile bacteria; it can localize almost anywhere in the body and cause trouble. The leptospires may infect the entire body, causing one or more of these symptoms: fever, depression, loss of appetite, lethargy, jaundice. Pregnant mares usually abort. The body in general is usually able to combat the disease (horses rarely die from lepto) but even though the horse recovers, the leptospires may cause further problems in the kidney, the eye, or the joints.

The disease can be transmitted to a susceptible horse through contact with any secretions or excretions from infected animals (carrier animals, or from horses that are in the early, active stages of the disease). This includes urine, nasal and eye discharges, saliva, and so on. Leptospires can live for many days in damp or saturated soil, but die within a few hours on dry soil. These bacteria can't stand dehydration or sunlight. This is one reason that lepto is not highly contagious in dry climates and in dry seasons; the leptospire is very fragile and dies within a few hours if the temperature and moisture conditions are not suitable for it. In a warm, stagnant pond it may last from seven to thirty days. The leptospires that are secreted in urine usually die within a few hours (urine is too acid for them to live in) unless the urine is diluted, as in a pond or marsh or stream or very wet soil. Leptospires can live in wet soil for three days, in saturated soil for 100 days, but in very dry soil they last only about an hour.

Vaccine is successful in preventing the disease in horses, unless the horse contracts a type of lepto he was not vaccinated against. For instance, *L. pomora* vaccine will not protect a horse against *L. canicola, or L. grippotyphisa,* and so on. If an unvaccinated horse contracts the disease, streptomycin and broad-spectrum antibiotics are helpful in combating most cases of active infection.

Blood tests will tell if a horse has been exposed to the disease (showing that the horse has been vaccinated or has had the disease at one time) but are not necessarily conclusive in determining whether or not the horse has an active infection presently. Only the isolation of leptospires from the blood or urine or other body fluids is positive proof of the active disease, and this can be rather difficult;

the organisms must be cultured from a sample of one of these fluids. Blood tests merely tell the amount of reaction in the horse's blood serum to the lepto antigen. Many healthy horses which have never had any obvious symptoms of lepto show a blood titer, suggesting that they have come in contact with the disease at one time or another.

Another factor that makes blood tests an inconclusive test of active leptospirosis is vaccination. Any horse that has been vaccinated for lepto will have lepto antibodies in his blood stream, and perhaps a change or rise in titer.

Leptospirosis is one of the main causes of periodic ophthalmia in horses, a disease of the eye. There was literature written on this eye problem as early as the fourth century. The disease was called moon blindness because horsemen mistakenly thought the recurring attacks were related to phases of the moon. Periodic ophthalmia is the most common cause of blindness in horses. The disease is also called recurrent ophthalmia, recurrent irridocyclitis, and uveitis.

When a horse gets periodic ophthalmia, the first symptoms are usually a watering of one or both eyes. The pupil constricts, the eye is sensitive to light and there is conjunctivitis (inflammation of the conjunctivae). The cornea may become opaque—cloudy and bluish gray—at this time or at a later date. These attacks last for indefinite periods of time and then seem to get better, recurring again anywhere from a month or a year later or even longer. Each attack further damages the eye and in almost all cases blindness eventually results. Periodic ophthalmia is rare in foals, and occurs most often in horses from one to eight years of age.

Some years back periodic ophthalmia was blamed on riboflavin deficiency, but in recent years more emphasis has been placed on other causes. Current thought is that there may be several causes, including leptospirosis, streptococcus infections, and parasitic larvae (*Onchocerca cervicalis*). Periodic ophthalmia is not very contagious, and seems to be caused by one or more factors acting together or separately.

The most important cause seems to be leptospirosis, either as an active infection or as a second or third exposure. Yet some horses with periodic ophthalmia do not show blood titers for lepto, indicating that they have never been exposed to the disease. So, lepto is definitely not the only cause. One theory is that repeated attacks of any or several of the causative factors result in a progressively stronger sensitivity and a greater reaction in the eye. Migrating worm larvae may lower the resistance of the eye, allowing the leptospires to work.

Many outbreaks of leptospirosis in horses have been seen in the United States and in other countries, but often a horse with lepto has no eye involvement. Periodic ophthalmia is just one of several side effects that can result from the disease. Usually, lepto is a fairly mild disease in horses, and it is self-limiting; the horse will get over it completely, and when he does recover from the active infection, he is no longer able to spread the disease. But it is not known exactly how many days this takes, and he should be isolated while he has the disease. A mare or stallion should not be used for breeding purposes for a while after having lepto, if there is any question that he might still be able to spread the disease. The horse is not a carrier, as far as we know at this time. A recovered horse's blood will, however, have antibodies, and a blood test will show a titer for lepto, showing that he has encountered the disease at one time.

A horse usually recovers easily from leptospirosis. But periodic ophthalmia is a different story. There is no real cure for periodic ophthalmia; an attack can only be treated to relieve the horse's pain. Effective treatment can help postpone blindness. The horse should be put in a dark barn or stall until the attack subsides, to rest the eye and help relieve the pain caused by bright sunlight. If the horse has lepto, high levels of streptomycin and broad-spectrum antibiotics will help control that aspect of his problem. Isolate him so he won't spread it to other horses that might come in contact with him or his excreta or with water he has contaminated. Your veterinarian can direct you to a program for treatment of the horse.

I hope that more horsemen will become aware of this disease and the threat it can pose. In our part of the country we have vaccinated our cattle for lepto for many years, but I had never heard of the disease in horses until one of our mares provided us with first-hand experience during 1966 and 1967. Since then I have seen more cases, and several breeders in the area have lost large numbers of foals because of leptospirosis abortions. Any horse breeder is running a big risk if he doesn't vaccinate his mares. Often lepto cases go unnoticed unless they cause abortion or periodic ophthalmia. And I have seen several cases of periodic ophthalmia in which the owner of the horse did not realize what the problem was or that lepto might have been the cause.

Our mare, Nikki, had an irritation in her left eye during the summer of 1966, two different attacks that lasted only about a day or two each. The eye watered and she held it partly closed. I thought perhaps a branch had flipped her in the eye. I had no idea that this might be the early stages of periodic ophthalmia, for I had never heard of it. The eye irritation cleared up and we thought no more about it until the spring of 1967 when her left eye suddenly became cloudy. I put her in a dark barn, treated the eye, and a veterinarian diagnosed the problem as leptospirosis.

Throughout that summer I doctored her for lepto and for periodic ophthalmia, giving her hundreds of injections (antibiotics, vitamins, particularly vitamin A, and injections of nonspecific protein treatment which was to help clear up the cloudiness in her eye). The eye varied in its usefulness that summer; sometimes she had good vision in it, sometimes not. The eye bothered her very much at times and I made a blinder on a halter to cover that eye so sunlight wouldn't bother it. By late summer she started to improve, and I began riding her again that fall. The eye cleared up fairly well and to this date she has not had another attack. Because of our experiences with Nikki we've tried to learn all we can about lepto and periodic ophthalmia, to know how to treat her and how to protect our other horses. It was not easy to find accurate information and it was frustrating to be confronted with vastly differing opinions. One vet thought my mare should be destroyed, claiming that she would be a carrier for the rest of her life. So I was led on a desperate search for accurate information and have been grateful to several veterinarians who were able to give me information and advice at a time it was sorely needed.

Since then, I have run across one other piece of information that may prove valuable to anyone trying to treat a horse for periodic ophthalmia. The information was in a letter to *Western Horseman* magazine in November, 1969, written by Philip Brown, M.D. He suggested using a preparation containing sodium sulfacetamide and prednisolone acetate, putting two or three drops in the eye twice daily for ten days or two weeks. He reports that a dramatic improvement is usually seen within two to five days. He also suggests using a drop of atropine (ophthalmic) in the eye at beginning of treatment, to keep the pupil dilated. The horse should be in a dark barn during treatment.

Because one of the most important side effects (besides abortion in pregnant mares) of leptospirosis is periodic ophthalmia, eye irritations in horses should never be overlooked. Proper care and treatment can make much difference in the future usefulness of the horse. When in doubt as to the cause of an eye problem, contact your veterinarian immediately for proper diagnosis and treatment.

EQUINE INFECTIOUS ANEMIA

Equine infectious anemia, commonly known as swamp fever, is a contagious disease of horses caused by a virus. It is an ancient disease and one that is still a problem. It has occurred on all continents, in all countries where horses are raised, and in nearly every state of the U.S. In North America it was first recognized as a specific horse disease in Canada in 1880. It has existed in the United States for more than 60 years and was a major problem 40 years ago. Members of the horse family are the only known natural hosts and horses of all ages and breeds are susceptible.

The disease gets the name swamp fever from the fact that the disease occurs most frequently in animals that are pastured in swampy places. Most cases occur during the late summer and early fall, a circumstance that fits in with the idea that many infections occur from the bites of blood-sucking insects that spread the disease from horse to horse.

The appearance of infectious anemia at several race tracks in recent years has made it a subject of vital interest and concern. The occurrence of the disease is reported more and more frequently; it may possibly be on the increase.

Several of the recent outbreaks have actually been man-made; the disease had been spread from carrier horses to susceptible horses by carelessness—by the use of unsterile equipment, especially hypodermic needles. One major outbreak occurred in New Hampshire in 1947, and many Thoroughbreds were lost, but the outbreak was controlled by destroying affected animals and applying strict sanitary measures. Several major outbreaks in more recent years have been studied and many were found to be caused by the use of unsterile needles or by transfusions of blood. Outbreaks in 1965 attracted national attention, occurring among horses at race tracks in California, Illinois, New York, Florida, and Maryland.

Symptoms of affected horses vary. Infectious anemia may appear in acute, subacute or chronic form. The disease is usually characterized by high, undulating fever, sweating, stiffness and weakness (especially in the hindquarters), depression, anemia, jaundice, edema and swelling on the legs and lower body, loss of condition, and unthriftiness—even if the horse has a good appetite.

When a horse suffers an acute attack, symptoms develop suddenly. The horse may have a temperature of 105 to 108 degrees. He appears dejected, refuses feed, becomes anemic, and has congested and jaundiced mucous membranes. He may sweat profusely in warm weather. Sometimes there is a watery discharge from the nostrils. Attacks may last three to five days or longer. Then the animal may appear to recover. Other affected animals become progressively weak, dying within two to four weeks, after several fever cycles. Some horses recover with a lingering type of anemia that leaves them weak and useless for the rest of their lives.

Chronic cases are the most common. Horses with chronic infectious anemia have intermittent attacks of fever (usually a fluctuating low fever), suffering from anemia and progressive loss of weight. They become weak during attacks and may show signs of staggering or paralysis. Other symptoms which may appear are jaundice, hemorrhage of the mucous membranes, swelling of the legs and belly, and the horse may urinate frequently.

Many chronic cases show no symptoms over long periods of time, but their blood contains the virus and is a source of infection for other horses. Most chronic cases develop anemia, heart action becomes irregular, body and leg swellings appear and disappear, muscular weakness varies from slight to so much the horse cannot stand or walk, or has a wobbling or uncertain gait. Such horses become thin and emaciated in spite of good appetite. Sometimes the horse will have diarrhea. Horses that seem to recover from the first attack may be free of symptoms for days, weeks, or even months, but sooner or later other attacks follow. These relapses often coincide with periods of stress (hard work, disease, and so on). In any of these attacks the horse may die.

Horses that survive become carriers for years; after a few acute attacks they develop a type of immunity and seem to be in good health for many years. But it is believed they remain carriers of the virus for the rest of their lives. Most infected horses usually die of the disease eventually, but there are rare cases in which horses eventually become free of the disease and the virus, after carrying the active virus for many years.

Infectious anemia is caused by a virus that can withstand ordinary disinfecting methods (even boiling). Contaminated instruments and hypodermic needles must be cleaned and boiled for at least 15 minutes to destroy the virus; a short period of boiling will not kill it. Alcohol and other antiseptic solutions will not destroy any viruses and definitely won't kill equine infectious anemia virus. This virus is destroyed by sunlight, but can persist for many months at room temperature in urine, feces, dried blood, and so on.

The virus is present in all tissues, secretions and excretions of the infected horse—milk, urine, saliva, semen, etc.—and the persistence of the virus in the bloodstream is astonishing, even in horses that appear normal and healthy (chronically affected horses that are carriers). There is record of one case which, after several acute attacks lived in apparent good health for 14 years and then suddenly developed acute symptoms and died. During the apparently normal period 18 other horses were innoculated with blood from the affected horse taken at intervals of about a year, and all but one of these innoculated horses developed acute symptoms of infectious anemia.

The virus in the bloodstream is quite resistant to heat and chemicals; the virus in dried blood retains virulence for many months if protected from sunlight. But thin layers are quickly inactivated by sunlight, so it is thought that infected secretions such as urine do not remain virulent for very long out on pasture.

The disease is spread from a carrier horse to a susceptible one through transmission of tiny particles of blood carried by biting insects or on instruments which puncture the skin, such as hypodermic needles. Incubation period 12 to 15 days, sometimes shorter and often longer, even as long as 90 days in some cases.

There is evidence that the disease can also be spread from one horse to another by simple contact without the intervention of insects. The disease can also be spread through infected secretions; thus horses that are kept in close contact can pick up the virus at common watering places and from contaminated feed and pasture. Mares can become infected when bred by a stallion that carries the disease. Intra-uterine infections have been known to happen. Foals occasionally become infected by drinking milk from infected dams.

The disease has always been widespread in the United States and recent studies indicate that it is still a problem. Reseachers at Texas A & M have tested over 7000 blood samples from horses of all breeds in 35 states and the results show that infectious anemia is still found in all parts of the country.

Much work is being done to try to find out more about the disease in an attempt to find a way to prevent it. In 1904 two Frenchmen proved that equine infectious anemia is caused by a virus. Ever since then scientists have been trying to unravel its mysteries and to discover a way to grow the virus in pure form, to try to develop a system in which the virus could be cultured without contamination from other viruses. But progress was slow, and a good method for control of the virus was not achieved until researchers were finally able to grow the virus in white blood cells from horses.

But there were still disadvantages. More white blood cells had to be taken from horses repeatedly and this meant that contamination could occur with each introduction of new white blood cells. Then after seven years of hard work, a method was developed in which scientists could keep the horse's white blood cells growing continually in the laboratory, at last providing a constant supply of "pure" blood cells to use in the culture of the virus.

It was found that equine infectious anemia virus would grow in the laboratory white blood cell and that the cultured virus could produce the disease in susceptible horses. Now scientists know they are on the right track in culturing the virus in its pure form and it is now possible to study the virus and learn its characteristics. Further studies can now be made to try to adapt the virus into a vaccine which, hopefully, will produce antibodies in the horse's blood stream and prevent the disease. The next few years may bring good news for horsemen, and many concerned horsemen are watching the progress of research with interest, with hopes for a future weapon against the dread disease of equine infectious anemia.

ANTHRAX

Anthrax is a disease that can affect most warm-blooded animals (including man). Herbivorous animals are most susceptible. The disease is sudden and runs its course quickly, characterized by septicemia (presence in the blood of bacterial toxins) and sudden death. Tarry blood oozes from the body orifices of the dead animal. Anthrax occurs all over the world, but most commonly in tropical or warm climates. Ruminants are most susceptible, followed by horses and swine; the disease is usually fatal except in swine.

The anthrax bacilli are not usually transmitted directly from one individual to another. The organisms are present in the tissues of affected animals and are eliminated in secretions and excretions for a short while before death. If an animal's carcass is left unopened, putrefaction tends to destroy the bacteria. But if the carcass is scattered by birds or animals of prey, soil and water may become contaminated. When material containing anthrax bacilli is exposed to the air, spores are formed which can exist for some time, often contaminating the environment for long periods. The organism is often spread during flood time, and then contaminates grasses that are grown in soil containing the bacilli. Grasses grown in flooded lowlands may be infected, and animals fed hay grown on these lowlands may contract the disease as well as the animals that graze the lowlands. In hot climates the disease is sometimes spread by blood-sucking insects, particularly the large horseflies of the genus Tabanus. Close grazing of scratchy stubble (causing injuries to the mucus lining of the mouth) and confined grazing in contaminated areas (such as around water holes) are sometimes predisposing factors in the spread of the disease. The introduction of infection into a new area is usually through contaminated animal products such as bone meal, fertilizer, hair, wool, hides, etc., or contaminated feeds.

Anthrax in horses is always acute. If the horse contracted the disease by eating contaminated feed or drinking contaminated water, the horse suffers septicemia with enteritis and colic. If the infection was contracted through the bite of an insect, the horse develops hot painful swellings around the throat, neck, thorax and abdomen, and prepuce or mammary glands. There is a high fever, the horse is severely depressed, and he may have difficulty breathing because of swelling at the throat, or colic due to intestinal irritation. The disease is usually fatal within 48 to 96 hours.

Antibiotics and antianthrax serum are most commonly used in treating the disease; if treated early enough, the animal may recover. In areas where anthrax is a problem all cattle and horses should be vaccinated. Spring is the usual time for the annual vaccination of cattle, horses, mules, sheep, and sometimes swine. Occasionally a few animals develop severe or even fatal reactions to the vaccination. If an animal develops a reaction he should be treated with antibiotics and antiserum. Carcasses of animals that die of anthrax should not be opened, but completely burned or buried at least 6 feet deep, covered with a good amount of lime.

PIROPLASMOSIS

Piroplasmosis is a disease caused by protozoa, and spread by blood-sucking ticks. This disease is also called babesiosis and redwater fever. Similar parasites cause disease in cattle, sheep and pigs. Two kinds of protozoa have been recognized in horses—*Babesia caballi* and *Babesia equi*. Originally the disease appeared only in southern Europe, Africa, southern Asia, and South America. Then the disease was discovered in Central America, and in 1961 the first confirmed case of equine piroplasmosis was diagnosed in Florida.

The disease is characterized by high fever (often intermittent, with the afternoon readings being higher than the morning), loss of appetite, depression, weakness, increased heart and respiration rates. As the disease progresses, there is severe jaundice, and urine is dark colored. Anemia is pronounced. Fever lasts about a week, and the total course of the disease is about three weeks. If the horse is pregnant, abortion is common. If the horse recovers, it takes a long time to overcome the anemia and emaciation.

Incubation time is about 7 to 20 days as the protozoa multiply in the bloodstream. Death is usually due to lack of oxygen in the tissues, from anemia. If the animal survives, it becomes a carrier for about a year (unless it becomes reinfected). There are a few drugs that seem to be effective in treating the disease. Eradication of the disease depends mostly upon the elimination of the vector tick; in new, uninfected areas, prevention of the disease depends upon effective quarantine and other measures to prevent the introduction of the blood-sucking tick that carries it.

PNEUMONIA

Pneumonia is a disease that can be caused by several things. In the horse, pneumonia can be caused by virus, or bacteria (sometimes as a secondary invader) either as a primary disease, or the aftereffect of some specific disease (such as glanders or strangles, or any disease that lowers the horse's resistance and is accompanied by extra stress factors), or caused by breathing in moisture when "drenched" with medication, or from eating Crofton weed.

Pneumonia is inflammation of the lungs, usually accompanied by inflammation of the bronchioles (finer branch subdivisions of the bronchial tubes), and pleurisy (inflammation of the membrane around the lungs, causing difficult and painful breathing). There is an increase in respiration rate (rapid shallow breathing), cough, and, in most bacterial pneumonias, toxemia (bacterial toxins in the blood stream). In later stages of the disease there is difficult breathing because much of the lung tissue has become useless. Nasal discharge may or may not be present.

There may be fever, loss of appetite, depression, increased pulse rate and reluctance to lie down. The horse will usually stand with his front feet wide apart, or if he lies down he may lie on his breastbone for easier breathing. In severe cases he may try to breathe with his mouth open.

With adequate treatment during the early stages, bacterial pneumonia usually responds quickly and completely, but virus pneumonia may not respond at all, or may respond just temporarily because of control of secondary bacteria. A horse with pneumonia should be isolated, kept in a warm barn, and given plenty of fresh water and light, nourishing food. During convalescence he should be kept protected from bad weather.

Infectious equine pneumonia (IEP) is a highly fatal disease of horses characterized by severe pneumonia, high fever, and septicemia (presence in the blood of bacterial toxins). This particular disease is now uncommon, but used to be quite frequent when large numbers of horses were brought together from various sources and kept under poor conditions (inadequate nutrition or improper housing). It was a common occurrence among horses on board ship. This disease is infectious but the rate of spread is rather slow at first. Primary cause of this disease is a virus, but the serious nature is due to secondary bacterial invaders such as streptococci, staphylococci, pasteurella, and pseudomonads. Predisposing causes seem to be exposure and chilling, fatigue, poor nutrition and overcrowding. The disease can be spread by contaminated feed or infected droplets in the air, inhaled by a healthy horse.

Corynebacterial pneumonia of foals is an infectious disease of young foals that localizes in many organs but especially in the lungs. Foals which survive the acute disease are often affected for the rest of their lives, seldom developing normally. Infection from the mare (via the uterus) is very likely an important cause, as is infection via the navel soon after birth. Cases can occur up to 6 months of age. The disease may be acute (with signs of pneumonia, fever, loss of appetite and sometimes arthritis) in very young foals, or subacute in older foals, the pneumonia developing slowly with cough, difficult breathing and "rattling." The foal continues to suck and temperature is normal, but he becomes emaciated. Severe cases usually die within one or two weeks.

There are many other diseases that can affect horses. These mentioned here are some of the most common and most dangerous ones—those you should know about and vaccinate your horse against, if they are a threat in your part of the country, and if there is a vaccine available. For some diseases there are no vaccines. Good health is very important, for without it a horse is useless. The conscientious horseman works closely with his veterinarian on disease prevention, control and treatment. For most diseases, prevention is the best cure. But if you ever do have a sick horse, accurate diagnosis, prompt and proper treatment is a must, for the sake of that particular horse and for those around him that might be susceptible.

10

MISCELLANEOUS HEALTH PROBLEMS

THE PREVIOUS CHAPTER DEALT WITH SOME OF THE MAJOR diseases of horses, most of them caused by a virus or bacteria. In following chapters we shall look at internal and external parasites, unsoundnesses, and colic. So for this present chapter, let's examine some of the other health problems that your horses may encounter from time to time.

The health of your horse is a very important matter, and any serious problem should always be taken care of by a veterinarian. But the conscientious horse owner strives to learn as much as he can about the various ailments and diseases, in order to better care for his horse—in order to prevent as many problems as possible and to know how to recognize them when they do occur, and also how to best treat the horse and speed him to recovery. Proper care and management are usually the best prevention for many ailments, but occasionally injury or misfortune occurs even with the best of care, and a horse owner should know how to cope with a problem whenever it does happen.

Let us take a look at some of these problems.

FOUNDER

"Founder" is a word used when speaking of the changes that take place in the horse's feet as one of the results of a severe shock to his system. This problem is also called laminitis. The hoofs can become deformed, the extent of damage and deformity depending upon the severity of the case. Founder is often an unnecessary disaster that could have been prevented by proper care, for it usually results from carelessness, neglect, or ignorance on the part of the horse owner.

To understand why a horse founders, let's take a look at the structure of the horse's feet. All his weight is carried by those feet, the front feet bearing more of the weight than the hind (the horse's head and neck make him somewhat "front-heavy"). The bones and tendons of the horse's legs terminate inside the hoofs in an area that serves as a cushion and an attachment between the bones and tendons and the outer horny shell of the hoof wall which is protective and insensitive. This cushioning area, the intermediary tissue between bones and hoof wall, is called the sensitive laminae.

When the horse suffers severe shock that upsets his body chemistry, many changes take place within his body. Often there is an increase in blood pressure, creating congestion of blood in certain vessels. Circulation speeds up, blood pressure increases, and the horse has a fever. Usually this congestion and fever do not cause serious or permanent damage in other parts of his body, but in the hoofs (which are encased in solid walls) the congestion, the swelling and the increase in pressure causes terrible pain and permanent damage to the sensitive inner tissues of the hoof, the laminae. This is why founder is also called laminitis. When this happens, the pain makes it difficult for the horse to walk, and in severe cases the horse does not want to stand up on his feet, spending his time lying down or rolling around. Lameness often shows up first in the front feet because they carry more weight. There is heat and pain around the coronary band.

Founder seems to be caused many times by an allergic reaction to protein, especially when the horse overeats such feeds as grains. Founder occurring in mares that retain the placenta after foaling may also be a manifestation of allergy to protein—the protein derived from tissue breakdown or from the streptococci that invade the uterus.

After the original cause of the problem is corrected and the fever leaves, the congestion and pressure and circulation of the blood return to normal, but usually permanent damage has been done to the feet and they become deformed. The laminae separate from the hoof wall, the sole drops down, the hoof wall spreads and develops rings and ridges. The slope of the front of the hoof becomes concave. Sometimes the separation of the hoof wall from the sensitive laminae is complete and the hoof wall comes off. In most cases the severe lameness and pain eventually disappears, but the horse is clumsy and goes lame easily; he may suffer from repeated mild attacks of founder.

Founder may develop as an acute severe attack and be followed by either recovery or the persistence of a chronic state of deformity and lameness, or it may develop in a mild form from the beginning. In acute cases there is usually pain in all four feet, but sometimes only the front feet are affected. In acute cases the horse is in great pain and shows much anxiety and distress. He shivers, sweats, breathes rapidly and shallowly, and has a fast pulse rate. He usually stands with all four feet bunched together under his body, with his back arched up and his head hanging down; he doesn't want to move and if he does move his gait is stumbly and shuffling because of the pain in his feet. It is hard for the horse to get up and down and he may make several attempts to lie down before actually accomplishing it. Once he is down, it is hard to get him up again and he may lie down, often flat on his side, for long periods of time.

Most of the causes of founder are the results of strain or stress or imbalance that upsets the horse's body chemistry. Some of the most common causes are listed below.

Overeating, especially of grains and concentrates, is perhaps the most common cause of founder in horses. The founder results from an allergic reaction to the protein in these feeds.

Drinking cold water when overheated can also cause founder. A good rule to follow in preventing this kind of founder is to *never* let a horse take a big drink when he is working hard or is hot and sweaty. When he takes a big drink of cold water, the blood rushes to his stomach to try to maintain an even body temperature (to warm the chilled stomach), leaving the muscles that have been working hard. The horse's muscles are in desperate need of a constant blood supply bringing fresh materials and repairs, and if these muscles and extremities get neglected because the blood rushes to the stomach, the horse's whole system is thrown into a state of confusion.

The same thing can happen any time the horse's temperature is drastically upset—whenever there are rapid changes from one extreme to another, such as when a hot horse is washed with cold water, or left to stand and cool rapidly after hard use on a cold, windy, or wet day, or has to stand in a cold wind when he is hot and sweaty. Sudden chilling of large areas of his body sends blood rushing to these parts to try to keep them warm, but there is only so much blood and some part has to suffer. Any time the horse's system has to make a drastic and rapid change, his carefully balanced body-control mechanisms are thrown into confusion and founder may occur as one result of the congestion.

Grass founder is another example of the horse's system being upset by too sudden a change. Grass founder can happen when a horse on dry feeds is suddenly allowed to eat a large amount of new spring grass or is changed suddenly from dry feed to lush green pasture (especially pasture containing clover or other legumes) any time of year. Horses should be changed *gradually* to the new green feed, allowed to graze only a few minutes the first day, a little longer the second, and so on until they can be safely left out on green pasture by about the fourth or fifth day as their systems have adjusted to it. Horses on lush green pasture (especially clover) must be closely watched for signs of founder, especially if they are fat and not getting much exercise. Fat ponies out on pasture and getting little exercise often develop chronic founder.

Concussion caused by riding on hard surfaces such as pavement or hard-packed ground is another cause of founder, especially if the horse is fat or heavy or carrying a heavy load.

Increased weight on the feet may cause founder, as when a horse is lame in one foot and tries to carry most of his weight on his good foot; he may founder in his good foot. Horses have also been known to founder just from shifting their weight back and forth continually, like when traveling at sea. Horses should never be required to stand continually. If a horse must be stationary for long periods of time, like when traveling a great distance in a van or trailer, he should be taken out of the trailer periodically and exercised, and allowed some rest stops in which he can move about freely and relax. Extremely overweight horses are more prone to founder than horses in proper physical condition. An overweight mare with the added burden of being pregnant is susceptible to founder, particularly if she gets no exercise and stands around for long periods of time.

Exhaustion can be another cause of founder, especially if the horse is out of condition. Severe exhaustion may not only cause founder; it can send the horse into shock, and if it is severe enough he may die.

Serious infections that go untreated can also be causes of founder, infections such as pneumonia, or infection of the uterus after foaling. Heat stroke can also be a cause. Founder may also follow the use of drastic laxatives that upset the horse's system.

To keep a horse from foundering, here are some good rules to follow:

1. Always feed a well-balanced ration and always introduce new feeds slowly, over a period of several days. Never feed him (or let him eat) excessive amounts of any kinds of feed.

2. Condition a horse gradually, never working him to the point of exhaustion.

3. When a horse is hot and tired, limit the amount of water he drinks to just a few swallows, and do not feed him very much at all until he is completely cooled (heartbeat, respiration, and temperature back to normal) and dried off.

4. Always cool a horse slowly after using him—by walking him, blanketing him, or rubbing him vigorously to increase circulation. Dry him and protect him from cold breezes until he is completely cooled and dry.

If a horse ever becomes foundered, prompt treatment should be given to relieve his pain and to try to keep the feet from suffering permanent damage. If the laminitis is caused by overeating of grain, give the horse a mild purgative to hasten the elimination of toxic substances in his stomach.

Keep his feet cold and wet, and relieve the strain on his feet by giving him soft footing. A wet mud pit is ideal. This "soaking" treatment has been used for foundered horses for many, many years and is still an effective way to help relieve pain, but there are now drugs (antihistamines and cortisone) that can relieve the horse's pain more quickly. Your veterinarian can prescribe them.

Many horses with deformed feet from founder are not too seriously affected and can still be useful if shod correctively with special shoes after proper foot trimming. These horses with mild cases of founder are fine except when barefoot; then they go lame rapidly as the hoof wall wears away and the weight is borne on the abnormally flat sole.

Recent research and experimentation have shown that the hoof changes in foundered horses are caused by alterations in the hoof horn during the early stages of laminitis—the formation of epidermal horn, resulting in a weakening of the bond between the hoof wall and the sensitive laminae. If, during the treatment of early founder, the horny hoof wall can be kept from bearing weight (by rasping away or removing excess horn and transferring the horse's weight to the sole of the foot) the horny wall will tend to regrow normally. Thus the deforming hoof changes can often be prevented in many early and mild cases of founder. Completeness of the recovery depends of course on the severity of the case and the length of time elapsed before treatment was started. There have been several cases of acute founder in which the horse lost a hoof and in time regrew a completely normal one. This led researchers to believe that weight borne by the hoof wall during the acute stage of the disease was part of the cause of deforming

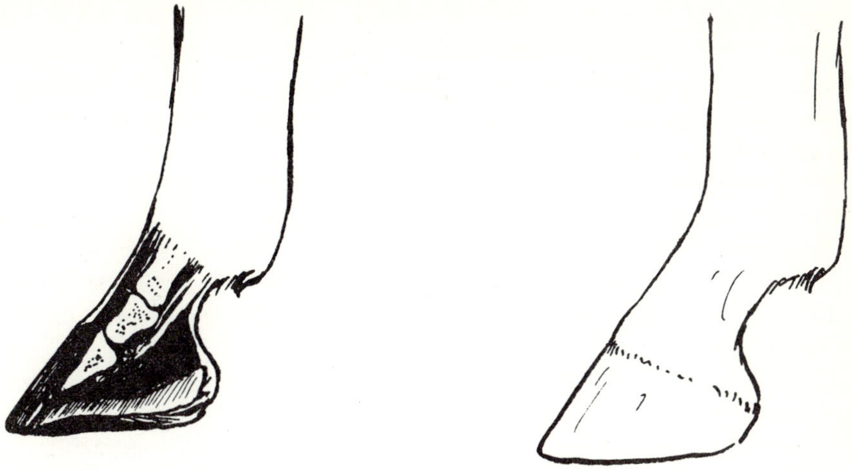

Top, **normal foot;** *bottom,* **foundered foot (note the dropped sole and downward rotation of the third phalanx or coffin bone)**

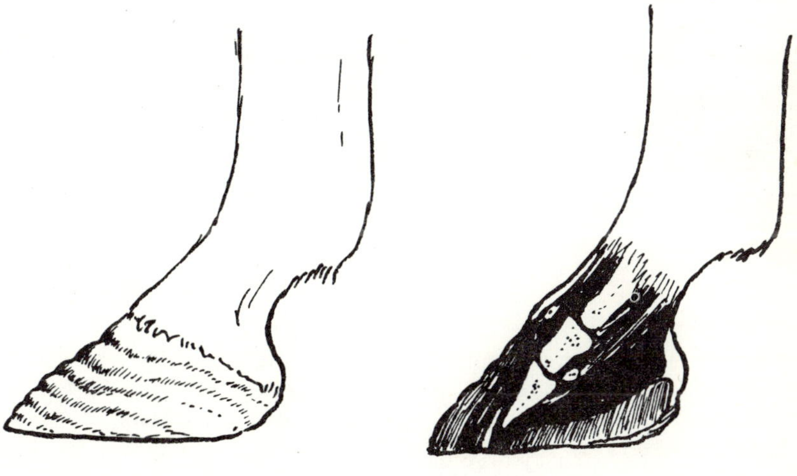

changes, and eventually led to the discovery that the hoof wall will regrow normally if enough of it is removed in the early stages of laminitis to prevent its bearing the weight of the horse.

As with many other ailments, prevention of founder is by far the best cure. Using and caring for your horse wisely is the best insurance against founder.

HEAVES

Heaves is a disease of the lungs (chronic pulmonary emphysema in horses). If a horse has heaves he will have it for the rest of his life. With good care he may not get rapidly worse, but he will never get better.

A horse with heaves has an abnormal distension of the lungs, caused by the rupture of the alveolar walls; the millions of tiny air sacs in the lungs break down. This creates the condition called emphysema. There is a reduction in actual air space; the capacity of the lungs becomes limited. And because the lungs lose their elasticity (especially their ability to recoil, or push out the air) the horse's breathing becomes more forced and labored and it is hard for him to expirate the air from his lungs. Because the horse has to force the air out when he breathes, over a period of time he develops a big chest and a pot belly because the diaphragm muscles enlarge. The horse is short of breath and has no stamina.

There can be several things that cause or aggravate heaves. Chronic heaves, the kind that comes on rather gradually and gets progressively worse, is usually associated with prolonged feeding of poor quality roughage or dusty hay. Some horses seem more susceptible to heaves than other horses. With some it seems to be an allergy problem, caused by dust or mold that irritates the respiratory tract which is extrasensitive to these things. Dusty or moldy hay is the worst kind of feed for any horse, because this type of feed can start respiratory irritation that can lead to heaves. If a horse already has heaves or a tendency in that direction, dusty or moldy feed can make his problem become rapidly worse. Heaves is virtually unknown in horses that run at pasture all their lives, never having to eat dry, dusty or moldy feeds.

Heaves can also follow infections such as chronic bronchitis or pneumonia. But usually heaves is caused by the horse being sensitive or allergic to something he breathes, such as dusty or moldy hay, dusty bedding, or a dusty corral.

Heaves can be acute or chronic. When a horse suffers from acute emphysema, rest is essential. He may even need oxygen to aid his breathing. He should be kept in an environment where his respiration will not be impeded by humidity, high temperature or foul air.

Chronic heaves is more common. It comes on gradually and becomes progressively worse. The symptoms appear gradually and in the early stages they are only evident during exercise. A cough is usually the first noticeable symptom —a short, weak cough which becomes more pronounced and wheezing when the horse exerts or is exercised. Instead of getting better after few days (as most coughs do which are caused by colds or minor irritations) this cough persists and gets worse. The horse may have coughing fits after exercise, after drinking cold water, or when coming out of a warm barn into the cold air.

As the disease progresses, the horse becomes unable to get rid of all the air in his lungs. The residual carbon dioxide from the incomplete evacuation of his lungs stimulates him to try to breathe more deeply, but even then he doesn't seem to get enough air. He must breathe much more strongly than normal in order to get enough air for a normal exchange of carbon dioxide and oxygen, because the capacity of his lungs is so limited. This is why exercise is so hard on a horse with heaves; the added exertion and subsequent need for oxygen and more air puts a strain on his breathing, upon his efforts to get enough air. Hard breathing develops and the metabolism of all of his body tissues is reduced.

A horse with heaves can only get progressively worse. But even though there is no cure for heaves, a careful horseman can help control the advance of the disease. He can put the horse out on green pasture (a horse on pasture will have fewer symptoms than one fed hay). If you must feed hay, avoid legume hay for it tends to contain more hay particles (dust) than well-cured grass hay. Feed good bright grass hay and sprinkle it with water to settle any possible dust. Pelleted feeds are usually less dusty than hay and are often beneficial for a horse with heaves. In some localities you can also buy a specially prepared ration that is made specifically for horses with heaves.

If you ever have a horse with heaves, try to keep his environment as free of dust and mold as possible. Don't bed him on straw or any other material that might be dusty or moldy. Wood shavings are much better. If you ride him, never work him on a full stomach; if you keep him away from feed and water for a few hours before you ride him his stomach will not be so full and his lungs will have more room to expand. And never overwork him. Don't overexert his lungs with too much running or climbing of hills. Not only the lungs, but the heart, too, is usually under greater stress when a horse has this problem. The blood flow into the chest cavity is somewhat restricted. Animals with advanced emphysema may die of heart failure, but in horses this is not common because most horses are disposed of (due to inability to work) before they get this bad.

There are several drugs in use now that can help relieve the suffering of a horse with heaves. Your veterinarian can prescribe products that may be beneficial if you ever have a horse with heaves.

This is a disease which should never be neglected. If a horse develops a cough, don't use him until it clears up, and if the cough becomes persistent or severe, have the vet check him. Early cases of heaves properly cared for may not progress rapidly; good care may help to assure many more years of usefulness for the horse.

AZOTURIA

This is a functional disorder (a metabolic problem) which is also called paralytic myoglobinuria. It occurs during exercise after a period of inactivity on full rations and is characterized by the presence of myoglobin (a substance that gives muscle its color and acts as a store for oxygen) in the urine and muscular degeneration.

The usual cause is feeding the horse full working ration during a short time that he is idle—feeding large amounts of good hay and grain without giving the horse sufficient exercise to utilize this nourishment. It is not a matter of the horse becoming fat, but of the blood and muscle becoming overloaded with nourishment that is not being used. If this kind of feeding is continued, without exercise, the horse adjusts to it and grows fat. But if a horse is being used semiregularly and fed a good diet to give him the energy he needs for this exercise, and continues to receive this same kind of nourishment when he is idle, his body doesn't adjust rapidly enough to the change.

Large stores of glycogen have been laid down in the muscles during a period of idleness on full rations, and when the horse suddenly exercises this glycogen is rapidly metabolized to lactic acid. If lactic acid is produced faster than it can be removed in the blood stream, accumulation occurs, causing hard, painful swelling of the large muscles in the horse's body.

The system of the horse which was overloaded with an accumulation of "reserves" cannot handle and eliminate the broken-down by-products of sudden exercise. These by-products accumulate, backing up in the horse's system because they cannot be eliminated fast enough.

Azoturia has been called "Monday morning disease," because in the days of the draft horse it was a common occurrence in horses that were grained heavily and worked hard all week and then rested on Sunday. If their grain ration was not reduced on their day off, azoturia sometimes hit them when they went back to work on Monday morning. The pleasure horses of today have a problem that is similar to that of the draft horse of yesteryear. More and more horses are being used for pleasure or for show, maintained in good physical condition for riding or showing, but only exercised a few days a week or basically only on weekends. This tendency to fully feed a horse but only use him part time can have serious consequences. In most cases of azoturia the horse has been idle for two or more days just before being exercised. An attack in not quite as likely if the horse's period of rest has been very brief (one day) or as long as two weeks. Horses taken out of training or track work because of minor injury or illness are often kept on full rations and then suffer attacks of azoturia when put back to work.

Symptoms of azoturia usually develop within about 15 minutes to an hour after the beginning of exercise. The attack is usually quite sudden. The horse has been lively, but suddenly he falters and comes to a stiff halt. He begins to sweat, becomes stiff, and is reluctant to move. He may even go down. There is usually a tight feeling of the muscles over the horse's hips and shoulders, and these large muscles are quite sore. In most cases there is a stiffness of gait or a tendency to knuckle over at the pastern (especially hind legs). Membranes of the eyes and nose may look rather dark instead of pink. There is sometimes a peculiar odor from the horse's breath and urine, and his urine will be dark in color.

Symptoms may disappear within a few hours if the horse is given complete rest immediately (he should not be made to move at all), but in severe cases the horse may die. Often the horse goes down, the horse first sitting down on his haunches, then lying down on his side. When he is down, he may be restless and in pain, struggling and trying to get up again. His respiration is rapid and his temperature may rise to 105 degrees in the later stages of a severe case. If exercise is stopped as soon as lameness begins, the horse may recover in two to four days, especially if he remains standing. But if he goes down, chances of recovery become less. Being down is usually followed by a fatal uremia (the presence of urine in the bloodstream), due to the suppression or deficiency of urinary secretion.

Emergency treatment for azoturia, while waiting for the vet to arrive, consists of blanketing the horse, trying to keep him relaxed and warm—warm enough to sweat freely. Let him have water if he will drink. You are trying to hasten the elimination of the toxic wastes in his blood stream. Avoid movement. Care for the horse right where he is or take him home in a trailer and confine him. Keep him on his feet if posssible, even if you have to use a sling. He should be given fluids, intravenously or orally, to maintain a good flow of urine. If it is difficult for him to urinate, a catheter should be used. The vet may prescribe certain drugs and hot packs to help ease his pain.

To avoid azoturia, the horse should always have his feed (especially grains) cut down on days he is not exercised. If he has been rested for several days, the workout should begin gradually, allowing the horse to warm up gradually, and the workout kept light, only gradually increasing on successive days. After exercise, the horse should be cooled out slowly and completely. Proper care will usually prevent azoturia.

CANCER

Cancer can affect any kind of animal, and horses are no exception. Horses can die of cancer. But some types of cancer in horses can be treated and cured if detected early enough. This is not always easy however, for in its early stages, cancer may be painless and may not bother the horse; his owner may not notice it until too late.

A cancer is a malignant tumor—a growing mass of abnormal tissue. It can grow until it interferes with the organs around it, or it can send out cells in the animal's blood stream to take root in other places in the body, finally killing the animal. Some types of cancer can kill the horse within a few weeks; other types take years. Let's take a brief look at some of the more common types of cancer found in horses.

Cancer of the eyelid (squamous cell carcinoma) is fairly common, horses with unpigmented eyelids being the most susceptible. Light-skinned or white-faced horses are the usual victims (albinos, paints, palominos, and horses of any breed with white marking around one or both eyes). Cancer of the eyelid is caused by the irritating effect of sunlight upon the eyelid that is unprotected by dark coloring. Skin pigment, called melanin, protects against harmful effects of sunlight. Cancer of the eyelid forms most often on the inner part of the lower lid, looking at first like a runny sore. The eye is irritated and waters excessively. This kind of cancer will not clear up without treatment and will continue to grow and spread, eventually killing the horse. If caught early enough it can be cured by operating or cauterization.

Horses with unpigmented eyelids should be protected from excessive sunlight by keeping the horse stabled or in a shady pasture during the daytime, or painting a dark substance around the eye to help absorb the sun's rays. My first horse, Possum, had a white face and one blue eye surrounded by unpigmented skin. That eye watered profusely in the summertime until I painted black shoe polish around it

Old Possum had a white face and one blue eye that was greatly irritated by bright sunlight

periodically. It made him look a little funny, but it gave him great relief from the bright, irritating sunlight. Mascara or black marking pencil—anything dark that will stay on a while (but be careful not to get any in the eye itself) will work as well.

The same kind of cancer, squamous cell carcinoma, can occur on other unpigmented areas, especially those that are thinly haired, such as under the tail, around the mouth, or on the sheath. The sheath and penis of a horse are common locations for several types of cancer, and keeping them clean may help to prevent some types.

Melanoma is another kind of cancer that occurs in horses. This kind is a black malignancy. It can occur in any horse, but is most common in old gray horses, often under or close to the root of the tail and appears as a black lump. This type of cancer can spread slowly or rapidly and is hard to get rid of.

Leukemia, a cancer of the blood, can also occur in horses. There are other types of internal cancer, too, and they are often difficult to diagnose and treat. Signs of cancer include any unusual type of sore or abnormality—a growing lump, persistent sore, bleeding or discharge from any body opening, chronic cough, persistent weight loss, and so on. If there is something unusually wrong with your horse, have a vet check him, and don't wait until it is too late. Skin cancer seems to be the most common kind of cancer in horses, and many skin cancers can be cured if treated early, so don't neglect any abnormalities that look suspicious.

SARCOID

Some horses develop tumors of the skin called equine sarcoid. These are fibroid tumors, consisting mainly of connective tissue. They occur singly or in groups and sometimes have the appearance of rather large warts. They are moderately malignant in that they often grow back after removal, but they do not spread to other parts of the body. These tumors most commonly appear on the head, neck, shoulders, chest and legs. In general, there are two forms of sarcoid tumors. Some look like flat gray warts. Others, especially those that occur on the legs, look more like proud flesh, growing very large in size and proving very difficult to treat. Wounds on the legs sometimes develop sarcoids. A virus seems to be involved in causing these unsightly tumors. Sometimes they are best left alone, for removal only makes them grow back, often in a larger form. A few promising cases have been treated by freezing and by anticancer chemicals and these new techniques may bring some answers to the problem.

PHOTOSENSITIZATION

This is an unusual skin disease caused by an allergic reaction to light —the sensitization of areas of lightly pigmented skin to strong light. The result is dermatitis (inflammation of the skin). Certain plants sometimes contain chemicals that are

Melanoma under the tail of a 25-year-old gray mare (note the black lumps). This particular case was a slow-growing type and did not bother the mare for several years

absorbed by the horse's bloodstream and carried to the skin after the horse eats the plants. In the skin, these chemicals produce a serious reaction to sunlight, but generally only in areas of unpigmented skin which are not protected from the sun's rays by melanin. The affected areas look sunburned, developing blisters and cracks through which fluid oozes. In severe cases the horse is quite sick, and patches of skin may actually die and slough off. The horse should be removed from the pasture or feed that has caused the problem, and put in a dark barn. Laxatives can be given to help the horse eliminate the toxic substances he has eaten. Antibiotics may be necessary in severe cases, and antihistamines will help stop the reaction. Soothing ointments can be applied to the affected areas to help them heal.

WARTS

Warts are unsightly, but usually not a serious problem. They are usually found on young horses, most often around the muzzle—on the lips and nose. Occasionally warts appear on other parts of the body, or even inside the ear. Since warts are caused by a virus they can be spread from horse to horse. The virus can be spread by infected grooming tools such as curry combs and brushes, tack, and anything else that comes in contact with several horses. The best treatment for warts is time; if left alone, warts usually disappear without treatment in a few months. If they become a real problem (if they interfere with a horse's breathing, for instance) they can be removed surgically. If they are a major problem in a stable or wherever a great number of horses are apt to get them, a special vaccine can be prepared to immunize young horses against them. This is made by your vet after he takes a sample from one of the warts to find out exactly what kind it is.

RINGWORM

Ringworm is a skin disease caused by a fungus, and most kinds of ringworm are contagious to other animals and man. The ringworm fungus sends out spores, which start new infections when rubbed into the skin. With horses, ringworm can easily be spread by infected brushes and curry combs, cinches and saddle blankets, and so on. Young horses, and horses deficient in vitamin A seem to be somewhat more susceptible to this disease.

After the spores enter the skin, lesions are seen within a week to a month later, appearing as circular areas in which the hair falls out or breaks off. Sometimes thick crusts form. If only one or two horses in a group or in a stable become infected, the fungus can easily be spread to the rest, so precautions should be taken to keep the other horses from getting it. Saddle blankets, grooming tools and everything that may be used on more than one horse should be treated and disinfected, and in some cases the entire stable should be thoroughly cleaned. A horse with ringworm should be thoroughly treated to prevent spread; there are several medicines you can get from your vet that are satisfactory for treating ringworm and keeping it from spreading.

SCRATCHES

"Scratches" is a term applied to sores (bumps and open ulcers) in the skin that occasionally appear on the lower legs of horses, most often affecting unpigmented areas—horses with white feet seem to be the most susceptible. This problem is a dermatitis, or inflammation of the skin, and seems to be caused by several things. It can be an individual allergic reaction to ingredients in the pasture (either from contact with the plants, walking through them, or from eating them), sometimes coupled with photosensitization, or a reaction to constantly walking through mud or from the feet being wetted. The most common cause of scratches seems to be a fungus, *Sporotrichum schenki.* Some horses are affected more readily than others, and horses with dark legs (pigmented skin being tougher) are not nearly so likely to have this problem as horses with white feet and legs. This disease, also called sporotrichosis, is contagious, but spreads rather slowly. The fungus lives in organic matter and enters the horse through breaks in the skin, either directly from contact with the discharge of infected animals, or indirectly from the contaminated pasture, especially wet, swampy areas.

The dermatitis that results is basically an inflammation of the deeper layers of the skin, sometimes involving the blood vessels and lymph vessels. The most common site of inflammation is the pastern and fetlock area. The involved skin becomes warmer, reddish, and thickened. Then the skin surface becomes scabby and cracked, and if neglected, becomes badly cracked and oozing. Infection may spread to the inner tissues. Severe cases can cause lameness, succeeding

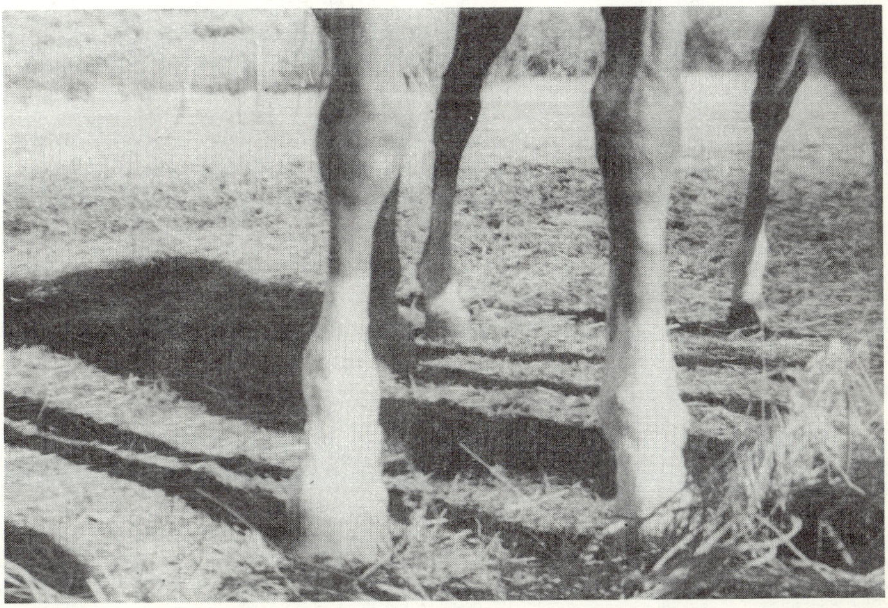

Scratches. Note the rough hair and enlargement around the pasterns and fetlock joints

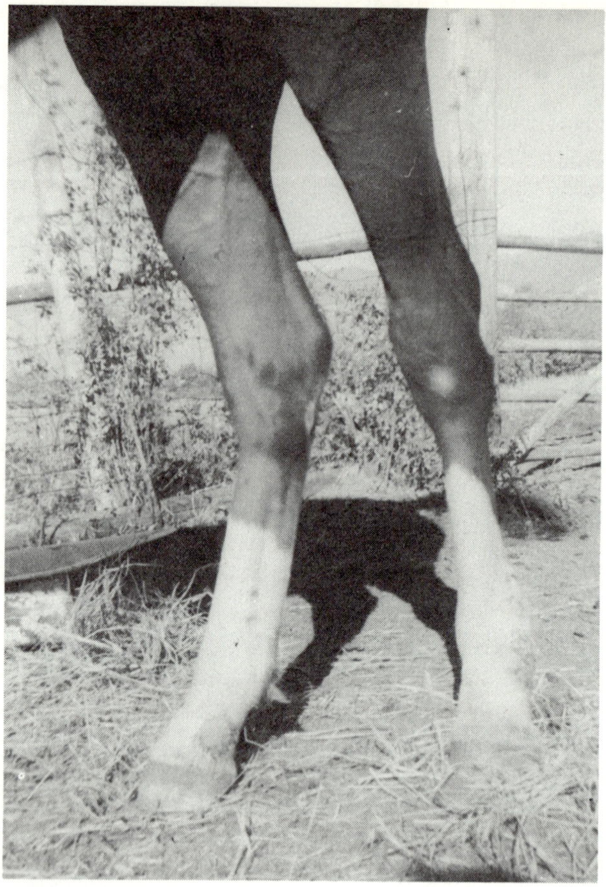

Note the round, crusted lesion on the inside of the pastern, and the enlarged, rough area on the outside of the near fetlock joint

crops of lesions may leave the horse affected for several months. The best prevention of this condition seems to be the use of clean, dry pastures, avoiding the use of swampy, contaminated areas for pasture, keeping the horse away from places where he must walk through deep mud or filth. If the horse's feet are kept clean and dry, and if all abnormalities or scrapes or injuries are promptly taken care of, scratches should not become a serious problem. In contaminated pastures, scratches is most prevalent during the wetter seasons of the year or during the summer when parts of the pasture are swampy. The problem receeds in drier times of the year, or in winter when the ground is frozen.

Several of the Arabian mares we were boarding in 1971 had cases of scratches when they arrived, having come from a contaminated pasture. Our standard method of treating the worst cases was to wash the feet and legs daily, applying a thick lotion to the scabby crusts (a lotion composed of nine parts glycerin to one part iodine). If the skin was broken, we also applied an additional protective dressing such as Turkol or Scarlet Oil. Astringent preparations such as methelene blue, pitan blue, "Blue-Lo" etc., are also helpful in treating scratches.

Scratches appear most often around the pastern, above the coronary band and in the heel area, especially above the heel and in back of the pastern where the foot bends. It also appears around the fetlock and sometimes above the fetlock joint. The thickened skin may come off, leaving bare spots covered with rough skin.

EAR PROBLEMS

Occasionally a horse will suffer from an ear infection or irritation, usually from ear ticks or fungus or some other abnormality. If a horse ever becomes suddenly head-shy for no apparent reason or dislikes having his ears handled, you should have them examined by a vet to determine the cause of the problem. Ear ticks are parasites that invade the ear canal and cause the horse discomfort. The horse may refuse to have his ears handled, may shake his head, or lay back the effected ear. These ticks can be killed by putting the proper medication into the ear. If the horse is secreting too much ear wax, because of the irritation by ear ticks, or because of infection, a special solvent can be put into the ear to dissolve excessive wax so that the horse can shake it out.

Warts in the ear generally do not bother a horse and are usually best left alone unless they become a serious problem. Surgery can correct several problems of the ear, including splits and tears from injury or wire cuts. Surgery is also used

Ahmahl. Note the warts in his ears; these did not seem to bother him

to correct a draining hole at the base of the ear caused by an "ear tooth" (dentigerous cyst)—a congenital deformity that occasionally occurs among horses.

There are some other problems that sometimes occur in horses—problems that are indirectly related to the ear, so we shall mention them here. Parotitis is swelling and inflammation of the salivary gland below the ear, and diseases of the guttural pouch can also cause swelling under the ear.

Parotitis, or inflammation of the salivary glands, is usually caused by infection elsewhere in the body (carried to the salivary glands via the bloodstream), or by irritation from grass seeds or other particles of feed caught in the salivary duct. Pain from the inflammation may interfere with chewing and may also cause the horse to carry his head in an abnormal position. Antibiotics are usually necessary to cure this problem, and if there are abcesses they may require draining.

Inflammation of the guttural pouches can cause swelling beneath the ear. The guttural pouch is a blind sac in the horse which opens into the eustachian tube of the inner ear. This pouch may fill with pus if it becomes infected, causing the unusual swelling. In foals, this pouch sometimes distends with air. These problems will need treatment and you should consult your veterinarian. We'll take a closer look at inflammation of the guttural pouch in the following section.

CHRONIC NASAL DISCHARGE

A persistent discharge from one or both nostrils (sometimes accompanied by coughing or snorting) can be caused by allergy, tumors, a foreign body in the nasal passage, or a chronic infection, but the most usual cause is infection of either the facial sinuses or the guttural pouch. The sinuses are hollows in the bones of the skull and they can sometimes become infected as an aftereffect of some type of respiratory infection (colds, influenza, strangles, etc.) or from an injury to the head, or from an abscessed tooth. Mild cases usually respond to a prolonged treatment of antibiotics, but serious and advanced cases usually need to be drained surgically.

The guttural pouch is an air sac on each side of the throat which connects with the inner ear and also the respiratory system. Occasionally one or both of these sacs become infected and fill with pus, usually after a respiratory infection. Infection of the guttural pouch can cause a chronic nasal discharge, and if the infection is severe the swollen pouch may interfere with swallowing, cause noisy breathing, or make the horse carry his head crooked. Sometimes the infected pouch will bleed. The only real cure for this condition is surgery, requiring a delicate operation under the ear. Any persistent nasal discharge in which you are uncertain of the cause, should be discussed with your vet. Chronic discharges should never be neglected, for this drainage is usually a symptom of something seriously wrong within the horse's nasal passages or respiratory system.

SWEENY

The term "sweeny" actually applies to any muscle or group of muscles that is withered or atrophied, regardless of their location, but in general horseman's parlance, the term has more or less applied to the atrophy of the shoulder muscles, caused by paralysis of the suprascapular nerve. Atrophy of a muscle results from disuse or from the loss of nerve supply, often through injury. In the shoulder area, the cause of injury to the suprascapular nerve is usually a blow on the point of the shoulder, or in the case of draft horses sometimes too much pressure from an ill-fitting collar or overwork. In a sweenied horse the shoulder joint is abnormally prominent and the shoulder area appears flattened. The spine of the shoulder blade is quite obvious, protruding from the withered muscles. If the problem is due to nerve injury, treatment is of little value, for the nerve damage is often irreparable.

WOUNDS

There are basically five types of wounds—bruises, abrasions, incised wounds, lacerations, and punctures. A bruise is a surface injury that doesn't break the skin. A horse may receive this type of injury from a fall or a kick or running into something solid. Bleeding under the skin, and lymph, causes swelling. Small bruises are best left alone for they will resorb in a few days and swelling will go down. Very large bruises may occasionally need veterinary attention, if swelling is profuse. A bruised leg (muscle and/or bone) can often be helped by soaking in ice or very cold water; this cold treatment reduces swelling and can sometimes keep a big knot from forming on a bruised bone.

An abrasion is scraped skin. It may bleed a little or ooze fluid and may be slow in healing. Usually the best treatment for an abrasion is to thoroughly clean the scraped area and then apply an antibiotic ointment, to combat infection and to help keep the area soft as it heals.

An incised wound is a clean cut, from an object such as sharp tin or metal, or broken glass. The usual problem, if the cut is at all deep, is excessive bleeding (a clean straight-across cut bleeds worse than a jagged cut). A pressure bandage may be necessary to stop the bleeding. Your first step in a wound of this kind is to try to stop the bleeding. A cut vein flows steadily and the blood is deep red; a cut artery flows bright red and the blood spurts each time the heart beats. In the case of a cut artery, apply pressure between the wound and the heart. With a cut vein, apply pressure on the side of the wound away from the heart, for the vein is bringing blood back to the heart. In extreme cases, when the horse is losing a lot of blood from a cut artery, a tourniquet may have to be used. But before going this far, try a pressure bandage; a tourniquet can do permanent damage by cutting off blood supply to the limb if applied too tightly or too long. If you must use a tourniquet, do so with care, tightening slowly until the flow of blood stops, and no tighter. In twenty minutes or less, loosen it and let the blood flow for a moment or so before tightening it again. Failure to release the tourniquet at intervals will kill the tissues from lack of blood. The tourniquet should be applied between the wound and the heart to slow a cut artery.

A pressure bandage can be made by placing a roll of bandage or a hard object such as a rock or piece of wood (wrapped and padded with clean cloth) against the cut, then bandaging firmly with bandage or strips of sheet. Continue wrapping until the bleeding stops; it will take many layers. Old sheets torn in three-inch strips are good for bandaging.

A laceration is a cut with torn, irregular and jagged edges; a barbed-wire cut is an example. Healing is usually slow because of the extensive tissue damage, and there is usually more danger of infection. Bleeding should be stopped and the wound thoroughly cleaned. Any dirt, hair or dead tissue should be removed. If the wound is bad and you fear chances of infection, a routine shot of antibiotics should be given to the horse. A tetanus shot is also a must when a horse suffers any wound that breaks the skin, unless his booster shots are up to date. To be effective, a tetanus shot must be given within 24 hours after the wound's occurrence.

The medication you use on a wound will depend somewhat on the type and seriousness of the wound. Remember that the wound must heal from the inside out. It should be cleaned and treated daily (preferably twice daily), and kept soft. With proper care serious wounds can be healed up with very little scar tissue to mar the horse's looks. In deep cuts on the lower legs (below the hock and knee) proud flesh may develop if tendons are damaged. Proud flesh is exuberant granulation —an unsightly rubbery growth—that sometimes results from an injured tendon growing profusely to try to repair itself. In this type of injury, consult your vet. He will usually tell you to use a healing medication for the first few days, then give you picric acid or some other preparation to use that will retard the growth of the proud flesh until the wound is healed. Once proud flesh gets started it is very hard to get rid of. It can be done but it usually takes much time and trouble.

Some wounds should be sutured; others should not. Deep lacerations usually need to drain and there may be some sloughing away of damaged tissue. Most wounds are best left unsutured. But sometimes a cut muscle can be sutured successfully, and wounds of the face and over bony areas are often sutured. A deep cut in a joint area may need veterinary attention. If the joint itself has been injured and has lost its fluid, the joint will need to be injected. Otherwise it may heal improperly and become permanently stiff.

A puncture wound should always have special attention. A puncture is sometimes hard to see and hard to treat. This type of wound usually does not bleed much. It tends to heal first on the outside, leaving a pus pocket that may abscess and send infection all through the horse's body. To treat a puncture wound you should first open up the hole and clean it with hydrogen peroxide or some other good disinfectant (not iodine, unless it is a puncture wound in the foot) recommended by your vet. Then put a piece of sterile cotton soaked in antiseptic in the wound. This dressing should be changed daily until the wound has healed from the inside out. If an old, neglected puncture wound begins to abscess, your horse may need antibiotics.

A puncture wound anywhere in the body is always serious because it can lead to tetanus (see the discussion of tetanus in chapter 9), but a puncture wound in the foot is doubly serious because it may damage the inner tissues of the foot, possibly causing problems such as inflammation of the bone, a fracture, decay of the pedal bone, or decay of the digital cushion. A puncture in the middle third of the frog is very serious because it may puncture the navicular bursa. A puncture wound in the foot is not always easy to locate. Cracks in the sole or black spots should be trimmed or probed to their full depth to see if any of them lead into sensitive tissues and are the cause of infection (if a puncture wound is not noticed and treated immediately, the horse will usually go lame, infection will be started and the foot may swell part way up the leg). Old shoe nails or nail holes should also be checked out. The frog should be trimmed back; a puncture in the frog is often hard to see if the foreign matter that made the hole (such as a nail) is no longer embedded in the wound, for the frog is spongy and closes over the hole. If the wound has introduced infection into the foot and this infection has no avenue for drainage, it will usually break out at the coronary band near the heel and begin to drain. The punctured area should be cut away so that infection will have adequate drainage, the foot soaked in epsom salts (magnesium sulfate solution) dissolved in hot water (but not so hot

The protective rubber boot we made for Khamette's foot from an old inner tube and some pieces of old harness leather

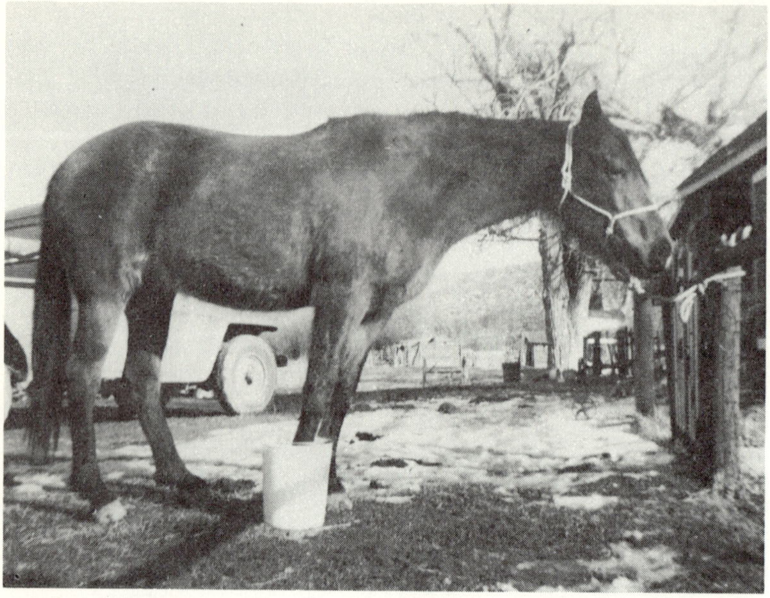

Treating a puncture wound in Khamette's right front foot. Soaking the foot in a hot solution of Epsom salts

We made a protective rubber boot for her

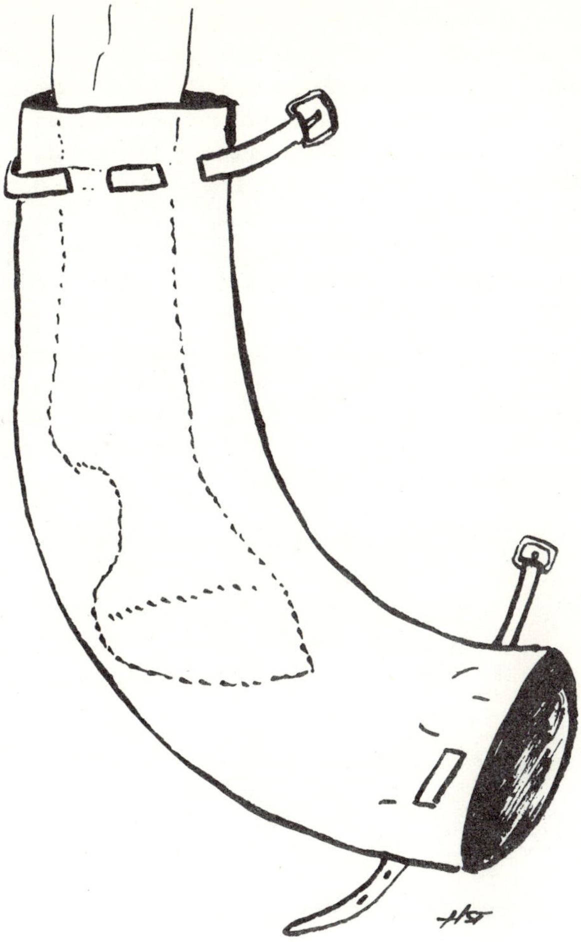

that the horse resents it; a rubber bucket or soaking boot can be used for this purpose). Then the wound can be disinfected (iodine is all right when used in wounds in the foot), and the hole plugged with some sort of wick that will allow drainage. A puncture wound in the sole should be opened so that there is at least a one-quarter inch hole into the infected area, with the walls of the drainage hole widening toward the ground surface so that it won't become obstructed. When the puncture is in the frog, the frog should be trimmed away at the site of the puncture until adequate drainage is established.

Once healing has begun it is not necessary to soak and bandage the foot every day—every two or three days is sufficient—but it is important that the wound be kept clean and dry until it has fully healed. A rubber boot designed to fit the horse's foot can be used.

Listed below are some items that you might want to keep on hand at all times as a sort of first aid kit for wounds. You should have a sulfa powder or other antibiotic powder or sulfa-base ointment (for many wounds ointment is better—less drying—than powder), boric acid for use as a mild eyewash, a fly-repellent dressing for summertime wounds, Scarlet Oil or some similar heavy liquid dressing, picric acid or some other suitable medication for retarding growth of proud flesh, old sheets torn into strips and rolled for bandages, squeeze bottles to use for squirting warm saline solution into a wound to clean it (horse's tissues tend to water-log when washed with ordinary water), or for applying powdered or liquid medications, hydrogen peroxide for cleaning wounds, iodine for disinfecting puncture wounds of the foot. *Never* use iodine or other strong medications such as methylene blue on an open wound. These burn the tissues; the flesh may subsequently slough away, leaving a scar. A deep wound that has been treated with iodine or methylene blue or some other strong disinfectant cannot be closed or sutured or it will fester and abscess. Iodine should be used only for treating thrush, puncture wounds of the foot and ringworm.

STRAINS AND SPRAINS

These are relatively minor injuries, involving pulled joints, stretched ligaments and tendons that support the leg and hold the joints together, but injuries that are often made worse by improper treatment. The muscles and tendons of the leg are

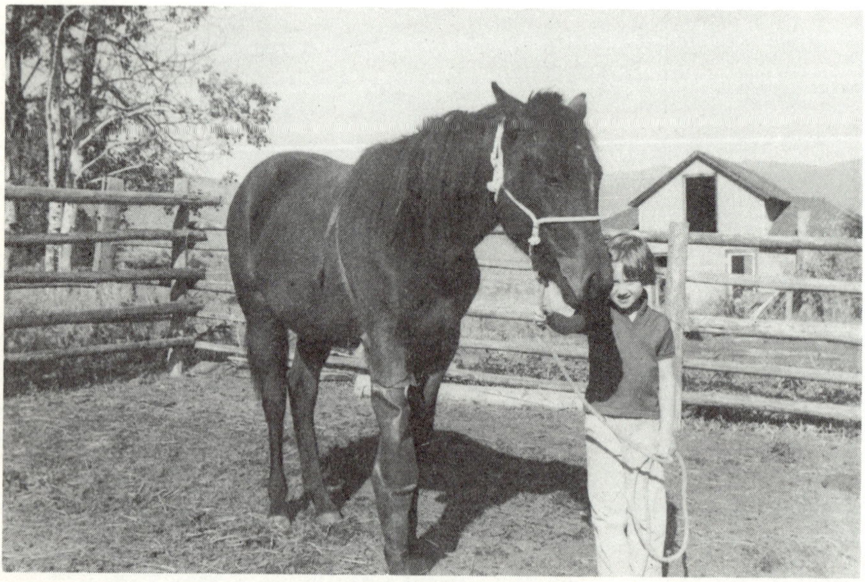

This rubber boot was made for Nikki to keep her from licking and chewing at a knee injury when she was a weanling. A similar type of boot, enclosing the foot, can be made for soaking a foot or a leg in ice water to reduce swelling caused by a strain

sore from overwork or twisting or sudden effort, and there is pain, heat, swelling, and lameness in the limb. Most strains and sprains will heal all right with no treatment at all (especially if the horse is rested and not worked), but they will heal faster and be less likely to leave the horse with a weakness or permanent swelling if treated properly. Rest and cooling are the first steps; for the first 24 hours the injured part should be cooled to help prevent excessive swelling. Then heat and liniment can be applied if you wish, and the injured part massaged to increase circulation. But never use heat on a fresh injury—keep it cold for the first hours to decrease circulation and keep the swelling down, and help prevent accumulation of fluids.

Resting the injured part is vital. Put the horse in a stall if necessary to keep him from moving about. A plaster cast can be used for complete immobilization if the injury is a bad one. A bandage will help immobilize a less serious injury sufficiently. Bandaging can also help keep swelling down, but never bandage too tightly or you will cut off circulation.

To cool an injured leg you can use water from a hose if the water is cold enough (below 55°F.), or ice-water in a "boot." A simple boot can be made from a piece of old tire's inner-tubing by sealing one end. Put it on the horse's leg and fill it with cold water and ice deep enough to cover the injured part, and keep adding ice from time to time. This can be done for a couple hours at a time, several times during that first day. The horse should show a marked improvement by the second day. If he doesn't improve, you should have a vet check the injury and make a proper diagnosis. An X ray may be needed to check for bone injury. The vet can also give the horse drugs that will help relieve his pain and reduce the swelling and inflammation.

BROKEN BONES

Most broken bones in horses necessitate the humane destroying of the horse. This is because the horse is one of the poorest patients when it comes to the setting and healing of a leg bone or joint. The horse cannot spend a great deal of time lying down and must either keep most of his weight on his good leg (which then often goes bad from the extra stress and strain) or be confined to a sling. Many horses will not tolerate a sling, and others that do often develop pressure sores from the sling. But new developments in equine surgery and new materials used for casting have helped to make the picture a little brighter. A higher percentage of horses with broken legs today can be saved than could be saved in the past. Some types of fractures can be taken care of more easily and simply than others. If you ever suspect a bone fracture (usually the swelling is greater, spread over a larger area than with just a sprain or strain, and the horse does not respond readily to treatment) have a veterinarian X-ray the limb to determine the extent of damage. If you are lucky, you may be able to save the horse with proper care and setting of the fracture or removal of the bone chips. But many times the procedures involved in repairing a broken leg cost much more than the average horse owner can afford. If a horse is extremely valuable as a breeding animal, expensive treatments are usually undertaken. But the average horse owner is usually faced with the sad decision of destroying the horse if a serious and expensive surgery and convalescence is involved.

BACK INJURIES

The horse has a strong, well-constructed back, but nature really did not intend for the horse to carry weight at high speeds, during fast starts, stops and turns or over jumps. It is amazing that the horse holds up as well as he does. The back can be injured from stress and strain (from the weight carrying) or from accidents such as falling down, slipping, being bitten or kicked, and so on. Or, the horse may get a sore back from an ill-fitting saddle; this type of back problem will be discussed more fully in the next section.

If a horse has pain in his back he may be noticeably lame and incapacitated, or just "grumpy" about doing certain things, depending upon the extent and seriousness of the injury. He may be reluctant to work, may not want to jump, turn in a certain direction, or perform some other specific maneuver, because it hurts him. He may hump up and protest when being saddled, or seem "cinch bound." He may try to rear or lie down. Or he may travel stiffly on his hind legs. If the back injury is serious, he may be reluctant to move at all, holding his tail out stiffly, trying not to move his hindquarters, or he may drag his toes when he walks.

The backbone consists of vertebrae, spinal cord, and branching nerves. The bones come together in a serious of joints and are separated from one another by discs, which are cushions and shock absorbers. The backbone is somewhat flexible and can bend up or down or to either side when certain muscles are put into play. Thus the horse's back is an elaborate and complicated system, composed of many different structures. Any one of these structures can become injured. Bones can be broken or dislocated, discs can rupture or become crushed or squashed, ligaments holding the whole system together can stretch or tear, as can the muscles that help hold and move the back. The spinal cord can become pinched, causing pain or partial paralysis (in severe cases, complete paralysis) of that part of the horse's body that is controlled by the particular nerves that are pinched off. In some back injuries, the joints between the vertebrae become damaged and calcified (arthritis of the back), leaving the horse somewhat disabled. For an injured back that is strained or wrenched, rest and confinement help the injured parts heal. Your vet can prescribe medications such as cortisone and phenylbutazone to help promote healing and relieve the horse's pain. Most back injuries caused by strain occur in the lumbar region—right behind the saddle area.

It isn't always easy to diagnose a back problem. The horse's back is difficult to X-ray. Other problems can often produce similar symptoms to back injury. It also isn't easy to tell just exactly what is causing the horse to act the way he does; it's sometimes difficult to tell whether the horse's pain originates in the back or elsewhere. We found this to be true when one of the fillies we were boarding came up hurting one winter day. She was reluctant to move, and her symptoms plus the fact that there was ice and slippery places in the pasture seemed to point to a back injury. The young horses often galloped and played and she could very easily have fallen down. But she had a urinary tract infection (probably in her kidneys) and we nearly lost her before we got her well again. We gave her antibiotics for four days (for good luck, because we weren't absolutely positive what was wrong with her) along with the medications for an injured back, and she improved. Then about the time we were going to put her out of the barn where she was confined, she got worse again, lying down most of the time, reluctant to move when she was up, droopy and cold-legged (failing circulation). And there was pus in her urine. So we immediately started her on antibiotics again, took a urine sample and the vet cultured it to discover she had a very bad staphylococcus infection which was resistant to most antibiotics. After trying 10 antibiotics on the cultured bacteria, the vet at last found one that would combat the infection and we switched to that one, with better results; the antibiotic we had been giving her merely knocked off the secondary infections and temporarily saved her life—it was not eliminating the primary kidney infection.

SORE BACK

The most common back problems are those caused by the equipment we use on the horse, by chafing or by pressure from an ill-fitting saddle or from the way we ride. There are several types of saddle sores and related problems, so let's take a closer look at them.

Fistula of the withers is caused by bruising of the withers, often by an ill-fitting saddle. It is an infected and draining bursitis (a bursa is a sac filled with fluid, situated in the tissues at places where friction would otherwise occur, such as around moving parts) at the withers. If not treated and cleared up in the early stages, it usually leads to a deep and nasty infection (often infected with the bacteria that cause brucellosis). By the time it reaches this stage it cannot be cleared up without radical surgery.

Many saddle sores are caused by pressure, either from an ill-fitting saddle or from dirty saddle blankets or from something under the saddle blanket. Once a sore starts, it is hard to heal unless you quit riding the horse and give it a chance to keep from being irritated. If a sore is repeatedly injured by riding the horse, it may become a permanent problem because of the development of thick scar tissue. Most sores will heal without treatment if the horse is allowed to go without being ridden until healing is complete. Sometimes a soothing dressing is helpful to keep the sore soft while it heals, but actually the only really necessary treatment is proper medication if the sore is infected. Your vet can prescribe a good medication.

Chafing can cause sores; for instance if a cinch ring is rusty or not smooth, or if it is situated too low, it can rub the movable skin in the horse's girth area. This can be prevented or corrected by moving the cinch ring, adjusting it so that it contacts the horse in a different place, and padding it.

A dirty or nonporous saddle pad or blanket can cause a heat rash on the horse's back. This can be prevented by always using a thick, clean, porous type

Heidi rode Khamette bareback during the summer of 1971—moving cattle, range riding, and conditioning for the trail ride—to allow an old saddle sore to heal

of pad or blanket that lets air through it, allowing some air under the saddle; this keeps the skin under the saddle somewhat cooler from sweat evaporation. On a long ride, it is wise to take off the saddle, or at least loosen it occasionally and lift it a bit to let some air under it. But a word of caution. Any time you remove the saddle after a long ride, do it gradually. After any long ride there has been a lot of constant pressure on the horse's back and the blood has been somewhat pressed out of that area. If you jump off and immediately take off the saddle, the blood rushes back into the area, and can cause lumps and bumps (heat bumps). These can become permanent lumps if this is done to a horse very often. To prevent heat bumps, dismount, loosen the cinch, and lead the horse a bit, perhaps lifting the saddle now and then to let air underneath it and decrease the pressure on the back so that blood begins to return to the area. Then after a few moments you can go ahead and unsaddle.

Pressure causes most saddle sores—too much pressure in one spot. A good thick saddle pad and well-built saddle evenly distribute pressure so that sores do not occur. Too much pressure in one place kills the skin cells because it deprives them of blood. A sore results if enough skin cells die. This tissue death is called pressure necrosis. A "sit-fast" is a plug of dead tissue caused by pressure necrosis under the saddle, or under the collar on a work horse. This is a hard skin lesion similar to a corn on a human toe. It can be caused by an ill-fitting saddle, too small a saddle for the weight of the rider (saddle bars too short), or uneven weight distribution (as when a rider sits crooked or slops in the saddle, or rides

Fahleen and her "cool back" saddle pad—made from synthetic fiber. This is the same material used in hospitals under bedridden patients to prevent bed sores. It is very soft and cushiony—and very durable

back too far on the cantle, or if the rider is carrying baggage that is unevenly distributed, like a rifle in a scabbard, and so on).

A saddle gall should be cleaned and cold-packed when fresh, and the horse not ridden again with a saddle until it has healed. If it is re-irritated it may become a permanent lump or sore. A deep gall can cause the horse great pain. A gall in the center of the back, over the spine, should be ice-packed for several hours as soon as you discover it, to help keep it from causing a permanent enlargement on the backbone. Your vet can prescribe medication to help it heal. If you have to keep riding the horse (though it's best if you don't) you should change saddles so that pressure will not come in quite the same place, and protect the gall with medicated powder (baby powder works) and use a saddle blanket that will not rub it. Some success has been found by making a special felt pad with a hole cut in it at the location of the gall (make sure it is large enough to account for any movement of the pad as the horse moves) and then putting regular blankets or pads over the felt pad.

In the fall of 1970 my mare Khamette (ridden by my sister Heidi and my husband) developed a saddle sore beside her backbone (on both sides) in the cantle area from a combination of too much riding (gathering cattle, every day) the prossure of the bars in my sister's youth saddle which she had gotten too heavy for, and old hair pads that were a little too stiff and dirty. Khamette had the whole winter to rest and heal, but the first two rides the next spring re-irritated the sore on one side and made a small spot raw again. Heidi had her heart set on riding that mare in a competitive trail ride that summer, so she rode the mare bareback all spring and summer, and we made a special pad for her to use on the trail ride—consisting of a foam rubber pad encased in a wool army blanket (doubled), with a very very soft thick wool baby blanket on the bottom side next to the horse's back. The combination of a very very clean back, and a clean soft, nonirritating saddle pad, and a new (bigger) saddle for Heidi, kept the partially healed and partially haired-over sore from getting any worse and it continued to heal all right even though a saddle was used several other times the rest of the summer. The special pad (foam rubber taken out) was washed after each use so that it was always perfectly clean and soft. There are now some new saddle pads you can buy which are made of the material used in hospitals to prevent bedsores on bedridden patients. These pads are very soft and can prevent saddle sores and help old sores to heal.

The saddle you use should always be large enough for your weight and should be constructed so that it places your weight evenly on the horse. The fork should fit the horse's withers, distributing your weight over as large an area as possible. The cinch should be wide and soft. Pads should be clean, soft, and porous. The horse's back should always be clean. Don't cinch the horse too tightly. If he hasn't got good withers and doesn't hold a saddle well, use a breast collar and crupper to help keep that saddle in place. Sit straight in the saddle—never carry your weight to one side; don't slop. Stay forward over the horse's center of gravity, not back on his loins. Carry some of your weight in the stirrups, transmitting it to the whole saddle tree. Keep in proper balance; ride with your horse. And on a long ride loosen your saddle a time or two. If you always think first of your horse, you're less apt to make him sore-backed.

ARTHRITIS

Arthritis, stated simply, is an inflammation of a joint. It can involve the bones, the cartilages, the joint capsule, or the ligaments holding the joints together, or several or all of these parts. The joint capsule contains an outer fibrous layer and an inner layer that secretes the synovial fluid which lubricates the joint. If there is inflammation in the joint, more fluid is secreted, causing swelling of the joint capsule. If there is injury to the joint in which the capsule or ligaments are torn from their attachments on the bone, new bone growth may form as a result of periostitis (inflammation of the lining around the bone).

Arthritis can be primary (caused by trauma to the joint) or secondary (the result of bone disease or the localizing of a disease in the joint, or the result of poor conformation which puts extra stress upon a joint). Enlargements of a joint in arthritis can be due to enlargement of the bones forming the joint, thickening of the joint capsule, too much joint fluid, or swelling of the tissues around the joint.

Acute arthritis may either clear up or develop into chronic arthritis. Chronic arthritis usually leaves the joint permanently damaged. Inflammation of a joint can be caused by trauma to the joint capsule, infection in the joint due to a wound that penetrates the joint or from a systemic infection that comes to the joint through the blood stream or lymph system (such as in navel ill, strangles, etc.), degeneration of the bone and cartilage and thickening of the synovial membrane from old age, and so on. Rickets in young horses can cause inflammation of the joints as can metabolism disturbances in mature horses (usually from lack of calcium or phosphorus). Arthritis is not uncommon in old horses, and can result from a "wearing" of the joint cartilages. There is degeneration of the bone and cartilage and a thickening of the synovial membrane. In old horses arthritis can become a chronic problem.

In acute arthritis, sedation or local anesthesia can be given to reduce the horse's pain. The inflamed joint should have absolute rest; in some cases a plaster cast can be used to advantage to keep the horse from using the joint, or a heavy bandage. Corticosteroids are sometimes useful for reducing inflammation. In infectious arthritis the joint may need to be drained and antibiotics given. In serious cases of arthritis you will need veterinary assistance.

OLD AGE

Horses usually live to be in their twenties, with some horses staying healthy and active into their thirties. Many horsemen feel that a horse is "past his prime" when he gets into his late teens, but these older horses can continue to be sound and useful with good care. The old horse needs a good diet containing sufficient vitamins and protein. He needs regular worming. Over the years bloodworms can greatly damage a horse's intestinal arteries (see the chapter on internal parasites and colic). So an old horse may have partly damaged intestinal arteries, but good care can help him out a little. Feed him more frequently but give him smaller

The old horse needs special care to keep him sound and healthy

portions. Don't feed or water him when he's warm from working. Keep his diet a little on the laxative side, feeding some alfalfa hay, and perhaps an occasional bran mash.

Take good care of his feet and legs, keeping his feet well trimmed. Use extra padding under your saddle to protect those sharp old withers and backbone. Use good judgment when riding him and spare him all you can. Warm him up gradually, and give him a rest when he needs it, don't ride him fast on hard surfaces or through the rocks; his old legs and joints are not as tough and resilient as they used to be. Bones get brittle as he gets older. Don't expose him unnecessarily to cold and dampnesss; if you ride him in winter make sure he is dry before you turn him loose. Pamper him a little over the winter; he may need grain and a little extra care to stay in good condition.

If he is a gelding, keep his sheath clean so that accumulated secretions won't build up and cause irritation or infection. Have his teeth checked periodically. If an old horse starts to lose condition, teeth are one of the first things you should suspect are not quite up to par. Horse's teeth keep growing a little throughout their lives. Sometimes as a horse grows older his teeth don't quite match up and a tooth may grow a ragged, jagged edge or a sharp point because it is not worn down properly by the opposing tooth. These sharp points and edges may injure the horse's gums or cheeks, causing pain, and interfere with proper chewing. If his food isn't chewed properly he cannot get as much nutrition from it, and the masses of rough food may make him prone to impaction and colic. Much of the improperly chewed food passes on as waste. If the irregular teeth cause him pain he may be reluctant to eat. Some old horses actually starve to death just because it is hard for them to eat and they refuse their feed. The irregular points and edges can be smoothed down with a "float"—a special rasp for "floating" teeth. Your vet can do this for your horse if teeth become a problem. As a horse gets very very old, he may lose some of his teeth, and eating coarse roughage becomes a problem. Green grass and soft feeds can help if eating hay becomes difficult. Your vet can help you with some individual problems of old age. Of course, some problems are incorrectable and the horse is better off to be humanely destroyed than to linger and suffer, but many problems of old age can be helped by proper care and treatment, allowing the horse perhaps several more years of usefulness.

While we are on the subject of age in horses, and as a sort of postscript to this chapter, let's take a quick look at the horse's teeth. His teeth are a fairly accurate standard by which to tell how old he is. Stallions and geldings have 40 teeth while mares usually have but 36, lacking the tusks or tushes. Occasionally a mare will have these.

When telling the age of a horse, it is usually only necessary to look at the lower front teeth, called the lower incisors. There are six of these; the two middle ones are central incisors, the two on either side of these the laterals, and the outer ones on each side the corners. Back teeth are grinders or molars. In the male horse, the canine teeth, or tusks, appear between grinders and incisors.

The horse gets two sets of teeth, the first teeth being temporory milk teeth. There are 24 of these, twelve in each jaw (six incisors and six molars, three molars on a side). The second set, the permanent teeth, are 40 in the male—six incisors, twelve molars, and two canines, in each jaw. At birth the foal usually has the central incisors in both upper and lower jaws (four front teeth) but sometimes these don't appear for a week or ten days after birth. At eight to ten weeks of age the lateral incisors on each jaw come through, and then when the foal is about ten months old the corner incisors appear. The molars (temporary) are present at birth. So at ten months the foal has all his temporary milk teeth.

At ten months of age the foal gets his fourth molars, his first permanent teeth (coming in at the same time his corner temporary incisors appear). At twenty months to two years the fifth molars come in. Since there is very little change in the incisors at this time, the presence or absence of the fifth molars help you tell the age.

The next change in the teeth is the shedding of the milk teeth and the appearance of permanent teeth. The permanent teeth push up under the temporary teeth, pushing them out. When the horse is about two years and nine months, to three years old, the temporary central incisors are shed, replaced by permanent ones. In about three months from their first appearance, the permanent central incisors are grown enough to be in wear. These permanent teeth are much larger than the temporary milk teeth, so there should be no mistaking them.

At three and a half to four years, the lateral incisors change to permanent. Thus at four years, the horse has four large permanent incisors in each jaw and two small temporary corner teeth. At five he gets his permanent corner incisors

Eight days old—central incisors (nippers)

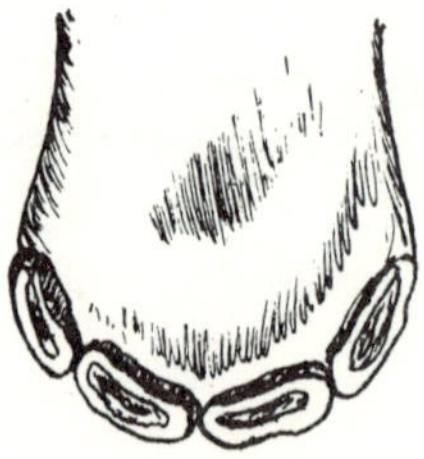

Two to three months old—centrals and laterals

Eight to ten months old—the "corners" appear

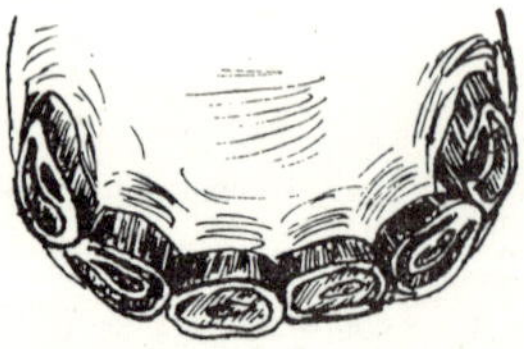

Two years old—centrals and laterals are wearing

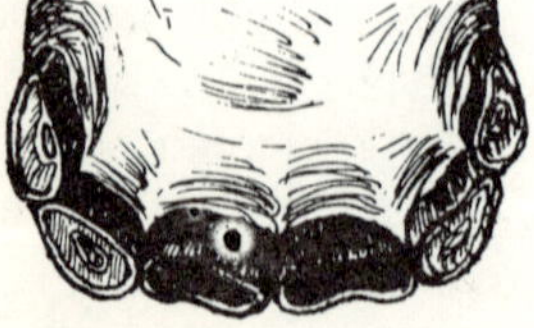

Two and a half to three years old—permanent centrals are coming in; no wear on them yet

Three and a half to four years old—centrals are fully in, showing wear, and the permanent laterals are coming in

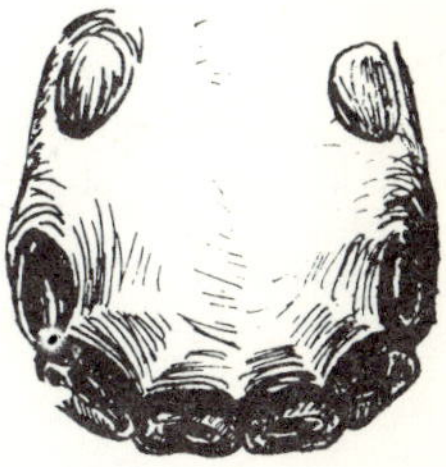

Four and a half to five years old—all permanent incisors are in. Male horses have canine teeth by this time

and now has all his permanent teeth, the permanent molars having replaced the temporary molars during the last two years and the sixth molars and canines having come in during the last year. The horse now has a full mouth. From then on the age of the horse is judged by how the teeth wear. When a tooth is fairly young there is a cone-shaped cavity (cup) in the grinding surface, and as the tooth wears away this cup becomes smaller and shallower. Since the teeth come into the mouth at different ages of the horse, the cups in the different teeth wear away at different ages. Around the cup is a thin layer of very hard enamel, and since it is harder than the rest of the tooth it wears away more slowly and protrudes above the rest of the tooth surface.

When the horse is six years old the cups in the lower central incisors are becoming small and shallow. The cups in the laterals are still of fair size. At age seven the cups are gone from the central incisors and becoming shallow in the lateral incisors. At age eight the cups are gone from the laterals and getting shallow in the corners. At age nine the cups are gone from the corner incisors. The horse is said to be smooth-mouthed. From this time on (approximately age 9 or 10) the age is harder to determine with the same degree of accuracy.

When the cup in the tooth is gone, the enamel that surrounded it will still be visible as a thin, whitish layer, forming an oblong ring in the center of the grinding surface. When the cup is first worn away, the ring surrounding it is oblong and quite large, but as age advances it becomes smaller and rounder, and finally disappears. Then the surface will feel smooth to the touch.

At age 10 the ring of enamel in the lower central incisors is getting small and round, but not entirely gone. At 12 the enamel is gone, or nearly gone, from the central incisors and is small in the laterals. At about 14 the enamel is gone from

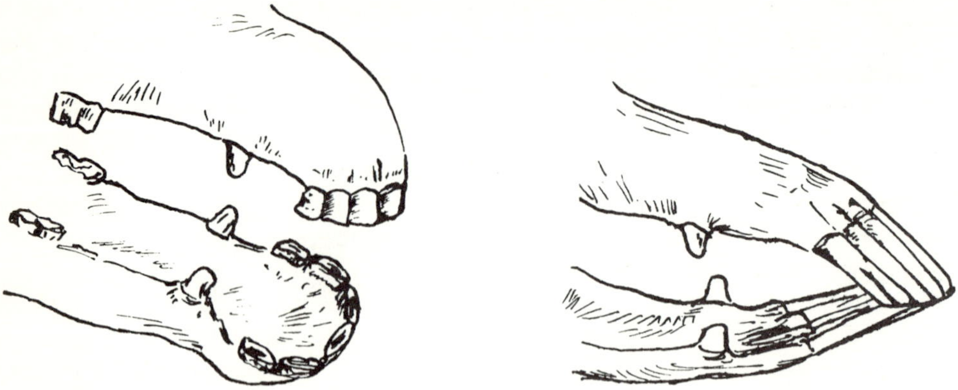

Angle of a horse's teeth; ***left,*** **a 5-year-old horse;** ***right,*** **a 25-year-old horse**

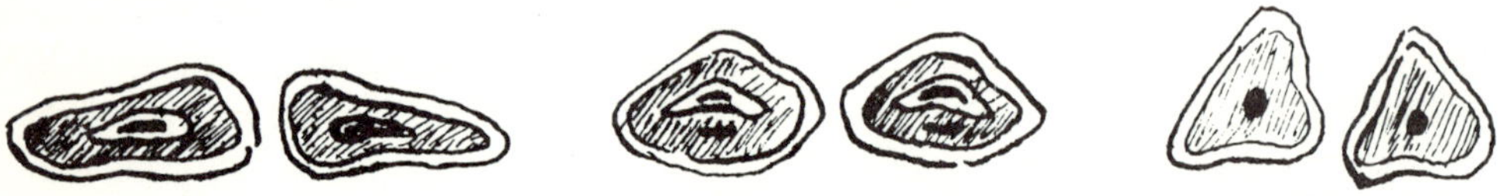

A cross section of a horse's teeth; ***left,*** **at 5 years old;** ***center,*** **at 10 years old;** ***right,*** **at 15–20 years old**

The outer surface of the corner incisor at various ages; Galvayne's groove appears at age 10, reaches the lower edge at 20, and gradually disappears (from the top first) by the age of 30

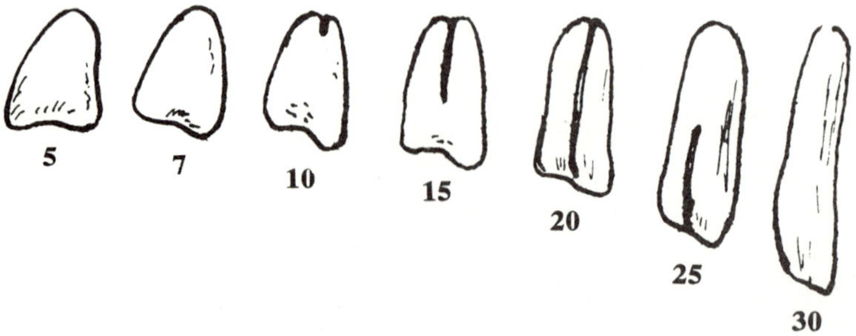

the laterals and small in the corners. At about 16 the enamel is gone from the corners. These rules are fairly reliable when the lower and upper teeth meet perfectly and wear evenly. If the upper teeth project farther than the lower, the lower teeth will not wear away as fast as normal. Sometimes one side will wear faster than the other; one corner tooth might indicate the horse to be 15 or 16 while the other tooth indicates only 12 years. Look at all the teeth when forming an opinion as to age.

There are a few other factors that can help you determine a horse's age in the later years. As a horse grows older, the shape and angle of the teeth change. In a young horse (five, six, seven) the upper and lower teeth are nearly straight up and down, the front surface of uppers and lowers together presenting a straight line. Then as the horse grows older, the teeth protrude forward, forming an angle, the angle becoming sharper as the horse gets older.

In the young horse the teeth are broad from side to side and narrow from front to back. As he grows older the teeth get narrower from side to side and wider from front to back.

Another aid in determining a horse's age is Galvayne's groove on the upper corner tooth. This groove appears at the gum line at about 10 years of age, advancing down the tooth with age. At age 15 the groove is halfway down the tooth, going all the way to the wearing surface at age 20. Then it retreats downward, appearing only on the bottom half of the tooth at age 25 and disappears altogether by age 30.

Other indications of old age, besides the teeth, are deepening of the hollows over the eyes, gray hairs over the eyes and around the muzzle, low hanging lips, sharpness of the withers and sinking of the back (sway back) and so on. But if a horse remains sound and healthy, with good teeth, he may be able to live a useful life for many years. My first horse, Old Possum, lived to be nearly thirty, dying one winter of kidney failure. An old mare that my husband's family had, Old Boots, lived to be 32 years old, and would have gone on longer if not for the fact that she'd lost most of her teeth. She managed all right that last summer on green grass, but was humanely destroyed before the winter because she could not eat hay.

11

INTERNAL PARASITES

ALMOST ALL ANIMALS HARBOR SOME TYPES OF PARASITES, INTERnally and/or externally. In this chapter we will take a quick look at internal parasites of the horse and in the next chapter we will deal with the horse's external parasites.

The domestic horse seems to be subject to all the various types of internal parasites found in his wild relatives. No horse is entirely free of internal parasites; all are infested to some degree, but several factors play a role in determining the degree and seriousnesss of the infestation. Nutrition and environment are very important factors. Poorly fed and undernourished animals are, as a general rule, more susceptible to the effects of internal parasites—less able to throw off infestations and thus more likely to carry heavy infestations. But some kinds of worms pose just as great a problem in well-fed animals as they do in undernourished ones.

Pasture management is another important factor. In the past, in underdeveloped areas and countries where poor farming methods were used, depleting the soils, horses were apt to carry large numbers of worms because of inadequate nutrition. But today, heavy infestations often occur just as commonly in areas where there is excellent fertility and productivity, because livestock are too confined—they can be grazed and maintained in smaller areas. Thus the contamination of pastures with worm eggs from manure is greater. Almost all intestinal horse parasites enter the digestive tract via the mouth. All of these, except the horse bot, are propagated by eggs which are passed in manure. The more horses that graze the same area, the greater the chance of one infested horse spreading worms to all the others.

Horses at pasture with a large area in which to roam (a large dry-land pasture or range, for instance) may have only light infestations provided the pasture was not previously contaminated (and the horses do not gather at water holes, swampy areas and so on where manure build up and conditions favorable to worms exist) and the horses introduced to the pasture were fairly worm-free at introduction.

In confined conditions (barns, stalls, corrals, and so on), horses should be fed in bunks or tubs that are kept clean, and never fed on the floor or ground where they can pick up worm eggs from manure. In a barn or stall, manure should be cleaned out daily and the bedding changed.

Most parasites thrive in an environment that is warm and moist. Sunlight, heat, and cold tend to destroy most parasites in the larvae stage, but the worm eggs are quite resistant to extremes in weather. Cold winters, dry summers, exposure to sunlight through the practice of harrowing fields and pastures, all help to keep parasite numbers down, but never eliminate them because of the hardiness of the worm eggs. Thus, pastures that have been previously grazed by horses can continue to be a source of infestation for other horses.

Worming horses regularly is very important, perhaps the most important aspect of proper horse care after proper nutrition. Far too many horsemen neglect this aspect of their horse's care, because they can't see the worms, and because if a horse is fat he looks healthy. All horses (unless they have just been recently wormed) have worms, and these worms rob the horse of vitality and nutrition. They can also damage arteries, cause colic and intestinal obstructions and irritate the digestive tract. They can interfere with the proper growth and development of young horses. And they can cause lung damage. Because worm eggs and larvae are passed in manure, and can live for some time in old manure, horses in confined areas are constantly subject to reinfestation. It doesn't do much good to worm a horse and then put him back in the same dirty corral. When we worm our

Yearling filly heavily infested with worms. She is thin, pot bellied, and rough looking. Even though it is August, she has not yet fully shed her long hair from the winter before

horses (and we like to do it several times a year) we try to then put them into a new, fresh pasture, preferably one that hasn't been used by horses for awhile. We like to alternate cattle and horses in our pastures.

Keeping worm numbers under control is a constant battle. Heavy worm infestations cause deaths in horses, but the greatest loss is in feed and vitality. The parasite depends on his host for survival, and robs the host of food in order to live. Worms do not generally kill their host, but they do add to the feed bill, rob the horse of energy, and perhaps affect his looks (leaving his coat dull or making him more pot-bellied than normal).

In the United States, there are several kinds of internal parasites that occur commonly in horses. Most of these are worms, but the horse is also subject to infestation from several kinds of fly larvae. Most internal parasites of the horse are in the digestive tract, usually picked up while the horse is eating, or from an insect that acts as an intermediate host.

Let's look first at the stomach parasites. Two types of stomach worms can be found in the horse, a small worm (about half an inch to one inch in length) called *Habronema* spp. and a tiny hairlike worm (less than a quarter inch long) called *Trichostrongylus axei.* The larger worm, *Habronema* spp., is usually present in small numbers, partially embedded in the stomach wall. This worm irritates the lining of the stomach and can cause ulcers and tumors. These worms (also called large stomach worms) are found all over the world but are generally a problem only in warm climates and wet areas. The three species which infest North American horses are dependent upon the stable fly and house fly as intermediate hosts. The larvae are passed in manure, where they enter the larval maggots of the flies which breed in the manure. Within the fly maggot, the worm larvae develop, then enter the horse when the fly feeds on the lips or wounds of the horse. Swallowed larvae reach maturity in the horse's stomach; larvae deposited in wounds of the horse cause cutaneous habronemiasis, commonly called summer sores. These lesions are most common in the skin between the horse's front legs, the face and abdomen—places where the horse has trouble removing flies easily. The larvae penetrate the skin, the horse loses hair there, and the area becomes encrusted with fluid that oozes from the skin. These sores often regress during colder months only to reoccur the following summer.

The smaller stomach worm, *Trichostrongylus axei,* is usually found in large numbers in the horse's stomach, and is acquired by grazing pastures contaminated with worm larvae that have hatched from eggs passed in the feces of infested horses. Infestation with this type of worm always occurs in conjunction with infestations of more pathogenic strongylid worms.

Another parasite that lives in the horse's stomach is the bot (gasterophilus), a grublike creature that is not a worm at all, but the larvae of a fly. These larvae can cause irritation of the stomach and occasionally they perforate the stomach wall. The adult bot fly is brown and hairy, about the size of a bee. It does not feed, and lives just a few days—only for the purpose of laying eggs. The fly is most prevalent in the summer months, the different species (there are five) appearing at different times. The three major species found in the U.S. are *Gasterophilus*

LIFE CYCLE OF THE STOMACH WORM HABRONEMA

A. Worm larvae are passed in manure

B. The worm larvae enter the larval maggots of flies that breed in manure

C. *Habronema* larvae are deposited in the wounds of horses when the matured flies feed upon the wounds

D. The larvae also enter the horse when the flies feed upon the horse's lips

E. The larvae are swallowed and mature in the horse's stomach; they are partially embedded in the stomach wall

intestinalis, G. nasalis and *G. haemorrhoidalis*. The fly lays her eggs on the horse's coat, one to a hair, where they can be seen as little yellow specks. These eggs should always be scraped off with a razor blade or sharp knife if you happen to observe them. The eggs hatch into tiny larvae which are transported to the mouth by the horse licking or nosing at his legs. The most common type of bot fly (*G. intestinalis*) lays its eggs on the horse's front legs, while other types lay

LIFE CYCLE OF THE BOT FLY

A. The fly lays her eggs on the horse's coat, on the hairs of the legs, or under the jaw

B. The larvae migrate to the horse's mouth or are picked up from the leg hairs when the horse noses or licks himself. The larvae burrow into the gums and undergo a molt

C. Then they emerge and pass to the stomach where they attach to feed and grow for 8 to 12 months

D. They pass out with the droppings in early summer

E. And migrate into the ground to emerge as adult flies

F. After 3 to 5 weeks, they begin the cycle again

their eggs around the jaw or cheek and those larvae migrate to the horse's mouth after hatching. The eggs which are laid on the horse's legs are picked up by the horse when he licks himself. After reaching the mouth, they burrow into the gums and undergo another molt. After about a month they emerge and pass to the stomach to feed and grow for 8 to 12 months, producing a small ulcer at the site of attachment. The larvae pass out in the droppings in spring and early summer, migrate into the ground and emerge as adult flies after 3 to 5 weeks to begin the cycle again. Occasional bot larvae migrate to abnormal sites within the horse, such as the brain, heart, lungs or ovaries. Bots in large numbers cause unthriftiness and poor coat, occasional mild colic and lack of appetite. Adult bot flies irritate and startle horses by flying and darting around the horses' heads.

Let's take a look now at some of the more common intestinal worms. Two types of roundworms are usually found in the horse's small intestine. The horse ascarid (*Parascaris equorum*) is the largest roundworm of the horse and the largest intestinal parasite found in the horse. These worms resemble an earthworm in shape, are white, and may get up to 12 inches in length. The adult worms are usually located in the small intestine, where a single female may lay over a quarter of a million eggs per day. These eggs are passed in the feces and must usually remain on the ground for two or three weeks before maturing enough to be infective. When mature eggs are eaten by a horse, the larvae hatch and begin migrating through the horse's tissues, developing into their adult form after this migration is complete. These worms are most common in young foals, robbing the foal of nutrition, and causing tissue damage during the larvae migration. The larvae can cause lung damage, a cough, and nasal discharge because of the lung irritation.

When first swallowed and hatched, the larva penetrates the intestinal wall and enters the blood stream. It passes through the liver, where it develops and grows and is then carried by the blood to the lungs. Then it passes through the small vessels in the lungs, breaking through the alveoli (air sacs), passing into the bronchi. In nine or ten days the larva is coughed up the trachea and is then swallowed, returning to the digestive tract. It then matures and develops into an adult in the small intestine, and begins laying eggs. The interval between ingestion of the egg until the development of an egg-laying adult is called the prepatent period, and takes about two to two and a half months. During this time of development and migration the host horse may undergo serious damage from the migration through the liver and lungs. Adult worms use up nutrients from the small intestine and occasionally become so numerous that the intestine becomes partially or completely blocked, causing the death of the host. Occasionally a worm will penetrate the intestinal wall and cause peritonitis. Signs of ascarid infestation are slow growth and poor hair coat, potbelly, lack of energy, digestive disturbances and sometimes diarrhea. Because it takes two months or more for the worms to mature and start passing eggs, fecal examinations for parasite eggs are usually negative until the foal is at least three months old. Therefore a diagnosis must often be based on the appearance of the foal. Most pastures that have been previously grazed by horses, especially foals, are heavily infested with roundworm eggs, so most foals should be wormed (treated for ascarids) beginning as early as two months, or even earlier in heavily infected cases.

Another roundworm sometimes found in the small intestine, but not as common as the ascarid, is *Strongyloids westeri,* a worm that is quite different from all other nematodes of horses. These worms in the larva stage penetrate the skin of horses as well as being ingested with feed. Only female worms develop from the larvae that penetrate the skin. These larvae migrate to the small intestine through the body, doing a certain amount of damage to the host. The females are less than one and a half inches in length. They attach to the wall of the intestine and produce

This weanling filly is infested with parasites, particularly with ascarids (roundworms). She is very thin but has a large belly.

fertile eggs without the presence of male worms. The most serious effect of this worm on the host horse is the acute diarrhea produced by the adult worms.

There are three kinds of tapeworms that can infest horses, two of which can appear in the small intestine, and one which invades the large intestine. The two that can invade the small intestine are *Anoplocephala magna* (known as the large horse tapeworm) and *Paranoplocephala mamillana* (known as the dwarf tapeworm) and have been found in a few horses in the U.S and Canada. Both of these tapeworms are rather small compared to those of other animals. They are picked up by the horse when he eats the intermediate host (mites) along with his grass. In large numbers, these tapeworms can cause problems, but this is rare. Usually no ill effects are noticeable in horses because most infestations are light. Adult horses are very rarely affected, and heavy infestations are necessary in the young horse before adverse effects are noticed. Symptoms are diarrhea, emaciation, and anemia. Blood-stained feces may be the only symptom in lightly infested animals.

The third and most common (and most serious) of the horse tapeworms is *Anoplocepha perfoliata.* They are found in a very narrow area near the ilio-caecal valve (between the small intestine and the caecum) and can cause hemorrhage, ulceration, edema, and granulation. Occasionally they can block the tract, causing a fatal colic, or may cause rupture of the intestine with resulting fatal peritonitis, but these cases are rare.

Tapeworm eggs are passed in the feces either singly or protected by a tapeworm

LIFE CYCLE OF THE ROUNDWORM (ASCARID)

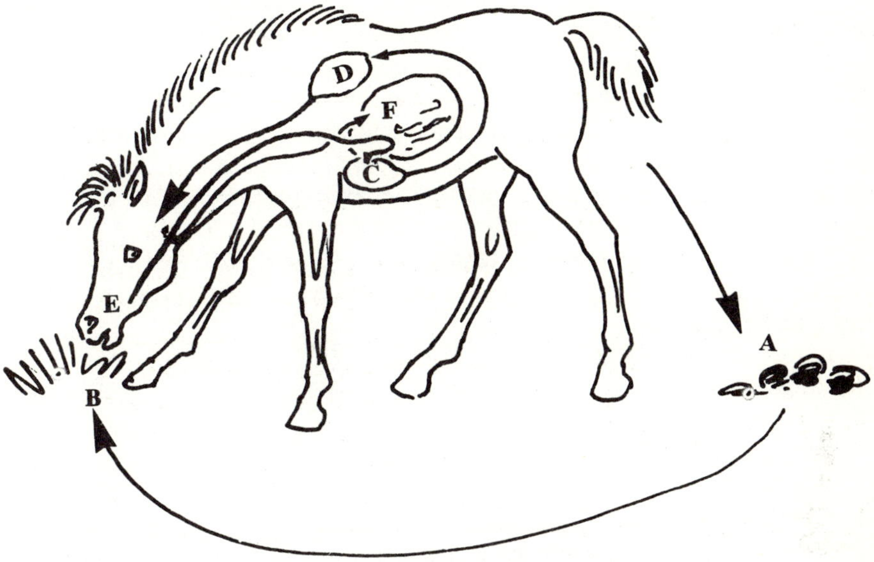

A. Eggs are passed in the feces and remain on the ground two or three weeks before maturing

B. The horse eats mature eggs with his grass or hay

C. The eggs hatch when swallowed, penetrate the intestinal wall, and enter the bloodstream—traveling to the liver where the larvae develop and grow.

D. The larvae are then carried by the bloodstream to the lungs

E. In 9 or 10 days they are coughed up, then reswallowed

F. Then they return to the digestive tract where they mature and begin laying eggs

segment or egg capsule. An intermediate host (such as the mite) must eat the eggs before the intermediate stage of the tapeworm can develop. Mature tapeworms develop only after the primary host (such as the horse) eats the intermediate stage.

Bloodworms (strongyles, also called red worms, or palisade worms) commonly cause debility and anemia in horses. These worms occur all over the world and nearly all horses are infested. There are three species of large strongyles (bloodworms) which affect the horse: *Strongylus vulgaris*, *Strongylus edentatus* and *S. equinus.* The bloodworm is the most widespread parasite of the horse and the most serious and damaging. The largest bloodworm, *S. equinus*, is one to two inches long. The three species mentioned are generally referred to as large strongyles. They are active blood suckers and their grey color is usually obscured by redness because of the blood they drink.

Eggs of the bloodworm are passed in the feces of the host horse, and in a suitable environment they hatch to produce two nonparasitic larval stages, and then go into an infective third stage in about 5 to 7 days. These larvae are resistant to drying and low temperatures and usually live for about three months. Under ideal conditions they may survive for a year or more. In warm, wet weather the larvae migrate onto forage plants, where they can be eaten by another horse. When they are swallowed by a horse, they penetrate the intestine and begin extensive migrations through the horse's body before they reach a length of about one or two inches in the caecum or large intestine. *S. equinus* becomes mature and starts laying eggs in about 260 days, after traveling through the liver and then back to the intestine. *S. edentatus* usually migrates through the liver for about two months, causing damage to the liver, and then to the intestine again. This species becomes mature and begins laying eggs in about 300 days. Larvae of *S. edentatus* and *S. equinus* have been found in many different tissues of the horse's body.

The most serious problem, however, is caused by the larvae of *S. vulgaris*, a smaller bloodworm, some of which may lodge in the walls of the blood vessels that supply the intestine with blood, and in the walls of the large arteries leading away from the heart.

Too many bloodworms in a vessel may cause ruptures or blockages. While the bloodworm larvae are traveling through the arteries, they damage them, often to the extent of causing illness or premature death of the horse. Too many larvae in a blood vessel cause inflammation, scarring and roughening of the inside vessel wall, and weak spots in the wall. This damage can lead to the formation of clots or of aneurysms. The vessel wall is no longer smooth, and clots can build up on the uneven surfaces. If a clot breaks loose (embolus) and travels down the blood vessel, it eventually comes to a place where the vessel is too small for it to pass, and it promptly blocks the vessel, shutting off the blood flow to the area supplied by that vessel. Although any artery in the horse's body can be involved with bloodworm larvae, *S. vulgaris* seems most often to damage the mesenteric arteries (which supply the intestine). When this happens, a section of intestine suffers a decrease in blood supply. This can cause abdominal pain and colic. A severe decrease or blockage results in the death of a section of intestine, and the subsequent death of the horse. Colic caused by *S. vulgaris* can be mild if only a very small artery is blocked, or fatal if the artery is a large one that supplies more intestinal tissue.

Another common site of arterial blockage caused by bloodworm larvae is where the aorta (the large artery leaving the heart and running under the backbone) branches into the iliac arteries which supply the hind legs with blood. If blockage occurs here the horse will be lame in one or both hind legs, and the affected leg or legs will be cold. If blockage is complete and permanent the leg tissue will die from lack of blood, killing the horse.

Artery damage can cause aneurysms, which are dilated sacs, bulging out from the vessel wall. Where the artery wall is weakened, it may balloon outward. If the aneurysm gets large, or if the blood vessel is quite weak, it may suddenly rupture, causing sudden death of the horse due to internal hemorrhage. The arteries have

LIFE CYCLE OF THE BLOODWORM—LARGE STRONGYLE, *S. vulgaris*

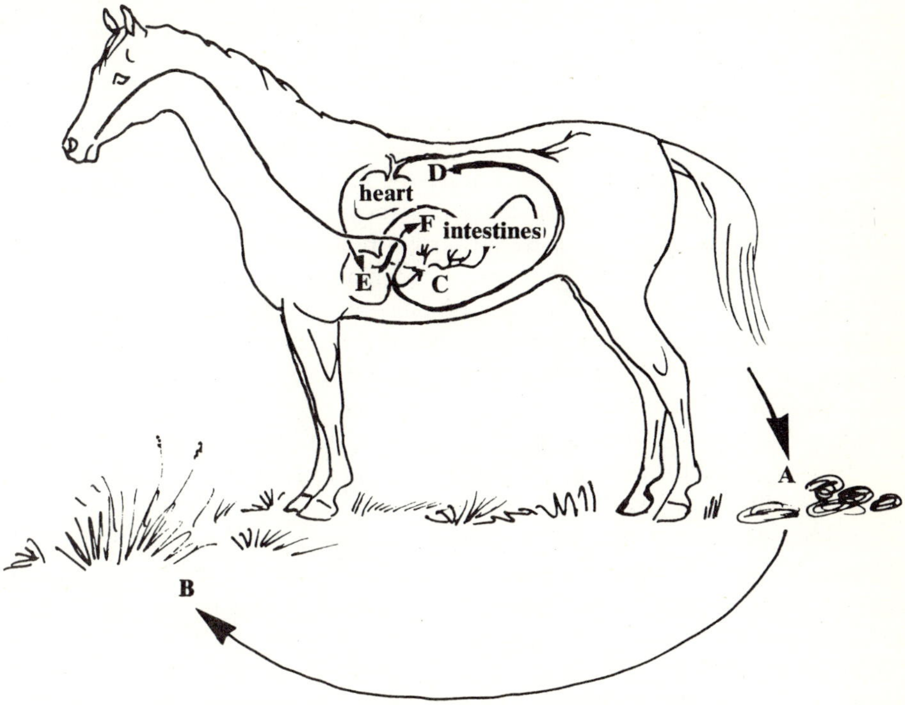

A. Eggs are passed in feces, and undergo three larval stages

B. In warm weather the larvae migrate onto forage plants and may be eaten by the horse

C. They are swallowed and they penetrate the intestine, where they enter the bloodstream and travel to various body sites such as the arteries that supply the intestines, the aorta, and the iliac arteries

D. They travel up to the liver

E. When fully mature, the bloodworm ends up back in the intestine

F. There, she lays eggs

quite a bit pressure within them, for blood surges through them with every beat of the heart. If an artery is brittle with scar tissue and weakened with aneurysms, it is not unusual for it to rupture, often happening when the horse's blood pressure is elevated because of increased activity or excitement.

Once a blood vessel has been damaged by bloodworms, it cannot be repaired. The only thing that can be done is to keep the horse regularly wormed, to insure that there will be no further damage. There are several drugs available that are effective against bloodworms, and they can be given in the horse's feed or by stomach tube. But these drugs only kill the mature worms in the intestine and are not effective against the larvae that are migrating through the body. Therefore it will take several repeated wormings to rid the horse of bloodworms, even if the horse eats no more contaminated feed. To keep a horse fairly free of bloodworms, he should be wormed several times a year, and put into a clean pasture. A fecal sample can be studied by your vet (observed under a microscope) to determine if your horse is host to bloodworms. Mature bloodworms in the intestine fasten to the intestine wall and can cause diarrhea, anemia, weakness and unthriftiness. Other symptoms of a horse suffering from bloodworms may be unexplained bouts of colic, or mysterious lameness of the hind legs. Anemia, mild or severe, is a factor that should not be ignored. Bloodworms deprive the horse of nutrients and blood, often causing young horses to be slow in growing, and causing working horses to be inefficient—lowering their energy levels and cutting into their endurance. There is a direct correlation between the ability of a racehorse to perform and his hemoglobin level. If your horse has bloodworms, he would no doubt be able to perform more creditably if he were free of them and their drain upon his system.

Small strongyles can also cause debility and anemia. They are blood suckers that infest the large intestine, but generally grow to be only ¼ to one inch in length. There are several kinds of small strongyles, including *Trichonema* and *Triodontophorus* (there are more than 40 species of *Triodontophorus*). The small strongyles do not migrate through the tisssues like the large strongyles, but do their chief damage in the intestine (principally in the caecum), ulcerating the mucus lining. In heavy infestations of these worms there may be very little normal area in the intestinal lining. The ulcers made by these small strongyles are deeper than those made by the large strongyles, and can occasionally cause fatal hemorrhage.

Pinworms (*Oxyuri sequi*) in horses are not usually serious, and are primarily a problem in horses confined to a small area. The main problem with these worms is the intense irritation they cause, and the subsequent disfigurement of the horse as he rubs out his tail This irritation of the region around the anus is annoying to the horse and he rubs and bites at his tail, causing loss of hair and sometimes physical damage to the surrounding tissues. The affected horse has a rat-tailed appearance. Sometimes a secondary bacterial infection may accompany this problem. The irritation from these worms may be so great as to cause the horse to be constantly restless to the point that he loses weight.

The mature pinworm is gray. The male worm is about one half inch long, but the female grows up to as long as 6 inches, with a long threadlike tail. Mature females crawl out of the large intestine and lay their eggs around the anus, in yellow patches of egg clusters. The eggs hatch in about three days and the horse becomes reinfested either by biting at the area or by eating contaminated feed. The eggs are resistant to drying and may remain alive in stables for long periods of time.

The donkey is the natural host for *Dictyocaulus arnfieldi,* a lungworm, and can harbor heavy infestations without harmful effects while serving as a source of infection for the more susceptible horse. This lungworm is a thin threadlike worm about two to three inches long. As an adult worm it lives in the bronchi. It can produce bronchitis and in heavily infested horses it produces symptoms of diarrhea, rapid shallow breathing, cough, nasal discharge and fever. The affected animal loses weight and may continue to have labored breathing for several months, if he survives. Light infestations of lungworms cause a hacking cough, but few other ill effects.

Occasionally a two-inch filarid worm called *Setaria equina* inhabits the peritoneal cavity of the horse. This worm is acquired by its host through the bite of a mosquito and does not seem to do much harm in the abdominal cavity. Sometimes it is found in other locations such as the eye, other body cavities, liver or digestive tract.

Infestations of another type of filarid worm (*Onchocerca*) occasionally produce problems in horses. *Onchocerca cervicalis*, (also called *O. reticulata*) causes local swelling and nodules in the neck ligaments, sometimes causing an impaired gait and lameness. *Onchocerca cervicalis* is thought to be important in causing fistulous withers and poll evil. This worm has also been found in the cornea of many horses and seems to be involved in some cases of periodic ophthalmia (see "leptospirosis," in chapter 9). These worms are also involved in occasional cases of dermatitis (skin inflammation). *O. cervicalis* is thin and threadlike, about 6 to 7 inches long. As in the case of most filarid worms, the microfilarie in the blood vessels of the horse's skin are picked up by blood-sucking insects (such as midges, sandflies, or black-flies) and deposited in the skin of other animals, then migrate to other sites in the body of the new host.

Screwworms and cattle grubs occasionally are found in horses. Like the stomach bot, the screwworm and the cattle grub are parasitic fly larvae rather than worms. The screwworm has been nearly eradicated from the U.S. with only an occasional outbreak. There are two types of screwworm and both are introduced to the horse as eggs laid on an open wound by a fly. The larvae of the primary screwworm *Callitroga hominivorax* dig into the tissue, enlarging the wound. Migration of the larvae, causing great tissue destruction, can sometimes result in the death of the horse. The secondary screwworm, *C. macellaris*, is more a scavenger fly, eating dead tissue, and is not so destructive.

Once in a while cattle grubs (*Hypoderma* spp.) will develop in horses instead of cattle even though cattle are the normal host. They are not very common in horses, and rarely do much damage. The late larval stages can sometimes be found in the backs of horses, in cysts or warbles. The lumps may cause some irritation if a saddle is used on the horse, and may detract from the horse's appearance. The grub comes on out when it is mature and ready to become a fly. The fully developed larvae are about one-third inch thick and about one and a quarter inch long. They perforate the skin, wiggle out and fall to the ground where they pupate for about five weeks and then become flies. The adult flies are active during the hottest part of the summer, laying eggs on the hairs of cattle (and occasionally on horses). These hatch in about four days and the emerging larvae burrow through the skin and migrate through the tissues toward the animal's back, the migration taking most of the fall and winter. In early spring they come to rest under the skin of the back where they develop a breathing hole to the outside and continue to grow until they are ready to wriggle out (about a month later). The area heals with very little scar tissue if no secondary infection has occurred.

There are a great variety of internal parasites that can infest the horse. Most of these are nematodes (roundworms or threadworms, a class of worms belonging to the phylum Nemathelminthes). Some don't seem to cause much problem, while others can do great damage, causing the death of the host. The most common harmful effects of internal parasites in horses are tissue destruction, irritation of tissues, loss of blood, robbing of nutrients, mechanical blockage and loss of function of specific areas of the body. Secondary bacterial infections are always another possible undesirable side effect of internal parasites.

The most common parasites occurring in horses of all ages and in most geographic areas are strongyles and stomach bots. The most common parasite in young horses is the ascarid. Some geographic areas may be more highly affected with some of the other parasites than other areas. The types of internal parasites, their life cycles and their susceptibility to various drugs can vary greatly from one area to another.

A regular worming program for your horses can help prevent serious parasitism problems and keep your horses healthier. A horse that is kept under clean and uncrowded conditions can be kept relatively free from internal parasites if he is wormed three or four times a year. Worming in spring and fall should control bots as well as strongyles and bloodworms. One of these wormings should come after the first killing frosts in the fall. Wormings given between the fall and spring wormings should be aimed mainly at strongyles and roundworms; worming for bots all year round is usually a waste of money. If you are planning to breed your horse or to get that horse in good condition for showing, training, or any type of riding that requires him to put forth extra energy or effort, you should begin your conditioning program by worming him. He'll be able to utilize his feed better and get in shape faster. Your vet can advise you on a specific worming program. He knows what time of year in your area to eradicate stomach bots—when they are all in the stomach but not yet fully developed. He can tell you which of the three kinds of bots is in your area, where they lay their eggs on your horse, and how you can remove the eggs. He can tell you which drugs work best on the worms in your area, and which drugs they have become immune to. He can examine fecal samples of your horse under a microscope to see what parasites your horse has and to determine which drugs would be most effective against them.

There are three ways you can worm your horse. The most reliable and efficient way is to use a stomach tube; the tube is passed through the horse's nostril to the back of the throat (where he swallows it), down the esophagus and into the stomach. The worming preparation, in liquid form, is poured down this tube. The horse receives the exact dosage for his size, age and condition. Your vet can do this for you; putting the tube down into the proper place is a little tricky unless you know how. We use the stomach tube quite often on sick baby calves as an aid in combating dehydration from diarrhea, feeding the calf extra fluids and medication. But we do not like to use the stomach tube on our horses. Tubing is undoubtedly the most effective way to worm a horse, but it can also be a very trying ordeal with a high-strung, sensitive horse. Some horses are more easily tubed than others and most of ours we prefer to worm by feeding a medication in their grain. The only ones we tube are the few who do not readily eat worming preparations in their feed.

Boluses can be given by mouth, and powders or granules can be fed with grain. The preparations that can be put in the horse's feed work quite well on the parasites they are designed for, but the main problems here are that many horses resent the strange taste and smell of the preparation and eat it so slowly that the proper concentration of medication in the stomach is never reached, or they refuse to eat it at all. With some horses this problem can be gotten around by feeding the horse molasses in his grain for several days, increasing the amount of it until the horse eats his grain readily even when fairly saturated with molasses. Then when you mix the worming preparation with his grain he will not be so apt to smell or taste it because it is disguised by the strong smell and taste of the molasses. Some preparations give off less odor when chilled; we kept some in the freezer prior to feeding in the grain and it tends to cut down the offensive odor if fed immediately after it is taken out of the freezer.

Some preparations are not so offensive to the horse, but they are also milder; they do not kill all types of parasites. You may want to use a combination of drugs (at different intervals) that your horse will eat, in order to get all the parasites. Your vet can advise you and help you with worming problems. Another method of worming is to feed a constant low-level preparation in the grain; this is a preparation that must be fed daily. This method usually works best as a follow up after a horse is initially wormed by tubing or with some other very effective medication. Tubing, waiting 10 days, then feeding a daily low-level wormer for 21 days, can be very effective in breaking the worm cycle. The problems here are that some horses will refuse the medication, and the danger that the parasites will become immune to it. You may find that one of these worming methods, or a combination of several, provide the best worming program for you. But whatever method you use, do it often enough to be effective in keeping the parasites under control; you are just wasting your time if you worm your horse inadequately and he becomes reinfested, carrying a heavy population of worms for more of the year than not. Be careful when using worming preparations; always read the directions and use the medication accordingly. These preparations are deadly; after all, they are meant to kill worms. They are designed to be used in a dosage that will be deadly for worms but not significantly harmful to your horse. All of these preparations can produce toxic reactions in your horse if used incorrectly, so read the label, use the proper dosage, and administer as directed. Your horse will be healthier for it. And your feed bill may go down a little when you say goodby to those worms!

12

EXTERNAL PARASITES

THERE ARE SEVERAL KINDS OF INSECTS THAT PARASITIZE THE horse externally, feeding upon the horse and causing irritation and discomfort, loss of body condition, and sometimes diseases. Flies are the most common and troublesome external parasites of horses; during the warmer months of the year no area seems to be free of these pests. We'll take a closer look at some of the various types of flies later on in this chapter.

Lice are sometimes a problem on horses. Lice are tiny insects that spend their entire life cycle on their host, although some species can survive for up to two weeks away from the host. The eggs, known as nits, are laid on the hairs next to the skin and there are three nymphal stages before mature lice emerge to feed upon the horse's skin. The life cycle of most species varies from 2 to 4 weeks under good conditions. Lice are transferred from one animal to another by direct contact but objects that come in contact with an infested animal can also carry lice to another horse. Blankets, grooming tools, harness, saddle blankets and so on may remain infective for several days after being used on an infested horse.

There are two kinds of lice that infest horses, and these are *Hematopinus asini* (sucking louse) and *Damalinia* (*Bovicola*) *equi* (biting louse). Sucking lice feed upon the horse's blood, and in case of severe infestations they can actually cause serious anemia from the blood loss. Biting lice cause skin irritation and itching; as a result the horse rubs and bites at the affected areas. Large patches of skin may become hairless and raw. Nits and lice are very small, but they can be seen on the affected areas with close observation.

Lice are most active in winter, and multiply most rapidly on animals that are in poor condition. Some animals seem more susceptible to heavy infestations than others. Lice rarely cause much problem during warmer months of the year. Affected animals can be treated with insecticides—with sprays, dips or dusts. Lice can develop resistance to insecticides; the same type of product may not always be effective in ridding your horses of lice. When using insecticides, follow the directions on the label closely and treat the horse in an open area where there will be no danger of contaminating feed or feeding areas.

Ticks are occasionally found on horses, especially if they pasture in brushy areas. Ticks are a dangerous parasite because they can carry several diseases as well as causing death from anemia in severe infestations. Once an area becomes infested with ticks, they usually become an increasingly serious problem, so all ticks found on your horses should be destroyed.

Many ticks are active blood suckers. Some species cause tick paralysis. As with louse infestations, heavy tick infestations irritate the host animal enough to interfere with grazing and feeding, and may lead to loss of condition.

The life cycle of the tick may vary. Some species spend their whole life on one host while other species pass different stages of their life cycle on successive hosts. Other species are parasitic only during certain stages. The eggs are laid in the soil and the larvae attach themselves to a passing host on which they may develop through one or more nymphal stages before becoming mature. Adult females suck blood or lymph and then drop to the ground to lay their eggs.

LOUSE (GREATLY ENLARGED)

TICK—**very flat, with a tough shell**

ENGORGED TICK **(its head embedded in the skin)—much larger than the empty one; soft, fat, and gray**

MITE

Ticks are the main vectors (carriers and spreaders) of many protozoan diseases, the protozoa in most cases surviving from one generation of ticks to successive generations by infecting their eggs. One protozoan disease of horses that ticks can spread is equine babesiasis (also called red water fever and piroplasmosis; see piroplasmosis in chapter 9). Some ticks also spread diseases caused by bacteria, viruses and rickettsia. One example of a viral disease of horses spread by ticks is encephalomyelitis (sleeping sickness); ticks can be vectors of the western-type virus.

When only occasional ticks are seen, they can be squirted with a tick bomb (those made for dogs are fine to use), or dabbed with some screwworm smear. Then they will die and fall off eventually. Where there are heavy infestations, the horses should be sprayed or dipped with an insecticide solution, preferably one that the ticks in the area have not already become immune to. Toxaphene or lindane are usually good insecticides for eradicating ticks; so are chlorinated hydrocarbons (such as DDT) or organic phosphates (such as Malathion).

One species, *Otobius megnini,* is hard to get rid of, for the nymphs drop off to moult and lay eggs in protected spots, necessitating the spraying of buildings, fence posts, feed troughs, tree trunks and so on, when heavy infestations occur. Other types are also difficult to control because the nymphs and adults attach to feed for only short periods. Complete eradication of ticks is often difficult because many kinds of ticks can live on several hosts, including wild animals, and adult ticks are capable of living for long periods of time apart from a host.

Ear ticks are very irritating to horses because they live deep down inside the ear, causing discomfort. Horses with ear ticks may become quite ear shy and head shy and hard to bridle because they resist any handling of the ears. They may shake their heads, rub their ears, or cock the head to one side.

Ear ticks can be killed by putting a mixture of mineral oil and lindane into the ear. There are some commercial mixtures available for this purpose also. The horse may have to be restrained and twitched before he will stand still for you to put anything into his ears, for the ticks have made them very sensitive. If the ticks have been present in the ear for some time, there may be a great deal of debris and secondary infection in the ear, requiring further medical attention.

Mites are another type of parasite that can bother horses, causing dermatitis (inflammation of the skin). Trombidiform mites are usually harvest or grain mites, primarily attacking harvested grain and infesting animals only secondarily. These mites can also transmit diseases. The natural hosts for trombidiform mites are small rodents. Larvae drop off the rodents and develop into a nonparasitic stage which live in grain and hay. These larvae become most active in the fall at harvest time and may cause dermatitis in animals grazing in pastures or being fed newly harvested grain. Horses are usually affected around the face and lips (in white-faced horses the problem may be mistaken for photosensitization); affected areas are itchy and scaly.

Another type of mite, *Demodex equi*, infests the hair follicles of horses, causing demodectic mange (also called follicular mange). It occurs most commonly in late winter and spring, especially among animals in poor condition. Invasion of the hair follicles and oil glands by the mites leads to chronic inflammation, loss of hair, and sometimes the development of secondary infection (Staphylococcal pustules or small abscesses). Because of its deepness in the tissues, demodectic mange does not cause as much itching as some other kinds of dermatitis; the most important symptom is the appearance of small nodules which are about one-eighth of an inch in diameter, feeling to the touch like small particles of birdshot in the horse's hide. In severe cases there may be a loss of hair and thickening of the skin. In horses the area around the face and eyes are most commonly affected. This problem is sometimes confused with ringworm. The life cycle of this type of mite is completed on the host. Transmission from one horse to another can occur with the use of grooming tools and other equipment that is used on several horses.

Sarcoptic mange, also known as barn itch and red mange, is caused by the mite *Sarcoptes scabiei.* This type of mange is a severe, itching dermatitis and is difficult to treat. Affected animals are constantly itchy and are often too restless to graze, losing weight and vitality. Horses in poor condition are most susceptible, as are horses that are overcrowded. The disease is most active during cold, wet weather. The mite eggs are laid in tunnels in the skin made by adult mites. The hatching nymphs burrow in the skin and mature in about 17 days. Mite infestation can be spread by direct contact from horse to horse or by materials such as bedding, saddle pads, grooming tools, clothing, etc. In moist, protected sites the adult mites can live as long as three weeks away from the host.

The affected areas of skin contain small red elevated areas and general redness. The area is intensely itchy and may become injured from the horse's constant scratching and biting. As the disease progresses, hair is lost and thick brown scabs form over the raw surface, with a thickening and wrinkling of the surrounding skin. The most commom site on horses is the neck.

Psoroptic mange is caused by *Psoroptes communis equi* in horses and this type of mite tends to migrate to all parts of the skin, preferring areas that are covered with hair. The adult mites puncture the skin to feed on lymph and cause local irritation and itchiness. The skin "weeps" and forms crusts which tend to spread, as the mites are most active around the edges of the crusts. In horses the large thick crusts are found on those parts of the body with long hair, such as at the base of the mane and the root of the tail, and on hairless areas such as the udder.

Chiroptic mange is also called leg mange, and is the most common form of mange in both cattle and horses. It can be a source of great annoyance to horses, causing inefficiency and nervousness. The mites that cause it (*Chorioptes equi* on the horse) are most active in winter. Transmission is by direct contact or by infested bedding or equipment. In horses the parasites occur almost entirely in the long hair on the lower parts of the legs, and are not often found on other parts of the body. These mites cause severe irritation and itchiness, the first sign usually being violent stamping of the feet and the rubbing of the back of the hind pasterns on wire, poles, stumps or anything else that is available. The horse may resent the handling of his feet. In cases that have continued for long periods of time the skin may be swollen, scabby, cracked and exceptionally greasy. Scrapings from the affected area usually contain large numbers of mites. The disease may be related to greasy heel, except that pain is more common in greasy heel, and itchiness the major symptom in leg mange.

Any itchy rashes on a horse should be examined by your veterinarian. He can usually differentiate between the various skin diseases by close examination and by taking skin scrapings to observe under the microscope. There are several types of sprays and dips that can be applied to eliminate the mites that cause mange, and your vet can advise you on an effective eradication program after diagnosing the skin problem, if mites are the offenders. Affected horses should be quarantined so that the problem will not spread.

Now let's take a look at the various types of flies and mosquitoes that parasitize horses. Flies are a major problem during warmer months of the year, because of the irritation and worry they cause, and because of the diseases they can spread.

The stable fly, *Stomoxyx calcitrans,* occurs in most countries of the world. This fly is about the size of a housefly and is often called the biting housefly. It is gray in color and can be distinguished from the housefly by the seven dark spots on its gray abdomen and by the long slender proboscis that sticks up bayonetlike in front of the head. The stable fly usually sits with its head upward. Adult stable flies commonly rest on vertical surfaces such as fences, walls, trees, or any structures that are near to livestock. They go to the horse only long enough to get a blood meal, usually feeding on the legs and lower parts of the horse's body. Bites from this fly are quite painful, causing the horse to stamp its feet and lick at the wounds. It takes only five to ten minutes for a stable fly to drink its fill of blood, but it may puncture the skin several times in the process. These bites often bleed freely when fresh. Both the male and female stable fly take a blood meal two or three times a day.

The stable fly lays its eggs in rotting hay and straw (especially when contaminated by urine) and in horse manure. A complete life cycle takes about 30 days, the larval and pupal stages taking place in the rotting organic matter. The eggs hatch in two days, the larvae complete their development in about ten days, and the pupal stage takes four to five days. After the adult fly emerges, the female may be ready to lay eggs in five or six days. Several generations can develop during a summer season. Warm, moist conditions are favorable to multiplication of these flies and in bad seasons the large numbers of stable flies may cause anemia (from loss of blood), worry and loss of grazing time. In some cases of extremely heavy infestations, deaths may occur. In some areas stable fly infestations may exceed 25 flies per animal. This does not seem serious at first, until a person realizes that the number of flies observed on a horse at any given time may be only about 2 to 3 percent of the total number of

Standing under a shade tree fighting flies or just out in the pasture swatting, stamping, and head tossing, horses suffer a great deal of torment and discomfort from external parasites

flies in the area that could potentially feed on that horse during the day.

Stable flies spread a number of diseases, including anthrax, and equine infectious anemia. This fly acts only as a mechanical vector of these diseases. The stable fly also acts as the intermediate host for *Habronema microstoma* (a stomach worm of horses; see chapter 11, Internal Parasites). In northern climates the stable flies spend the winter in the larval or pupal stage. In warmer climates the length of each development stage may increase during winter months, but on warmer days flies emerge and breeding is generally continuous off and on throughout the year. Control of this fly involves removal of rotting organic matter which is used as breeding sites. Repellents are helpful in keeping the flies away from livestock, and the spraying of barns and stables and foliage where the flies rest can help reduce fly numbers.

Horse flies and deer flies belong to one of the most annoying groups of biting flies that attack horses and other livestock. These flies belong to the family Tabanidae. The horse flies belong to the genus *Tabanus* and the deer flies to *Chrysops, Hematopota,* and *Pangonia.* In most geographic areas these tabanids are the number one biting fly problem affecting horses. These flies may be black or brown, yellow or gray, varying in size from 3/8 inch to over an inch in length. They cause a great deal of worry to large animals; the broad bladelike mouth parts cut a deep painful wound. The blood flows freely from the bite and the fly laps up blood with a fleshy lobe at the tip of its proboscis.

Only female tabanids take a blood meal; the males feed on vegetable sap or juices of soft-bodied insects.

Tabanids are found in all parts of the U.S., especially where there are large areas of permanently wet ground. Females lay large masses of eggs on the leaves of plants growing near water or on other objects that project over water or moist ground. The eggs hatch in 5 to 7 days and the larvae drop into the moist soil where they burrow down and feed on organic matter or the juices of other insect larvae or earthworms. They stay in the larvae stage during summer, fall, and winter, migrating in the spring to drier areas of soil where they pupate. This pupal stage lasts two or three weeks; then the adult fly emerges. Most species complete just one generation each year.

Soon after emergence, the female fly needs a blood meal. Deer flies attack mainly around the head, neck and shoulders of the horse, while horse flies attack principally on the legs and abdomen. Because the bite is very painful, horses try to dislodge the flies with swishing tail or mouth or stamping feet. But this type of fly is very persistent and worries the animal until a blood meal has been gotten. Large numbers of flies may leave the horse's hair crusted with blood and there may be streams of blood coming from the bites. Horse flies can drink more than their own weight in blood at each feeding. Daily blood loss can range from three to ten ounces per horse, not counting the blood that flows from the bite after the fly has fed. Some horses will run frantically when attacked by horse flies. The tabanids are often mechanical vectors of diseases, due to their habit of feeding on one animal and then immediately attacking another. Control of these flies is difficult unless wet swampy areas can be drained or horses kept away from these areas when the flies are most active. Some kinds of fly repellent are helpful.

The horn fly, *Lyperosia exigua*, is one of the most plentiful and widespread of the biting flies affecting horses and cattle throughout the U.S. and Canada. Although they are primarily parasites of cattle and cannot survive apart from cattle, they can be a nuisance to horses and man in bad seasons. When horses are used in working cattle, or when they are pastured near cattle, horn flies may be a major source of irritation, especially when infestations on the cattle are heavy. Often when the cattle are being treated with insecticides, the horn flies will leave the cattle and go to the nearby horses. These flies are blood-suckers but are not known to transmit diseases. Their primary fault is the worry they cause, interfering with an animal's grazing. Heavy infestations can cause serious loss of condition and deaths in cattle, but horses usually attract fewer flies and extensive blood loss from the horn fly is not common.

Adult horn flies are only half the size of the common housefly and are commonly found on the withers, back, neck, shoulders, underline and around the eyes of the host animal. When disturbed by a swish of the tail or a toss of the head, black clouds of flies swarm up (on heavily infested cattle) to immediately settle back down again. Horses are more sensitive to these flies and even as few as 20 of them may be undesirable. Horses may develop sores around the eyes and under the belly, providing sites for infestation of the larvae of the stomach worm, *Habronema megastoma* (these infestations are called summer sores; see chapter 11). Infestation of skin wounds of horses by these larvae causes the formation of extensive granulation tissue, and invasion of the conjunctivae in the eye may lead to development of a granular conjunctivitis.

Adult flies stay on the host most of the time, leaving it only to lay eggs in fresh cattle manure. The eggs hatch within 24 hours and burrow into the manure pat, completing their development within four or five days. Then they crawl to a drier part of the pat or into the soil beneath to pupate, taking about five days. Then the adults emerge and seek a host, laying eggs about four days later. Low temperatures or lack of moisture inhibit the growth of these flies. In cold climates the flies spend the winter as hibernating pupae. Cattle feces has the proper moisture content for the development of the larvae, and although mature flies can live equally well on cattle or horses, the eggs are generally laid in cattle manure. Horn flies may be carried long distances by strong winds, but are generally spread by the movement of cattle. Insecticides applied to the host animal are very effective in getting rid of the horn flies since these flies spend almost all their time upon the host.

Hippobosca equina (the horse louse fly) is a common fly and a parasite of horses and cattle in most countries with a warm climate. This is a flat, glossy, reddish brown fly, a little larger than a housefly. It is a blood sucker and spends most of its time on the host, especially between the hind legs. Eggs are laid in dry soil where they develop and mature into adult flies.

Biting midges are members of the family *Ceratopogonidae* and of the genera *Culicoides*. These tiny flies are bloodsuckers, causing irritation and annoyance, and can transmit infectious diseases. They also act as intermediate hosts for filarid worms of the *Onchocerca* genera (see chapter 11). Hypersensitivity to the bites of this midge is the cause of allergic dermatitis (Queensland itch) in horses in Australia. The midges are plentiful in warm months and most active at dusk and in the early morning.

Black flies, buffalo gnats and sandflies are small grey to black flies that belong to the family *Simulidae*. In North America the important ones seem to be *Cnephia pecuarum* in the southern states of the U.S. and *Simulium arcticum* in northern Canada. These very small flies live in most parts of the world, and with the exception of *S. arcticum* they are a problem only in warm climates. Large numbers of those flies are present after periods of flooding or in areas with many streams; the larval stages are passed in flowing streams. The flies congregate in swarms and attack all animals, biting mainly around the legs, on the belly and around the head. The irritation is so great that many animals stampede or mill about nervously, often injuring one another or trampling young ones. These flies transmit filarid worms (*Onchocerca spp.*)

Mosquitoes can be a serious nuisance to horses, especially in swampy areas where permanent pools of water or frequent flooding create breeding sites. The most annoying species of mosquito belong to the genera *Aedes, Psorophora,* and *Culex*. In most types of mosquito, the female must drink a blood meal before she can lay eggs. The males feed on plant juices and do not bite animals. Eggs are laid on the surface of water or on the ground in low places that will be flooded. Several broods may hatch each season. Strong winds can carry mosquitoes many miles. Some species of *Culex* breed in ponds of water that collect in irrigated pastures and fields. Several species of *Aedes*, the snow mosquitoes, appear in great swarms in the high mountain and northern regions, causing annoyance to livestock in the early spring. These mosquitoes spend the winter in the egg stage and begin developing in pools of water when the snow melts in the spring.

The mosquito plays a major role in the spread of disease. *Culex tarsalis, Aedes dorsalis*, and *A. nigromaculis* transmit equine encephalomyelitis. The filarid worm *Setaria digitata* is also spread by mosquitoes.

Mosquitoes can be controlled by elimination of breeding sites (filling in or draining swampy areas), destroying mosquito larvae by treating breeding grounds with the proper larvicides, and chemically destroying adult mosquitoes. Some insecticides will kill adult mosquitoes on contact, but there is no material that can be applied directly to horses that will give long-lasting relief from mosquito attack. Some preparations can be applied daily and give partial protection. Animals should be covered thoroughly with the spray, but if oil solutions are used, the skin should not be wetted.

The face fly, *Musca autumnalis,* is a small fly that has become a problem in the U.S. only recently. This fly (indigenous to Europe and Asia) first appeared in North America in 1952 and is now present over large areas. The face fly resembles the housefly, and congregates on the face of cattle, feeding on nasal and eye discharges and saliva. These flies also irritate horses. Several species of Muscidae act as intermediate hosts for *Thelazia* worms which occasionally infest the conjunctival sacs and tear ducts of domestic animals.

Insecticides for the controlling of external parasites can be purchased in a variety of forms, the most common being dusts, wettable powders, emulsifiable concentrates, and oil solutions. Always be certain to use only the insecticide preparations that are specifically designed for use on livestock; other types may be quite harmful to the horse.

Dusts are applied directly to animals in dry form, sprinkled on and then rubbed or brushed into the hair. Wettable powders contain a dispersing and wetting agent which allows them to be diluted with water and used as sprays. The particles of insecticide are then suspended in the water and must be kept agitated to keep from settling out. Some wettable powders can also be applied dry, but many can't, so be sure to read the instructions before using.

Emulsifiable concentrates are probably the most common type of preparation. These are solutions of insecticides in petroleum oils or other solvents, with an emulsifier added so that the solution will mix with water. Occasionally this kind of preparation will separate into its various parts after standing in storage for a long time, and if this happens it should not be used.

Oil solutions are insecticides dissolved in oil, usually ready to use as packaged; these are commonly applied as sprays or low-volume mist sprays for livestock.

If insecticides are used improperly they may be harmful to man and to livestock. They should be handled with care and used only when needed, and kept in closed, well labeled containers in a dry place. Empty containers should be disposed of and not used for anything else. Do not breathe the dusts or mists. Don't spill any on your skin or clothing; wash it immediately if you do. Apply insecticide on a still day so that air currents will not carry particles to other areas.

Insecticides can be applied to horses as sprays, dusts or wiped on. Because most horses are frightened and excited by the sound of spraying, dusts or wiped-on preparations are usually preferred. When sprays are used, it is best to use the wettable powders, because the oils and solvents in other types of sprays may cause hair damage or skin irritation. If the horse sweats much after the insecticide is applied (as he would do if ridden on a warm day), the effects of the insecticide will not last very long.

In a control program, you should be familiar with the life cycle and habits of the insect you are trying to eradicate, so that you can attack it at a stage when it is most susceptible to the control measure you are using. In areas where a certain pest will be a problem throughout the summer, the best control can be obtained if you make an attempt to eradicate the pest early in the season before those flies are noticeable rather than waiting until they have become annoying and widespread. This is especially true when trying to control stable flies. If you are using insecticides, repeat the treatment before the flies become numerous again. This way there won't be much chance for a large breeding population to develop. And remember, when using insecticides, always read the label and use according to directions, taking special note of precautions and warnings. With proper care, you can help to keep your horses' external parasite problems to a minimum.

13

FEEDING HORSES

FEEDING IS PROBABLY THE MOST IMPORTANT ASPECT INVOLVED in caring for a horse. What he eats—the quantity and quality of his feed—is what determines how well he will perform, and to a large extent, how healthy he will be. And if he is still young and growing, the feed he eats will make all the difference in how well he develops. A horse inherits certain potentials, certain tendencies toward height and size, form and appearance, and inherits potentials of ability (for speed, strength, endurance, agility, and so on). But how well these latent potentials develop depends a great deal upon the feed the horse eats, as he is growing and maturing, and after he is mature. A horse with good breeding may still develop poorly if fed inadequately. He may have potentials for size and strength but end up weedy and stunted as the result of improper feed. His bones may have had potentials for being straight and good, but he may end up with rickets, splayed feet, splints or other problems because of inadequate nutrition. And a well-developed horse with potentially great ability may perform poorly on insufficient or improper feeds, lacking proper energy to perform as he should, or suffering such metabolic problems as founder or azoturia.

The horse is unique among large domestic animals in that he is used for his athletic ability. He is kept, raised and trained to perform such feats as carrying a rider, traveling long distances, performing fast starts, stops and turns, jumping, and so on, until he falters in old age. He must be fed for a long life of usefulness, fed to stay healthy and fit for many years. Other large animals are fed for maximum production of meat, milk, or wool. A beef animal is fed for maximum pounds in a minimum time. The horse should be fed for energy instead, and not fed excessively to the point where he becomes fat. A horse should not carry too much extra flesh or his health will be impaired and he won't be able to perform as he should.

The horse is a roughage eater, but unlike the cow, he has but a simple stomach and only chews his food once. He also has a smaller stomach than the cow. Because of the differences in their digestive systems, the eating habits of the cow and the horse differ greatly. The horse eats small amounts, but eats more or less continuously throughout the day or night, stopping to rest or to switch flies in the shade, or to dream awhile, then going back to graze for a bit again. The cow, on the other hand, can eat more at one time, and eat it more quickly. Then she finds a nice place to lie down, and rechews her hurriedly eaten meal at her leisure.

The horse is a roughage consumer by nature, but this fact has been overemphasized so much that he is often expected to eat grass or hay of poor quality. The horse is actually less efficient in his digestion of roughages than a cow or a sheep. This is because digestion takes place faster in the horse; he has only a simple stomach, does not chew his cud or reprocess his food. The feed passes through his system more quickly than feed passes through the digestive tract of a ruminant, and some types of feed are less fully utilized. Coarse feeds are often not broken down enough and can lead to digestive disorders and colic (see chapter 14).

The horse has only a simple stomach, like man. The reason he can digest grasses and hay and man can't is because he has a special type of large intestine. The roughages a horse eats are not thoroughly processed until they pass through the caecum and large intestine.

The horse needs carbohydrates, fats, proteins, vitamins, minerals, and water. On a natural diet of grasses that are grown in fertile soils, he gets all his nutrient requirements if he also has access to water. But today many of our feeds are not grown on fertile soils and many of our horses eat unnatural diets—hay or pellets part or most of the year instead of grasses, and concentrates such as grains. Because the horse's diet is partly unnatural, the conscientious horseman must make sure that the diet is balanced and that the horse gets the proper amounts of the various things he needs. The nutrition in that feed, and the way the various nutrients are proportioned will make a difference in the horse's appearance and performance.

Let's take a closer look at the way the horse eats. As he grazes he uses his lips to pick and sort his feed—to push away any sticks or rocks, and to place the grass or hay into position to bite it. The horse bites off the grass (and can cut it very short; sheep and horses are the closest grazers and can both abuse and damage a pasture if allowed to overgraze it), while the cow, who has no upper teeth (just a hard palate), gathers the grass with her tongue and breaks it off with a motion of her head.

The horse uses his molars to chew the grass or hay into fine particles, mixing saliva with it to make it easier to chew and to swallow. The salivary glands in a horse's mouth produce a great deal of saliva to assist in his eating. This is one reason the horse drinks lots of water, usually 8 to 12 gallons a day, and more in hot weather when he sweats. The horse and cow both need quite a bit of water to aid in the digestion of their feed, because they eat so much roughage. Both will take several drinks while eating a goodly portion of hay. The cow generally drinks more than a horse of equal body size, because the cow's digestive tract contents must be more fluid; the cow's end product (feces) is always more soft and moist than that of the horse. The horse's feces, in comparison, is generally firm and fairly dry.

The digestive tract of the horse is a very long tube, consisting of esophagus,

The horse sorts the grasses with his lips and bites them off with his teeth

stomach, small intestine, caecum, large colon, small colon and rectum. Unlike man, a horse cannot breathe through his mouth very easily, nor can he readily cough up or vomit material from his esophagus or stomach. If a horse tries to vomit, the feed is generally forced out through his nose (he will snort or sneeze), for it cannot be ejected easily through the mouth. Material coming back up into the throat region may go into the windpipe, causing choking, or pneumonia if the foreign particles get into the lungs where they will cause irritation. A horse's stomach may rupture from the pressure created by trying to vomit. Because of the horse's inability to vomit to unload his stomach of bad feed, the horse should always be fed the best quality feeds. Spoiled or moldy feeds that might lead to digestive upsets can be disastrous.

The horse's stomach is small, holding only about three to five gallons. This is one reason that a hay-fed horse should be fed several times a day. The daily ration should be divided into two (and preferably three) parts and fed at intervals, the largest portion being fed in the evening so that the horse can work at it at his leisure during the night. A cow can be fed only once daily (as most ranchers do when feeding hay to their beef cows during the winter), but if a horse is only fed once a day, much of it will go through the stomach undigested because the feed portion is too large for the horse's small stomach to handle at one time. This is quite a waste of valuable feed, to say nothing of the fact that it does the horse very little good. Many horsemen used to think that watering a horse during or directly after feeding would wash feed out of the stomach and on into the intestines, causing loss of nutrients and the passing of undigested feed. But under natural conditions a horse waters periodically as he grazes. And when eating hay, the horse needs adequate saliva and water to moisten the dry feed and will drink several times before he is done eating it. Hurried feeding (too large a portion at once; more than the horse's stomach will adequately hold) is more to blame for the wasting of feed that goes on into the intestines undigested. The only time a horse should not be watered directly after eating is when he is fed whole grain. If he drinks water after eating whole grain the kernels will swell a great deal (a portion of the whole grain is not chewed because it gets in the hollows of the horse's teeth and is not ground up but swallowed whole) and may cause him discomfort. Whole grain is not a good type of grain to feed a horse because some of it is never chewed up, and passes through the digestive tract whole and undigested, wasted. If you examine the droppings of a horse fed whole grain, you will see quite a few grain kernels that have passed through intact.

The stomach begins the digestion processs, breaking down the feed a little and starting the digestion of proteins. The small intestine continues the digestion of proteins begun by the stomach and also breaks down starches and sugars, and to some extent the fats. The pancreas and liver supply digestive juices to the intestines. The portion of feed that is digestible is broken down, absorbed through the

Digestive Tract of the Horse

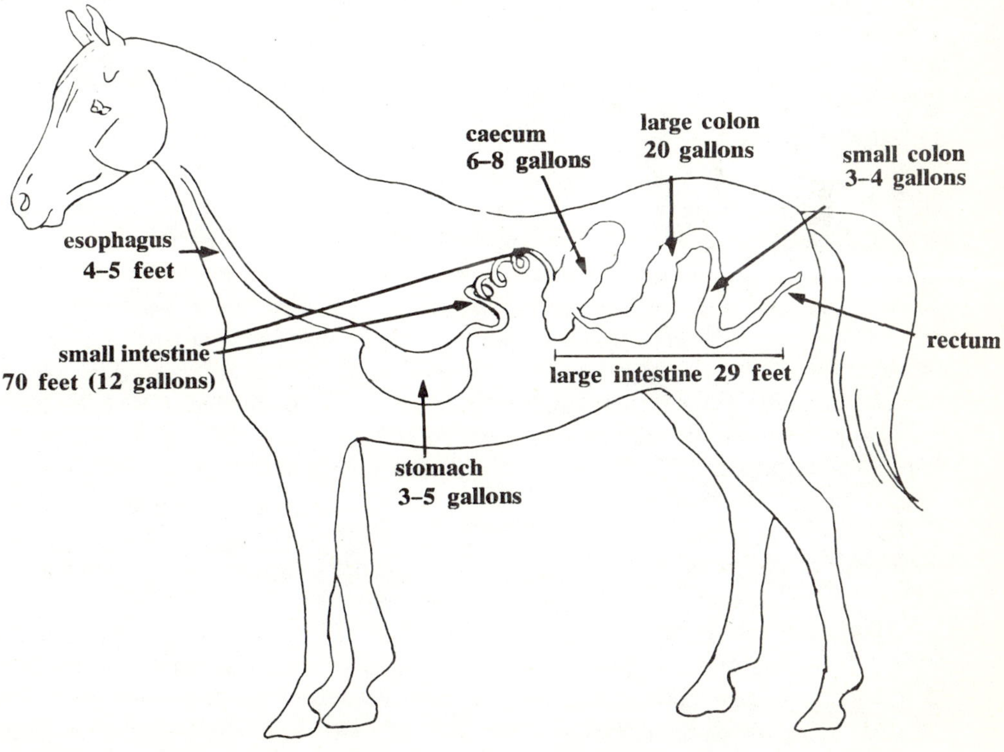

intestine and carried throughout the body by the bloodstream. The horse has no gall bladder (in other animals the gall bladder stores up bile to be released into the intestine for breaking down fats). In the horse, bile is secreted continually in the small intestine, but not in large enough amounts at once to thoroughly digest the fats. Therefore these must wait until they reach the large intestine.

The large intestine digests the roughage more thoroughly. Actually, no animal (by himself) can digest the cellulose, the fibrous part of roughage. The cellulose is broken down by helpful bacteria and other microorganisms that live in the horse's caecum and large colon. These microorganisms change the cellulose to energy-producing fatty acids, and also manufacture amino acids (from which the horse gets usable proteins) and vitamin B. These various B vitamins are essential, and would not be produced without the microorganisms. Digestion is finished in the large colon, then the remaining excess water is absorbed from the residue in the small colon, forming balls of feces, which are then passed out through the rectum and anus. The horse normally passes 30 to 50 pounds of manure daily. It takes a lot of roughage to provide the necessary nutrients of such a large herbivorous animal. Grains (considered to be concentrates) provide more nutrients per pound than hay or grass.

Carbohydrates, the major source of energy for the horse, can include sugars, starches and cellulose (the carbohydrates are derived from cellulose as it is broken down). Oats, barley, and corn may be as much as 60% sugar and starch. Cellulose (found in all roughage) is less easily digested and must be eaten in greater amounts to produce the same amounts of energy.

Horse rations contain very little actual fats, for the horse does not digest them well. The small amounts of fat eaten, however, do aid the horse in the absorption of fat-soluble vitamins. If a horse is fed too much carbohydrates or fats, they will not be used up as energy and will be stored as body fat.

Protein (made up of amino acids) is necessary to the life and well-being of the horse. Protein is necessary for growth, for the building and repairing of muscles and any other body tissue. Protein helps to build bones, blood, skin, hair, hoofs, and muscle. Lack of protein in a horse's feed will impair his growth if he is young and still growing, and will interfere with hoof development. A horse lacking protein will also have a poor hair coat. Horses need extra protein when they are growing, and a mare needs extra protein during the last third of pregnancy (when the fetus is making most of its growth) and while she is nursing a foal. Legume hay is a good source of protein, and so are soybean, linseed, and cottonseed meals. A little extra protein in a horse's ration in the spring of the year can help him shed his winter coat (as the new hair comes in) more quickly and easily.

Vitamins are organic compounds that are required by all animal bodies in small amounts to function normally. Vitamins help to regulate chemical reactions within the body, keeping the body operating smoothly and healthily. There are two kinds of vitamins, the fat-soluble vitamins (A, D, E, and K) and the water-soluble ones (B vitamins and vitamin C). All good green forages that a horse consumes in a good pasture are rich in most vitamins that the horse needs. A horse on good pasture will need no supplementation, no additional vitamins. This is why good pasture is so valuable for horses; it provides adequate body requirements in a natural and inexpensive form. But if there is no pasture, limited pasture, or poor pasture (grown on poor soils or containing poor types of grasses), the horse should be fed good quality hay. Vitamin supplementation has become a common practice among horsemen, especially for breeding stock and growing young stock and horses under heavy stress in training or performance, and it is a wise practice when the horses involved do not get an adequate amount of natural vitamins at pasture. But there can be dangers in oversupplementing a horse with vitamins, and vitamins can be costly if the animal does not really need them. Don't go overboard on the latest wonder feeds or secret formulas. Natural feeds of good quality (if available) are always best. Carefully analyze your situation—your pastures, your feeds, your horse's condition and use—and then decide just how much supplementation he does or does not need. Our ranch horses do very well (often traveling up to 50 or 60 miles in a hard day of range riding, or putting in a strenuous day sorting or moving cattle) on pasture or good hay alone with no supplementation of any kind except salt, and very little grain. Khamette (half-Arab mare) went straight from a couple months of range work to enter a competitive trail ride with no additional conditioning or feeding and placed second in her division as well as winning the trophy for best Idaho horse. Don't underrate that good pasture!

Let's take a quick look at the vitamins needed by the horse. Vitamin A is manufactured by the horse from the carotene in various plants. Green pasture is a very good source of carotene. Green, leafy alfalfa hay is another good source. Grass hay is not very high in this vitamin. Because vitamin A is fat soluble, it can be stored in the body. Horses on good pasture during the summer months usually store enough of this vitamin to last them through the winter. But if a horse is never on pasture, or is under a special stress during winter months (such as a pregnant mare) vitamin A should be given consideration when making up the horse's ration. Lack of vitamin A can cause abortion or the birth of weak or poorly developed foals. Lack of this vitamin can also make a horse prone to digestive upsets and colic, respiratory ailments with cough, colds or runny nose. Vitamin A is important to reproduction, to the health of the eyes, and the development of hoof growth.

Vitamin D is produced in the skin if the horse is out in the sunshine. This vitamin is usually not lacking if a horse is kept out at pasture. Sun-dried hay is also a good source of vitamin D. Under normal conditions a horse should not lack this vitamin. But if a horse is kept stabled most of the time he may need supplementation. Vitamin D plays an important role in the conversion of calcium and phosphorus into body tissues such as bones. Lack of vitamin D can cause rickets in growing horses, and osteomalacia in mature ones. Osteomalacia is a condition in which the bones become soft, with pain, tenderness, muscle weakness, lack of appetite, and loss of weight.

Vitamin E in horses is necessary for the development and proper functioning of muscle. This vitamin is normally adequate in natural feeds. But in periods of stress, you may wish to supplement your horse if natural feeds are not available.

Vitamin K is essential to blood clotting and is manufactured by the horse in his intestines. All vitamins are required in very small amounts, and the fat-soluble ones (A, D, E, and K) can be stored in the horse's body tissue. Feeding excessive amounts of any of these fat-soluble vitamins can be dangerous for your horse, resulting in throwing the horse's system out of balance. Too much vitamin D can result in excessive calcification of soft tissues.

The B vitamins are created in the horse's caecum and large colon by bacteria—thiamine, riboflavin, niacin, pyridoxine, pantothenic acid, biotin, choline, folic acid and B12. Vitamin C is also synthesized by the horse.

Minerals are inorganic elements (not produced by living things). They are required in small amounts for the health and normal function of the body. Under natural conditions, the horse gets the proper minerals he needs by eating plants that store them from the soil. The major minerals needed by the horse are calcium, phosphorus, magnesium, sodium, chlorine and potassium. Others needed in only minute (trace) quantities are iodine, cobalt, copper, iron, zinc and manganese. Iodized salt should be fed in geographic areas that are lacking in iodine, the "goiter states." Lack of iodine in a pregnant mare's diet can result in weak foals, dead foals, or foals born with contracted tendons. Other trace minerals, such as iron, are also lacking in some areas. If your area is deficient in some of these minerals, feed your horses mineralized salt that contains these trace minerals.

Under good conditions of production the horse will normally get the minerals he needs in a good diet of natural grasses and grains. But where feed is grown in deficient or depleted soils, the horse will need these additional minerals in his diet. Calcium and phosphorus make up the major ingredients of bones and teeth. Magnesium is also found in the bones. Lack of these minerals can impair the growth of the young horse, or cause unsoundnesses. Roughages are usually high in calcium and grains are high in phosphorus. A mature animal needs about twice as much calcium as phosphorus, while a growing horse needs the same amount of calcium as phosphorus. Vitamin D is essential in the proper relationship of these minerals; if there is not enough vitamin D, the calcium cannot be absorbed, throwing the calcium-phosphorus balance off. Too much grain in a young horse's diet can interfere with the proper absorption of calcium; too much phosphorus (in the grain) also throws the balance off and, because of the improper absorption of calcium, soft bones may result. This can lead to serious problems such as improper bone development, crooked legs and unsoundnesses. Never overfeed a horse, and never oversupplement with minerals. Salt is one exception; for it can be fed freely.

Sodium chloride is ordinary salt. A horse needs more salt than most animals because he sweats (cows do not sweat). In the summer, or when he is working hard, a horse may consume quite a bit of salt to replace that which he sweats out of his body. A horse should always have access to salt, especially in hot weather or when he is working hard. It can be given in block form or granulated. Our horses seem to prefer the blocks. A horse will not eat too much salt (unless he has been deprived of it for long periods of time, and even then he generally won't eat too much unless the salt is in loose form), he eats only the amount his body needs, so salt can be kept available to him at all times. Sodium chloride and potassium help regulate the water in the horse's body and assist the passage of nutrients into the body cells.

The feed a horse needs will vary a great deal, according to what the horse is doing. If he is idle, he needs less than if he is working. Maintaining an idle horse requires the least feed, for the idle horse requires only enough for bodily functions and for a limited amount of activity—grazing, getting up and down, moving about. A good maintenance ration for an idle horse is 2½ percent of the horse's body weight in good hay. Thus a 1000-pound horse would need about 25 pounds of hay daily. If the hay is poor quality, more pounds of hay must be fed, or a small portion of grain added to the ration. Idle horses often grow too fat on good pasture unless they have a good amount of exercise. An idle horse that is an easy keeper will also grow too fat on a limited amount of good hay and may need his ration cut down a little.

A growing horse needs an adequate ration for the growth needs of his body. The young horse gains about 45% of his mature body weight by the time he is 6 months old, 58% at a year, and 87% in two years. For proper bone growth and muscle development proteins, vitamins, and minerals are very important for the young growing horse. Good pasture and a grain ration are usually adequate. If the young horse is not on good pasture, alfalfa hay can be fed; alfalfa hay contains more protein, minerals and vitamins than grass hay. Calcium and phosphorus are usually adequately supplied by roughages and grains.

Young horses should always be fed well for proper growth, but never overfed.

The idle horse needs only enough feed to meet the needs of his bodily functions and a limited amount of activity

Fahleen at five months. The growing horse needs an adequate ration to meet the needs of his growing body. He gains about 45% of his mature weight by the time he is six months old

Bones, tendons, ligaments and joints are more easily injured when the horse is carrying excess weight, especially when these tissues are still young and growing.

Foals and young horses fed too much grain or too many vitamins and minerals can have their body chemistry upset by this imbalance, as mentioned earlier. Some nutritional diseases are caused by too much supplementation of vitamins and minerals. And too much grain can interfere with the proper absorption of calcium, causing soft bones (the phosphorus in the grain upsets the phosphorus-calcium ratio). Soft bones and excessive weight can cause permanent damage in the growing horse. Calcium deficiency in these young horses cannot be corrected by feeding calcium; they cannot absorb the calcium that is already passing through their systems until their grain ration is drastically reduced. Some horsemen today (especially in the stock horse breeds) try to push their foals and young horses to fast growth and extra pounds in a hurry. This is not natural or healthy. The horse is not a beef animal. Too many people today are in too big a hurry to get that young horse "grown up" to where they can ride or race him. A good horseman has time to let a foal grow up and mature properly before he starts to ride it.

On our ranch our young horses have always been fed grain, but the mainstay of their diets has been natural feeds, preferably the dry-land grasses (western blue-bunch wheat grass, Agropyron spicatum; Sandberg bluegrass, Poa secunda presl; and Idaho fescue, Festuca idahoensis) when possible, especially in the spring and summer when these grasses are green and growing, and

Four of our ranch mares on pasture

Fahleen as a weanling. Our young stock are fed grain during the winter

the best hay and pasture available when those young horses are kept in closer confinement. During the winter our young horses are always fed grain and good quality hay as well as winter pasture. When we stop to consider the fact that thousands of good horses have been raised in the past on nothing but natural feeds and the grains available, we are not so apt to rush out and buy the newest brand of special feed for young horses which the advertising claims will speed them on to better and faster growth. Please don't misunderstand me. Young horses should always be fed well and fed properly, but feeding can be overdone, to the detriment of that growing young horse.

The pregnant mare has nutritional requirements above those of the idle horse. Most mare owners recognize this fact, and feed a pregnant mare more, but they tend to overestimate the needs of the pregnant mare and underestimate her needs when she is nursing the foal. Pregnancy increases the feed requirements of a mare, but increases it significantly only during the last fourth of her gestation period. Too many horsemen seem to think a mare must be fed extra as soon as she becomes pregnant, and as a result mares are often overfed and too fat at foaling time. A mare's real need for extra feed is during that last quarter of pregnancy and after she foals and is producing milk.

If a foal at birth weighs about 100 pounds, about 80 pounds of that are just water, (a foal's body weight is 80% water), so that means that the mare

Scrappy, dam of Khamette. The pregnant mare needs more nutrition than the idle horse, but her needs don't increase significantly until the last part of her gestation period

Sansih and foal, Arabians owned by Jim Hopkins of Twentynine Palms, Calif. A mare's real need for extra feed is when she is nursing a foal. She needs three times as much nutrition as she would for body maintenance alone

puts only about 20 pounds of protein, vitamins, minerals, and energy into that foal in her 11 months of gestation. And since about 80% of the foal's growth takes place in the last four months, the mare's requirements for the first six or seven months of pregnancy are not much more than just maintenance. If the mare is too well fed and is excessively fat at foaling time, she may have trouble foaling, and she may also have difficulty settling when rebred.

A mare's real need for extra feed is when she is producing milk for her foal. To illustrate this, we can compare the figures using total digestible nutrients (TDN). Total digestible nutrients are that portion of the ration that is actually utilized by the horse, not including that portion of the roughage that passes on as waste. A 1000-pound mare in the last quarter of pregnancy (in which her feed requirements are increased greatly above maintenance requirements) needs about 7.3 pounds of TDN each day. The same 1000-pound mare will need 14.4 pounds of TDN daily when she is nursing her foal. That's about twice as much. A lactating mare needs nearly three times as much nutrition as she would need for body maintenance alone. The average 1000-pound mare gives 5 to 6 gallons of milk per day. A high-producing mare gives more milk than a dairy cow. This is quite an accomplishment and requires a lot of feed. Milk production is usually highest by the time the foal is three months old, then begins to taper off until the foal is weaned. The foal will grow and develop best if his mother is fed properly so that she will milk well. There is no substitute for that mother's milk. You can creep feed a foal, feeding him extra grain

Fahleen. Good foals can be raised on mother's milk and a bit of grain

A foal can be raised even on mother's milk alone if the mare is adequately fed. These mares and foals are on excellent pastures of crested wheat

and supplements, but you are actually better off to just feed that mare well so that she will produce the milk he needs. A foal can be fed grain to advantage while he is still nursing, but it is better not to overdo it. Feed that mare (a nursing mare should always be fed grain; a lactating mare may not be able to meet her requirements on roughages alone) and she will put the pounds on that foal. Of course, if a mare is not producing enough milk for some reason, the foal should be fed supplementary feeds, but the average mare will raise that foal properly if she herself is well fed. Mare's milk is adequate (regardless of certain foal feed advertisements to the contrary) if that mare is properly fed. All of our own foals have been very adequately raised on nothing but their mother's milk and a little additional ordinary grain. But the nursing mares were on good, fertile pasture and receiving a good grain ration themselves.

A stallion should be fed well, for his work during the breeding season is equivalent to being worked hard under saddle, as far as energy output is involved. Stallions used heavily during the breeding season may breed as many as 60 to 100 mares during a period of five months, breeding about six mares per week, or sometimes even as many as two or three a day for a very short time. A stallion used this hard will lose a great deal of weight, and he must be properly fed to be energetic and fertile. Excessive fat is detrimental to his fertility, but so is lack of adequate nutrition. The right amount of protein, vitamins and minerals are essential for the production and quality of sperm. It is usually a good idea to feed a stallion a good vitamin-mineral supplement during the winter and spring.

A working horse being used under saddle needs additional feeding, according to how much work he is doing. As his work (or ridden hours) increases, he should be fed more grain, for that will give him more energy. Feed requirements of individual horses may vary quite a bit, but in general a horse doing light work, two to three hours a day, requires about 50% more energy than one doing no work. A horse doing medium work, or being ridden 4 to 5 hours a day, uses 70% more energy than one that is not working. And a horse doing heavy work, 5 to 8 hours or more, will require even more energy. But these are general figures. They will vary according to the type and volume of work (a horse ridden 3 hours for pleasure will not need as much energy as a horse ridden 3 hours sorting cattle, galloping, stopping and turning), the condition of the horse, the time of year, and so on. Horses need more feed in winter than in summer, for they have to produce quite a bit of body heat as well as other functions.

The U.S. Cavalry had a good general feed rule for horses doing steady work. These horses were fed one pound of grain and 1.1 pounds of hay for every 100 pounds of body weight. Of course, on days the horses were not worked, the grain feeding was cut down considerably. Thus on a working day, a 1000 pound horse would receive ten pounds of grain (split into several feedings) and about eleven pounds of hay, the hay being fed primarily in the evening. With the average horse, the harder he is worked the more feed he needs, and he can handle the extra feed better in concentrate rather than bulk form.

The nursing mare who is pregnant again should be very well fed, for she has the hardest job of all

Sur Amir. The breeding stallion must be well fed to keep him fit and fertile

The horse doing light work (the pleasure horse, the horse in early training), or two to three hours a day, needs about 50% more nutrition than one doing no work

The horse doing medium work (about four to five hours a day)—in this instance the rope horse and ranch horse—needs about 70% more nutrition than one that is not working

A horse that is an easy keeper may stay in good condition on pasture alone even when ridden a great deal, but he will have more energy and endurance if fed grain. Our half-Arabian mare Khamette, stays fat on pasture alone and even on rather poor-quality feeds. But she goes much farther in a day if fed grain. We get the best performance from this mare when she is kept off pasture on a limited amount of hay (to keep her from getting too fat) and a fair portion of grain to give her the needed energy for strenuous rides. But during the summer of 1971 she did very well in a competitive trail ride, conditioned on pasture alone, with about a pound of grain on days she was ridden.

No two horses are alike. You have to discover the amount of feed that seems right for each. Some are gluttons, some are nervous and picky eaters. Some eat slowly, others eat too fast. Some need more feed than others. Some need more roughage, others more concentrate. Some will eat anything; others are more fussy.

Rations should always be changed gradually, over a period of several days or a week. Regularity of feeding is important, too. Abrupt changes in a horse's feed can cause digestive disturbances. Small amounts given regularly are far better than a huge amount given all at once. This is especially true of grains. Too much grain can founder a horse. And remember the horse's small stomach capacity; his system can utilize only a certain amount at one time and the rest passes on as waste.

When feeding grain it is best to split the ration into two portions and feed it twice daily. Horses that are being fed a large daily ration of concentrates (horses doing strenuous work, mares nursing foals, young stock, etc.) can be fed three times a day, so that none of the separate feeding portions is very large. The horse will get much more good from the feed that way. Grain should always be cleaned up within 30 minutes, or less. If a horse takes longer than this to eat his grain, he is being fed too much, or he has bad teeth. If a horse lacks appetite or takes a great deal of time to eat his feed, have your vet check his teeth; they may be hurting him.

Never overfeed a horse. A horse that is too fat cannot perform well and will tire easily. If he has a crammed stomach from eating too much, he will not feel like working. Fat horses are more prone to founder than are horses in good condition. Too much feed, especially when a horse is not working, can produce serious metabolic disorders, such as azoturia. In azoturia, large stores of glycogen have been laid down in the muscles during a period of idleness on full rations, and then when the horse does exercise, this glycogen is rapidly metabolized to form lactic acid. If lactic acid is produced faster than it can be removed in the blood stream, accumulation results, causing hard painful swellings in the large muscles. Severe cases may be fatal. To prevent such disorders, cut down the horse's grain ration considerably on days he is not ridden or exercised. Our ranch horses are kept on good pasture or hay and are only fed grain on the days we ride them.

Let's take a quick look at some of the various feeds for horses. Horse feeds can be classified into various categories based upon fiber content and food

A horse doing hard work—moving cows all day in steep country, sorting cattle, or conditioning for endurance riding or for any other event which takes a great deal of energy and stamina—requires a lot of good quality feed

value—the total digestible nutrients they supply. Concentrates are low in fiber content and high in TDN, (such as grains and grain by-products). Roughages are high in fiber (more than 18%) and are relatively low in TDN (such as hay, straw, silage, and pasture grasses). Protein content is also a factor to consider when categorizing a feed. Most grains are less than 16% protein. In the roughages, the legumes are usually higher in protein than the grasses, depending partly however, on the stage of maturity. Grass hay cut while it is still green and growing, may actually have a higher protein content than a legume (such as clover or alfalfa) cut too late in its growing season.

The horse needs digestible nutrients in his diet, but he also needs roughage for proper function of his digestive tract. This roughage comes in the form of extra bulk, or fiber—the cellulose; stemmy parts of hay and grasses and the hulls on grain. A bulky feed is one that has relatively little weight for its volume. A quart of oats weighs about one pound while a quart of corn weighs about one and three-fourths pounds; therefore oats are bulkier than corn. The corn weighs more for its volume. The more weight, the more energy available in the feed. Hence the name concentrates—they contain more TDN for their volume.

While we're on this subject, it might be good to mention that horses should be fed by weight, not by volume of feed. Not all grains weigh the same. Feeding by quarts or gallons can be confusing. As mentioned a moment ago, corn weighs more than oats, and a quart of corn may give the horse twice as much energy (or twice as much potential fat) as a quart of oats. Weigh your feed, find out how much your scoop or bucket really holds in terms of weight. And recheck it when changing feeds. Making a change in the horse's ration without adjusting the weight of the new feed may lead to a digestive problem or colic. As a general rule a quart of oats weighs about a pound, a quart of barley weighs a pound and a half, and a quart of shelled corn weighs about a pound and three-fourths.

Oats weigh about 32 to 40 pounds to the bushel, are about 70% TDN, and are 9.4% digestible protein. Oats contain about 11% crude fiber. Oats are an old standard horse feed, palatable and nutritious. They are also safe to feed, (the safest of all grains for horses), even in fairly large quantities. This is because they are about 30% hull (outer covering around the kernels), providing some bulk. Shelled corn, vegetable oil meals, and wheat (heavy concentrates without much bulk) can tend to form a doughy mass in the horse's stomach, which can cause serious problems such as colic, whereas oats form a loose mass in the stomach which can be easily digested. The bulky oat hulls help to keep this feed moving on through the digestive tract at a normal rate. Crimping or rolling the oats increases their food value because they can be more easily chewed and digested.

A bushel of barley weighs about 48 pounds and barley is about 78% digestible nutrients. It is about 10% digestible protein and only about 5.9% fiber. Barley is the most common grain fed to horses in Arabia and in many parts of Europe. Barley hulls, like oat hulls, provide some bulk to aid in digestion, but not as much. A portion of oats or wheat bran is often added to barley to increase the bulk. Barley kernels are very hard, and are best if rolled or steam rolled. No grains should be ground for use as horse feeds, for ground grains are too dusty (and can lead to respiratory irritations) and too fine, tending to dough up in the horse's stomach.

Corn, with no hulls, is high in energy content and low in bulk. Because it is rich in energy, it should be fed sparingly, or it can make the horse too fat. Corn is also very heating and therefore should be fed sparingly in the summer. It is low in protein, so if it is fed to young growing horses or pregnant mares, it should be supplemented with legume or some other feed that is high in protein. Corn weighs about 56 pounds to the bushel and is 80% digestible nutrients, but only about 6.7% digestible protein. Corn has only 2% crude fiber. It takes about 15% less corn than it does oats to keep horses in good condition, when the ration is properly balanced with some additional protein. Changes from oats to corn in a horse's diet should always be made gradually. Ear corn is safer to feed than shelled corn, for horses eat it more slowly and chew it better. Corn should never be ground finely (coarsely cracked is all right, however) because the fine meal forms a doughy mass in the stomach that is hard to digest. Corn and cob meal can be made by grinding up the ears, cob and all, which adds bulk. Molasses should be added to this meal to reduce the dustiness.

Wheat is usually not fed to horses because of its high cost, but in some wheat-growing areas it may sometimes be economical. It should be rolled or crushed to make it easier to chew and digest, fed only in moderate amounts (usually less than 20% of the concentrate part of a horse's ration), and mixed with a bulky concentrate such as oats or wheat bran to avoid colic. Wheat is a high energy feed that is also high in protein. It weighs about 60 pounds to the bushel, and contains about 80% digestible nutrients. Wheat is about 11% digestible protein (one of the highest grains in protein content), and only 2.6% crude fiber.

Wheat bran is a by-product of the milling of wheat, consisting of the coarse outer coating of the kernels of wheat. It can be mixed with concentrates for additional bulk, and for its laxative effect. Wheat bran can be added to the daily ration (about 10 to 15% of the ration is plenty) or given once a week as a bran mash. A bran mash can be made by mixing hot water with dry bran and letting it stand and soak up for half an hour before feeding it. Because of its laxative quality, many horsemen like to feed a bran mash to a mare after she foals, to help keep her from being constipated. The first few bowel movements after foaling are sometimes a bit difficult for the mare and she will have less problems if fed a laxative diet for a few days. Wheat bran contains about 67% digestible nutrients and about 10% crude fiber. Wheat bran often contains more than 13% digestible protein.

Sorghum grain, also called milo, is a good horse feed in areas where it is grown. This is a hard, round grain that should be rolled or crushed, and because it is lacking in bulk it should be mixed with bran, oats or barley or some other bulky feed when used as a horse feed, to avoid constipation. Milo weighs about 56 pounds to the bushel and is about 80% digestible nutrients and only 2.3% crude fiber. Milo contains about 8.5% digestible protein.

Molasses, a by-product of sugar refining, is a good source of carbohydrates for horses. This product is often mixed with grains to reduce dustiness and to increase the energy content of the ration. Molasses contains about 54% digestible nutrients. The proportion of molasses should not be more than 10%. Too much molasses will make a horse have loose bowels.

Linseed meal, soybean meal and cottonseed meal are by-products of the extraction of oil from seeds. These meals are high in protein and are often called "protein supplements." Linseed meal is 31% protein. This meal is produced from flax and has been a standard horse protein for many years, being useful for building up run-down horses and putting a bloom on a horse's coat of hair. Linseed meal is somewhat laxative and should never be fed in large quantities. One pound per day is usually plenty.

Soybean meal is a by-product of soybean oil, and contains 42% digestible protein. This is an excellent protein supplement for growing horses, and broodmares and stallions during the times of year that they need extra protein. It is often a cheap source of digestible protein. This is a very heavy, concentrated feed and you should feed less than a pound of it per day.

Cottonseed meal is 34% digestible protein, and when added to a ration that is low in protein, it can be a good cheap source of protein. This is a heavy feed and should not be fed in amounts exceeding a pound per day. If fed in excess it can cause digestive problems and toxic reactions. Horses do not like this feed very well and it should be mixed with better-liked (and more bulky) feeds such as oats, bran or corn-and-cob meal. When starting horses on cottonseed meal, feed only a small portion at first, mixed with other grain, and gradually increase it up to about a pound or less.

There are many commercial horse feeds today that are mixes of various grains, proteins, vitamins and minerals, often with molasses added. They are handy to feed, but often more expensive than mixing a ration yourself, especially if you have many horses to feed.

Now let's take a quick look at roughages—pasture, hay and silage. Straw should never be used as feed for horses. If cut as hay (while still green and growing, before the grain heads are mature) oat hay is palatable and can be used as part of a ration, but mature straw is too coarse for horses. Most other types of grain have rough "beards" and if cut for hay they are not suitable for horses because the rough pieces irritate the mouth and throat.

Mature straw has little feed value because it is coarse and not efficiently digested. Coarse feeds such as straw or poor quality hay can also lead to colic because the particles of coarse feed can sometimes partially or completely block the digestive tract.

Pasture is the most ideal feed for horses because it is usually cheap and, if grown in fertile soils, nutritious. Pasture is usually only half as expensive as hay, and only one-fourth as expensive as grain. Pasture is the natural environment for the horse for he gets sunshine and exercise.

If horses are kept on pasture, the pasture should be large enough to support them, taking into consideration that horses may beat out the grass in certain areas (such as near fences or watering places, under shade trees, and so on), will graze only their favorite types of grasses and leave others (also leaving grass that grows in the shaded areas or wet areas), and will not graze where they have defecated. The size of the pasture which will support a horse or group of horses depends on the type of grasses in it, the fertility of the soil, and the amount of moisture it gets. Some types of grasses must be irrigated in dry climates to get any production from them. It will take more dry-land area to support a horse than good irrigated pasture (there is more regrowth on the irrigated pasture), but some of the native dry-land grasses are excellent horse feed. Irrigated pasture grasses sometimes have too high a moisture content; horses get full before they've eaten enough volume to provide them with the proper nutrients. There is more food value in the dry-land grasses by volume. A good pasture is one that is large enough for the horses that graze it, carrying several types of good palatable grasses. A mixture of clover in with the grasses makes a good horse pasture, for clover is high in protein. White clover, red clover, sweet clover or alfalfa are all satisfactory forages for horses, if mixed with grasses. But if portions of the pasture go predominantly to clover, especially white clover, horses may eat too much and suffer from colic or founder.

The greatest single factor in determining the quality of pasture grasses and hay is its stage of maturity. The older and drier the grass, the less palatable it is, and the less food value it has. Mature grasses are coarse and not as easily digested as green and growing ones. Some of the hay grasses —timothy, smooth brome, intermediate wheatgrass, etc., are good horse feed

Horses are very selective grazers and will practically eat their favorite grasses into the ground (note grass at right) while leaving others untouched (note grass at left)

when pastured, if grown in fertile soils and grazed while young. Kentucky bluegrass also makes a good pasture if it is mixed with other palatable grasses or a suitable legume. Any of the pasture grasses or dry-land grasses (except downy brome, often called "cheat grass") are preferable to swamp grass, for swamp grass is very coarse. Pasture land should always have adequate drainage to avoid swampy soil, for not only does permanently damp soil grow inferior types of grasses, but also serves as breeding place for horse flies.

A good hay for horses provides bulk, energy, protein, minerals, carotene (which the horse converts to vitamin A) and several B vitamins. Hay should make up the major portion of a horse's diet if he is not on pasture, for hay is a cheaper source of energy and nutrition than grain. But hay must be of good quality to supply the nutrition that the horse needs. Hay should be cut before it is fully mature (just before the bloom stage or the heading out of the grasses)—while the grasses are still young, tender, and growing. Good hay is green in color (not yellow or brown), leafy, with small, fine stems. Coarse, stemmy hay is hard to digest. Clean, bright, sweet-smelling hay is most nutritious. Dry, dusty hay is not only unpalatable, but can also give a horse respiratory problems. Mold or mustiness can cause colic.

Legume hay should always be fed sparingly because horses like it and tend to overeat. Mixed hay, containing both grasses and legumes, is a good type of hay for horses because the legume adds extra protein. If you are feeding straight legume hay, don't feed more than one pound of alfalfa hay per hundred pounds of body weight; fill in with another type of roughage. Too much straight alfalfa can cause digestive problems, loose bowels and excessive sweating. Grass hay can be fed more freely.

If well cured (dry, and not damp and moldy, but not *too* dry or it will be sun bleached and have lost much of its vitamins and feed value) both grass and legume hay make good horse feeds. Grass hay is less likely to be dusty or moldy than alfalfa and other legume hay. Moldy hay is dangerous and should never be fed; it can fatally poison a horse and can cause pregnant mares to abort. Discard all mold, and try to avoid dusty hay. If hay is a little dusty, you can shake it to remove some of the dust and then sprinkle it with water to settle the rest; but sprinkled hay should be eaten right away —if it sits for any length of time wet it can begin to mold. Extremely dry hay is usually dusty; the leaves have shattered. It doesn't hurt the horse to eat dusty hay; the damage is done by his breathing the tiny particles into his lungs. These hay particles cause irritation and can lead to infection or allergy. The result is a cough and a runny nose, or even worse—heaves or pneumonia.

Good, leafy, early cut alfalfa hay is 12.8% digestible protein, good leafy-red clover hay is about 9.2% digestible protein, while mixed hays (legume and grass) are about 7.2% digestible protein. Timothy hay, if cut before the bloom stage, contains about 6% digestible protein, while mature timothy hay may be as low as 2.4% digestible protein. Good Kentucky bluegrass hay is about 4.7% digestible protein.

There are several ways to feed hay. Overhead racks are handy in a barn or stable if the hay is dropped down from a loft above, but it is unnatural for a horse to reach up for his food, and dust and chaff often fall into his eyes. Mangers and feed bunks are a better way to feed hay, for they are low, allowing the horse to reach down; he needs to exercise his neck muscles as though he were grazing. The only disadvantage to mangers and feed bunks is that they must be cleaned out periodically. It is not good to feed a horse on the stall floor or on the ground; he may pick up worm eggs from manure and become infested with internal parasites. He also scatters some of the hay and wastes it. Hay that has been urinated on will not be eaten. When feeding hay on ground that is sandy or gravelly, the horse may swallow sand when picking up the last wisps of hay or leaves of alfalfa. If this continues over a period of time the sand can accumulate in his caecum and cause impaction and perhaps fatal colic.

Lawn clippings can be used as horse feed, but are best if dried a bit before gathering. Lawn grass is young and growing and therefore highly digestible. But like any new kind of feed, the horse should be changed to it gradually, replacing only a small portion of his hay at first. Because of its palatability, horses will overeat this type of feed. Freshly cut clippings are too high in water content and tend to give a horse loose bowels. They can't provide enough nutrients per volume. They may also spoil and mold if piled for any length of time; dried clippings are much better.

Pelleted feeds are handy to use. There are basically three types of pelleted horse feed: pelleted concentrates (grains), pelleted hay, and "complete" pellets that contain hay, concentrates, vitamins, and minerals. The "complete" pellet has advantages and disadvantages as a horse feed. Advantages: rations can be easily measured, the contents are uniform, the pellets are easy to store and to transport; there is no waste, and the ingedients are nearly all utilized and very little passes on as manure, making less clean-up problem in stables. Pellet-fed horses are trim (no hay-belly) and old horses, horses with bad teeth, and horses with heaves seem to do better on pellets than on hay.

The main disadvantage to the complete pellet is the lack of bulk. Pellet-fed horses usually chew wood because they crave extra roughage. Pellets that are a complete horse feed contain some roughage in the form of hay, but it is all finely ground. The horse eats less total bulk and may not feel full. The horse normally requires a moderate fill of roughage to satisfy his normal body functions. The bulky fiber of roughage has practically no nutritional value in itself, but it stimulates the digestive process and keeps it moving at a normal pace. Roughage assists in the digestion and assimilation of the digestible matter the horse eats. Most horses chew wood badly when first on pellets and then taper off somewhat after they have become adjusted to the new ration and their stomachs have shrunk in size. Another disadvantage to pellet feeding is that if a horse accustomed to nothing but a diet of pellets gets into

Feeding hay on the ground is not a good practice because the horse may inhale dust from the ground (causing coughing or even heaves), eat bits of dirt or sand that may accumulate in the digestive tract and cause colic, or may pick up worm eggs and become infested with internal parasites

a pasture accidentally or finds access to grass or hay he may be in danger from colic or impaction. His stomach and intestines have shrunk somewhat and he cannot manage the extra bulk that he is unaccustomed to.

Pellets are expensive if you can get good hay or pasture that is cheap or home grown. We have never fed any of our horses a complete pellet for we have good hay and pasture and we feel that horses need roughage. But if you live in an urban situation where your horse is stabled instead of pastured and you are buying all of your horse's feed anyway—hay, grain, and supplements—pellets might be practical. This is one reason more and more owners of pleasure horses will probably use pellets; and as scientists and feed manufacturers learn more about proper horse feeds and needs, the pellets will undoubtedly be improved further. Pelleted feeds can cost anywhere from $40 to $100 a ton, depending on the pellet and upon your locality (how much freight you have to pay). That seems expensive until you think of the small amount you need to feed, the lack of waste, the fact you don't have to buy any other feeds. The controversy over proper roughage still rages, but pelleted feeds are here to stay and you might find them practical in your locality and individual situation.

Silage can be used as horse feed. Good corn or grass silage can be used as a substitute for one-third to one-half of a horse's hay ration, but should be introduced into the ration gradually, and it should never be the only roughage fed. Silage has a high moisture content and it usually takes about three pounds of silage to equal the food value of one pound of hay. Silage that is spoiled, moldy or frozen should never be fed to horses because bad silage can cause digestive upsets or even poisoning. Horses are much more susceptible to poisoning from spoiled silage than are cattle or sheep.

A few last parting suggestions on feeding. Feed each horse as an individual, taking into consideration his condition, his feeding habits and his keeping ability. Keep feed boxes clean so that moldy or spoiled feed won't accumulate in the corners. Make sure a horse has plenty of fresh water, salt, and exercise. Exercise is just as important as feed when it comes to keeping a horse fit and in good condition. "Feed the horse to stand the riding, then ride the horse to stand the feed," is an old horseman's saying and it is certainly true. Muscle and fitness are never built by feed alone.

14

COLIC

COLIC IS ONE OF THE MOST COMMON CAUSES OF DEATHS IN horses. Just what is colic? In this chapter we will take a quick look at various types of colic and its symptoms.

The word "colic" means abdominal pain. So in a sense, colic applies more to the symptom than to the disease or to the cause of pain. There are many things that can cause abdominal pain in horses, so there are many different kinds of colic. Generally speaking, colic refers to a problem that originates within the digestive tract. True colic refers to pain within the digestive tract (stomach and intestines) and that is what we'll be discussing here. The term "false colic" refers to abdominal pain that originates outside the digestive tract, from abdominal organs such as the liver, kidneys, bladder, and so forth.

Colic, or digestive tract disturbance, is often fatal because of digestive tract blockage, twisted intestine, or some other abnormality. Pain in the digestive tract means that something is seriously wrong. Not all cases of colic are fatal, but a horse with signs of abdominal pain, even though mild, should be constantly watched, and attended to if symptoms worsen. A vet should be called immediately for serious symptoms of colic; he may be able to save the horse if called in time. Until the vet arrives, the horse should be kept on his feet.

Horses are more susceptible to colic than are many other kinds of animals, partly because they cannot vomit as easily to unload the stomach of harmful contents. And the horse does not easily burp to relieve himself of pressure caused by gases. If the stomach or intestines become blocked for some reason, pressure builds up and causes pain. If the tract remains blocked, the pressure can become so great that the digestive tract ruptures and the horse dies.

Digestive upsets which cause colic may be mild or severe. The horse may be merely uncomfortable, or in great pain, depending on the seriousness of the problem. Mild colic may merely make the horse restless. He might switch his tail, nose at his belly, paw, stamp his feet or lie down. Severe colic may make the horse more noticeably uncomfortable, and if he is in great pain he may get up and down, staggering on his legs, throwing himself to the ground to roll violently or thrash about.

Abdominal pain (colic) can be caused by many types of problems—inflammation in the digestive tract, gas pressure, spasms, obstructions and so on. In all types of true colic, the pain is caused by distention of the stomach or intestines. This may be persistent and static when the digestive tract is filled with too much feed, gas, or fluid, or the distention may be intermittent when local and periodic pressure occurs because of spasm and increased activity of certain parts of the digestive tract. The persistent and stationary accumulations are called "physical colics" (due to obstruction or improper feed, etc.) and require physical treatments. The intermittent distentions are "functional colics" and proper treatment is to relieve the spasms. Let's take a closer look at some of these problems.

Spasmodic colic, in which the digestive tract works in spasms, can be due to excitement and sometimes occurs in horses that have an excitable disposition. The intermittent cramps of spasmodic colic can also be caused by eating spoiled or moldy feed, by a sudden change in diet (taking the horse from dry to green feed, from grass hay to alfalfa hay, suddenly introducing a large amount of concentrates to the diet, or eating a large amount of any kind of food that the horse is not used to). Drinks of cold water when the horse is hot and sweating can also cause it. With this type of colic the horse suffers from short attacks of abdominal pain. He may roll, paw and kick for a few minutes, then shake himself and stand normally for awhile until the next attack of pain hits him. Sometimes noisy intestinal sounds can be heard, even some distance away from the horse.

Symptoms of spasmodic colic often disappear without treatment; the horse may be fine in a few hours. But if his attacks of pain do not subside, he can be given medication (usually in the form of an injection) to help relax his digestive tract. Atropine is often given as treatment for spasmodic colic.

Another type of colic, which is similar to spasmodic colic in that it is a functional colic and manifested by increased activity of the digestive tract, occurs when there is inflammation of the intestines. Because of the inflammation the horse does not absorb fluids as he should and he suffers from abdominal pain and diarrhea. This can happen when the horse has too many parasites within his digestive tract, or other irritations such as sand or poison within the tract, or can be caused by bacteria or virus infection. In acute cases the horse suffers severe abdominal pain and rolls and kicks at his belly. Bowel movements are soft and watery and have a very unpleasant odor, sometimes containing blood and mucus. In sand colic the bowel movements will contain bits of sand or gravel.

Sand colic is caused by an accumulation of sand or small gravel that the horse has swallowed over a period of time. This can happen if the horse's feed gets blown full of sand during a windstorm, or when a horse is fed hay on sandy ground; he picks up sand when eating the last wisps or leaves of hay. It can also happen if he is on sandy pasture land where grass pulls up too easily and he swallows some of the sand clinging to the roots. The sand irritates the digestive tract and over a period of time can accumulate in places within the tract. The sand is heavier than the rest of the material eaten, and instead of passing through the digestive tract it tends to lodge in the caecum or in the lower loops of intestine. After a time it can cause a serious type of colic that is often fatal. Treatment can be undertaken if a horse begins to have colicy symptoms and sand colic is suspected. Accumulations of sand or dirt in the digestive tract can best be removed by repeated treatment with a bland and oily purgative; mineral oil is about the best. Daily doses of one or two pints of mineral oil may be necessary for more than a week to clear up a chronic case. Acute cases are best treated with a washing out procedure involving giving the horse two or three gallons of warm saltwater and administering medication (arecoline) at intervals for about eight hours.

Gastric dilation (dilation of the stomach) is an occasional cause of colic in horses. There is too much fluid in the stomach, causing great distention and pain. This can happen when the horse eats too much grain or has a pyloric obstruction (the pyloris is the opening from the stomach to the intestines), or drinks too much fluid such as whey. The overeating of large amounts of feed such as grain is the most common cause of this type of colic in horses. Abdominal pain is severe and the horse may sweat, roll, kick at his belly or sit on his haunches. His heartbeat and respiration rates are greatly increased.

Chronic dilation of the stomach sometimes occurs in horses if a tumor obstructs the pylorus, or if the stomach wall loses its muscle tone because of old age or debility, or if the horse is often fed coarse and indigestible kinds of roughage. Ulceration of the stomach can sometimes occur if a horse is a wind-sucker or cribber (he swallows splinters of wood, that can irritate the stomach lining) and this ulceration can cause pyloric spasm and subsequent stomach dilation.

This kind of colic can result in fatal dehydration. Also, local damage to the stomach lining may cause shock. In some cases where the stomach wall is already weakened, the stomach may rupture. In chronic cases of stomach dilation, the stomach loses its muscle tone and its activity is decreased, indigestion is commonplace and the horse becomes undernourished.

Flatulent colic is caused by a formation of gas in the intestine. This build-up of gas may be due to an intestinal obstruction, or may be caused by eating highly fermentable feeds such as lush green feed or too much clover or alfalfa. All cases of excessive gas in the small intestine are due to some kind of intestinal obstruction. Gas in the large intestine can be caused by fermentable feeds, or can be caused by a narrowing or obstruction of the passage.

Impaction is another cause of colic, and occurs when the intestinal tract is partly or completely blocked by a mass of food. This is most common in the large intestine and causes moderate pain, constipation and general depression. The intestine is distended with an accumulation of food which cannot get past the blockage. Usually impaction is a dietary problem and is caused by eating coarse, low-grade roughage. Bad teeth and improperly chewed food can be a cause. Exhaustion of the horse, or lack of fluid in the digestive tract can also predispose a horse to impaction. Long intervals between feedings can also be a cause. Overfed fat horses and heavy eaters are susceptible to recurrent attacks of impaction. If the blood supply to the intestine is interfered with, as sometimes happens when a horse is infested with bloodworms, or if there is lack of muscle tone in the intestine because the horse is debilitated, the intestinal tract may not be able to move the bulk of feed at a normal rate and impaction occurs.

The affected horse suffers moderate abdominal pain which may continue for three or four days or for as long as two weeks. The attacks of pain are moderate (the horse is not violent, but rather merely uncomfortable and restless) and occur at intervals. The horse is constipated, and when feces are passed they are hard and small and covered with mucus. Most cases respond to treatment, but impaction of the caecum is difficult to correct and may be fatal (the caecum lies between small and large intestine). Fatal cases of impaction of the large intestine are usually due to rupture of the intestine or from exhaustion of the horse after a long period of colic. Most effective treatment of impaction is to administer up to a gallon of mineral oil along with a drug to relax the intestines.

Impaction of the ileocaecal valve (between the small intestine and the caecum) is usually fatal, and is just as serious as acute intestinal obstruction. The most common cause is from feeding a horse low-grade, finely chopped roughage. The roughage passes through the stomach undigested and collects at the end of the small intestine at the valve that leads to the caecum. The horse suffers only mild pain for about 8 to 12 hours, pawing or rolling or nosing at his flank. Then the pain becomes more intense. The horse is severely depressed, sweats sporadically, stands with his head hanging down or sits on his haunches, then rolls or struggles violently. Pulse, respiration, and temperature go up. The small intestine and sometimes even the stomach become tightly distended with gas and fluids, and death usually occurs within 36 to 48 hours from the time the first symptoms were noticed.

Intestinal obstruction is very similar, causing acute abdominal pain, severe shock, constipation, and sometimes the passage of blood and mucus. The most common causes of obstruction are intestinal accidents, twisted intestine, prolapse of one part of the intestine into the cavity of the adjoining part (the most common cause of intestinal obstruction in foals), and strangulation. Twisting of the caecum or large colon may occur while the horse is rolling, especially if the rolling is violent. This can happen when the horse is rolling and thrashing about because of pain from spasmodic colic or from impaction.

Intestinal obstruction causes severe pain and shock and the horse usually dies within 12 to 24 hours unless the obstruction is surgically removed or corrected. Occasionally a displaced loop of the intestine will correct itself spontaneously if the horse is exercised vigorously or hauled over a rough road. The sign of recovery is the immediate passage of great quantities of feces from the large intestine.

One other type of colic that should be mentioned is thromboembolic colic caused by strongyles, the bloodworms that live as parasites in nearly all horses. Bloodworms damage the intestinal arteries, interfering with blood supply. If there are too many bloodworms they may block the arteries or cause clots to form, which block the arteries. When this happens, part of the intestine dies from lack of blood, and fatal gangrene of the bowel is the result. Sometimes there is merely a chronic, low-grade impairment of blood flow to the intestines, causing lack of muscle tone in segments of the large intestine. Sometimes a blood vessel will balloon out in a weak spot caused by the bloodworms, creating an aneurysm. These occasionally put pressure on a nerve and may be a cause of spasmodic colic. Recurrent attacks of colic may also occur when a secondary bacterial infection becomes established in the aneurysm.

When any of the foregoing digestive tract problems occur, the horse tries to relieve the pain in his abdomen. At first he may just be restless. Then as the pain increases, he may lie down, roll back and forth, or kick and bite at his belly. He may strain to have a bowel movement, stretch himself, or break out in a sweat. If the pain continues to worsen, he may grunt and squeal and thrash violently on the ground.

If a horse seems restless, watch him closely to determine whether or not he may have symptoms of colic. If he seems to get more restless, or starts rolling, call a vet immediately; don't wait around to see if he gets better or worse. Colic can worsen rapidly and a quick call to the vet may save the horse.

Often when a horse has colic, he tries to relieve the pain by rolling or thrashing about on the ground. This can be dangerous, as the violent rolling may twist an intestine

If a horse has colic, lead him around. Keep him moving and on his feet; the mild exercise of walking may help him pass a bowel movement

Until the vet arrives, keep the horse on his feet. Walk him slowly around and don't give him a chance to lie down. If he gets down on the ground he might roll so violently trying to relieve the pain that he might twist or rupture an intestine. Gentle exercise may relax the muscles of the intestines enough that the horse can pass a bowel movement, which may relieve some of the pressure and pain.

One of the fillies we boarded in 1971 had a heavy infestation of intestinal worms when she arrived, and upon worming her she passed so many worms at once that she suffered a blockage. An hour after worming she showed symptoms of colic—discomfort and rolling. We caught her and walked her for about six hours, keeping her on her feet when the bouts of pain hit and she wanted to get down and roll. A mild dose of medication to help relax the intestinal tract, and the constant gentle exercise of walking helped her finally to have several bowel movements, passing large quantities of dead worms. Then she was all right again.

Nikki frightened us with the only serious case of colic we've ever had among our own horses, one night in the summer of 1972. She was about three months pregnant (and being pregnant seemed to change her body chemistry—she was more allergic and sensitive to things then than at any other time in her life) and I had ridden her quite strenuously for about eight hours that day, moving cattle. Upon arriving home that afternoon she was tired, but still perky and energetic. I let her graze at pasture with the other horses for about 45 minutes, after she had rested for a couple of hours. Nikki was very sensitive to clover that summer and I always rationed her time at pasture, never letting her graze for more than 30 to 45 minutes at one time. I brought her in from pasture then and gave her her evening hay. Two hours later she was suffering from severe cramps, staggering around and lying down to roll. It was already dark and I wouldn't have known she was ill except that I just happened to go out to give her one more flake of hay for the night. She staggered a few yards, weaving on her hind legs, then went down.

I urged her up off the ground and took her out into the pasture where there was lots of room and the ground was soft—in case she did go down again. There I walked her, around and around, Heidi following her with the whip to keep her moving when she got shaky on her legs and wanted to go down. We gave her injections to help stop the cramps (she was suffering from spasmodic cramps of the small intestine, probably from the clover she ate when she was tired), and had to repeat the treatment five hours later. We walked that mare all night long, 11 hours in all. She would seem better for short periods, and we would let her stand and rest a few minutes—she was nearly exhausted because she had worked hard moving cattle most of the day, and had to walk constantly most of the night. But she wasn't able to stand and rest very long; after a few minutes she would crouch down on her hind legs, or buckle at the knees, trying to go down to the ground as the pain came again.

But she tried her best to please us; she seemed to know that we wanted her to stay up off the ground. When we were walking her, we urged her on every time

Jirre, the suckling filly who suffered colic after worming. After a few hours of walking she passed several bowel movements and was all right again

she got staggery, and she soon began to move faster on her own accord when the pain got worse, trying even to trot while her hind legs were staggering. She no longer needed someone behind her with the whip; the person leading her could usually keep her moving with a word or a touch of the whip on her hindquarters. Somehow she sensed that there was some relief of pain to keep moving, even though she was very tired, and she tried very hard.

We made a pot of coffee to keep ourselves awake, and spelled each other off in the leading, so we each got time to rest, sitting or lying there in the dark pasture, lending encouragement to the one doing the footwork. At 5 a.m. Heidi went to bed and I carried on until 6:00 when Lynn got up and took over.

Finally at about 8:30 a.m. Nikki's pain subsided and she was able to stand, resting, without wanting to go down. We let her stand quietly for about half an hour, then gave her an enema, using a four-foot hose (put into her rectum about 18 inches) and a homemade funnel fashioned from a plastic bottle with the bottom cut out. We gave her 1½ gallons of warm salt water (one teaspoon per quart) at body temperature (100 degrees). It went in easily, for her large colon was rather empty; she had passed manure 11 times during her walking. The saline solution would be partly absorbed by her large intestine and help to keep her from being dehydrated. She was to have nothing to drink for four more hours and nothing at all to eat for 12 more hours, to help insure that the cramps would not begin again.

She was a very tired mare, but the crisis was over. We were tired and bleary eyed, but elated just the same, for our night of worry was over and our mare was going to make it. We fed her very sparingly that evening and gradually increased her feed during the next day. She was now strickly on hay and a little cut grass—no more clover for Nikki that summer, for it seemed to be somewhat toxic to her.

When feeding a horse it helps to know his individual problems. Horses are all individuals and some feeds will disagree with one horse and not another. Some horses need to be limited in the amounts of certain kinds of feed, while other horses do not. Colic can be a very serious problem, so good judgment in feeding is always very important.

Most cases of colic can be prevented by proper care of the horse and conscientious feeding. To avoid colic, always make dietary changes gradually, and always feed the horse clean, palatable, good quality feeds. Never overfeed a horse. Never feed a horse moldy or coarse hay. Coarse hay or straw is likely to cause impaction. If a horse is to be changed from dry to green feed (from hay to pasture, for instance), allow him only a little green feed the first day, longer the second day, and so on until the horse has been changed completely to the green feed in a gradual process of about four or five days of adjustment. And even then, if the horse is on lush pasture containing clover or other legumes, he should be watched closely because overeating of these feeds can cause colic, as we experienced with Nikki. If your horse tends to overeat on clover, limit his time at pasture.

If a horse is not on pasture and is being fed hay or pellets, feed him two or three times a day, regular amounts, at regular intervals. Irregular feeding can cause a horse to eat too much at once if he is hungry due to long hours between feedings. And a hungry horse may eat poor quality, unpalatable feeds (that might cause impaction) that a properly fed horse would not even look at. If you feed hay or pellets, feed in a manger or feed bunk and not on the ground where the horse may pick up sand or dirt or worm eggs.

Worm all your horses regularly. They will stay healthier and in better condition, utilizing their feed more efficiently. And they will be less likely to have colic from bloodworms, or colic from passing too many worms at once when you worm them sporadically with long intervals between wormings.

Older horses and horses with bad teeth are sometimes likely to have colic because they do not chew their food properly and some of it passes through the stomach undigested. If a horse's teeth become jagged or irregular, they should be floated (filed smooth).

Don't feed large amounts to a horse that is hot, tired, or chilled. His blood supply is busy in other areas of his body than his digestive tract, refreshing tired muscles or trying to keep him warm. His stomach and intestines are not properly supplied with blood, and to feed him at this time may create digestive disturbances or colic and upset his whole system (perhaps causing founder as well) as blood leaves other parts of his body to come to the digestive tract. Don't feed grain when the horse is tired. After a long ride it is much better to give him a little hay to nibble on until he has rested somewhat and his body is back to normal; his system can handle a little hay or green grass much more easily than grain at this point.

The horse seems to be more susceptible to digestive disorders than some other animals because of the way his digestive tract is built and the way it functions. The horse's intestine is large in some places and narrow in others. Masses of rough food material (coarse food or unchewed food) sometimes lodge in the narrower places, causing impaction. Ruminants (cows, sheep, deer, etc.) have a stomach with four compartments in which feed is processed and digested by millions of bacteria and yeasts. The ruminant's food is mixed and processed and then rechewed, passing into the intestines practically already digested, where it is then absorbed. But the horse has just a simple stomach. He chews his food just once and it goes from his stomach to his intestines in a rather rough state, not too well digested. His food is not processed by microorganisms until it reaches the caecum, toward the end of the small intestine. This is one reason that some of the food the horse eats passes on as waste, undigested. Cattle can handle coarser roughage than horses, usually, and get by. Horses need the best quality, most palatable feeds that are fine and tender. The horse's digestive tract is not as efficient as that of a cow, and he needs everything in his favor. His natural feeds in the wild state were native grasses, not straw or coarse swamp grass hay. The old-fashioned notion of some "economy-minded" ranchers of feeding their horses their worst hay, saving the better hay for the cattle is one of the reasons that colic and digestive disorders in horses have been so prevalent. A good horseman feeds his horses the best quality hay he can get.

Poor quality coarse hay is very poor economy when it may cost you the life or health of your horse.

Good feed, good care, and some individual attention are the best preventative measures for guarding against colic. A good horseman should know his horses well enough to be able to tell when one of them is not feeling well, and he should check on them often enough to be able to know something might be wrong. If a horse is seen periodically during the day, if he is doing something unusual or out of the ordinary for his personality, he can be checked more closely and carefully observed. Colic can be a serious matter; it is never something to take a chance on. Good care is the best prevention. And immediate attention and treatment is the best way to deal with colic should your horse ever experience it.

15

HORSE CARE YEAR ROUND

CERTAIN ASPECTS OF CARING FOR A HORSE HAVE ALREADY BEEN discussed in earlier chapters. In this chapter we will look at the horse in winter and in summer to discuss specific problems that occur with the seasons, and the various details involved in proper care of the horse during these varying seasons.

Winter care for horses will vary somewhat in different parts of the country, depending on the severity of the weather. It will also vary under individual circumstances—such as whether your horse is kept out on winter pasture or in a barn or stable.

Good winter care of a horse involves adequate feed, shelter, care of his feet, and, if you are riding him in winter, additional feed and care to provide the needed energy and to keep him healthy and comfortable. In the winter a horse must have adequate feed of proper nutritional contents to keep him warm and in good flesh. He should have access to water even if this means breaking ice daily for him during the coldest weather. He should have shelter of some kind if your winter weather is severe. He should be properly wormed so that he can utilize his feed and won't harbor internal parasites all winter. His feet should be properly trimmed at intervals so that they won't grow too long during the winter; his shoes should be taken off if he is out on winter pasture. He should have adequate foot care if you are riding him during the winter and should be shod with special shoes if you are riding on ice. He should be properly groomed if you ride him in winter; he should always be cooled out and dried off after a ride to make sure he doesn't chill.

Cold weather in itself, for horses out on winter pasture, is not a serious problem, but it can become a problem when accompanied by wind or storm. Some geographical areas, due to the flatness of the land, have a lot of wind. The northern parts of our country which are not protected by mountains (especially the Great Plains) often have severe winters for livestock. In these windy places horses must have some kind of winter shelter because the wind makes the temperature seem much colder than the thermometer registers. This can create a real survival problem if low temperatures are accompanied by strong winds. Accompanying this chapter is a wind chill table which shows the equivalent temperature under various wind conditions. It feels colder when the wind is blowing; it *is* actually colder. The wind chill table will give you accurate figures, but a rough rule of thumb to give you some idea of the difference made by the wind is this; each mile per hour of wind is approximately equal to a degree or more lower in temperature. You can see why livestock sometimes freeze to death on the western and northern Great Plains when temperatures drop down to ranges of about 40 below zero, accompanied by 50 and 60-mile-an-hour winds. Twenty below zero, accompanied by 40-mile-an-hour winds, would be the equivalent of 87 degrees below zero on a still day. Forty below zero, with 50-mile-an-hour winds would be the equivalent of 120 below zero on a still day. And so on.

Cold weather without wind is not hard on a horse. His winter coat of hair is good insulation, but wind can penetrate winter hair. Dry, still cold is much more bearable. Besides wind, the weather that seems to bother a horse most is that which comes in the fall or spring or during a freak warm weather winter storm—when he gets soaked with rain or melting snow and then becomes chilled when the temperature drops before he has a chance to dry off. These are the times when a horse seems most susceptible to colds and respiratory infections.

If you live in an area that has cold winters, there are several factors that are of importance in caring for your horse during the winter months. First, you should be concerned with the condition of your horse as winter begins. A thin horse will not winter as well as a fat one, for he has no reserves. A horse should never be too fat, but he should be carrying enough extra flesh at the beginning of winter to remain healthy and fit and to have some reserves for body heat and energy, especially if he is kept out on winter pasture and gets no supplemental feeding except during severe weather conditions. Even if kept confined and fed well all winter, a thin horse will not gain weight as fast as he would if he had been gotten into good condition before cold weather set in. As a general rule, horses need more feed in winter than in summer, for it takes more energy just to keep warm.

Old horses and young horses winter harder than mature horses in the prime of life. Old horses often have poor teeth and cannot chew their food as well as they should, therefore getting less good from it. If an old horse stays in poor flesh in spite of good feed, have his teeth checked; sometimes floating his teeth (smoothing off the rough edges of his molars) will help him chew better and utilize more of his feed. The condition of an old horse when he goes into the winter may very well determine whether he will survive or not. If he goes downhill in flesh too badly during the winter, he may get too weak and may die, or if he makes it through the winter it will take him a long time to get back into condition the next summer. Unless an older horse is fat and a good keeper, don't turn him out to winter pasture unless you can check on him often and feed him extra if he needs it.

Mature horses on winter pasture do not need grain if they have good pas-

Nell and Khamette, wondering when the snow is going to stop. Their "hot" wire fence is nearly buried

A freak snowstorm, April 1971. Nikki, Fahleen, and Bambi enjoying a break in the two-day storm

Fahleen, knee deep in snow

ture or hay. But young horses—weanlings, yearlings, two year olds and sometimes even three year olds—do much better if fed grain, and a good portion of protein in their diets as well (such as alfalfa hay or a protein supplement). They are less apt to be stunted or set back in their growth if fed liberally during the winter months. Young horses need sufficient nutrition for growth as well as for energy and body heat during cold weather. They need more feed than mature horses (proportionately), yet it is good for young stock to be out on winter pasture if possible. They need room to exercise and move about freely, developing surefootedness, muscle, bone and wind, but if they are out on winter pasture, grain them daily. If they are wintering with mature horses, fix a pen or corral in which you can feed the young horses separately. If your winter pasture is not conveniently located for you to feed young horses daily, it is better to keep them close at hand so they can be adequately fed during the winter months.

Another factor to be considered in caring for horses during the winter is shelter. The horse is a rugged, hardy animal and one of the most adaptable creatures in the animal world, for he seems to thrive in various extremes of climate. Most horses in the northern climates winter out-of-doors in spite of cold and inclement weather; they never see a barn and are none the worse for it, but this does not mean they don't have shelter. Shelter is important. Many winter pastures in these colder northern areas have valleys and canyons, draws and gullies where livestock can get out of the wind. Brushy or timbered areas along the draws and creek bottoms provide a great deal of shelter, not only as wind-

WIND CHILL TABLE

Actual Temp.	35	30	25	20	15	10	5	0	5	−10	−15	−20	−25	− 30	− 35	− 40	− 45
Wind Calm . .	35	30	25	20	15	10	5	0	− 5	−10	−15	−20	−25	− 30	− 35	− 40	− 45
5	33	27	21	16	12	7	1	− 6	−11	−15	−20	−26	−31	− 35	− 41	− 47	− 54
10	21	16	9	2	− 2	− 9	−15	−22	−27	−31	−38	−45	−52	− 58	− 64	− 70	− 77
15	16	11	1	− 6	−11	−18	−25	−33	−40	−45	−51	−60	−65	− 70	− 78	− 85	− 90
20	12	3	− 4	− 9	−17	−24	−32	−40	−46	−52	−60	−68	−76	− 81	− 88	− 96	−103
25	7	0	− 7	−15	−22	−29	−37	−45	−52	−58	−67	−75	−83	− 89	− 96	−104	−112
30	5	− 2	−11	−18	−26	−33	−41	−49	−56	−63	−70	−78	−87	− 94	−101	−109	−117
35	3	− 4	−13	−20	−27	−35	−43	−52	−60	−67	−72	−83	−90	− 98	−105	−113	−123
40	1	− 4	−15	−22	−29	−36	−45	−54	−62	−69	−76	−87	−94	−101	−107	−116	−128
45	1	− 6	−17	−24	−31	−38	−46	−54	−63	−70	−78	−87	−94	−101	−108	−118	−128
50	0	− 7	−17	−24	−31	−38	−47	−56	−63	−70	−79	−88	−96	−103	−110	−120	−128

(Winds greater than 40 mph have little additional chilling effect.)

It feels colder when the wind is blowing because it *is* actually colder, as shown by this wind chill table. Wind chill charts have been in use for many years, especially by the U.S. Army, but were recently given "official" status by the Weather Bureau a few years ago.

breaks but also providing some relief from driving snow or rain. In pastures where there are no natural windbreaks (trees and bushes) or shelter of any sort, a simple three-sided open shed will suffice. Horses seem to prefer being out of doors even in the coldest and nastiest weather but they do appreciate a good windbreak.

A good general rule to follow concerning winter shelter during storms and cold weather is to keep a horse in an environment to which he is accustomed. If you keep your horse in the barn every night and he is used to it, by all means bring him in to the barn during stormy weather. But if he is used to being outside, don't bring him inside just because of foul weather. If he is accustomed to a blanket, keep it on him and keep it dry, but if he is not used to a blanket, leave it off. Bringing a horse into a barn and blanketing him when he is not used to it may actually bring on a respiratory infection; the horse may become overheated under the blanket and start sweating, then become chilled because he is damp.

A horse in good condition is best left to fend for himself—outside. Horses were meant to be outside; it is their natural environment. It isn't necessary to cover them or bring them inside, so long as they have a chance to toughen themselves naturally and gradually as winter comes on. They grow thick hair as part of their protection from wind and cold. If you keep a horse blanketed and indoors in order to prevent this extra growth—to keep him sleek and smooth and in "show shape," he will not grow a winter coat and can't be expected to stand outside in a winter storm. It is not wise to clip and blanket a horse in preparation for shows if he will have to spend some of his time outside exposed to winter weather.

Don't take a horse from one climate to another without taking the necessary precautions. If you are traveling to a colder area, use a covered horse trailer, blanket the horse when traveling, and when he is not being exercised, until he has had a chance to become accustomed to the cold (preferably over a period of several weeks, letting him outside at first only on nice days) and has started to grow a protective coat. Traveling in cold climates will not bother a horse that is already accustomed to cold weather and has grown a good winter coat of hair.

If you ride a horse in winter and he sweats, make sure he is thoroughly dried off after you ride him, before you put him away. If you blanket a horse in winter, don't let his blanket get wet from rain or melting snow, or from his own sweat. Even in cold weather, a horse can become too warm under a blanket and start sweating, and this moisture under the blanket will not evaporate as quickly as the moisture on an uncovered horse; the blanketed horse may chill.

If you must bring a wet horse into a barn because of sickness, injury or some other problem, make sure you get him thoroughly dry. Rub him with towels or clean feed sacks until you have gotten all the moisture you can, then groom and brush him vigorously. Rubbing and brushing help dry him and stimulate his circulation, making him warmer.

Never expose a horse to drafts, especially if he is damp. Drafts of moving air can cause chilling and make the horse susceptible to respiratory infections.

The horse is better off outside in the open than inside a drafty place. The cold air will not bother him outside, but a light gusty bit of air moving around his stall or through the barn can chill him.

Make sure the horse has water. Horses drink less water in cold weather than in warm weather, but they still need it. Horses will eat snow, and if snow conditions are right (not frozen or crusted) they can get by on snow alone for quite awhile, but they still will appreciate some water now and then to get sufficient moisture for their bodily functions. Occasionally a horse will die from constipation on winter pasture with nothing but snow for moisture. If your horse is out on winter pasture and the weather gets cold, you may have to check his watering place every day or so and break ice if it freezes over. Some horses are very reluctant to walk on ice and you may have a problem getting them to drink if they must step out on ice to reach a water hole. You can combat this problem by spreading dirt or gravel over the ice, to give the horse safer footing. If you water your horse in a tank or tub, make sure your water supply does not freeze up. In very cold weather you may have to empty all the ice out of the tank periodically. Where it is feasible, warm water is best for the horse; he will drink more and it won't freeze up so fast. No matter what kind of water system or natural water your horse has, make sure he has access to water and is drinking. Don't just break a water hole in the creek for him every day without checking for his tracks to the water and occasionally staying by long enough to watch and see whether he is really using it. If the approach is steep or icy he may not be going to water.

Khamette and Fahleen, waiting for a handout

Nell walking across her pasture after a heavy snowfall. After snow like this, a horse needs hay to give him energy to paw for grass

If a horse is on good winter pasture he usually won't need hay unless the snowfall is exceptionally deep. Even in mountainous terrain where there is heavy winter snowfall, there is usually grass that can be found on the ridges or south slopes and the horses don't have to do much pawing through the snow to get at it. The south slopes receive more winter sun and sometimes melt off several times during the winter, whereas the north slopes stay more shaded and the snow keeps piling up deeper and deeper. Wind usually keeps some of the snow blown off the ridges. A winter pasture that is basically on a south slope is better than one situated on a north slope.

After a bad storm or unusually heavy snowfall you might want to take hay to horses out on winter pasture. Even in deep snow a horse will paw through to grass, but he needs strength and energy to do it. If you give a horse some extra feed right after a heavy snowfall, it will give him the energy to go out and paw for grass and in a few days the snow may not be so deep. In some parts of the country in years when snowfall is unusually heavy, horses have to be brought in from winter pasture and fed hay.

Don't overfeed a horse during a storm. You may be tempted to feed him extra while the weather is bad because he looks so miserable, but a heavy feed at this time is the worst you can do for him. He needs his blood circulating to keep him warm, and if you feed him a lot the blood will go to his stomach and intestines instead. If most of his blood is concentrated around his digestive tract, he may chill. Wait until the storm is over and then feed him extra.

These horses are in good flesh and ready for winter

If you feed hay in winter, make sure it is always good quality and dry. Hay that is freshly wet from a snowstorm is all right to feed (if it will be cleaned up right away) but hay that has been wet for more than a few hours (especially alfalfa hay) can begin to mold.

If a horse is being turned out to winter pasture, his shoes should be removed so his feet can have a chance to wear normally, with frog and bars bearing some of the weight and aiding proper circulation of the foot. Going without shoes for a time can be a healthful rest for a horse's feet. And there will be no chance of a shoe catching in a wire fence or some other hazard and entangling him. Snow will not ball up as badly under a bare foot as it will under a shod foot, making slippery footing. A barefoot horse can get around a little better on ice and snow than a horse with ordinary shoes, for snow tends to pack and ball up in shoes. On sheer ice a horse will be more surefooted if he is shod with special sharp shoes (with toe and heel calks, for traction) or shoes with borium welded on, but in snow he will get around much better barefoot.

A horse should be wormed before winter. If he harbors a heavy infestation of internal parasites all winter, he may lose weight and become rough and thin by spring. He will utilize his feed much better if he is free of internal parasites. A good thorough worming is best done after the first killing frosts in the fall; at this time there is usually a heavy infestation of bots in the stomach and these can be eliminated. The horse will not become reinfested with bots at this time because the mature flies are killed by cold weather.

If you ride in winter, you must dress properly for cold weather, wearing warm but lightweight clothing. If you are too heavily bundled you can't move freely. Wool is good insulation against the cold and can absorb a lot of moisture from perspiration and still keep the air next to your skin fairly dry. Gloves should be warm, yet lightweight and flexible for a good feel on the reins.

On windy days you'll want a scarf or knitted face mask to protect your nose and cheeks from frostbite. The danger of frostbite is much greater when the wind is blowing than when the weather is calm, for as mentioned earlier, wind makes the actual temperature much colder than the thermometer registers. You might expect the danger of frostbite at 20 degrees below zero and be prepared for it, but you are just as apt to suffer frostbite at a warmer temperature if there is wind. Don't forget that when you and your horse are moving, it will have the same effect on your exposed skin as if a wind were blowing. If you and your horse are traveling 10 miles per hour on a still day, exposed skin is subjected to a 10 mile per hour wind. Even though the temperature given by the thermometer is a pleasant 15 degrees, the effect of that 10-mile-per-hour wind will make it the equivalent of two below zero on your exposed flesh.

If you ride on cold winter days, remember that the temperature will suddenly drop after the sun goes down. Winter days are short, and evening usually comes sooner than you expect. Plan your rides so you'll be done before the temperature gets quite cold so you can cool out your horse and put him away *dry* so he will not chill. If you do end up riding after dark on a cold night and your horse is wet (as ours sometimes are after a long day moving or sorting cattle), you must take

Warm clothing is a "must" for winter riding. Dad on Khamette, wearing chaps, Lynn on Bambi, wearing coveralls

Heather on Nikki, Lynn on Bambi, checking through the cattle on winter pasture

the extra time necessary to dry him and cool him before you turn him loose.

A horse being ridden steadily in winter should be well fed, receiving grain as well as good hay or pasture. His feed is where his energy comes from and he can handle the extra feed better in concentrate form. In winter a horse needs more feed than in summer because he has the added necessity of keeping warm, and a horse being ridden in winter needs more feed than a horse being ridden in summer, for it takes more calories in winter just for maintenance.

This is the kind of winter pasture where we usually winter our horses

Heather, Nikki, and Pepper on a cold December day

When you ride on cold days, warm the bit in your hand for a few moments before you put it in the horse's mouth. An icy-cold bit on a warm tongue is not very pleasant. Before you saddle the horse, brush any snow or moisture from his back. A dry back is less apt to gall and get saddle sore than a wet one.

When riding in winter, use caution and good judgment in where you ride. Beware of slippery places or deep snow drifts. Keep the horse well collected when he is unsure of himself; he is less apt to slip and slide or to panic and plunge if he is collected, with his weight balanced and his actions controlled and smooth. Collected does not mean rigid restraint; let the horse use his head if he needs to. He will need free use of his head and neck for balance if he is on unsure footing.

Don't ride on sheer ice if you can avoid it. Many horses will not walk on ice because they are afraid of falling. Unless a horse has special shoes (sharp shod or shoes with borium pieces welded on) he has very poor traction on ice and can slip and fall. When riding in snow, remember that it can ball up under the horse's feet. A wet, packing snow under a shod horse will soon have him walking on four balls of ice—very slippery footing indeed. Under these conditions it is good to grease the bottom of his feet with something that will keep the snow from clinging and packing. Special pads for snow can be used in the shod hoof that will help keep snow from packing in the hoof. The Nature Plate shoe is designed to simulate the natural bare foot and tends to have the same self-cleaning effect in snow; thus this new type of horseshoe supposedly does not collect snow quite as readily as an ordinary horseshoe.

Be careful when riding on trails or roadways, for packed snow is always more slippery than undisturbed snow. Beware of packed roadways or trails, especially when going downhill. Horses are more apt to slip and slide going downhill than on the level or going uphill. This is because of the balance problem; horses tend to be front heavy. Beware of frozen ground even if there is no snow. Frozen ground can be slippery, especially on side hills. If you are riding in hilly terrain, watch out for shaded places that might be slippery because of ice or frozen ground. North slopes are usually more frozen than south slopes, and they are slicker even when not frozen because there will be moisture and mud on north slopes even after the south slopes have dried out.

One of the most important things to consider in winter riding is to make sure your horse is always cooled out and dry after you ride him. In winter the horse has a heavy coat of hair and if he is wet it will take longer for his normal body heat to dry him than it does in summer. His winter skin and hair are good insulation, letting very little body heat escape. The air is cold and it takes longer for moisture to evaporate. These three factors make him stay wet for a longer period of time in winter, and the cold makes him more susceptible to chilling. If you ride in winter and your horse becomes hot and sweaty (and he will get hot easily because of his heavy coat of hair), take time to cool him properly and dry him.

All too often a horse is unsaddled after a ride and given a short rubdown or some brushing for 5 or 10 minutes and is then turned loose into his stall or pasture. This attention is usually sufficient in summer, but is not nearly enough in winter. The horse becomes heated during any brisk exercise. In summer this heat is rapidly dispersed through the skin, but in winter the horse has a heavy coat of hair designed especially for keeping body heat in. Thus the radiation of body heat out through the skin is much less in winter than in summer, so it takes a longer time to properly cool the horse in winter. You

Working cattle in winter can be a tricky job, and dangerous unless you have a sure-footed horse

Here Nikki is bringing in a foxy cow. Though the cow dodges and doubles on a run, Nikki brings her ever closer to the gate. The cow may slip and fall on the slick snowy ground, but the horse is collected and always has a "spare leg" for any emergency

Lynn on Bambi, heading home after a winter ride. The horse will be thoroughly groomed and dried if sweaty

Bringing home the cows. Horses can paw through snow for grass, but cattle can't. After a heavy snowfall in late December, Heather and Nikki bring the dry cows down from the 320. They know there's hay awaiting them in the fields and they line out for home

can cool him slowly by giving him mild exercise such as walking, or by massaging his chest and legs and giving him a good rubdown. If he is breathing rapidly and shallowly or his pulse rate is still higher than normal, he is not yet cool and you must work with him further. When at rest, the horse's respiration is about 16 to 20 breaths a minute, his pulse rate is about 40 and his temperature is 99 to 100.5. During exercise the pulse and respiration can both get as high as 100 per minute, and the temperature becomes somewhat elevated. After exercise these rates begin to return to normal, and the horse is not cooled out until they do.

Sweating can be a problem in winter riding. Because of the horse's long winter coat, it is possible for him to still be wet after he is cooled off and his circulation, respiration, and temperature have returned to normal. He should never be turned loose wet at the end of the ride, for he may chill, especially if you finish your ride at the end of the day and the colder temperatures of night are approaching.

There are two things that should be mentioned about sweating. First, a horse in good physical condition whose muscles are in shape is less apt to sweat as much as a soft unconditioned horse. You can hurt a horse by riding him too hard when he is not conditioned for it, especially in winter when being wet and then chilling adds an extra shock to his system. If your horse sweats profusely when you ride, you may be working him too hard. If he has a heavy coat and the day is unusually warm, he will probably do some sweating even if he is in fairly good condition; in this case you will just have to spend a lot of time drying him off after you finish your ride.

Secondly, what you feed the horse can also make a difference in how much he sweats. If a horse is fed alfalfa hay (especially second or third cutting, which is mostly straight alfalfa with no grass) he will sweat a great deal more than if he is on a diet of grass hay or pasture, regardless of his conditioning. If you use your horse a lot in the winter and sweating is a problem, don't feed him much alfalfa hay unless you want to take extra time drying him off after you ride.

As winter moves into spring, there are often sudden changes in the weather. There are warm days, then colder ones again, and then some rain instead of snow. There may also be more wind than usual. Spring can be a hard time for horses. They are shedding their winter coats but there are still some cold and wet days before warm weather really arrives.

Some horses on winter pasture may get thinner at the first signs of spring. They prefer the new green grass to the dry grass or hay they have been eating, even though there is not yet enough new green grass to give them all the nutrition they need. Supplemental feeding (grain or pellets) can help combat this problem and should be given if spring is slow in coming and your horses seem to be losing some weight.

Spring is the time of year that most of the elk and deer that "winterkill" starve to death. They have survived the tough winter but are now so eager for green shoots that these are all they will eat. If they are thin and weak and spring is slow in arriving, the few green spears they find are not enough to support them and they starve. Horses are a little bit the same. So when winter seems about over and spring is on its way, don't neglect your horses. This may be the time of year they need good care the most.

Taking care of a horse in summer involves the same intelligent care that you would give the horse at other times of the year, but involves a few additional factors. Summer is a pleasant time because the weather is warm, and it is the time of year that most people are out-of-doors enjoying their horses. But summer and its warm weather bring a few problems as well, and these must be dealt with when caring for your horse. Some of these problems are heat, flies, parasites, disease, and in some instances bright sunlight and dryness of climate.

Let's take a look first at feeding and watering. In the summer, many horses are out on pasture. Green grass grown in fertile soil is the most natural and ideal feed for your horse, and if your situation permits, you will want to let your horse graze in pasture during the summer months. Good pasture is not only the most nutritious food for your horse, but usually the most inexpensive as well.

The ideal summer pasture is one with good palatable grasses, large enough in area for the horse to move about and not tramp out the grass, situated on well-drained land, and containing trees or some type of structure that will provide shade for the horse on hot days. The horse should also have free access to water in the pasture. A large pasture is always preferable to a small one, for in the large pasture the horse has plenty of room to move about and to choose the types of grasses he likes best. In a small, confined area, the horse soon becomes bored because he has little to do, and he also may run out of grass. Horses are very selective grazers, as you will notice when you look closely at the grazed pasture. Some of the grasses will be grazed close (these are the tender, palatable grasses that the horse likes best) and some grasses will not even be touched. A horse usually will not eat grass that grows in the area where he defecates, nor will he eat the coarser grasses that grow in wet, acid soils or in shaded areas. Because horses are selective in their eating habits, the pasture area should be large enough that the favorite grasses are not overgrazed. Rotating pastures can help this problem, changing horses to a different pasture before the favorite grasses get eaten down to the ground. Horses should be taken out while the grass is still about two inches high; it will have a quicker recovery than if horses are kept there until the grass is chewed clear down to the ground. If pastures get tall and coarse, they should be clipped; then the horses will eat the new regrowth for it will be more tender and nutritious. Rotating horses with cattle on pasture land is a good practice, because cows will eat some of the grasses the horses won't, and vice versa. Cows will also eat where horses have defecated, and horses will eat over cow feces. Rotating cattle and horses helps to get the best use from all the grasses in a pasture and also helps to break parasite cycles, for most internal parasites require a specific host. Horse parasites cannot survive in cattle, as a general rule and cattle parasites usually will not live in horses.

Water is very important, especially in summer, and should always be available to the horse except when he is hot from exertion. If there is no creek or pond or other natural water in his pasture, a trough, tank or tub will do, but make sure that the horse has as much water as he wants and that it is always clean and fresh. If you are watering him in a small tub, water him at least twice a day, and preferably three times a day in hot weather. The average horse will drink about 12 gallons of water daily in warm weather, and more than that if weather is quite hot or when he is working and sweating.

A horse should always have salt, especially in hot weather. Eating a little salt helps combat fatigue and helps keep his body chemistry in proper balance by replacing that which he sweats out of his body. Iodized salt should be fed if your part of the country is lacking in iodine. Salt comes in several forms. You can feed it as blocks, or granulated, or mixed with the horse's ration. Many horses like to lick and chew on block salt; it helps give them something to do. If rainfall is frequent in your area, or if there is no high spot in your horse pasture that is free from irrigating water, you will have to make sure the salt stays dry. Make some kind of container and cover for it; otherwise it will dissolve in the rain or irrigating water, killing the grass in the surrounding soil.

If you live in a dry climate, you may find that your horse's feet become brittle during the summer. In dry climates, the hoof tends to dry out and become harder, but if hoofs become too dry they get brittle and are more likely to split and crack. A little moisture in the pasture solves this problem. If the horse waters in a creek or pond and has to stand in moisture when he drinks, or has to walk through some water, it will keep his feet from getting too dry. If there is no natural moisture where your horse is kept, you can create a mud puddle at his water tank. Or you can use a hoof dressing. Some types of hoof dressing are helpful for dry, brittle feet. Your vet can probably advise you on these. If feet are brittle, keep them trimmed and smoothed often, to keep them from splitting.

Heat and bright sunlight can be a problem for horses in summer. If the summer months in your area are hot, make sure your horse has shade during the heat of the day. A few trees and bushes in his pasture will give him shade and a cool place to stand and fight flies on a hot day. If there is no natural shade, a shed or sun roof can be very simple and inexpensively made.

Horses with white markings or light-colored skin are sometimes bothered by bright sunlight. Black skin is the best protection against the sun's rays, but most horses have pink skin underneath white markings, and a few horses have light colored or pink skin all over their bodies. Pink skin will sunburn and peel and can make the horse very miserable. Sometimes a horse with light skin or white markings (with unpigmented skin underneath) suffers from photosensitization, which can be very serious (see chapter 10). Some horses have white markings around their eyes (bald-faced horses, for instance). These horses and horses with nonpigmented eyes (blue or "china" eyes) are often bothered by bright sunlight. Bright summer days and occasional bright winter days when the sun reflects off the snow make the horse's eyes water. Black skin absorbs the sun's rays. An eye surrounded by pink skin will be irritated by sunlight because pink skin reflects the sunlight—right into the eye. Ultra-

Make sure there is shade in your horse's pasture

violet rays can irritate and damage the eye and its surrounding tissues if accumulated in large doses. Cancer of the eyelid is not uncommon in horses that have unpigmented skin in that area (see chapter 10). If you have a horse whose eyes water from the sun's irritation or a horse that sunburns, consult your vet. If your horse has a white face, with white around one eye or both, you can reduce the irritation from bright sunlight by applying something dark around the eye, like black shoe polish or mascara; just be careful not to get any into the eye itself. The black coloring will tend to absorb the ultraviolet rays and they won't be reflected into the eye.

Insects are usually a nuisance to horses in summer. There are many species of biting insects that can cause annoyance and injury to horses. Some of the worst are horse flies, deer flies, stable flies, horn flies and mosquitoes. These insects have specialized mouth parts with which they can pierce the animals' skin and suck blood. This form of feeding can cause a great deal of annoyance and irritation to horses, and if the population of biting insects is very large, it can result in a considerable loss of blood. Horses are more thin skinned and sensitive than most other kinds of livestock and the insects bites are painful.

Horn flies bother cattle, and if there are horses near the cattle these flies bother the horses, too. Horn flies are about half the size of a housefly and are usually found along the neck, back, and underline of the animal host. The constant presence of these flies can be very annoying to horses, making them irritable and restless, and sometimes creating sores along the underline or around the eyes.

The stable fly looks a lot like the housefly but it has dark spots on its gray

Old Possum, swatting and kicking at flies

abdomen and a long slender proboscis. Stable flies are found in all parts of the U.S. but are most plentiful in the central and southeastern states, in irrigated areas and in some coastal areas. They usually occur in large numbers around stables or wherever livestock congregate, and where there are piles of wet straw or other kinds of decaying organic matter. They usually feed on the horse's legs or lower body, making the horse stamp his feet or lick and nose at the flies.

Horse flies and deer flies are perhaps the most annoying flies that attack horses. Deer flies can be black or yellow, with wings marked with brown or black. Horse flies are usually black, brown, or grey. Horses being attacked by horse flies or deer flies will stamp their feet, switch their tails and sometimes begin running in an attempt to get away from them. Horse flies consume more than their own weight in blood at each feeding, and when these flies are plentiful the horse may suffer quite a blood loss. Shelters or sheds of some sort are helpful in protecting the horse from these flies, for they usually do not follow an animal into an enclosed building or a dim place. If flies are bad you may want to stable your horses during the daytime. It also helps if you keep your horse in a well-drained pasture. A horse that lives in a wet pasture or near a swamp will be more bothered by horse flies because they breed in wet areas. Mosquitoes are also a major problem wherever there are swampy or damp areas.

Some of these horse pests can be fairly well controlled with insecticides. For mosquitoes and some species of flies, there is no long-lasting treatment, but there are preparations that can be used daily on the horse to give him some relief. Insecticides can be obtained in a variety of forms, the most common being dusting powders, wettable powders, emulsifiable concentrates and oil solutions.

When using insecticides on or around horses, be sure to use only those products which are prepared specifically for horses; others may be too strong or cause toxic reactions. Dusting powders can be sprinkled on, wettable powders can be diluted and used as sprays, and emulsifiable concentrates and oil solutions are usually used as sprays. Insecticides should always be used according to the directions, for when used improperly they can be harmful to man and animals. Store them in a safe place and use them wisely, avoiding contact with skin and eyes. Use them on a calm day so that air currents won't drift the insecticide to another area.

Insecticides for controlling bothersome flies on the horse can be applied as dusts, sprays or wiped on. Sprays will often startle a horse and must be used with care. If you do use a spray, use a wettable powder, for the oil solutions can damage the hair and cause skin irritation on some horses. Treatment for flies may have to be repeated often, especially if you are riding the horse and he sweats; the sweat washes away the insecticide, making it less effective and shortening the duration of its potency. See chapter 12 for more details on flies and fly control.

When grooming your horse in summer, take into consideration the fly problem before you cut off his mane and make his tail so short that it doesn't even hang down to his hocks. His mane and tail are his most effective weapons against flies. In some breeds a roached mane and a short wispy tail may be the popular style, but in areas where the fly population is large and their attacks on the horse intense, this kind of short hair styling may be almost cruel. A good horseman thinks first of his horse. Manes and tails should be well groomed and neat, but leave your horse enough to fight flies with.

Before we leave the subject of flies, there are two more villains that ought to be mentioned; the bot fly and the face fly. The bot fly lays its eggs on the long hairs of the horse's legs or jaw, or flanks. This fly is very irritating to the horse, especially one species called the chin fly that lays its eggs along the jaw and chin and greatly annoys the horse as it deposits them. Many horses become quite frantic when bothered by the bot fly and care must be taken when handling them at this time because they may jerk their heads up suddenly or strike out wildly with a front foot when the fly comes up under the chin. You can help reduce the bot problem for your horse by clipping off the hairs along the legs and jaw where the little yellow eggs have been deposited, and burning them.

The face fly is a cattle pest and also bothers horses. Some years the face fly will be more of a problem than in other years; this species seems to go in cycles. These flies constantly light on the face and around the eye of the horse, and can make the eyes matter and run from the constant irritation (this fly feeds on eye secretions). Face flies spread pinkeye in cattle and influenza in horses. The subject of flies which bother horses is more fully covered in the chapter on external parasites (chapter 12).

You can keep the face flies from irritating your horse's eyes by applying insecticides or repellents, but be sure you don't get any into his eyes. There are sprays you can use, but many horses are frightened by the sound of an aerosol

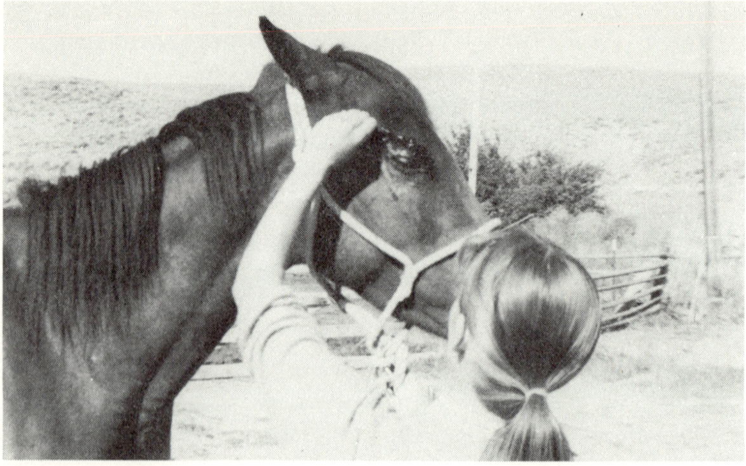

Putting black oil around a horse's eyes to keep the flies away

The black oil is applied with a feather. The horse closes his eye as the feather approaches

spray, especially when it is close to the ears and eyes. Also, it is difficult to keep a spray out of the horse's eyes. If you don't want to use a spray, another effective way to combat the flies is to apply a liquid repellent over the horse's entire eyelid and surrounding area with a feather. The horse will close his eye as he sees the feather approaching and you can apply the repellent with practically no danger of getting any into the eye itself if you are careful. Treatments every other day are usually sufficient to alleviate the face fly problem during the height of the fly season. Repellent sticks can also be used safely around a horse's head.

Perhaps one of the most important factors in caring for your horse in summer is the prevention of disease. Summer and fall are the times of year in which most outbreaks of communicable diseases occur, partly because of the warm weather (which favors the insects that carry and spread diseases) and partly because there are usually more chances for diseases to be spread by direct contact. Many horses become exposed to strange horses during these seasons, because of horse shows and fairs, play-days, trail rides, and other activities.

The old saying that an ounce of prevention is worth a pound of cure applies very well to horses and the disease problem. If you own a horse you should have some sort of yearly vaccination program for him. A great deal of trouble can be avoided by innoculating the horse against the most common contagious diseases. You can't afford to let a horse be sick. Treating a sick horse can be time-consuming, costly and heartbreaking. The horse may be useless to you during the months you need him most. Leaving a horse unprotected from disease is not worth taking the chance of having to see him suffer or even die.

Which diseases you vaccinate against will depend partly upon the area in which you live. Some diseases occur in nearly all localities (tetanus and sleeping sickness, for instance) while others may appear in some areas and not in others. Check with your vet to see which diseases you should include when you vaccinate the horse. On our ranch we vaccinate against sleeping sickness, tetanus, strangles, influenza, and leptospirosis. For many years we only vaccinated for sleeping sickness (tetanus is practically unknown in our area, but in recent years we've vaccinated for this disease also, just to be on the safe side). Our ranch is quite isolated and we've had very few disease problems, but now we occasionally take a horse off the ranch to a competitive trail ride or a show, where he might come in contact with diseases, so we vaccinate for strangles, influenza, and leptospirosis.

Vaccinating all your horses every year can be costly, but losing a horse or having an epidemic run through them all can be even costlier. It pays to vaccinate. You can save some time and expense by giving the vaccinations yourself.

Most shots are given intramuscularly—deep into muscle tissue. These are the easiest kind to give. Relatively few shots are given subcutaneously (just beneath the skin) or intradermally (between the layers of the skin.) There is now a vaccine for sleeping sickness that can be given intramuscularly; these are much easier to give to a horse than the old type of vaccine that had to be inserted between the layers of the skin on the horse's neck.

For intramuscular shots, use a long needle, an inch or so in length. An 18-gauge needle is usually about the right diameter size to use on horses. Needles for livestock can range in diameter size from the tiny 26 to 22 gauge needles we used for intradermal sleeping sickness shots to the large 16-gauge needles we use for giving shots to our cattle. Cattle have a thicker, tougher hide than horses and need a bigger needle.

Always use a needle that is sharp and straight. Dull needles cause more

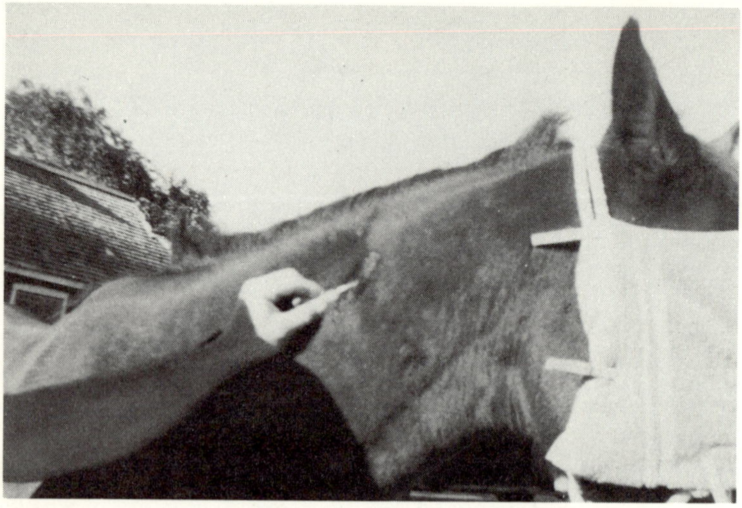

Giving a sleeping sickness shot intradermally, between the layers of the skin. Note the blindfold—a great help for making some horses stand still

pain than sharp ones. Bent needles can break. When giving shots intramuscularly, choose a thick muscle and don't hit any bones. Some of the sites we use on a horse are the chest muscle (on either side of the breastbone) and the thick flat muscle covering the croup (on either side of the backbone) and the muscle below the horse's buttocks.

We like to give shots in the chest muscle because it is a very convenient place. The horse seems to be less sensitive here than on some other parts of his body. The horse is accustomed to you working at this end of him, and when you are this close to his head he can be easily controlled. If a horse is nervous, upset, or disagreeable about having the shot, the injection is much easier to give in the chest muscle than in the rump, for he can't swing his front end around as much while he is being held, and you are less apt to get kicked. When you give shots in the chest muscle, just be careful not to hit the breastbone.

For large shots (anything more than 10cc.) we prefer the heavy muscle covering either side of the horse's croup. If a horse is touchy and nervous about shots in the croup we can split up a large shot into two or three doses and give it in the chest muscles at two or three sites, but the muscle tissue may swell. Large shots given in the chest area are more apt to swell the tissues than large shots given in the croup; the croup area has more muscle tissue to absorb a large shot. The amount of swelling resulting from a vaccination will depend on the type of shot (strangles shots often cause more swelling than some of the others) and will also differ with various horse's reactions. Swelling is unsightly and in some cases may make the horse a little stiff and sore for a while, but it usually will go down to normal within 24 to 48 hours. On small foals we give shots in the muscle just below the buttock, because the chest muscle and croup muscle are not always big enough or thick enough yet.

Giving a shot in the breast muscle

1. Inserting the needle

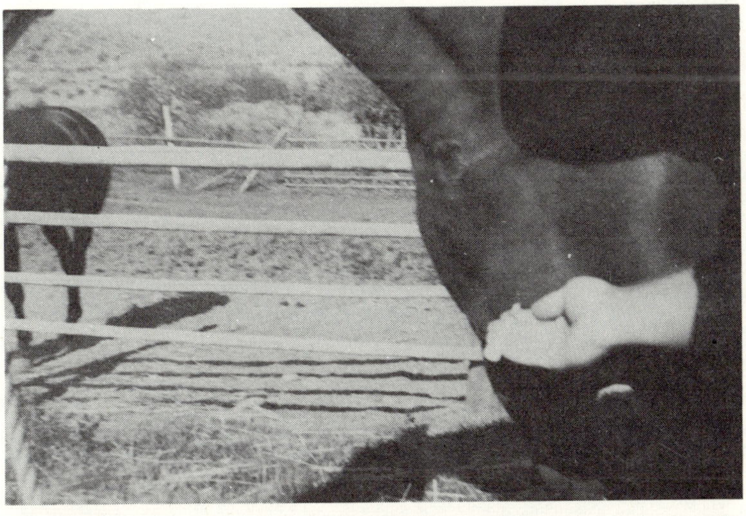

2. Syringe is then attached to the needle and the shot can be given

Here's how we give a shot: first we select the needle and syringe, choosing a size that is appropriate for the size of the shot. Syringes in our medicine chest range in size from one or two cc. (for sleeping sickness shots and for injecting a cow's eyelid with antibiotics to clear up a bad case of pinkeye) to 40 cc. (for giving large shots, such as shots of antibiotics to cattle with foot rot, etc.) The vaccinations we give our horses are usually in the 2 cc. to 10 cc. range. If we are giving a small shot, we use a small syringe. It will be more accurate and easier to measure on the individual calibrations than a larger syringe.

Always disinfect and sterilize the needle and syringe. Needles and glass-and-metal syringes should be boiled (they should be taken apart first) for 15 minutes. Plastic syringes will not stand boiling and are usually meant to last for only one shot. It is best to use syringes that can be boiled, to lessen the chance of contamination. After use, syringe and needle should be washed, well rinsed, boiled, and then kept in a clean place. We leave the plunger rather loose when storing the syringe so that it will thoroughly dry inside. Every so often you have to get new rubbers for the plungers when the old ones get worn.

Just before using a syringe, we usually rinse it out with a bit of alcohol, and then we also clean the area on the horse where we will give the shot. If the horse is really dirty, we brush him well. When his hair is free of dirt we clean a spot with alcohol. For this purpose we put some alcohol in a cup and make small cotton swabs, as many swabs as there are horses and shots to give, to put in the cup and take out for the vaccinating. Also, we like to use a different sterilized needle for each horse. This helps to insure that there will be no chance of contamination from horse to horse. One sterilized needle can be used to fill the syringe each time from the vaccine bottle.

To fill the syringe, insert the sterile needle through the rubber top of the bottle, tip the bottle upside down, then slowly pull the plunger out to create a vacuum in the syringe. This vacuum will be filled by vaccine from the bottle. Pull the plunger slowly; creating too much vacuum too quickly can break a glass syringe.

If you have a large bottle of vaccine and several horses to vaccinate, you will find that filling the syringe becomes harder each time because of the vacuum created in the bottle. If this happens, you can insert some air into the bottle from the syringe (fill the syringe with air, then inject it into the bottle); then the vaccine will come out more easily. We do this when we are vaccinating several horses or a group of cows and will be using up the entire contents of the bottle. But it isn't good to inject air into a bottle that you plan to keep and use again at a later date, for this can contaminate the rest of the bottle.

And speaking of keeping well, note the expiration date on every bottle of vaccine or antibiotic that you buy, and don't buy any that is getting old (never buy any that has already expired). Try to buy vaccine in a bottle size so that you can use it all at one time. Antibiotics will keep fairly well, but once opened, a bottle of vaccine should be used right away. Don't save fractional contents. Using part of it one day and the rest a few days later is all right, but trying to keep it several months or a year later is something else and should never be done.

When filling the syringe, try not to get any air into it. If you do have air in it when it is filled, hold the syringe in a vertical position (needle straight up) and press gently on the plunger until all the air has been pushed out. You can squirt this air back into the vaccine bottle and not risk wasting any vaccine if you accidentally push too hard.

When we give shots, we like to have the horse out in the open rather than tied to a fence or in a stall because there is nothing to get in your way if he moves around a little. Have someone hold him for you. On a few of our horses that are nervous about shots, a blindfold helps. A blindfolded horse will usually stand quietly because he's afraid to move (he can't see where he is going), and since he can't see what's going on, he won't be aware of the exact moment of injection. But don't try to blindfold foals or young horses; it can frighten them. Some young horses will plunge about blindly when they can't see, whereas an older horse will usually stand still.

On a really bad horse, we use a blindfold and also tie up one front foot to make him stand quietly without fighting and kicking. One other trick that can help with a really fussy horse is to put him in a trailer (our horses all feel at home in a trailer eating a little grain) and then injecting the shot in the croup muscle, standing in the other trailer stall with the partition between us and the horse.

If shots are given properly the horse won't flinch or jump. Swab the area with a piece of cotton soaked in alcohol to clean the place where you want to insert the needle. Then detach the needle from the syringe, and, holding the needle in one hand, thump the area soundly several times with that same hand, using the side of your fist. This thumping (not done too hard, of course, but hard enough to do the job) does not hurt the horse and tends to desensitize the area and mask the prick of the needle. If you try to jab the needle in with no preparation, the horse will jump every time.

After thumping a few times, jab the needle in with the same rhythm. The mistake a lot of beginners make is in not jabbing hard enough to push the needle clear in. After a few tries, you'll know about how hard to put it in—swiftly but not *too* forcefully—just enough to get it in all the way. The faster the needle goes in, the less it will hurt. With this method, if the horse does flinch or jump a little, you can then wait until he settles down before attempting to give the shot. The syringe can then be attached to the needle and the shot given. I've seen some people try to jab the needle in still attached to the syringe, usually not getting the needle in deep enough, and then they have to try to give the shot to a jumpy, fighting horse. This often ends up with the needle coming back out and spilling part of the vaccine.

After giving the shot, the syringe and needle can be removed together. Sometimes a shot may ooze a little after you remove the needle, but if you will press and rub the area with your finger this oozing will stop.

Another advantage to putting the needle in alone (besides the fact it's easier on the horse) is that you can tell whether or not you've hit a blood vessel *before* you inject the vaccine. Never put vaccine, antibiotics or any other

Lynn vaccinating Bambi

1. The area is cleansed with alcohol

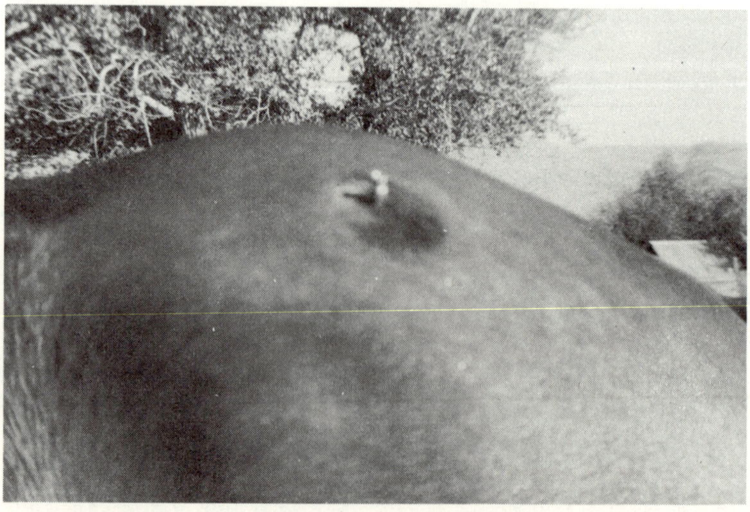

2. The needle is inserted

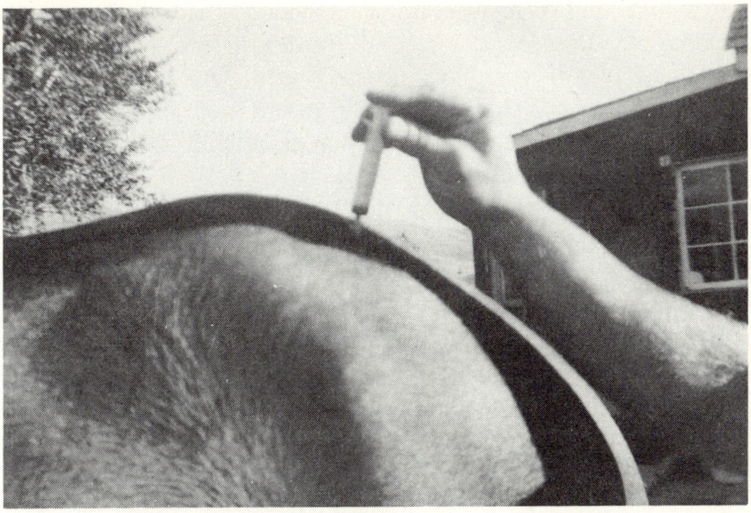

3. The syringe is attached, and the shot given

4. Syringe and needle are then removed together

intramuscular injections directly into the blood stream, or you may have a serious or fatal reaction. Wait a second or two before you attach the syringe to the inserted needle. If you have struck a vein (veins are nearer the surface than arteries; you are much more likely to hit a vein than an artery) the needle will begin to drip blood. If this happens, pull the needle out and try again in a slightly different spot.

Another way to tell whether or not you've hit a vein is to pull back a little on the plunger after the syringe is attached to the inserted needle. If blood appears in the syringe, remove needle and syringe and try again.

Teamwork helps when giving shots. It's not impossible to give shots by yourself, especially to a horse that knows and trusts you, but it helps to have someone at the horse's head, holding him still and talking to him, reassuring him and keeping him calm and quiet. The person at his head can make a big difference, distracting him from the shot and keeping him from moving at the wrong time.

When giving a horse his yearly vaccinations, you may be giving him quite a number of shots. Some of these can be given to him at the same time, with no ill effects, while others you may want to space out and give to him individually at a different time. You may want to discuss a good vaccination schedule with your vet to see when it would be best to give the various shots. It's usually wise to start your summer vaccinations early, before the various diseases (and the insects that spread them) become a problem; your horse will be better off to have immunity before he comes in contact with a contagious disease.

Once in a while a horse will have a reaction to a shot. There are two kinds of reactions, though in a sense you might say that the difference is one of degree. A person should be prepared to give the antidote or call the vet if a severe reaction occurs. A horse may have an allergic reaction to vaccine, which manifests itself in swellings or hives, and in severe cases congestion of the lungs and shortness of breath, like in an attack of asthma. A severe allergic reaction can be alleviated by proper medication (antihistamine). Allergic reaction is an individual horse's sensitivity to a particular type of shot and may occur even if the shot is given properly. All of Nell's offspring are somewhat sensitive to vaccine. Nikki is the worst. In 1972 she got severe respiratory congestion following her encephalomyelitis shot, her strangles shot made her legs swell, and her influenza shot produced a local swelling about the size of a grapefruit. But these reactions lasted only a few days.

Another type of reaction, which is more severe (but also more rare, for it usually only occurs when vaccine is given improperly), produces a severe shock to the horse's system. This can happen in the case of overdoses, injection of vaccine directly into the bloodstream, or in giving shots of certain types such as strangles vaccine to a horse that has already been sensitized (in the case of strangles vaccine, *never* give a shot of vaccine to a horse that has already contacted the disease and has the disease in the incubation stage, or to a horse that has already had a shot). This reaction, called anaphylactic shock, is caused by an extreme sensitivity to the substance being injected; the horse's system is already sensitized, either by the disease itself, or by previous vaccine. When a horse suffers from anaphylactic shock he has sudden attacks similar to severe allergy attacks and he may collapse and die (heart and lungs are restricted).

We have never had this happen to any of our horses, but we now keep an antidote on hand at all times just in case. We once had a calf nearly die when we gave him a routine shot of antibiotics during treatment for scours. That particular calf just happened to be allergic to penicillin, and within 15 minutes was flat on the ground turning blue and gasping for breath. We rushed him 12 miles to the vet and saved him, but now we keep the antidote on hand ourselves.

All vaccine bottles have a printed warning concerning the possibility of anaphylactic shock. The best way to avoid this reaction is to be sure to read the instructions for the vaccine and be sure to give the shot properly, in the locations described and in the proper dosage for that particular horse, and discuss the vaccination with your vet beforehand if you think the horse may have already contracted the disease; in some cases he'll tell you not to give the vaccine.

If shock ever occurs, call the vet immediately. If called in time, he can administer an antidote which may save your horse's life. If you are like us, with many horses (and cattle) and don't want to take any chances, talk to your vet ahead of time and pick up a bottle of medication to keep on hand for emergencies. You can usually get a bottle of epinephrine which will keep for many years. You may never need it, but one small injection may save a horse's life, if shock or severe allergic reaction ever do occur.

Giving shots is not difficult, and if you have several horses you can save a lot of money by doing it yourself. It really pays to vaccinate your horses, for just one loss will cost you more than you'll ever pay for vaccine for many years. In this day of trailers and vans and jet age horse transportation, there are many opportunities for diseases to spread. Vaccinating for the most prevalent diseases in your region is now a must, especially if you ride with other horsemen, keep your horse at a stable where he will come in contact with other horses, or if you take your horse to shows or other activities where there will be many strange horses. Even if you live on an isolated farm you can't really be sure that you are isolated from diseases. When a disease can be spread by insects rather than by direct contact (horse to horse) you can never really be isolated. To take a chance on not vaccinating your horse because you are isolated, is just that—taking a chance.

Your horse's health is extremely important. A sick or unhealthy horse is of no use to its owner. The conscientious horseman will do his best to keep his animals healthy and in good condition. You can work toward this goal by having a good yearly vaccination program for infectious diseases (usually vaccinating your horses in the spring or early summer), worming your horses regularly to keep them relatively free of internal parasites, and reducing the fly problem as much as possible during the summer months. Summer care also involves feeding a horse adequately and properly (especially if you are riding him), taking care of his feet (trimming them often, or shoeing him if he needs protection from excessive wearing, and resetting the shoes before his feet grow too long), giving him some form of regular exercise, and keeping that horse in the safest possible surroundings.

If you do your best to keep your horse comfortable and healthy and in good condition during the summer months, you can expect a pleasant summer working with him. If you have to cope with the problems of sickness, or debility from internal parasites, or colic from internal parasites or improper feed, foot problems due to neglect, or a horse that is out of condition, or uncomfortable and irritable because of external parasites, your summer with your horse will not be nearly so pleasant, and you may have to forego plans of riding, training, showing or competing with that horse. Summer in some ways is an easy time for horses because of green grass and pleasant weather, but in other ways it can be a difficult time as well, because of the various warm-weather problems mentioned in this chapter, and because of the various stresses of riding, training, and using the horse. If you use the horse regularly or strenuously, he must be given adequate care, proper feeding, and proper conditioning to be able to hold up under the steady use. And he'll need a helping hand health-wise to keep him free from parasites and infectious disease. Good summer care of your horse is essential if you hope to keep him sound and healthy and energetic throughout the riding season.

16

CONFORMATION—WHAT MAKES A GOOD HORSE

CONFORMATION IS PERHAPS THE MOST IMPORTANT ASPECT OF A horse's overall makeup. Conformation determines whether he will be a good athlete or a poor one—whether he will be agile or clumsy, speedy or slow, able to stop and turn quickly or unable to balance and handle himself with speed and smoothness when he changes directions. His conformation determines in large part whether he will be safe to ride (or whether he will stumble or trip on his own feet), whether he will stay sound with constant use, and whether he will be able to perform well or poorly at any number of jobs he may be asked to do. Conformation means how the horse is put together—how his bones, tendons, muscles and other parts are shaped and how they fit together.

Let's take a quick overall glance at the horse, and then take a more detailed look at his feet and legs, for feet and legs are the most important parts when it comes to determining whether the horse will be a good athlete or a poor one.

The horse is a four-legged hoofed animal that has a long neck which allows him to graze, and which he uses to help balance himself. Almost every part of the horse except the tail corresponds with a similar part of the human body. The stifle is the same as the human knee and has the same kneecap. The horse's "knee" (foreleg) is comparable to the human wrist and has the same seven bones. The hock joint in the hind leg is the same joint as the human ankle. The cannon bone in a horse is similar to the human middle finger, and the fetlock joint compares with the middle knuckle on that middle finger.

The horse has a large head. This large head helps him maintain his balance, for he is able to raise or lower it as needed to counterbalance whatever is happening at the other end of his body. His large head also allows him to have large, wide nostrils, giving him a large flow of air when he needs it. A horse cannot breathe through his mouth very readily (as a human does when he gets out of breath), so he needs to be able to take great gulps of air through his nose when he is running hard or trying to catch his breath from exertion.

The design of the horse's head and neck enables him to move his head quickly so that he can see in all directions. His wide head, with far-apart eyes, is also an advantage in this area, for he can see a different picture with each eye and can watch things going on around him (on both sides of him) much more easily than a human can.

In judging a horse, the horseman looks for small, finely formed ears (which are more attractive than large floppy ones), eyes set wide apart with good width of forehead, a finely formed face, and small muzzle. The face should not be too narrow or it may interfere with the horse's flow of air when breathing hard—due to the narrowing of the nasal bones. The head should be set onto the neck in such an angle that the horse can flex at the poll. A long neck is more desirable than a short one, for the horse with the long neck can usually balance himself better. A short, thick neck is usually a hindrance to a horse's balance and action. The neck and head should be compatible and complement one another: a horse with a long neck and heavy head will not have as good balance as a horse with a long neck and lighter head. A coarse heavy head requires a heavy short neck to carry it. The angle of a horse's neck may vary a little with his breed, but in any breed there are several undesirable types of necks—the short, thick pony neck (which is not very flexible), the upside-down ewe-neck, and the thick heavy crest.

A horse with good conformation is well built and well balanced, with a smooth, pleasing overall appearance. His body should be well proportioned;

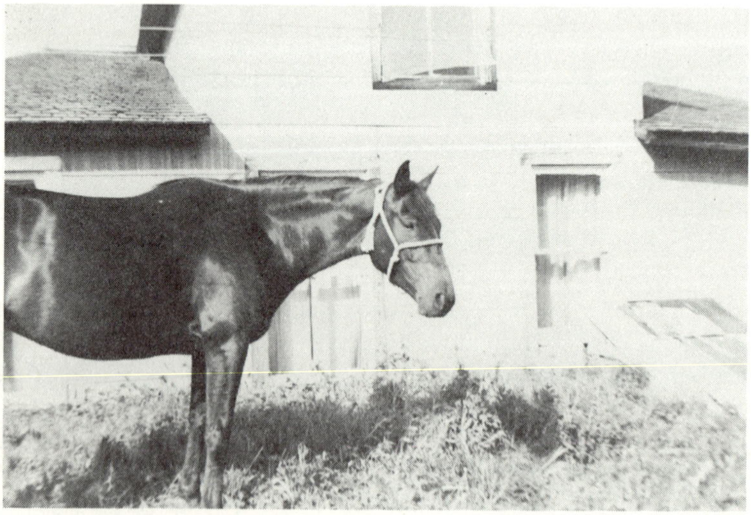

Nell in her younger years, about five years old. The judge of horses looks for small, finely formed ears, eyes set wide apart with good width of forehead, a finely formed face, and a small muzzle

Nell (riderless, at left) in middle age (about 10). Note the long neck, good withers, long shoulder, well-balanced body, long hip, deep girth, and good legs

even though our main concern will be his feet and legs, the upper structures of his body are important, too, for they play a part in the movement of his feet and legs. The proportions of his body can also be a factor in determining whether or not there will be limb interference when the horse is moving.

Body conformation varies with the breeding of the horse and this factor should be taken into consideration when evaluating him. For instance, an Arabian will have a shorter back than a Thoroughbred. The Thoroughbred generally has a longer, trimmer body than a Quarter Horse. The Quarter Horse is often shorter and heavier-bodied. But there are some good general rules for all breeds. Body and leg conformation should be compatible. A long-bodied horse should be long legged, and a short-bodied, short-backed horse should be shorter legged. If a short-backed horse has legs that are too long for the body, he will be more liable to overreach when he travels, striking his front feet with his hinds. A horse that is too long backed may have a swing to his gait that changes the movement of his legs so that he is not traveling perfectly straight, making him prone to speedy cutting or cross firing. When judging a horse's conformation, make sure his body conformation complements his leg conformation.

When you judge the body of the horse, look for a deep heart girth and well-sprung ribs. A horse needs plenty of room for heart and lungs and the inner workings of his body. Horses that are wide through the ribs and deep through the flank are usually better doers and do not draw up as quickly on long rides as horses that are "bacon sided" (flat sided, with very little spring of rib) and too shallow through the flank. Viewed from the front, the horse's barrel (rib cage) should be wide. The greater the width (curvature of the ribs), the greater the horse's chest capacity will be. The closer the rib cage is to cylindrical shape, the more volume it will have.

There should be sufficient width between his front legs, yet the chest and front end of a horse should not be too wide and heavy. A wide breast is a disadvantage in a riding horse for it cuts down his speed and agility. The horse should not be too front-heavy. He should be wider at his hind end than at his front, for he needs adequate strength of loin and quarters and it is best if his center of gravity is not so far forward, allowing him freer use of his front legs. Ideally, the horse's body should be wedge-shaped—narrow in front and wider behind. The sternum (breastbone) lies between the forelegs and carries the lower ends of the first eight pairs of ribs. If the breastbone is too wide and the pectoral (breast) muscles are too bulky, the horse will be too wide and heavy in front. Good muscling between the front legs should look like an inverted V.

Extreme heaviness of muscle on thighs, arms and forearms is not necessary and is actually a hindrance to free action. A heavily muscled horse may have a quick burst of speed, but he will tire sooner than the trimmer horse. Speed and endurance come from long, lean muscles, not bulky ones. The ideal quarter in a riding horse is long and level with sufficient width and long, level muscles. Long, slender muscles denote speed; round heavy muscles denote weight carrying and power.

A good horse has a long underline and a straight back. Another way to describe this is to say that the horse is close coupled. A long back is often a weak back, unless the horse is strong across the loin. Loins should always be muscular and slighty raised, never hollow. A mare usually has a slightly longer back than a stallion or gelding; this gives her room to carry a foal. A slack-backed or sway-backed horse usually has a weak back. Some horses become sway-backed with old age, particularly old mares that have produced many foals.

The backbone is made up of 7 cervical (neck) vertebrae, 18 dorsal vertebrae, 6 lumbar vertebrae and an extra bone, made up of several fused vertebrae, the sacrum. In young horses there is both lateral and up-and-down movement of the spine when the horse travels, but in mature horses there is little flexibility in the spine. The Arab usually has an extra dorsal vertebrae and one less lumbar vertebra. Because each dorsal vertebra carries a pair of ribs, most Arabian horses have 19 pairs of ribs instead of 18, giving the Arabian a little more total chest capacity; hence his great stamina and endurance. In any horse it is important that the last 11 pairs of ribs extend well back, for these 11 produce the greatest effect upon increase in chest cavity as the horse breathes.

The various breeds differ on the croup. The Arabian generally has a flatter croup than other breeds. But aside from breed differences, there are some good general rules for conformation in this area. The line from the point of the hip to the buttocks should be long. The angle made by this line and the line from the point of the buttock to the stifle joint should be very sharp. As in the shoulder, the sharper the angle, the more freedom of movement and better action in the

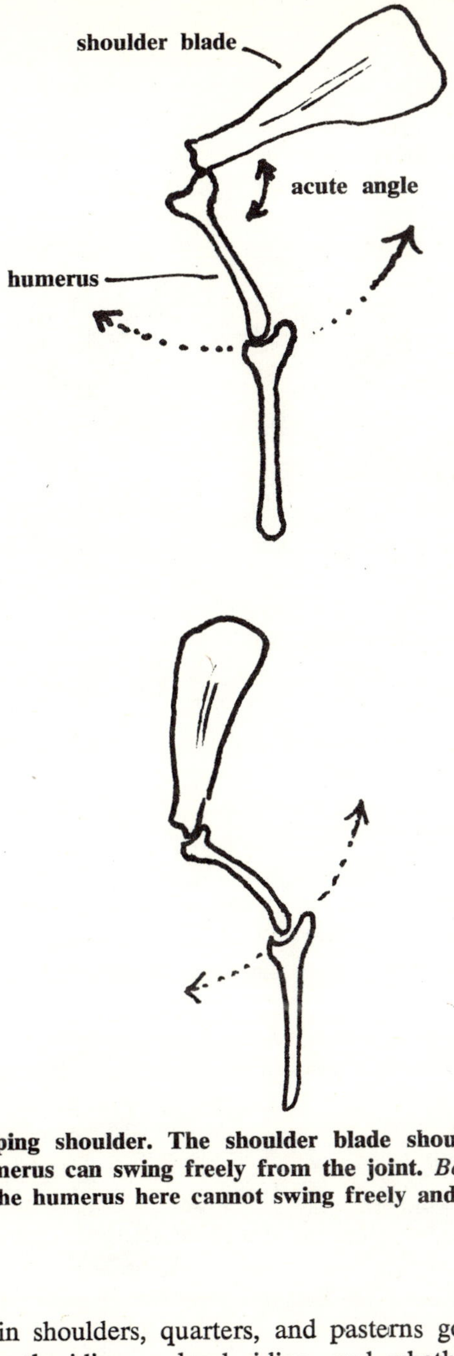

Top, **a good, sloping shoulder. The shoulder blade should slope back at such an angle that the humerus can swing freely from the joint.** *Bottom,* **an upright shoulder, short and steep. The humerus here cannot swing freely and the foreleg has a cramped movement**

legs. The angles in shoulders, quarters, and pasterns govern whether or not the horse will be smooth riding or hard riding, and whether or not he will remain sound and active with hard use. The angles in shoulders, quarters, and pasterns help determine how much concussion the bones of the feet and legs have to take, and concussion (too much) is a major factor in the breakdown of vital struc-

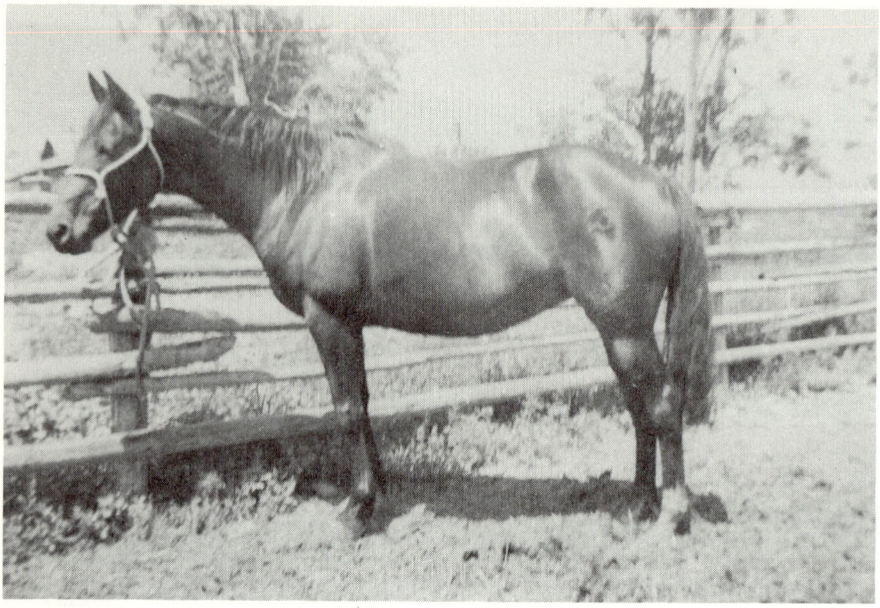

Nell in middle age after raising two foals (and after wire cuts on both front pasterns). She still looks good from any angle

tures of the horse's feet and legs, causing lameness and unsoundnesses.

These angles greatly determine the horse's action. Good action means freedom of movement of the legs. If the shoulder is too steep, it puts the front legs too far under the horse, making him unable to reach out in front as he should. An easy riding horse should be able to reach out past his nose with his front feet at the extended trot and at the canter. He should flex well at the hocks and knees; his action appears springy and not wooden-legged. A sloping, well-laid back shoulder is essential for speed (as well as for smooth action) because the humerus (bone of the forearm) can be flexed more completely and rapidly. A horse with an upright shoulder is cramped in his foreleg action and his gaits lack elasticity; his movements are less springy and there is more jar when his feet come to the ground.

The withers of the horse are composed of the spines of the third to ninth dorsal vertebrae. If withers are very thin and high, an ill-fitting saddle can bruise them and sometimes cause a fistula. When trying a saddle on a horse, you should always be able to get at least three fingers between the saddle and the top of the withers after you are mounted. If there is not enough room, your saddle tree is too wide and low for that particular horse and you need a lot more padding to raise the saddle up, or a higher-built saddle.

Good withers are important in a horse. A horse with good high withers holds a saddle well and you don't need a painfully tight cinch, double rigging and breast collar to hold the saddle in place. A round-backed (mutton-backed) horse with low withers must be cinched more tightly and still the saddle

Note well-balanced head, neck, body, legs—good angle in shoulder, hips, pasterns. Excellent legs and hock angle (a line dropped from buttocks that travels down the back of the hock and cannon)

Compare these two horses:

Ginger, a five-year-old grade mare that my brother and I used in years past as a 4-H project and cow horse. She is short necked with short, steep shoulders and pasterns, very little wither, apple rumped, with hind legs "out in the country" and a bit sickle-hocked. Ginger is higher at the hips than the withers; her back is too long.

Nell at age 19, after having four foals. She is long necked, long shouldered, with good slope to shoulder and pastern. She has good withers, a long hip, a better croup, is deeper through the girth and has better leg angles. Her hind legs especially are a contrast to Ginger's crooked ones

may turn with you when you mount. A tight cinch can be very irritating to a horse as well as inhibiting; it tends to cut down his breathing capacity, just as a tight belt or girdle cuts down yours. When traveling up and down steep mountainsides, or when changing directions suddenly (as when cutting cattle) it is difficult to keep a saddle in its proper place on a round-backed horse. Choose a horse with good withers and you'll have a minimum of saddle problems if you use a saddle that fits him. A narrow-fronted saddle is possible only on a horse with a narrow front and good withers. Most horses with a dash of Thoroughbred blood or Arabian blood have high withers and good backs for carrying a saddle that will not be pulled sideways. Double riggings and breast collars came into existence with the advent of the Quarter Horse, who in general (unless he carries some Thoroughbred blood in his veins) has a broad, muscular back that does not hold a saddle very well.

A horse with good high withers usually has a long shoulder that slopes and is well laid back. A shoulder that is too straight and upright limits the horse's freedom of action and gives a shorter, choppier stride. A horse with short, straight, upright shoulders usually also has a short neck and short, upright pasterns. Angles in a horse's body tend to be similar throughout the body.

Ideally, a horse should be the same height at the withers as he is at the highest point of the hips. In young horses (up to two or three years of age) the croup may be higher than the withers, for the withers often attain their full height later. If there is a choice, you should choose a horse with croup higher than withers rather than vice versa (if croup and withers are not the same height). A high croup is a sign of speed in a horse, if the muscles of the loin and quarter are well developed. The highest point of the croup is where the hind limb attaches to the spine. A horse with a good croup and proper hindleg bone angles will have good speed because he has a long swing of leg. A horse with a "droop croup" (a high croup and a sloping quarter) is called goose rumped.

A well-balanced horse has about the same distance from fetlock joint to his underline as from his underline to the top of his withers. In other words his front legs are about the same length as the depth of his body from withers to underline.

Now let's take a closer look at the horse's feet and legs. Remember that the horse should always be judged while he is moving as well as while standing still. He should have a good way of going; he should travel "straight and clean," not brushing or interfering in any way. A horse with good conformation has straight feet and legs that move forward in straight lines. Stand directly in front of the horse as he is being led toward you, at both the walk and the trot. Then have him led straight away from you at the walk and stand directly behind him. Some points of bad conformation of feet or legs will show up much more sharply when he is moving. Watch closely as he travels so thay you can see any possible faults or deviations from straight forward leg action. If his front feet paddle outward or wing inward, or if his hind feet come dangerously close to one another when in motion, or his hocks roll out-

Have the horse led toward you and away from you at the walk and trot

Nell. Note the straight leg action. Her legs move forward in straight lines; there is no winging, paddling, or interfering

ward, the horse does not have good conformation.

A good judge of horses always starts from the ground up, for any horse is only as good as his feet and legs. If they are weak, poorly formed or crooked, they are likely to go bad with steady use, and then the horse will be lame and useless. Poor conformation of feet and legs can contribute to some type of lameness and in some cases may actually be the cause of lameness and unsoundness. The horse is primarily an athlete and his main purpose is moving. How well he moves determines how useful or how valuable he will be to his owner, whether he is a race horse, roping horse, endurance horse, ranch horse, show horse, competitive games horse, or pleasure horse. Any horse is only as good as his feet and legs, for they are what propel him and carry him over the ground. If his legs are unsound from injury, or poorly built to begin with, he won't be a very good horse.

The various breeds of horse differ a little in some points of conformation (the Arabian has a flatter croup and higher tail carriage than the Quarter Horse, for instance, and a shorter back than the Thoroughbred), but these points are rather minor as far as the basic functions of the horse are concerned. The fundamentals of good conformation are the same for almost every breed, especially when we talk about conformation of feet and legs. Good straight legs that move forward in straight lines are always desirable over crooked legs that deviate from a straight line when in flight. Sickle hocks are sometimes tolerated in the Quarter Horse breed, a tendency toward cow hocks can be found in some Arabians, and low, thin heels are found in some Thoroughbreds, but as a general rule straight, well-built feet and legs are always preferred.

Feet should be well shaped and of a desirable size, proportionate to the size of the horse. A horse is a large animal and needs enough hoof to support his weight. If feet are too small, the horse will not hold up under strenuous use and the increased shock of concussion in the small area may cause lameness from such problems as navicular disease. A large or heavy horse needs feet of an adequate size to support his weight, while a small horse should have smaller feet. Selective breeding, particularly in the Quarter Horse breed, has resulted in some horses having feet that are too small for their body weight. Small feet may look neat and dainty, but a horse as heavy as the Quarter Horse needs sufficient foot structure to support his body. If the feet are too large, the horse will be clumsy. For good sound feet and maximum agility, the feet should be of appropriate size for the size and weight of the horse.

Hoofs should be wide at the heels, not narrow or contracted. Wide heels give and spring apart when the foot hits the ground, absorbing some of the concussion. Otherwise all the shock and jar of hitting the ground would be transmitted directly to the bones of the foot and leg and joints. Front feet are generally larger, rounder and stronger than hind feet, for the front feet support nearly two-thirds of the horse's weight.

The foot should be fairly deep at the heel. When shod, at least an inch of hoof horn should remain between the shoe and the back of the coronary band. The sole should be almost round, with deep frog, good strong bars, and definite grooves on each side of the frog. The wall of the hoof is replaced by new horn every eight to ten months. The hardest and toughest horn grows at the toe; the toe of the hoof helps to pull the horse along and it therefore suffers more wear and tear than the rest of the hoof wall. The bars of the foot help to take additional weight upon the heels and to act as wedges which keep the heels from caving in and contracting. Every time the foot takes weight, the frog forces the bars apart, lifts the plantar cushion and exerts a pumping action on the blood vessels within the foot. Contraction of the heel usually results from lack of weight upon the frog. Without the frog taking weight, there can be no pressure on the plantar cushion which spreads the lateral cartilages apart.

The sole should be slightly concave in the front feet and more so in the hind. Flat feet are not desirable in a horse, for the flat soles are more easily bruised. A flat sole is a bad fault, for the navicular bone, over which the flexor tendon glides, lies within the foot just above the center of the frog. A horse with flat feet is more likely to suffer from navicular disease than a horse with good concave soles. Horses with naturally flat feet, or horses that have dropped soles from foundering, often go lame unless they are correctively shod.

There are many individual differences in the foot structures of different horses. Some have much stronger feet than others. Ideally, the hoof wall should be fairly thick, pliable, resistant to drying out, and should grow normally. The sole should be thick so that it won't be easily bruised, and should shed normally. The bars of the foot should be strong and well developed, and the frog should be large and healthy, and should be exactly centered in the sole, with the point of the frog pointing to the toe of the foot. A pigmented hoof wall is preferable to a white foot, since the white hoof wall is not as tough. White hoofs are more brittle than pigmented hoofs—more easily dried out and subject to breaking and cracking.

If a horse has good leg conformation, his feet will usually wear evenly; the foot will break over squarely on the center of the toe, and he won't wear one side of the foot lower than the other or "dub off" one side of his toe more than the other. To a certain extent, the conformation of the horse's legs determines the shape of the feet, how the feet wear, and the flight of the feet—how they are picked up, swing through the air, and are put down.

The angle of the pastern should be the same as that of the hoof. If the angle is broken, for instance, if the toes are too long or the heels too long (throwing the foot off balance), there will be strain on bones and tendons. Proper trimming of the foot can help to correct a broken foot angle.

Ideally, the angle of foot and pastern should be about 45 degrees. This will vary a little with the individual horse and the breed. An Arabian or Thoroughbred will usually have a well-sloped foot and pastern and the angle may even be slightly less than 45 degrees, while a Quarter Horse may have a steeper foot and pastern and have an angle that is more than 45 degrees. Hind feet angles are normally a little steeper than front feet, so a horse with a 45-degree angle in front will usually have about a 50-degree angle in the hind foot; a horse with a 50-degree angle in front will have about a 55-degree angle in the hind, and so on. The foot and pastern angles should not be either too sloping or too steep. If it slopes excessively, the pastern will be weak. If it does not slope enough, the foot will not have enough give and the horse will have a choppy, jarring gait likely to damage the feet and legs from excessive concussion. If the pastern slopes, it transfers some of the weight to the tendons as the horse moves. If the pastern is too upright, a far greater amount of the concussion is transmitted directly to the bones.

When the horse moves, the foot should break squarely over the center of the toe, and the flight of the foot should be an even arc, the foot landing again squarely on the ground. In a normal foot of good conformation, the foot in motion reaches the top of its arc just as it passes the opposite leg. If the horse has a long toe and less than a 45-degree angle to the hoof wall, the long toe acts as a longer leverage point and delays the break-over of the foot. The added effort required to force the foot to break over makes the horse have a long sweeping stride. A horse with sloping pasterns, shoulders and feet will have a smooth ride and longer strides than a horse with shorter toe and steeper angles. When the horse has a short toe and a higher heel and steeper angles, he has a short choppy stride and increased concussion, making him more subject to ringbone, navicular disease, and traumatic arthritis of the fetlock joint. The foot breaks over quickly, comes to the ground at a sharp angle, creating quite a bit of jar.

Since the front feet bear two-thirds of the horse's weight, they are more susceptible to injury from concussion and stress. The front feet of a horse should be quite round and should have good wide heels. The sole should be slightly concave. The frog should be large, with a good cleft, and should divide the sole into two equal halves. Unequal size of the halves may show that the horse is base-wide or base-narrow in the front legs. Hind feet are more pointed and more concave than fronts. We will discuss abnormal conditions in the feet when we talk about unsoundnesses in the next chapter.

Poor conformation of the legs is in itself not an unsoundness (except for curby conformation), but it is a weakness, predisposing the horse to certain types of lameness and unsoundness.

Front legs should be well balanced and straight, with proper length of bone and proper angles so that there is no undue stress on any one part of the leg. Both legs should bear weight equally and they should be perfectly straight when viewed from the front. A line dropped from the point of the shoulder should go down the center of the leg, bisecting the forearm, knee, cannon, fetlock joint, pastern and hoof. Toes should point directly to the front, and the feet should be exactly the same distance apart as the forearms are where they come out of the chest. Crooked legs put a strain on the joints.

Knees should be large, flat, and well proportioned. They should face directly to the front and not point toward or away from one another. They should not be small and pinched up; a pinched-in knee crowds the bone, tendon and cartilage and hinders free, smooth action. A knee must never be too small. The front should be flat, and the cannon bone should be placed centrally. The elbow must rest squarely on top of the forearm and not turn in or outward. An inturned elbow will make the horse throw

A B C

A. Normal front foot and pastern with a 45-degree angle

B. Foot and pastern angle less than 45 degrees

C. Foot and pastern angle more than 45 degrees

Broken Foot and Pastern Angles

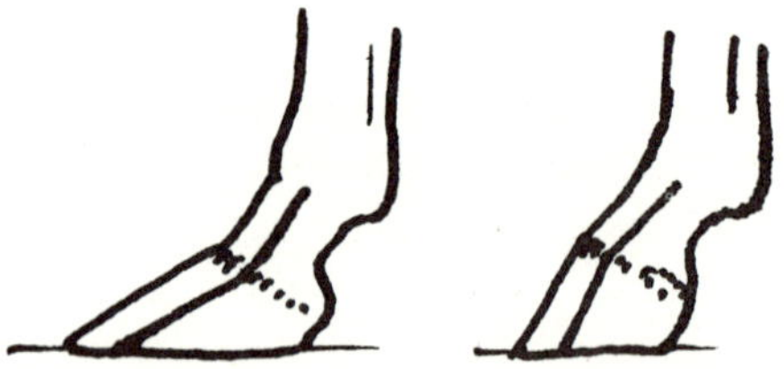

A B

the front foot outward when the knee is flexed. If a horse doesn't travel straight it is sometimes because the set-on of the elbow is not straight.

The cannon bone should be directly centered under the knee. If it is too far to the ouside the horse has bench knees (offset cannons) and will be more susceptible to splints because of the extra strain on the inner splint bone.

If the cannon bone is too far back (the head of the cannon lies behind the line of the radius and the front line of the foreleg looks concave at the knee) the horse is calf kneed. This is very weak conformation.

Cannon bones should be short, for they are mainly just for support. The horse has muscles in the upper parts of his legs; the muscles in front of the forearm are extensors which pull the lower part of the leg forward, and the muscles at the back of the forearm bend the knee. All of these muscles continue downward below the knee as long tendons; there are no muscles below the knee or hock. A long forearm allows longer muscles and shorter tendons. The horse will have more speed and agility if he has long forearms and short cannons, for it is the slim, lengthy muscles (operating through tendons in the lower legs on the lever system) which produce speed. Very few leggy horses with long

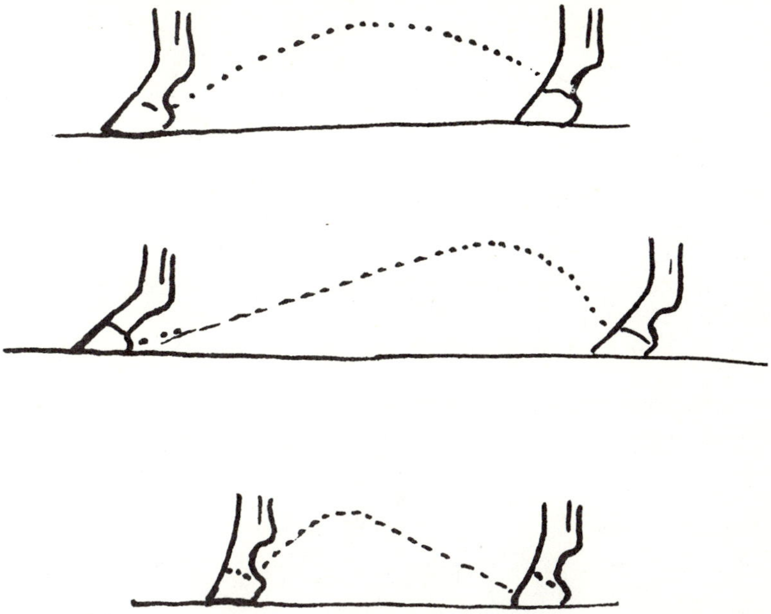

Flight of the foot; ***top,*** **a normal foot with an even arc;** ***middle,*** **small foot and pastern angles, creating a long, sweeping stride; large foot and pastern angles, creating a short, choppy stride**

cannon bones have much speed or stamina; a horse with long forearms and short cannons is a better athlete than the horse with comparatively short stubby forearms and long clumsy cannons. The cannon bone in the front leg is shorter than the cannon of the hind leg. The metatarsal bone of the hind leg is an inch or so longer, coming approximately to the height of the chestnut on the front leg (the chestnut is the horny growth on the inside of each front leg, above the knee.)

Viewed from the side, the cannon should have good flat bone. Flat bone is actually a combination of bone and tendon, being wide from front to back; the tendon should be set well back of the cannon bone, giving the appearance of depth and flatness. Thus, when viewed from the side the cannon should appear wide, while viewed from the front it is more narrow.

A common fault is being "tied-in" close to the knee or hock. This means that the tendon is too close to the bone, causing friction between bone and tendon, a condition that can cause wearing and other problems between the moving parts. Undesirable round bone occurs when the cannon bone and tendon are too close together all the way down the cannon. This type of leg will not hold up as well as a cannon that has wide, flat bone. The term "fine boned" means insufficient support under the horse—cannon bones that are too small in diameter. "Coarse bone" usually means a heavy, clumsy cannon. Good bone is short, strong, flat and proportionate to the horse's size.

Viewed from the front, a horse's legs should be straight

A. Front legs follow straight lines. Spacing is the same at the hoofs as at the chest.

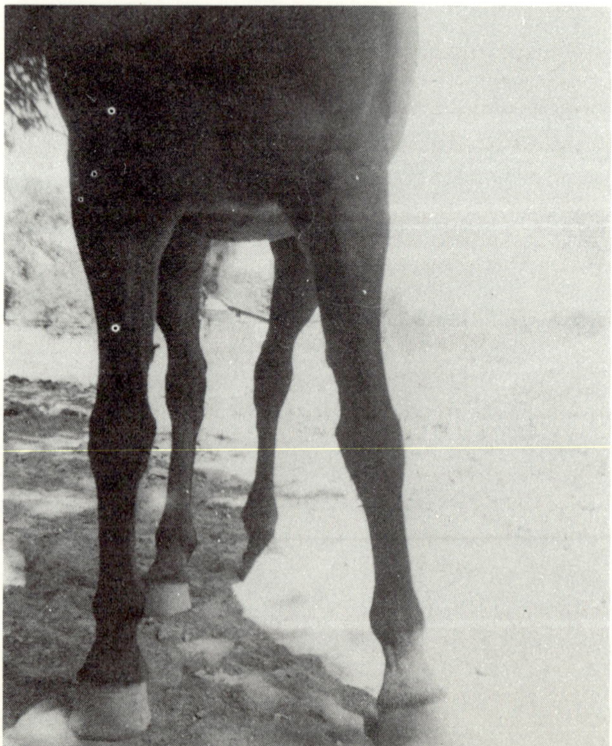

B. Front legs base-wide. The horse stands wider at the ground than his legs at the chest. This particular horse is also bench-kneed

A more obvious example of bench knees (offset cannon bones). Drop a line down this horse's front leg and you will see how far to the outside the cannon bone is placed. This is an extra strain on the inner splint bone

The fetlock should be broad from all angles, rounded a little at the front, but firm and flattened on all sides. The sesamoid bones should stand out prominently with the flexor tendons gliding over them. When viewed from either front or side, the forelegs should be perfectly straight. Undesirable deviations from this include calf knees, bucked knees, bowlegs, knock knees, bench knees, tied-in knees, cut out under the knees, standing under in front, and camped in front. Calf knees (back at the knees) are of very weak construction and the horse rarely stays sound under heavy work, for there is great strain on the knee bones and ligaments. Bucked knees (over at the knees) are not as bad as calf knees, but does put some extra strain on the leg. Sometimes horses that have been run too much or worked hard and long eventually become buck kneed.

When a horse is knockkneed his knees point inward toward one another. This puts added strain on the legs. Bowlegs (out at the knees) puts strain on the knees and are also often accompanied by base-narrow, pigeon-toed conformation. Bench knees (offset cannons) is a condition in which the cannon bone is offset to the outside of the knee. It is usually an inherited condition and is considered weak conformation because the inner splint bone is under greater stress than in a normal horse and splints are common. In a horse with tied-in knees, the flexor tendons are too close to the cannon bone just below the knee, inhibiting free movement. "Cut out under the knee" is very weak conformation in which there is a cut out appearance just below the knee in front of the cannon bone, when viewed from the side.

When a horse is "standing under in front" the entire foreleg is placed back too far under the body. This is not necessarily a conformational fault but is usually due to lameness or disease. The same is true of a horse that is "camped in

DEVIATIONS FROM NORMAL, STRAIGHT FRONT-LEG CONFORMATION

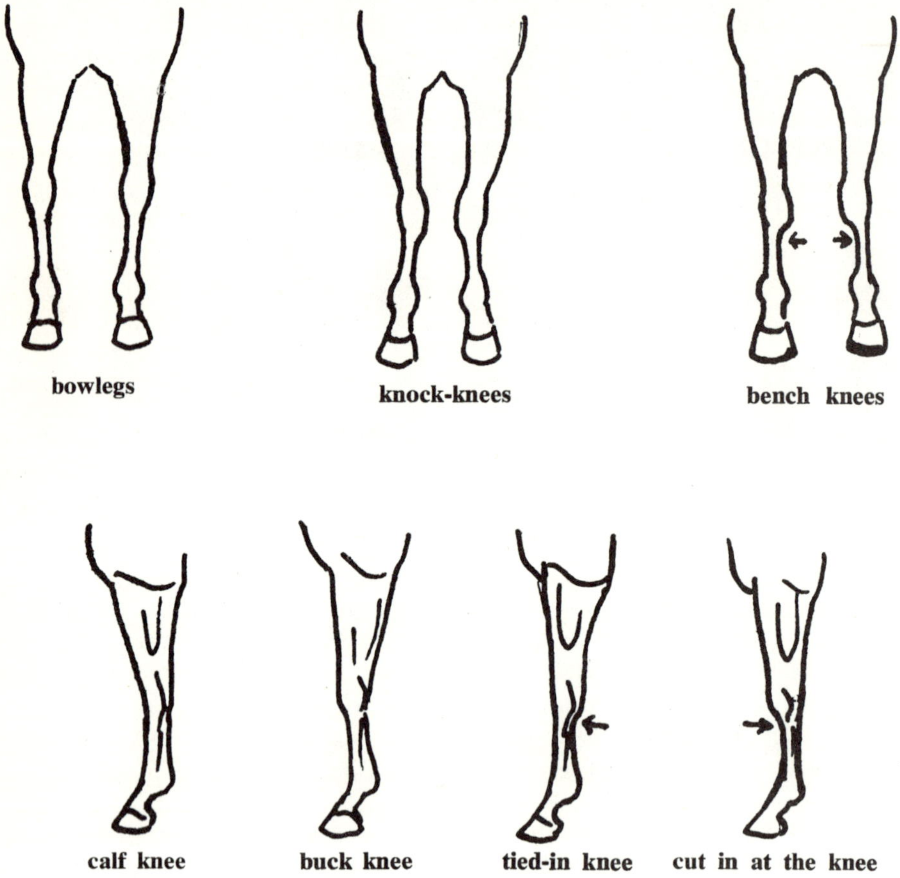

front"—the foreleg is too far forward and is not a conformational fault but can be due to pathological conditions such as navicular disease.

Shoulders and pasterns should be long and well sloped. A horse with sloping shoulders and pasterns gives a smoother ride and is less likely to become lame from problems due to concussion than is a horse with more upright and shorter shoulders and pasterns. Pasterns are shock absorbers. If they slope, they transfer some of the weight to the tendons. If they are too steep and short, most of the concussion is transmitted to the bones of the leg and the structures of the foot, giving a choppy, jarring ride and often causing such problems as ringbone, sidebone, splints, navicular disease and road founder.

The stride of the horse with sloping pasterns is usually longer and smoother than that of the horse with short, upright pasterns. Horses with short steep pasterns usually take short, choppy strides. Corrective trimming and shoeing can sometimes help to correct minor faults in the hoof, helping the horse to travel properly. Corrective trimming and shoeing can help straighten a horse that has toes too long or too short, or heels too high or too low. Corrective trimming can also help if the hoof has a tendency to turn in or out too much, helping the horse break over a little straighter. But if the conformational fault is higher in the leg—in the pastern, cannon, knee, hock, shoulder, etc—the correction will only be temporary and may even cause lameness by forcing bones, tendons, and muscles to work against the way nature put them together. How much success the correction has depends on the degree of conformational fault all the way up the leg.

Let's look now at some examples of poor conformation in the front legs besides the abnormalities we have already touched upon. Base-narrow is an undersirable type of front leg conformation and occurs when a horse's front feet are too close together. This is found most often in horses with large breasts and large pectoral muscles (such as the Quarter Horse). Very often base-narrow conformation is accompanied by pigeon toes (the horse's toes point inward toward each other instead of straight to the front) but not always. Base-narrow pigeon-toed conformation puts strain on the pastern and fetlock joints, commonly leading to such problems as windpuffs in the fetlock joint, and ringbone or sidebone on the outside of the foot. If you pick up a front foot and look at the bottom of the hoof, you will see that the hoof is worn excessively low on the outside edge, because the hoof tends to land on the outside wall instead of squarely flat. Basenarrow pigeon-toed horses "paddle" with their front feet, swinging them outward as they travel. The pigeon-toed horse will usually paddle whether he is base-narrow or base wide. Pigeon toes can usually be partially corrected by trimming the inside of the foot more than the outside, in order to keep the foot more nearly level.

A worse conformation than base narrow and pigeon-toed is base-narrow and splay-footed (toes pointing outward) for this puts even greater strain on the leg below the fetlock joint. Horses of this conformation rarely hold up under hard use. The feet of a splay-footed horse wing inward as he travels, often causing limb interference (one foot striking the opposite foreleg), especially in a base-narrow horse where the feet are so closely placed to begin with.

Base-wide is another type of poor conformation. Base-wide front legs are farther apart at the feet than at the forearms. This conformation is usually found in horses with narrow chests (such as the American Saddlebred and Tennessee Walker). Most base-wide horses are also splay-footed, which causes winging when they travel. The inside of the limb is under great strain in a horse that is base wide, and in horses that are both base-wide and splay-footed, because the ligaments of the fetlock joints and pasterns are constantly under tension. As a result, windpuffs, ringbone, and sidebone can occur.

Base-wide and splay-footed are generally considered to be weaker conformations than base narrow and pigeon-toed since winging is more dangerous to both horse and rider than paddling; there is the likelihood of interference.

Front-Leg Conformation and Action

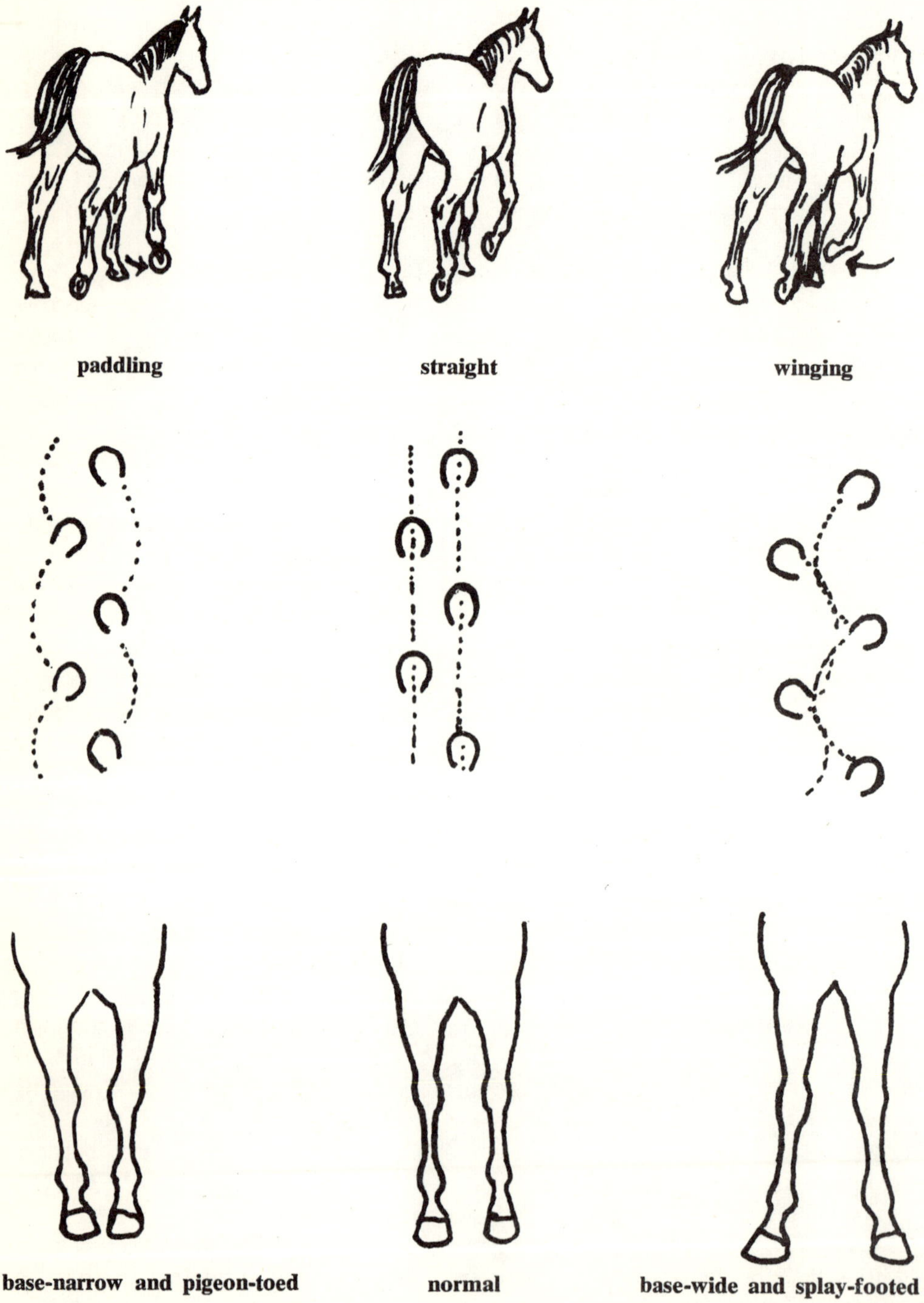

FRONT-LEG CONFORMATION

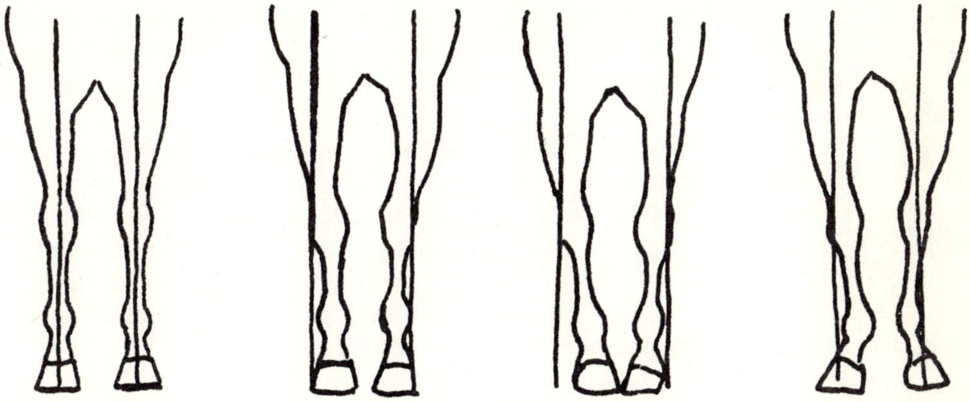

straight | base-narrow | base-narrow and pigeon-toed | base-narrow and splay-footed

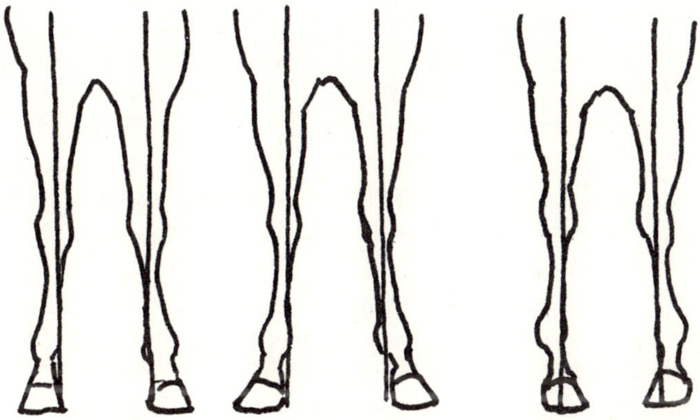

base-wide | base-wide and splay-footed | base-wide and pigeon-toed

Paddling is unsightly, but the horse does not hurt himself. Winging is dangerous as well as unsightly because of the possibility of the horse striking himself and tripping himself. Splay-footed horses usually wing to the inside, whether they are base-narrow or base-wide. If you pick up a foot and look at one of the front hoofs of a base-wide splay-footed horse, you'll notice that the foot is worn excessively on the inside wall. This is because the horse tends to land on the inside of his foot. When trimming this type of foot, lower the outside wall to make the foot more level. A few horses are base-wide and pigeon-toed. This is very weak conformation because of the additional stress to the limbs.

Now let's look at the hind legs. The hind legs are less subject to lameness than the front legs because they suffer less concussion and trauma, but conformation in the hind legs is very important because it plays a part in how the horse travels, whether or not he interferes, and whether he will be likely to get curb or some forms of spavin. The conformation of the hind legs also determines the speed and agility of the horse to a certain extent.

The hind legs have several functions—progression, holding the entire body weight in an emergency, propelling the body through the air when jumping or galloping, applying the brakes when stopping, and so on. Much of the horse's propelling power comes from his hind legs, so the hock joints must be sturdy. The hock joint does most of the work and must be firm but not fleshy, flat on the inner side and only slightly rounded on the outer side. This joint must be large and strong, to carry the weight of the horse, but smooth.

Proper angle is very important. Viewed from the rear when the horse is standing squarely, the hind leg should be well balanced and perfectly straight. A line dropped from the point of the buttocks should perfectly bisect the leg. A straight leg gives equal distribution of weight on various parts of the leg, and equal strain and pressure; no one part of the leg is under extra stress.

When viewed from the side, the back of the hock and the back of the cannon bone should be perfectly straight and perpendicular to the ground when the horse is standing squarely. The point of the hock should be the same distance from the ground as the chestnut on the forearm. If a line is dropped from the point of the buttocks, it should touch the point of the hock and follow the back of the cannon on down. This is the best degree of angle in the hock joint. If the hock joint is too straight or too angled it can cause serious problems.

Individual horses may vary a great deal in angulation of the hock joint. Viewed from the side, if the hind leg is too straight, a line dropped from the buttocks falls in back of and completely misses the cannon. This type of conformation can cause bog spavin because of too much strain on the front of the hock joint capsule (resulting in a chronic swelling and irritation due to too much joint fluid in the capsule). This type of leg is easily injured by hard work. Stifle and hock joints are synchronized and have the same degree of angle. A horse with too straight a hind leg (thus too straight a stifle) may be prone to dislocation and locking of the stifle joint (upward fixation of the patella—the kneecap). The patella slips upward, locking the joint. When the stifle joint locks, the leg is extended back behind the horse and the stifle and hock cannot flex. The joints may stay locked, or may just catch temporarily; the horse will hop along for a few strides with his leg extended behind him and then will travel normally for a few steps. A horse with hind leg joints too straight will usually have short straight pasterns as well.

Too much angle at the hock joints, creating a condition called sickle hocks, is another type of poor conformation. This type of hind leg structure predisposes the horse to curb, bone spavin and bog spavin. Sickle hocks is also called curby conformation because it puts great strain on the ligaments and hind parts of the hock, causing curb (injury to the plantar ligament and consequent swelling) if the

HIND-LEG CONFORMATION

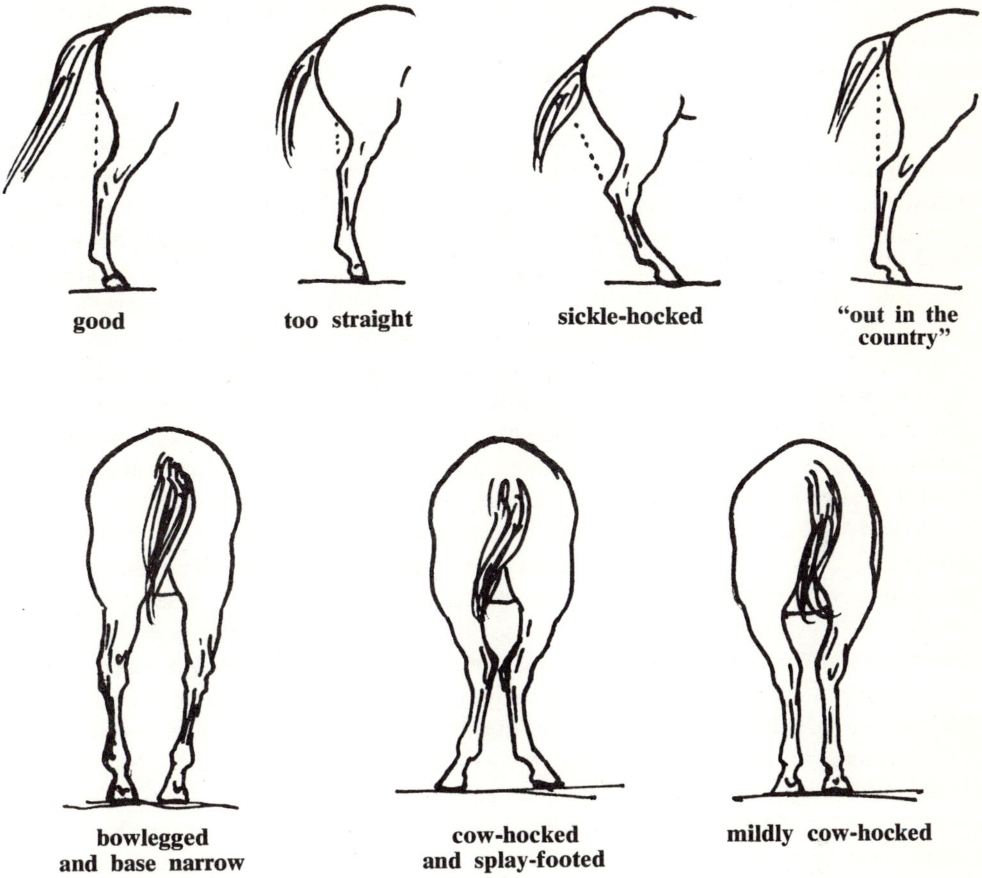

horse is worked hard. When a horse is sickle hocked, a line extended up the cannon bone is at such an angle that it completely misses the buttocks. The horse's feet are too far under his body; they are not directly beneath his hocks and buttocks. Some Quarter Horse breeders like a horse with sickle hocks because they feel that a horse with hind feet farther beneath his body can stop and turn more quickly and still keep his head low. This is true to a certain extent; it takes greater effort for a horse with normal angles in his hind legs to stop and turn and still keep his head low (cutting horse fanciers want that head kept low at all costs). But a sickle-hocked horse will not hold up as well as a horse with proper hind leg angles. Also, sickle hocks are a great disadvantage when it comes to wanting speed from a horse. Straighter stifles and hocks are much better than bent ones when it comes to producing speed; when there is too much angulation in the stifle and hock joints there is considerable loss of time in straightening the joint at each stride. A straighter leg can swing faster.

Some horses have hind legs that are "out in the country." The cannon is vertical (rather than angled as in sickle hocks) but it is too far back behind the horse. This conformation is not as weak as sickle hocks but it is not as desirable as good, well-formed hind legs of proper angle.

Other deviations from good hind-leg conformation are base-wide and base-narrow, bowlegs, and cow hocks. When viewed from behind, a base-narrow horse has his hind feet too close together. Bowlegged horses (with hocks too wide apart) are generally base-narrow. A base-narrow, bowlegged condition is most common in heavily muscled horses and puts great strain on the hock and fetlock joints, leg bones, and ligaments. The feet may be either straight or pigeon-toed. The legs might appear fairly straight down to the hocks, but become base-narrow from there on down. Base-narrow conformation in the hind legs is often accompanied by bowlegs. Bowlegged and base-narrow is very weak conformation; this type of horse often has trouble with interference as he moves. His hocks roll outward and his feet come too close together. If the condition is pronounced, they may strike one another. Also, even if a horse has good conformation in front, if he is base-narrow behind there are several kinds of interference that can occur between the front and hind legs (cross-firing, speedy cutting, etc.).

Base-wide conformation occurs in the hind legs when the hind feet are too far apart. The most common type of base-wide conformation in the hind legs is cow hocks. A tendency toward cow hocks is rather common .When a horse is cow-hocked, the hocks are too close together and the feet are widely separated. Mild forms, where the hocks are fairly close together but the legs are nearly straight from there on down, are not a serious fault and are a much stronger construction than bowlegged and base-narrow. Almost all hoofed wild animals—deer, elk, moose, antelope, caribou, buffalo, etc.—are cow-hocked. Straight hind legs are best, but mild cow hocks are preferable to bowlegs because the cow-hocked horse has less trouble with interference and cow hocks give more support to the horse than bowlegs. Severe cases of cow hocks should always be avoided; the hocks are too close together, pointing inward toward one another, and the feet are widely separated. This construction puts strain on the side of the hock joint and can result in bone spavin. Viewed from the side, this type of cow-hocked horse may also be sickle hocked. Some young horses are mildly cow-hocked and then develop straighter hind legs as they reach maturity.

Regarding legs and conformation, here is an old saying that is worth remembering "Bad hind legs can hurt the horse, but bad front legs can kill the rider." A horse with bad hind legs may not move well, and bad hind legs may brush and strike against one another as the horse travels, but bad front legs may make the horse stumble and fall.

Before we leave this chapter, let's take a quick look at the way the horse's leg structure distributes concussion and works to lessen the stress and strain on any one part of the leg, remembering of course, that this distribution system works best when the horse has good conformation. Poor conformation always puts added stress and strain on certain parts, which may eventually produce lameness and unsoundness when these stressed parts become broken down and injured from the excessive wear and tear.

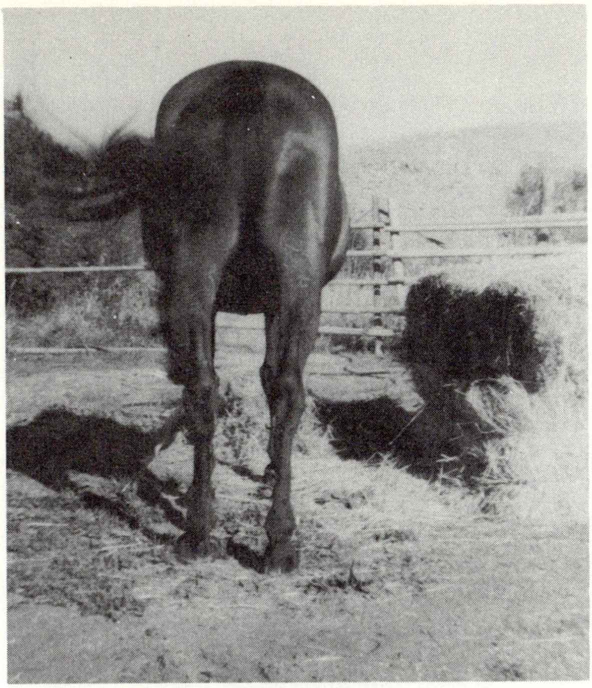

Good straight hind legs

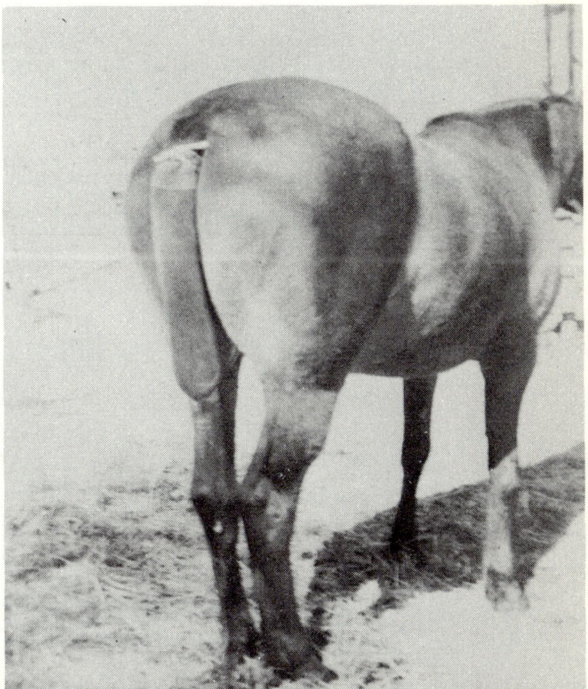

Cow-hocked

Cow-hocked and Splay-footed

In the front legs, several factors work together and help to overcome the effects of concussion. The horse's chest is suspended between the front legs entirely by muscular attachments rather than bony ones, and these muscular attachments are somewhat flexible and stretchy, absorbing some of the concussion which otherwise would be thrown upon the front feet and skeletal system. The shoulder blade is also attached to the body by muscles rather than bones.

Foot expanison and the buffering effect of the plantar cushion help to distribute the effects of concussion in the foot itself. Gently yielding joints (in combination with muscle coordination) help to distribute concussion in the leg and alleviate much of the jar that would otherwise occur in the bones and joints. Much strain is also transferred from the bones to the more elastic tendons. And finally, the shoulder blade of the horse glides over the ribs at each stride, transferring any remaining shock to the muscles of the body.

In the hind legs, the yielding of the joints and the increased arching of the sole (remember that hind feet are more concave than front feet) help decrease the effects of concussion. The pelvis is attached to the horse's spine at the sacrum, which is capable of only limited movement, so any concussion not absorbed within the hind leg itself has to be taken up by the disks between the vertebrae. But because the front legs bear more of the weight of the horse, the hind legs are not likely to be as apt to suffer from excessive concussion.

The suffraginis bone (first phalanx, or large pastern bone) and the navicular bone are perhaps the weakest links in the structure of the horse. But there is a unique arrangement of tendons and pulleylike surfaces created by the sesamoid and navicular bones, so that there is very little direct strain upon the suffraginis, *if* the horse has a good sloping pastern that is not too upright. Short steep pasterns put more strain on these bones. When the horse is standing, most of its weight falls upon the back of the joint between the lower end of the metacarpal bones (cannon) and the suffraginis, and upon the sesamoid bones. Extra weight is also put on the navicular bone if the horse gallops uphill (the heel is lowered and the toe is pointed upward), or if the horse has low heels and a long toe. Horses with low heels and long overgrown toes are more likely to develop navicular disease than are horses with normal feet.

Any horse is only as good as his feet and legs. Leg conformation is perhaps the most important factor to consider when evaluating a horse for any purpose, for the way his legs are built and the way they move make all the difference in his athletic ability—his agility and speed, his pleasantness and safeness to ride, and his ability to hold up under steady work without becoming lame.

17

UNSOUNDNESSES

AN UNSOUNDNESS IN A HORSE IS ANY DEFECT THAT INTERFERES with the usefulness of the horse, any physical problem which makes him unsafe to ride, or lame, or in any other way unable to perform as he should. An unsound horse will not hold up under strenuous activity, but will usually go lame.

Tendencies toward certain unsoundnesses may be inherited through poor conformation; some types of conformation make a horse more likely to have problems such as ringbone, splints, curb, or some other unsoundness. Some types of unsoundness are due to injury, such as a dislocated stifle, blindness from a blow on the eye, sidebone caused by a wire cut, and so on. And some types of unsoundness may be caused by too much strain and overwork—bowed tendons, certain types of splints, etc.

There is a difference between an unsoundness and a blemish. A blemish is unsightly—an old wound scar, a splint caused from injury—but it is not an unsoundness as long as it does not in any way interfere with the working ability of the horse. Several of our ranch horses have blemishes. Nikki has a thickened place on one knee that she has had ever since she ran a nail into her knee as a weanling. Nell has large scars on both front pasterns from old wire cuts. Khamette has splints and windpuffs. Ahmahl has scars on both hocks from the time the neighbor's dog put him through a fence when he was a foal, and Fahleen has a scar on one hind leg from an old injury. It's very hard to raise an unblemished horse when they must live within wire fences! But I guess we can be thankful that they are only blemished, and not unsound. Blemishes are unsightly, but do not make the horse lame or unsound.

Unsoundnesses can be mild or severe, depending partly on what the horse is being used for. There are many unsoundnesses which would hinder a horse used for racing or some other types of strenuous activity such as working cattle in the mountains, but which might not bother a horse used strictly for pleasure riding or as a child's mount. But the person selecting a horse as a pleasure horse or child's mount must bear in mind the fact that some types of unsoundnesses may get worse with use, and he must be doubly sure to guard against getting a horse that is unsafe to ride at faster gaits—because you never know when junior might decide to get a little speed out of "Old Dobbin." But if a horse is only mildly unsound, he may still be a safe bet for the person who never makes long or strenuous rides.

There can also be a difference between working soundness and breeding soundness. A horse may not be sound for work (he may have a bowed tendon, a bruised sole, a bad hoof crack, a saddle sore, or an injury which impairs his action, etc.) yet still be sound for breeding as long as his problems are not inheritable, or do not result from poor conformation which might be inheritable. A horse can also be just the opposite, being sound for work but unsound for breeding because of some fertility problem (like our mare Bambi was for seven years). But if a horse has an unsoundness that results from an inheritable problem (such as curb or ringbone resulting from poor conformation) he should not be considered sound to breed from because of the likelihood of passing on his poor conformation (and therefore the tendency toward unsoundness) to his offspring.

Good judgment is sometimes needed to differentiate between certain unsoundnesses and blemishes. For instance, a splint can be either an unsoundness or a blemish, depending upon where it is located on the cannon bone. A splint can

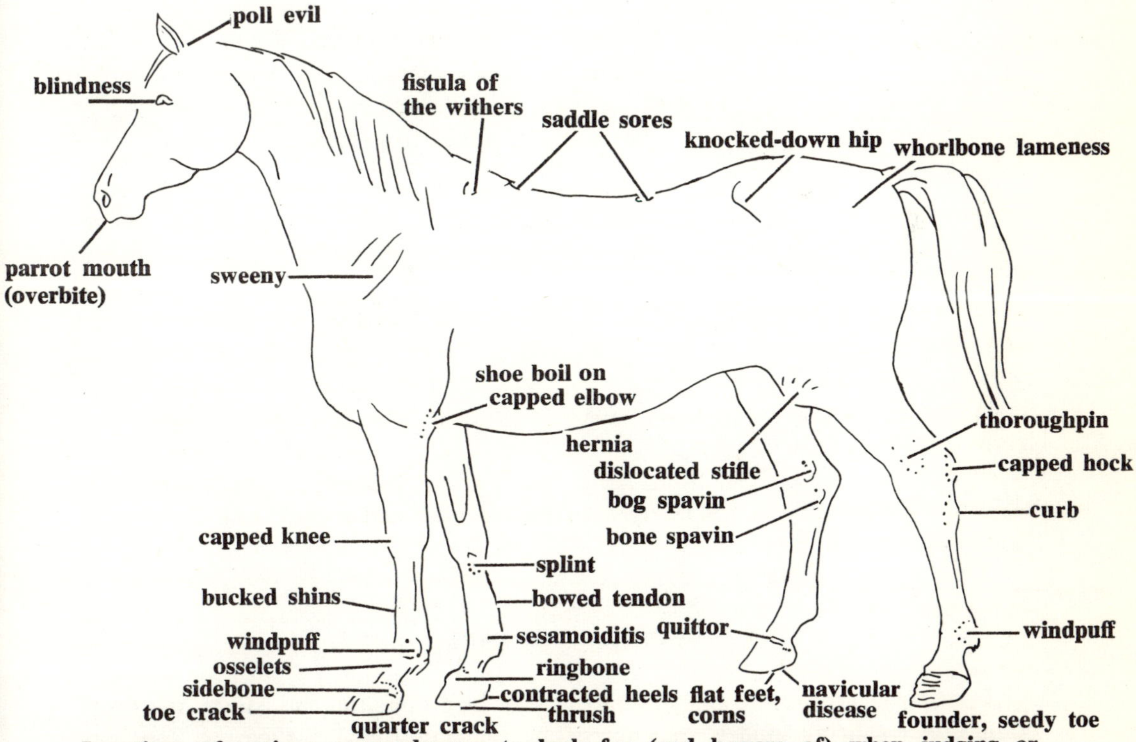

Locations of various unsoundnesses to look for (and beware of) when judging or selecting a horse

be a minor problem, merely a blemish, if it is not accompanied by poor conformation or lameness, and if it doesn't hinder any vital structure (if it is well away from any moving part such as the knee or back tendon). The same holds true for thoroughpins and windpuffs. If they were caused by injury and strain and are now healed and there is no lameness, they are considered to be merely blemishes, provided the horse has good conformation. But if they are caused by poor conformation, they should be considered unsoundnesses, for the condition is then inheritable. You would not select such a horse for breeding stock.

Let's look at the various points that should be covered when examining a horse for soundness. A good judge of horses always starts from the ground up, for a horse is only as good as his feet and legs. If they are bad or unsound, his usefulness is severely limited.

Feet should be well shaped and of a desirable size, proportionate to the weight of the horse. Beware of feet that are too small, for even if the horse is sound now, if his feet are too small and you use him hard he may become unsound and lame from any number of problems, such as navicular disease. Hoofs should be wide at the heels, not narrow and contracted. Wide heels "give" and spring apart when the foot hits the ground, absorbing some of the concussion. Contracted heels are more common in the front feet than the hind, partly because the front feet carry more weight and are more subject to stress and abuses than the hind feet. Some breeds of horses have more narrow feet than others. The Arabian has the roundest foot. A narrow foot is not necessarily a contracted foot; looking closely at the condition of the heel and frog will tell you if the foot is contracted. Foot contraction often occurs in show horses in the American Saddler and Tennessee Walker breeds because the hoofs are left to grow quite long as an aid to animated high action, but this eliminates frog pressure.

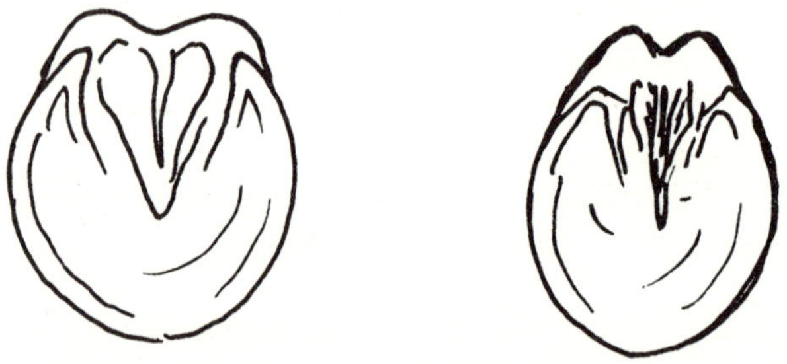

Left, **a normal foot;** *right,* **contracted heels and a small frog**

CONTRACTED HEELS

Contracted heels are usually due to lack of frog pressure, and can be the result of improper shoeing, injury, or unnecessary shoeing (leaving shoes on too long or keeping the horse shod all year 'round). Lameness from many causes can often make a horse put less weight on a foot, and over a period of time the lack of frog pressure can result in contraction. Once a foot is badly contracted, it may take more than a year for the foot to become normal again. A contracted foot is narrow at the heel and the frog is small and atrophied. Our oldest mare, Nellace, still has one front foot that is somewhat contracted, having become badly contracted many years ago when she was wire cut and lame for part of a year.

FLAT FEET

Flat feet is another undesirable condition in a saddle horse. A flat foot is abnormal and lacks the natural concavity of the sole. This condition is often inherited. Flat feet are more common in the front feet than in the hind feet, for hind feet are naturally a little more concave than the fronts. In a flat foot the sole is easily bruised, and lameness often results.

FOUNDER

Founder is another condition in a horse that can make him unsound for work. Founder (laminitis) is caused by fever and inflammation of the laminae of the foot (the area where the hoof attaches to the sensitive tissues of the inner foot). The fever that causes founder can result from infection, or severe upsetting of the horse's body chemistry (as can happen if the horse eats too much grain, drinks too much cold water when he is hot or exhausted, and so on) and acute cases of founder cause severe pain in the horse's feet.

A horse that has been foundered has a dropped, flaky sole, which makes him go lame very juickly from sole bruising unless he is correctively shod. The hoof wall will be ringed with ridges due to the inflammation of the coronary band; the hoof wall has grown more rapidly than normal because of the chronic irritation at the coronary band. The hoof may grow long toes which have a tendency to curl upward. Seedy toe (separation of the hoof wall at the white line) is common in foundered horses and may allow infection to get into the sensitive tissues of the hoof. Founder is covered more fully in chapter 10.

HOOF PROBLEMS

Cracks in the hoof wall can make a horse unsound if they are severe enough. Brittle feet, often the cause of cracking, may be due to dry climate and lack of moisture in the pasture. Brittleness is also more common in white feet than in pigmented feet. This is one reason we prefer our ranch horses to have dark feet.

Toe cracks, quarter cracks, and heel cracks (these are all called sand cracks) are those in the hoof wall that start at the ground surface and travel upward. Occasionally a crack will start at the coronary band because of some injury or defect in the band, and will travel downward, but this is more rare. Hoofs that are too long are the usual cause of cracks that start at the ground surface, for the excessively long hoof wall is often broken, chipped or split. The repair of serious sand cracks will be discussed in the chapter on corrective shoeing. Thin hoof walls can also be the cause of cracking.

Thin wall and sole are inheritable problems and can be considered unsoundnesses in some cases. The conformation of the feet may look normal, but the feet wear down too quickly and the horse goes lame because he walks upon his soles and then the soles are easily bruised. This condition is most noticeable at the heels. To check for thin wall and sole in a horse, pick up a front foot if he is barefoot and look at his heels to make sure they are not worn down excessively.

CLUB AND COON FEET

Other problems to look for in a horse's feet are club feet and coon feet. A club foot has a steep pastern with a hoof and pastern angle of more than 60 degrees. If just one foot is this way, it is usually due to injury; lack of use can cause this contracting and shortening of the tendons. But if both fronts (or both hinds) are this way, the problem may have been inherited or due to nutritional deficiency when the horse was young. Club-footed horses have a rough and stumbling gait and do not make good saddle horses. If a horse has a foot this way due to injury, you may still use the horse for breeding purposes, but this type of unsoundness will always limit the animal's usefulness for work, making him less agile and less surefooted, unsafe to ride.

Coon-footed horses have long, weak pasterns with a more sloping angle than the hoofs. This puts strain on tendons, bones and ligaments, and can also result in bruising of the fetlock joint when the horse travels at fast gaits over rough ground, because the weight of the horse on the weak pasterns pushes the fetlock joint down too low. A coon-footed horse may have an easy, soft ride, but he will not hold up under hard or strenuous use. Slope to the pastern is much desired in a good horse, but *too much* slope (more slope than the hoof) makes a weak pastern.

FOUNDERED FOOT

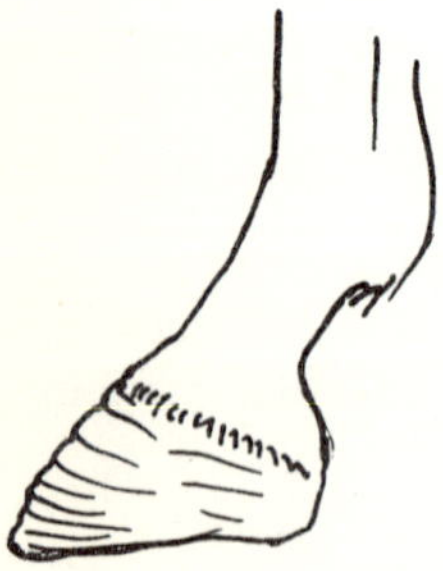

CLUB FOOT

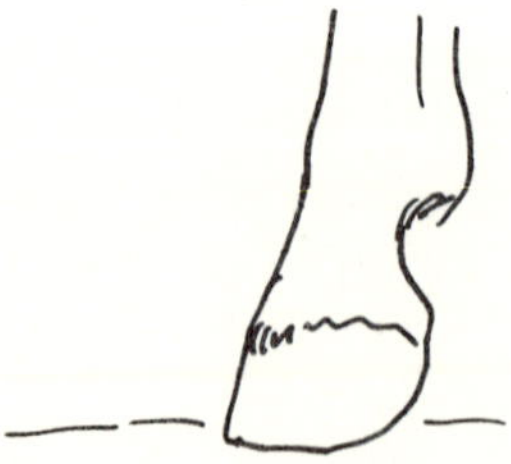

COON FOOT

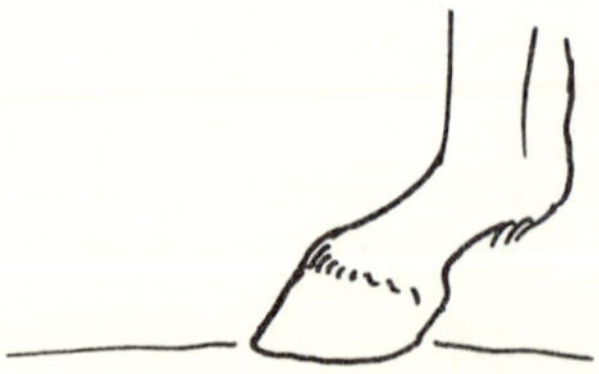

NAVICULAR DISEASE

Navicular disease is another problem that can make a horse unsound, for this problem is incurable. It usually begins as an inflammation of the navicular bursa (the fluid-filled sac around the navicular bone in the center of the hoof) which lies between the deep flexor tendon and the navicular bone in the foot. The navicular bone is a very small bone which lies in back of the coffin joint, inside the heel area of the horse's foot. The navicular bone serves as a kind of pulley and the big flexor tendon runs over it before it attaches to the coffin bone (third phalanx). Navicular disease usually affects only the front feet, except in cases of nail punctures in the hind feet, because front feet are more subject to shock and crushing from concussion.

Navicular disease is one of the most common causes of lameness. It may be due to inherited weaknesses (such as upright conformation—pasterns too short and steep—or weak navicular bones, or feet that are too small for the weight of the horse) or due to excessive concussion. Horses that are subject to a lot of strain and concussion from racing, working cattle and cow cutting, jumping, roping, barrel racing, and other sports where the horse must stop and turn quickly, are more likely to have the problem. A horse that is lame from navicular disease has a shuffling, stumbling gait, a shortened stride, and he tries to land on the toe of his foot instead of on the heel, to avoid frog pressure. The navicular bone is up above the middle of the frog, and pressure on this area of the foot is painful.

Navicular disease begins with an inflammation of the navicular bursa, caused by strain or concussion. As time goes on, the inflammation spreads from the bursa to the nearby bone and often to the tendon attachment. Once the bone becomes involved, it gradually degenerates and the horse becomes progressively lamer. The first symptoms of navicular disease appear as a mild lameness that keeps recurring and getting a little worse. It may take several years for all of the symptoms to appear (limping, particularly when first starting to exercise, standing with the involved foot pointed forward instead of placing weight on it, moving with the characteristic choppy, stumbling gait when worked, and showing progressive changes in the appearance of the involved foot). Over a period of time this kind of lameness will cause atrophy and shrinking of the foot, a contraction and raising of the heels. The foot becomes brittle, with an arched sole and shrunken frog. Sometimes you can tell just by looking at a horse's front feet that he has been lame from navicular disease, for one or both front feet may be somewhat contracted and smaller than normal, from disuse.

Navicular disease is incurable but sometimes treatment can be given to relieve pain. Corrective shoeing sometimes helps. In severe cases pain-relieving drugs can be given, or your vet can perform a neurectomy, cutting the nerves at the back of the heel so that the horse will not feel pain in the area of the navicular bone. Treatment is only a means to relieve pain and is never a cure. If the horse is used, the navicular disease will only become worse.

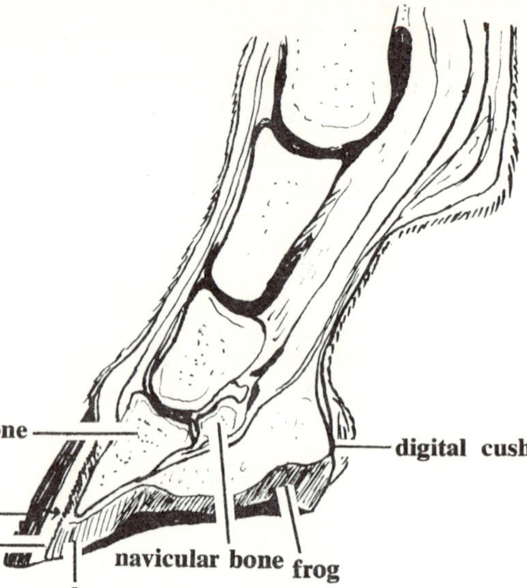

Cross section of the horse's foot showing the location of the navicular bone

THRUSH

When you check a horse for soundness, make sure that he is free from thrush or corns. Thrush is a type of hoof rot caused by lack of ventilation, usually because the hoof is packed with mud and dirt. This is common when a horse is kept in a swampy pasture, a dirty stall, or a muddy corral. Signs of thrush are black secretions around the white line between the hoof wall and the sole of the foot, and a penetrating bad odor; you can't miss it. The infection may go up into the sensitive tissues of the hoof. In this case the horse will flinch when his feet are cleaned, because the frog may be undermined with infection. Severe cases will make the horse lame, and there may be swelling in the lower leg, with the horse showing signs of infection. Several different bacteria can cause thrush, but the most common cause is the organism *Spherophorus necrophorus* (the same bacteria that causes foot rot in cattle).

Treatment for thrush is removal of the cause (putting the horse in cleaner surroundings if need be) and keeping the feet cleaned. Large areas of the frog may have to be removed. If there is infection in the foot it will have to be treated—drained and cleaned, and the horse given antibiotics. The foot should be cleaned daily and the cleft of the frog packed with cotton soaked with iodine, formalin or some other good medication. During treatment the foot may need to be protected in a boot of some kind. Improper trimming and shoeing sometimes cause thrush because of resulting lack of frog pressure. To remedy the situation a bar shoe may have to be used to exert pressure on the frog until it returns to normal. If thrush is treated before it causes extensive damage to the foot, the horse may fully recover. But if the sensitive inner structures of the foot have already become involved chances of complete recovery are not so good because there may be permanent damage and changes in the inner structures of the foot that may cause continual lameness.

CORNS

Corns are bruised areas on the sole of the foot, usually involving the sensitive and insensitive tissues at the angle where the hoof wall meets the bars. Corns usually appear as slightly reddened areas (especially noticeable when the feet are trimmed and the sole pared down) and are usually hot and tender in the early stages. Corns are most common in the front feet because they bear more weight than the hinds and are more subject to bruising. Flat feet make a horse more likely to get corns because a flat-footed horse is more likely to bruise the soles of his feet.

Corns are usually caused by improper shoeing (ill-fitting shoes) or by leaving shoes on a horse too long; the hoof wall begins to grow down around the outside of the shoe at the heel area and the shoe puts pressure on the sole at the angle between the hoof wall and the bars, bruising the sole. Trimming a horse's heels too low may also cause corns. When the horse's heels are too low there is increased pressure at the angle of the wall and the bar. Corns are rare in a horse that goes barefoot, but the barefoot horse may suffer sole bruising (creating a sore and tender area similar to a corn) from stepping on rocks. Horses with thin soles and horses suffering from chronic laminitis (founder) are especially subject to sole bruising. Bruises in the sole generally occur in the toe or quarter area rather than between the wall and bar.

A dry corn is bruised sensitive tissue in which there is red staining in the corn area (due to bleeding inside the sole). A moist corn is caused by severe injury to the sole resulting in seeping fluid beneath the injured horny sole. A suppurating corn is one that has become infected, the infection causing damage to the sensitive inner tissues of the hoof (the sensitive laminae, fibrous plantar tissue, or digital cushion).

Corns or sole bruises can make the horse quite lame if they are serious. If the problem is a corn, the horse will attempt to favor the heel and place more weight on the toe. If the problem is a bruise in the toe area of the sole, the horse will tend to land on his heel to keep weight off his toe. Hoof testers can be helpful in locating the corn or the bruise, for the horse will flinch when the bruised area is pressed. If the sole is pared down a little with a hoof knife, a reddish (or bluish, if an abscess is developing) discoloration can usually be seen in the area where the horse showed pain.

When corns are caused by improper shoeing, removal of the shoes may be all that is necessary for the healing of the corn. The horse should not be ridden and should not be reshod until the corn and lameness have fully disappeared. To make sure that shoes do not cause corns, the heels of the shoe should always extend well back and the shoe should be slightly larger than the hoof wall at the quarters. This allows room for hoof expansion. The hoof wall must always rest upon the shoe; if the wall extends out beyond the shoe when the foot expands, the shoe is not large enough.

If a horse ever gets a suppurating corn, the sole should be pared down at that area until the infection in the sensitive tissues can drain. The foot should then be soaked daily, just as you would soak it for a puncture wound, using magnesium sulfate solution (epsom salts) or a good antiseptic. The foot should be soaked daily for several days and then bandaged or put in a boot between soakings to keep the open area from being contaminated. If a bruised sole develops an abscess, it should be drained and soaked in the same manner. Corns and sole bruises are always a serious matter, for some cases tend to become chronic, eventually causing inflammation of the third phalanx (coffin bone) and permanent lameness; the foot bone becomes enlarged and painful.

RINGBONE AND SIDEBONE

Other unsoundnesses to look for on the horse's feet are ringbone and sidebone. These are bony enlargements (exotoses) around the coronary band where the hoof grows down from the hair line (at the coffin joint) or higher, at the pastern joint. This bone growth is usually caused by injury to the periosteum (the membrane lining around the bone) either from a blow that bruises the bone, or a wire cut, or from stress and strain. Malnutrition when the horse is young can later be a cause of ringbone. Horses with poor conformation (base-narrow and pigeon-toed, base-wide and splay-footed, and so on) are more likely to get ringbone and sidebone than are horses of good conformation, because of the constant strain on the limb below the fetlock joint. Short, steep pasterns also predispose a horse to ringbone because of the increased concussion to the pastern joint. Let's take a closer look at the problems of ringbone and sidebone.

Ringbone is new bone growth on the first, second, or third phalanges, as a result of periostitis, and can lead to osteoarthritis (inflammation of the joint) or the growing together of the pastern or coffin joints. Ringbone is very seldom found in Thoroughbreds; it is more common in other breeds. This is because most Thoroughbreds have fairly good leg conformation. Ringbone is caused by injury to the periostium (stimulating new bone growth) and the most common type of injury in this area is strain—damage to the ligaments attached to the bone and therefore pull and stress on the bone lining. Poor leg conformation vastly contributes to this type of strain, for a crooked leg is always under more stress than a normal leg, and work increases this stress. Ringbone has been thought to be inheritable, but what is actually inheritable is the poor conformation that can lead to ringbone.

Ringbone can either be high or low. "High" ringbone is around or near the pastern joint (between the first and second phalanx); if the joint itself is not involved, there may be no lameness. Low ringbone is around or near the coffin joint (between the second and third phalanx, usually at the extensor process of the third phalanx, at the front of the foot). Ringbone can also be articular (involving the joint surface) or periarticular (around the joint but not involving the joint surface). Periarticular ringbone is most common in high ringbone. Low ringbone is often more serious than high ringbone, because it is usually articular and because it occurs underneath the coronary band, inside the hoof. Thus it causes severe pain and lameness.

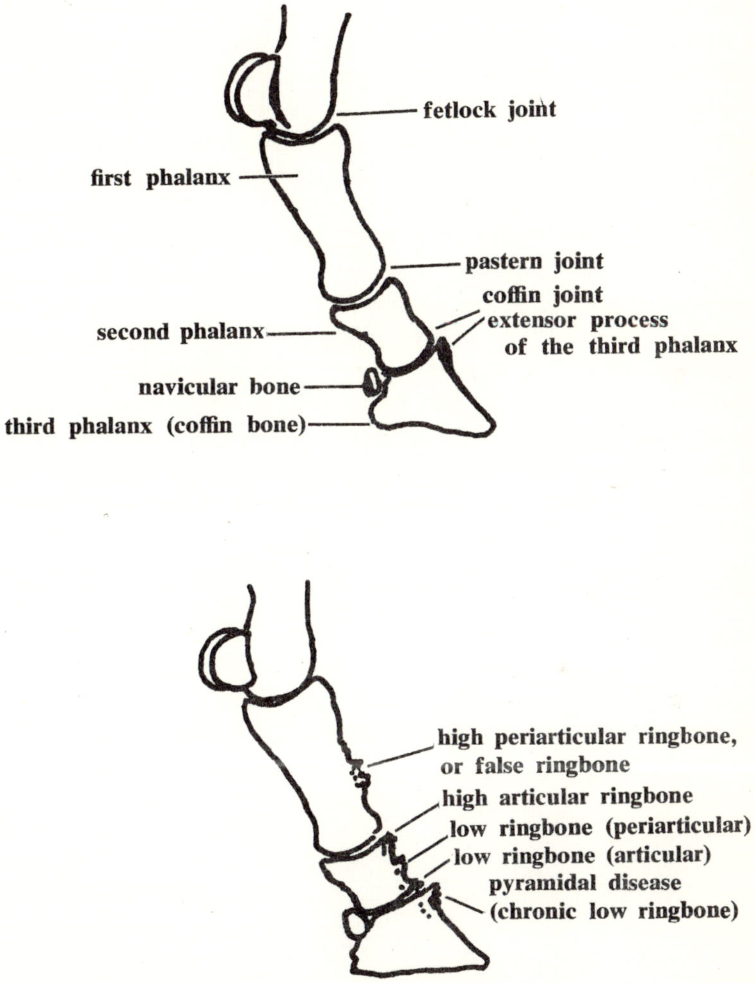

Ringbone; *top,* **normal foot and pastern;** *bottom,* **various locations of ringbone**

Chronic low ringbone is called pyramidal disease or buttress foot because of the bulging of the coronary band. The hair on the coronary band will stand up at the front of the foot and there will be heat and pain in this area. Low ringbone is usually more serious than high ringbone and chronic low ringbone (in which arthritis and new bone growth have already begun) is incurable. There will be a change in the shape of the toe of the hoof wall.

Ringbone can occur in either the front or hind feet but it is much more common in the front feet. The horse is usually lame and shows increased lameness when turning a corner. Heat and swelling will be evident at the area where ringbone is forming and the horse will flinch when you press the area with your finger. Cases in which the ringbone is periarticular (not involving a joint) may not show symptoms of lameness, and no heat or pain will be present after healing. Some cases of ringbone are hard to diagnose unless an X-ray is taken. Pigeon-toed horses are most likely to develop ringbone on the outside (lateral side) of the joints, while splay-footed horses are most likely to develop ringbone on the inside (medial side) of the joints, due to the extra stress placed on these areas because of crooked leg structure.

Very early ringbone can sometimes be treated and arrested in time to correct it. If the problem is diagnosed while the early ringbone is still in the periostitis stage, before arthritis and new bone growth have begun, it may have a chance of healing if the injured part is treated with corticosteroids and given strict rest. This is done best by limiting the motion of the joints by putting the leg in a plaster cast from the hoof wall to just below the knee joint. The leg should stay in the cast for at least four weeks, and the horse should be rested for an additional four months after the cast is removed. If the horse is put back to work too soon the problem may begin all over again. Ringbone can easily progress to the point where it is incurable, so early diagnosis and proper rest are very important.

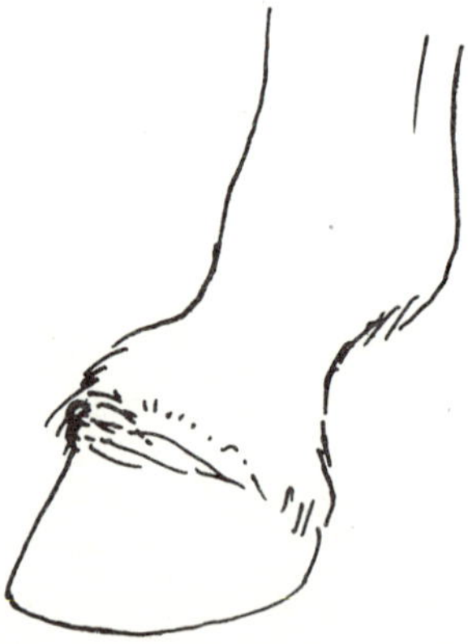

Buttress Foot (Pyramidal Disease)

High ringbone sometimes heals with a growing together of the pastern joint (first and second phalanges); when this fusion is complete, lameness disappears. But sometimes the joint will not fuse and merely grows massive deposits of new bone. Sometimes surgical fusion of a joint will help. Firing or blistering sometimes helps periarticular high ringbone (where the joint is not involved) but will not usually help an articular ringbone. Low ringbone, where the coffin joint is involved, almost always results in a permanently unsound horse. Neurectomy (cutting of the nerves) is sometimes done to relieve the pain, but this can be risky because when the horse loses the feeling in that area of his foot he may also lose part of his agility and may be unsafe to ride because of his tendency to stumble and fall. Occasionally the hoof wall will slough off when certain nerves are cut. Surgical removal of new bone growth (in cases where the bulge of new bone inhibits a tendon or some other vital structure) can sometimes be successful. Corrective shoeing, such as the use of a full roller motion shoe, can help to take some of the action from the fused joints, but rarely eliminates lameness completely. This type of corrective shoeing can be helpful, however, when first using the horse again after resting and healing an early case of ringbone; the roller motion shoe helps take some of the motion from the joints and may help to keep the horse from relapsing.

Sidebones are an ossification of the lateral cartilages on the sides of a horse's feet. The lateral cartilages are on each side of the horse's foot, back toward the heel and above the coronary band. In a normal horse these cartilages are firm but movable structures. They become sidebones if they calcify and turn to bone.

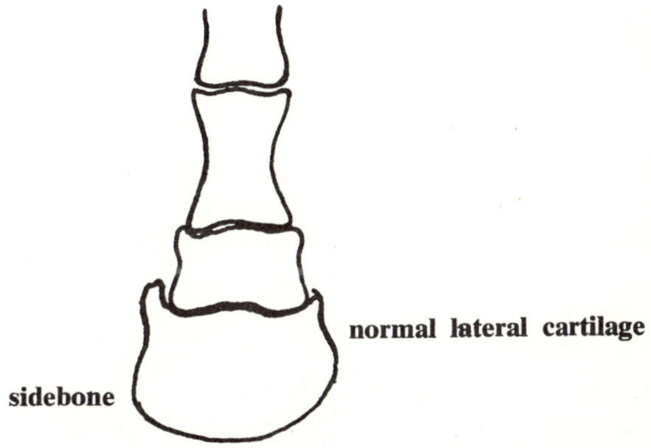

Concussion is a major cause of sidebones, especially in horses of upright conformation, for this causes trauma to the lateral cartilages. Splay-footed horses tend to get medial sidebone while pigeon-toed horses tend to get lateral (outside) sidebone, though it is not uncommon for sidebone to develop in both cartilages eventually. Lateral (outside) sidebones are most common. This is mainly a problem of the front feet, for they are more subject to concussion and trauma than hind feet. Wire cuts in the cartilage area can cause sidebone if they damage the cartilage. Poor shoeing can also be a factor: using long heel calks can increase concussion, and shoeing a horse off level can put more strain on one side of the foot than the other, increasing concussion to one of the cartilages. Foundered horses or horses with contracted heels are also prone to sidebone, because the feet in these horses are less able to absorb concussion.

Sidebone may or may not cause lameness. Lameness is often blamed on sidebones when it is actually due to other causes. Many horses with healed sidebones have no lameness. Lameness usually only occurs when the cartilages are in the process of becoming calcified and bony, for that is usually the only time inflammation is present (unless an old sidebone gets badly bumped and reinjured). If sidebones are the cause of lameness, one or both of the cartilages will show heat and pain when pressed. If sidebones are causing lameness, the horse should be rested until the inflammation goes away and the bone-forming process is complete. Sidebone, after healing, will usually not cause problems unless the bone-formation is quite large. A sidebone can join a ringbone and cause serious lameness. And sometimes a sidebone will fracture because the calcified area is brittle and less flexible than a normal cartilage. A fracture will cause sudden and severe lameness. A fractured sidebone often heals if given a complete rest, but sometimes surgery is needed to correct it.

Corrective shoeing can sometimes help a horse with lameness-causing sidebone. The quarters of the foot can be thinned or grooved (as for the correction of a contracted foot), aiding expansion of the hoof and relieving some of the pain and pressure over the sidebone. Full roller motion shoes help decrease the action in the coffin joint area, eliminating part of the motion that causes pain. Pain-relieving drugs can also be used in some cases.

QUITTOR

Quittor can be an unsoundness. This is a chronic inflammation of the cartilage area above the coronary band and it is characterized by the dying of some of the cartilage tissue, with drainage out through the coronary band. Injury to the foot is the usual cause of quittor—injury such as wire cuts or bruises above the coronary band, or damage to the cartilage by interference if the horse strikes himself with his other foot. The injury to the cartilage causes an abscess underneath the coronary band. Sometimes quittor can be caused by a wound through the sole of the foot in which infection has spread to the lateral cartilage, draining out through the coronary band. Sidebone or permanent swelling may result from quittor. Permanent damage and subsequent deformity of the foot may make the horse consistently lame. A few early cases of quittor can be corrected by removal of the dead cartilage tissue with a caustic substance such as silver nitrate (injected daily, followed by saline injections to neutralize the silver nitrate). Surgical removal of the dead tissue is often preferred, but this is a radical procedure, taking a long time to heal and sometimes involving damage to the coronary band, creating cracks in the hoof wall. Treatment of quittor is usually successful only if the problem has not yet progressed very far. If it has gone on for any length of time it may become a permanent unsoundness.

SPLINTS

Splints can also be an unsoundness. Splints are bony enlargements on the cannon bone, caused by injury, poor conformation, strain, or overwork. Sometimes splints are also caused by nutritional deficiency (or *excessive* supplementation of vitamins and minerals) when the horse is young.

Irritation of the periosteum (membrane surrounding the bone) either from injury such as a kick or a blow or from running into something solid, or strain from overwork when the horse is immature or strain on the legs from poor conformation, or from concussion (too much riding or fast riding on hard surfaces) are the major causes of splints. Many times young horses get splints from concussion and overwork while the same kind of work will not hurt a mature horse. This is because the splint bones are not firmly attached to the cannon bone until the horse is five or six years old—and until they are firmly attached there is possibility of getting a splint whenever there is too much movement between the bones (and irritation) due to overwork or stress and strain from too much fast stopping and turning, or trotting on hard surfaces, etc.

Whenever the periosteum is injured, circulation increases and the area is stimulated to make new bone growth, creating the unsightly bony lump. Poor conformation puts additional strain and concussion on the splint bones and makes a horse more likely to get splints. Two types of poor conformation can predispose a horse to splints—upright shoulders and pasterns (which give a horse more jarring gaits) and bench knees. When a horse has bench knees the cannon bones are set too far apart, to the outside of the knees, putting extra strain on the inner splint bones. Splints accompanied by poor conformation, or splints that interfere with moving parts (being too close to a tendon or to the knee joint) are considered to be unsoundnesses. A splint that interferes with a moving part will hinder the working soundness of the horse, and a splint accompanied by poor conformation is a problem that could be passed on to the horse's offspring.

Splints can develop on any part of the splint bone on either side of the cannon, but most occur on the inside of the front legs because the inner splint bones carry more weight than the outer ones and are more subject to stress. Most splints occur about three inches below the knee.

Splints usually cause lameness in the early stages. A few cases never cause lameness at all. When inflammation begins there might or might not be swelling, but careful examination of the lame leg will reveal the location of the soreness, for it will be hot, and painful to the horse when pressed. Lameness is most noticeable at the trot and most severe after the horse has been ridden or exercised on hard ground.

In time the splint calcifies and the hard permanent lump appears. Lameness then usually disappears and the splint is then considered a blemish rather than an unsoundness unless accompanied by poor conformation or in cases where the splint interferes with a moving part. Many splints are merely blemishes, but an old splint can flare up and become a problem if it is struck or reinjured.

Splints are almost impossible to remove once they have calcified (except

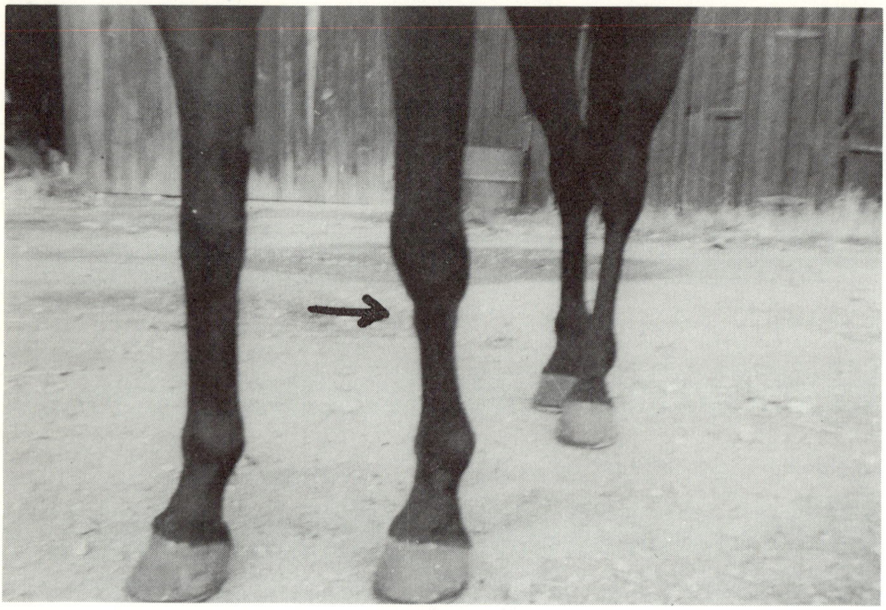

Splint

by surgery, and even then the bony growth returns in about half the cases) so it is important to diagnose them early. Old chronic splints are best left alone, though sometimes pegged splints (which go around the back of the leg, interfering with movement of tendons) can be surgically removed if they vitally hinder the usefulness of a valuable horse.

If a young horse (and by young I mean any horse under seven years old) turns up lame, strain or concussion may well be the cause if no other condition is obvious. Fracture of the splint bone is often confused with early splints, but with a fractured splint bone the swelling of the leg is usually spread over a large area and the horse is lame for a longer period of time. If you ever suspect a broken splint bone, have a vet check the horse and X-ray the leg. A fractured splint bone can be a serious problem and some of the treatments used for ordinary splints will only make the fracture worse.

The best treatment for a potential splint caused by strain or concussion is rest. Do not use the horse at all. Cold packs and astringents and anti-inflammatory drugs can also be beneficial. Resting the horse for at least 30 days while treating the splint will often be sufficient for the splint to heal.

Treatment for a splint that is already producing new bone growth (chronic splint) is more complicated, and effectiveness of the treatment varies. Veterinarians use several methods, such as firing or cauterizing the splint deeply, or injecting cortisone directly into the inflamed area, or using counter-irritants. But once new bone growth has started it is not always possible to stop the calcifying process. Needless to say, prevention is important. Horses should not be started too young, and should not have to progress to strenuous work or training until they are mature. When I was a girl I put splints on a three-year-old filly (Ginger) with too much long-distance traveling, trotting on hard ground. This filly had upright conformation and jarring gaits, making her prone to extra stress from concussion. A good rule to remember: never trot or gallop a horse on pavement, gravel, or hard ground. This goes for horses of all ages, for there are other problems besides splints that can be caused by concussion, including founder, sidebones, and navicular disease.

Several other problems (besides ringbone and splints) can be caused by periostitis (inflammation of the bone lining). The periosteum protects and nourishes the bone, serves as an anchor for ligaments and tendons, and contains a special layer of bone-forming cells which play a part in bone growth when the horse is young. But these bone-forming cells can also be stimulated to grow new bone when injured. Injury to the periosteum causes inflammation, swelling and heat. The blood supply increases, and the bone-forming cells are stimulated, forming excessive bone, creating an enlargement. We've already looked at the bone formation that occurs in ringbone and in splints. Now let's look at a few more.

BUCKED SHINS

Metacarpal periostitis, also called bucked shins, is a periostitis on the front of the cannon bone, usually on the front legs. It occurs most commonly in young race horses during their first weeks of training, and is due to strain and concussion. Mature horses are rarely affected this way because the bone lining is more mature and more fully attached. In young horses that undergo stress from intensive training, the condition usually occurs in both front legs. If only one front leg is affected, the cause is usually from direct trauma such as a blow. The periostitis causes painful swelling which is warm to the touch, and lameness. Lameness increases with exercise. If only one leg is affected, the horse will try to rest it, but if both front legs are affected he will shift his weight back and forth.

To recover from bucked shins, the horse must have complete rest. Most cases will heal without treatment if the horse is properly rested, but cold packs during the first day or two can be helpful. The horse should be rested for at least a month or more or the condition will recur. If the periostitis is serious or if the horse relapses several times because he is not given enough rest to heal completely, there will be permanent new bone growth on the cannons.

SESAMOIDITIS

Sesamoiditis (inflammation of the proximal sesamoid bones, in the fetlock joint) is usually accompanied by periostitis and new bone formation. This problem is usually caused by unusual strain to the fetlock joint. Sesamoiditis causes pain and swelling of the fetlock joint, and lameness, for the horse does not want to place weight on the leg; he tries to keep the fetlock joint from descending. After the problem becomes chronic, there will be new bone growth on the sesamoid bones and sometimes calcification of the suspensory ligament above the sesamoid bones. Treatment in the early stages is aimed at reducing the inflammation, and immobilizing the leg with a plaster cast. Chances for complete recovery are limited if there has been much periostitis and new bone growth.

OSSELETS

Osselets are caused by periostitis at the front of the fetlock joint, between the cannon bone and the first phalanx or on the front surfaces of either of these bones. Osselets are most common in young race horses, due to stress and strain. "Green osselets" refer to the early stages of this condition, before new bone growth has occured. The problem is most common in the front legs, often progressing to the point that the underlying bone is involved and the fetlock joint is inflamed (arthritis) with resulting calcification, stiffness and constant pain.

If both fetlock joints are affected, the horse moves with short choppy strides. If just one leg is affected, he shows lameness in that leg. There will be pain, heat and swelling over the front of the fetlock joint; pressure at this area will make the horse flinch. There is fibrous enlargement of the joint capsule. Some cases will have joint mice.

JOINT MICE

A joint mouse is a loose piece of bone within the joint capsule. Rest is essential for recovery. Ice packs can be used to reduce the acute inflammation. It sometimes helps to inject the inflamed joint capsule with a corticosteroid; if used soon enough corticosteroids can help prevent new bone growth as well as decreasing inflammation. Supporting wraps should be used on the fetlock area for about two weeks. If the osselets are chronic, firing can be used to create an acute inflammation that will usually heal leaving the joint free from chronic arthritis. The horse should be rested at least six months, even though he may begin to look normal sooner; putting him back in training too soon will only cause a relapse. If the horse has a joint mouse surgery may be necessary to remove it so that it will not catch between the cannon bone and the pastern bone and cause pain and lameness.

Chances for recovery are hopeful if periostitis has not resulted in new bone growth; sometimes chances are good even with bone growth if the growth does not occur on the moving surfaces of the joint. If it does involve the surfaces of the joint, or if a joint mouse is present, the horse may never be fully sound. The joint mouse can do permanent damage to the joint if it has been repeatedly caught in it. If a horse has steep pastern conformation, osselets may reoccur because of concussion.

BOWED TENDONS

Bowed tendons are an unsoundness which result from injury (usually severe strain) to the flexor tendons and the sheaths that surround them. There are two tendons in back of the horse's cannon bone, the deep flexor tendon and superficial flexor tendon, one lying on top of the other. The tendons are covered with a sheath and lubricated with synovial fluid. These tendons travel down the back of the cannon and attach to the hoof, flexing the hoof when the muscles in back of the forearm are contracted. When the tendons and their sheath and attachments are injured or torn, swelling is caused by hemorrhage and inflammation and we say that the tendon is bowed. The swollen area is very hot and painful. Bowed tendon is usually caused by strain and overwork of the part. Horses most susceptible to getting bowed tendons are horses with long, weak pasterns, horses that are too soft and out of condition for their work, horses undergoing rigorous training or strenuous exertion, horses with long toes, and horses that are too heavy for their tendon structure. Such a horse can easily bow a tendon while running hard, galloping uphill, turning quickly, bucking, or suddenly accelerating to full speed. A bad blow can also cause damage that can result in a bowed tendon. Once a tendon becomes bowed, it usually becomes a permanent weakness, limiting the horse's ability when performing at fast speeds.

When the tendon and sheath become injured, adhesions form, binding the tendons to the sheath and binding the two tendons together. Fibrous scar tissue forms between the tendon sheath and the surrounding tissue, and the bowed appearance is due to this fibrous scar tissue. Bowed tendon can occur high or low, and a severe case may involve the whole length of tendon. Bowed tendon is most common in the front legs, but when it does occur in the hind leg it is usually low. When a horse has a chronic bowed tendon he may appear sound at the walk and trot but he will quickly go lame when worked fast or hard.

A bowed tendon must be treated immediately after it occurs; the longer you wait the less chance of proper healing—the injury will become a chronic problem and permanent weakness. Regardless of the method of treatment, it takes at least a year for a bowed tendon to fully heal. If the horse is put back to work too soon (even though he may look completely recovered) the injury will recur.

Treatment consists of cold packs and a pressure bandage as soon as the injury is detected—to stop the bleeding inside the tendon sheath and to keep down the swelling. A local injection of corticosteroid into the tendon and tendon sheath should be given. The horse should be confined, and as soon as the cold pack treatment is finished, the leg should be put in a light cast (from coronary band to just below the knee) for about two weeks. Then the cast should be removed, the leg checked, and another cast put on if necessary. If there is quite a bit of improvement in the leg, it can be wrapped in supporting bandages for about a month rather than using another cast. Then the horse should be rested for a year. This kind of treatment, if given early enough, can sometimes help the leg to heal enough that the horse can be used strenuously again. Twenty percent of the race horses that bow tendons are able to race again, but always the leg must be supported by wraps. The injury can easily recur. Once a bowed tendon becomes chronic (once scar tissue is present), it becomes a permanent unsoundness, for it will never be as strong again.

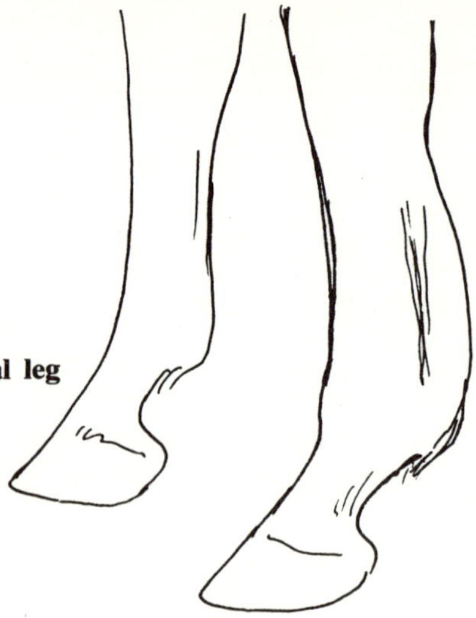

CURB

Curb is an enlargement at the back of the hock due to inflammation and thickening of the plantar ligament. This unsoundness is usually due to poor conformation—sickle hocks or extreme cow hocks—putting strain on the plantar ligament that runs down the back of the hock. Occasionally a horse with good conformation will also get curb from violent exertion or from kicking his hocks against something solid like a wall or a trailer tailgate. A permanent blemish results, but if the horse has good conformation he will usually be serviceable and sound after recovery. But in horses with poor hind-leg conformation curb is considered an unsoundness because the problem is likely to recur whenever the horse exerts and puts the hock under great strain, making him lame again; the hock is continually under greater stress than normal because of the poor conformation.

When curb occurs, there is inflammation, heat, and swelling at the back of the hock (lower portion) and the horse will be lame. He will stand with his leg resting, heel up. Exercise may make him lamer. In severe cases, periostitis will result in formation of a bony growth. Tissues around the area often become full of scar tissue, making a permanent blemish. Corticosteroids and cold packs can sometimes help reduce swelling and pain in the early stages. Once the condition becomes chronic, the swelling will usually remain in spite of treatment. If the horse has good conformation and is rested and treated, he will fully recover, even though he may always have a permanent blemish. But the horse with poor conformation will usually relapse.

SPAVIN

Spavin is a term meaning a disease of the hock joint. There are several kinds of spavin, most of them causing swelling in the area of the hock. A bone

SWELLING OF THE PLANTAR LIGAMENT

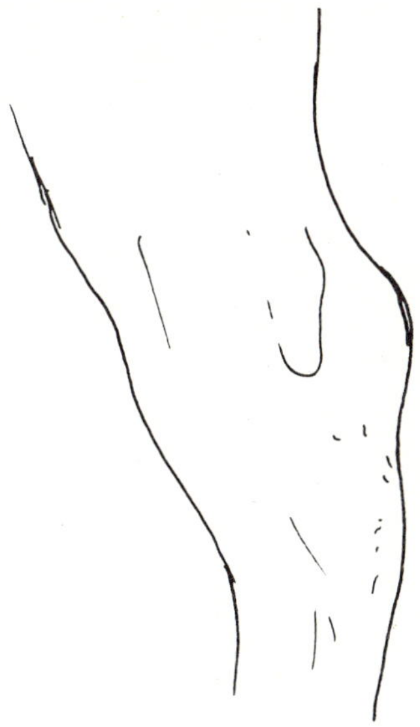

spavin is a periostitis which forms new bone growth. A jack spavin is a very large bone spavin. A high spavin is a bone spavin higher on the hock than usual. A blood spavin has no real definition but usually refers to an enlargement of the saphenous vein which angles across the front of the hock joint; it is usually associated with bog spavin. A bog spavin is a chronic swelling in the joint capsule of the hock. Occult or blind spavin is a term for typical spavin lameness without any of the obvious changes in the hock.

Bone spavin is the worst kind of spavin. There is inflammation of the bone, usually affecting the bones of the hock, causing immobility and a joining together of some of the moving bones in the hock joint. Bone spavin is usually due to poor conformation such as sickle hocks or cow hocks, which increase concussion and stress within the hock joint. Bone spavin tends to be an inheritable problem, through poor conformation. Bone spavin can also be caused by excessive stress and strain, as when a horse performs quick stops (as in roping). Mineral deficiencies, as in cases of rickets, may predispose a horse to bone spavin. Horses with narrow thin hocks are more likely to develop bone spavin than are horses with strong, well-developed hocks. Roping horses are less apt to get bone spavin if they have good hind legs (no sickle or cow hocks) and are not started at heavy roping until they are mature—at least four years old.

Bone spavin causes lameness and pain when the horse flexes the hock joint, making him take a shorter, lower stride. He tries to land on the toe of the foot, sometimes dragging it and wearing it excessively. Lameness from bone spavin is usually worst when the horse is first exercised; in mild cases the horse will tend to get over his lameness after he is warmed up. In severe cases the lameness will get worse with exercise.

Bone spavin creates an enlargement on the inner surface of the hock, sometimes detectable only by X ray. The spavin test is often used to help confirm the diagnosis, for most cases of bone spavin react positively to this test. The spavin test consists of keeping the hock flexed (by holding the hind foot up nearly to the horse's belly) for a couple of minutes and then trotting him. If the first few steps he takes show more lameness than he normally shows, he has shown a positive reaction. Bone spavin is a permanent unsoundness but some treatments are helpful. Corrective shoeing (forcing the foot to break over the center of the toe) helps some horses because they tend to wear their feet on the outside edge, attempting to place weight on the outside of the foot to relieve the pain of the spavin on the inside of the hock. If the spavin lies under the cunean tendon, surgical removal of part of that tendon may give the horse some relief from pain. Sometimes the involved hock bones can be fused surgically to help eliminate pain. Complete recovery should never be expected but sometimes the horse will be usable, though he may always be slightly lame when first starting the ride, especially in cold weather.

Bog spavin is a chronic swelling of the joint capsule of the hock (the capsule is filled with excessive synovial fluid) and is usually caused by poor conformation, especially if a horse has hocks that are too straight. It can also be caused by injury to the hock joint as a result of quick stops and turns or other stress, or caused by nutritional deficiency such as rickets.

When a horse has bog spavin, there is a large soft swelling that bulges on the front-inside area of the hock joint, and sometimes there are also two smaller swellings, one on each side of the hock towards the back. These should not be confused with thoroughpin, in which the swellings are higher. When any one of these three swellings (of bog spavin) are pressed, the other two will bulge out farther. Lameness is usually present only if the bog spavin is caused by trauma, and there will then be heat, pain, and swelling over the hock joint.

If the bog spavin is caused by poor conformation, treatment is of little value because the condition is likely to recur. When strain is the cause, corticosteroids can be injected to reduce inflammation of the synovial lining and help prevent excessive fluid formation. Elastic bandages around the hock will also help. The horse should then be rested at least three weeks. If bog spavin is caused by nutritional deficiencies (rickets), treatment is successful only if the horse's diet is also corrected. Bog spavin in horses with rickets is most common in young horses six months to two years of age. A large bog spavin is a serious blemish. If it is due to poor conformation, or if lameness continues, bog spavin is considered to be an unsoundness.

Occult spavin is the least common type of spavin and is a disease of the hock joint causing typical spavin lameness but no visible changes in the hock. Trauma is the most usual cause, probably damaging the joint cartilages—a condition which does not show up in X-rays. The lameness from occult spavin usually continues for the rest of the horse's life and therefore this type of spavin is a permanent unsoundness.

THOROUGHPIN

Thoroughpin is another unsoundness that is similar to bog spavin. A thoroughpin is a puffy swelling in the web of the hock, caused by distention of the sheath that surrounds the deep flexor tendon. It is usually caused by strain to this tendon. Like bowed tendons and bog spavin, once it is stretched and strained and swollen, it remains that way, becoming a permanent weakness.

STRINGHALT

Stringhalt is a problem that can only be detected when the horse is moving, so if you are checking a horse for soundness, have him walked and trotted. Stringhalt is an involuntary (spastic) flexion of the hock joint and can either be mild or severe, the horse jerking his foot up toward his abdomen. This jerking is exaggerated if the horse is making a turn. Usually only one hind leg is affected. As the horse lifts his leg to take a stride, it jerks up too high, and then when it goes back to the ground it hits the ground sharply. Stringhalt is worst when the horse begins exercising, then tends to disappear as he warms up. The jerking action is most noticeable when the horse turns or backs up, and is worse in cold weather. It also tends to get worse as the horse grows older.

The cause of stringhalt is not known, though it has been blamed on nervous diseases such as neuritis of the sciatic or peroneal nerves, or spinal cord problems. There is no treatment that will produce a cure, but many cases improve after surgical removal of part of the tendon that crosses the outside surface of the hock joint.

UPWARD FIXATION OF THE PATELLA

Upward fixation of the patella is a problem that can occur if a horse's hind legs are too straight. The stifle joint in the horse is similar to the human knee and has a kneecap (patella) at the front of the joint. When the patella slips upward out of place, the hind leg locks in an extended position out behind the body. The stifle and hock cannot flex but the fetlock joint can. In mild cases there is just a temporary catching of the leg for a second. In more severe cases the leg may remain locked for several hours or days. These cases are sometimes difficult to put back into place.

Conformation is an important predisposing factor, but debility and poor conditioning can also make a horse more likely to suffer upward fixation of the patella. Some cases occur when the limb suffers trauma while it is overextended. Once upward fixation has occurred, the ligaments are somewhat stretched and the problem is likely to happen again. Shetland ponies are more likely to suffer this problem than are horses of other breeds. Sometimes a foal will outgrow this condition if it is mild, but if the problem continues, surgery should be performed to correct it, preferably before the horse is two years old—before he is left with a permanent gonitis (inflammation of the stifle joint). The operation, called medial patellar desmotomy, is very simple, severing the medial patellar ligament. If this operation is performed before gonitis develops, the horse will probably fully recover, no longer suffering from upward fixation.

GONITIS

Other unsoundnesses of the upper hind leg are dislocated stifle and inflammation of the stifle joint (gonitis). The injured stifle is due to trauma, such as falling down, or being kicked, running into a gatepost or doorway, suffering periodic upward fixation of the patella, and so on. Arthritic changes take place in the stifle joint because of damage to the bones, the joint capsule, the cartilage and ligaments of the joint. This kind of injury often results in permanent disability. Years ago our black mare, Scrappy (dam of Khamette) was kicked on the stifle by another horse and suffered serious injury. She could not walk on that hind leg at all and hopped around on three legs for more than a year. But eventually the joint healed and she became sound again after about a year and a half. Time can do wonderful things—time and rest. But many times an injured stifle joint will never heal properly to let the horse become sound again. The degree of lameness with an injured stifle joint will vary with the severity of the injury. Gonitis is present if there is swelling in the joint capsule. If one of the ligaments of the joint has ruptured, treatment will do very little good and the horse will remain chronically lame. If there are bony changes or fractures in the stifle joint, chances of recovery are equally unfavorable. But if the gonitis is due to sprain and injury of the joint capsule or to ligament attachments (without rupture) a long period of rest (and injections of corticosteroids if no infection is present in the joint) may result in complete recovery. If infection is present it must be cleared up before the joint can heal properly.

BURSITIS

There are several types of blemishes that can also be unsoundnesses under certain conditions. Included in this category are capped elbow, capped knee, capped hock, fistula of the withers, and others. These are all cases of bursitis, caused by injury. A bursa is a lubricating sac that covers a joint. These sacs are found in many parts of the body, and they are filled with a lubricating fluid, serving to protect bony places and friction areas where tendons run over bones. When there is injury to a bursa, it fills with excessive fluid, creating a round swelling that is hot and painful. This bursitis can become complicated with infection, breaking and draining—sometimes forming a fistula that will not heal.

Fistula of the withers is an infected bursa at the top of the withers. This can be an unsoundness if it interferes with the use of a saddle so that the horse cannot be used. Poll evil is an infected bursa at the top of the head; usually caused when the horse strikes the top of his head (on a doorway, trailer roof, etc.).

Capped elbow is an enlargement (bursitis) on the point of the elbow. This can be caused by a blow on the elbow, or by the horse's shod heel pressing against his elbow when he is lying down. Sometimes a horse strikes himself in the elbow

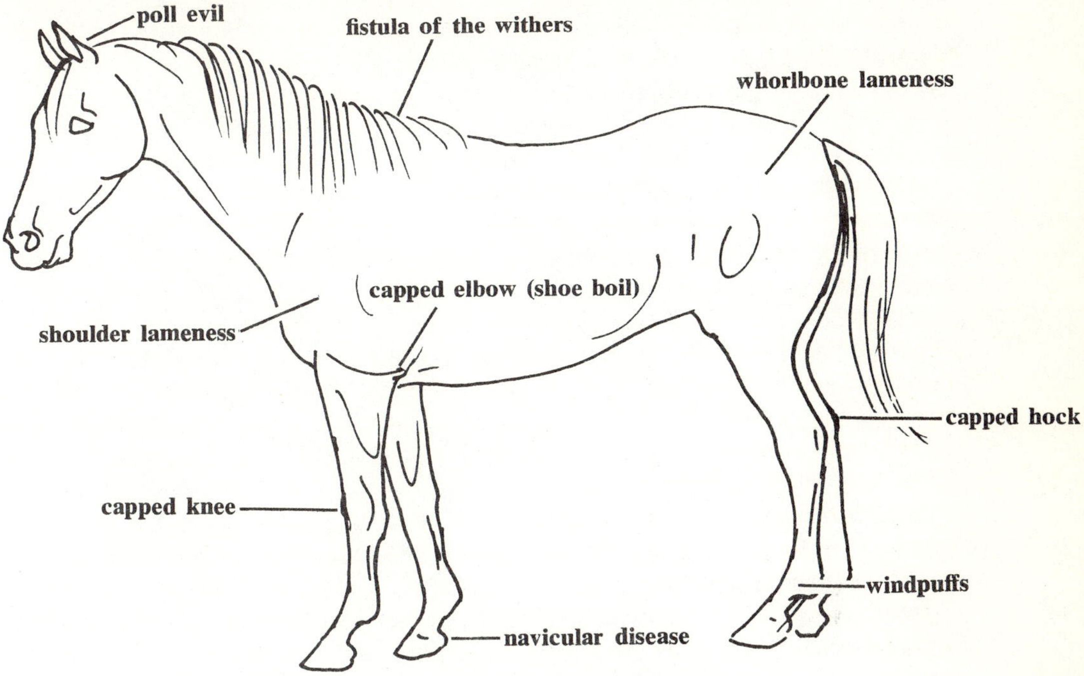

Some of the most common sites for bursitis

with his foot when stamping at flies. If infection develops in the swollen bursa, the fluid turns to pus, creating a shoe boil. This will burst and drain, often becoming a permanently draining fistula. The bursa may also fill with fibrous tissue.

Capped hock is an enlargement on the point of the hock, usually caused by injury such as a blow on the hock or by the horse kicking a wall or a trailer tail-gate. There is a firm swelling at the point of the hock, sometimes accompanied by curb, a swelling that is a few inches lower. Capped hock only rarely causes lameness, but the blemish is a permanent one, because of scar tissue within the bursa.

Capped knee is an enlargement at the knee (hygroma of the carpus) and is caused by injury to the knee such as from the horse falling on his knees, being thrown to the knees, having to get up and down on hard ground, or from a blow on the knee. If this bursitis is neglected, the swelling can become cystic. The cystic mass can eventually calcify, causing a permanent blemish and impaired knee action. Most bursitises cause permanent swellings. Often these are merely unsightly and do not interfere with the usefulness of the horse. But if they make the horse lame, they must be considered unsoundnesses.

Whorlbone lameness is caused by an inflammation of a bursa up in the hip region of the horse (trochanteric bursa)—a bursa that lies under the tendon of the middle gluteus muscle. This bursitis is caused by such things as bruising (if the horse falls on that side) or straining of the tendon during strenuous exertion. Bursitis in this area makes the horse lame, the stride of the affected hind leg being shorter than that of his good leg—making the horse travel dog fashion with the hindquarters moving toward the good side. If this condition is present in a horse for a period of time, the gluteal muscles atrophy and shrink.

Shoulder lameness can be caused by inflammation of the bicipital bursa (bicipital bursitis). This bursa cushions the movement of a tendon over the groove in the humerus (forearm). The most common cause of bursitis here is injury to the point of the shoulder, and occasionally it can be caused by injury to the shoulder blade or humerus. The same trauma may also cause sweeny (atrophy of the shoulder muscle) as a result of damage to the suprascapular nerve. Sweeny is also an unsoundness because there is no known treatment of any value for atrophy of the muscle due to nerve injury, and the horse is left with a permanently withered shoulder.

CARPITIS

Carpitis (popped knee) is inflammation of the knee joint, often involving the joint capsule, its bones and ligaments, due to concussion or trauma. Horses with poor conformation and horses that are soft and out of condition or horses that are worked too hard are most likely to develop carpitis. Occasionally a blow to the knee (or direct injury when the horse paws and hits his knee on something solid) will also cause the condition, causing new bone growth in the knee area. This type of injury can leave the horse permanently unsound, with an enlargement at the front of the knee and lameness at fast gaits. Treatment is not very successful if new bone growth has already started. In the very early stages, corticosteroid injections and a long rest (at least four months) may result in recovery, if the horse's conformation is good. Poor conformation (calf knees or bench knees) in the horse will make the problem more likely to recur.

WINDPUFFS

Windpuffs are a distention of a joint capsule, tendon sheath, or bursa in a horse's leg, and are usually caused by strain, especially when the horse is young and immature. Young horses that are ridden too hard or subjected to a heavy training schedule almost always develop windpuffs at the fetlock joints, in the joint capsules. Race horses, rodeo and contest horses, or any other horses that are worked excessively may develop windpuffs. Once a windpuff gets started, it will be a permanent blemish, with firm swelling in the fetlock joint just above and in front of the sesamoid bones. Chronic cases may harden. Windpuffs are usually just blemishes, but if they cause lameness they are unsoundnesses. They usually do not cause lameness unless accompanied by arthritis, bursitis, or tendonitis.

SADDLE SORES

Saddle sores, too, are usually just blemishes, unless they result in scar tissue that would interfere with the use of a saddle. Saddle sores are more fully covered in Chapter 10.

KNOCKED-DOWN HIP

Here are a few more unsoundnesses to look for when judging or selecting a horse. A horse with a knocked-down hip has a fracture at the point of the hip, usually from running into something or being kicked.

PARROT MOUTH

A horse with a parrot mouth has problems eating, because his upper incisors overlap his lower incisors. This overlapping of the upper teeth is considered an unsoundness because it interferes with the horse's ability to eat properly, and is also inheritable.

HERNIA

A hernia is another problem to look out for; it is a protrusion of a loop of tissue through an abnormal opening.

HEAVES

Heaves, a problem of the lungs, is another name for emphysema in horses. The symptoms of heaves are forced and labored breathing when the horse exercises, and coughing. Heaves is the result of inflammation of the lungs, either from infection (such as pneumonia) or irritation or allergy from exposure to dust or from being fed dusty or moldy hay. Heaves is a problem that can only get progressively worse, so if you are selecting a horse make sure he is free from this unsoundness. Heaves is covered more fully in chapter 10.

BLINDNESS

When checking a horse for soundness, don't forget to check his vision. Blindness in one eye may not be readily noticed unless you check for it. Blindness can be caused by injury to the eye or may be caused by certain diseases such as leptospirosis, the after effect often being periodic ophthalmia (moon blindness; see chapter 9), the most common cause of blindness in horses. Pass your cupped hand carefully in front of each eye and watch the horse for a response.

Whatever the reason you are choosing a horse, soundness should be one of your prime concerns. If you are going to use the horse for riding, you'll definitely want him sound—safe to ride and agile, and able to withstand the traveling and maneuvering that you'll have him do. And if you are choosing him for breeding purposes, you'll want him to pass on the best possible qualities to his offspring. If the horse has an unsoundness that might possibly have been caused or aggravated

A horse with parrot mouth; the upper incisors overlap the lower ones

by poor conformation, you don't want to take the chance of having the problem crop out among your foals. There are still a lot of good horses to choose from, so don't get stuck with a poor one. Look before you buy, and when you look, make sure the horse is sound.

18

CORRECTIVE TRIMMING AND SHOEING

NOT ALL HORSES ARE BLESSED WITH PERFECT FEET AND LEGS. Sometimes, due to poor conformation or accident, a horse may have crooked ones. The seriousness of the abnormality will depend upon the degree in which the horse's gaits are affected. If the horse merely paddles a little, his gait may be unsightly, but he can still be a useful, dependable animal. But if the faulty conformation of the feet is such that the horse strikes himself or is not sure-footed but clumsy, his usefulness and safety are limited. The horse is an athlete, and his greatest usefulness is in being able to move with speed, agility, balance, and safety. If he is limited in these aspects because of abnormal feet and legs, his value is therefore lowered.

Some abnormalities can be corrected with proper trimming and shoeing. Ideally, the corrective trimming should be begun when the horse is but a small foal. Regular and frequent trimmings can help to straighten legs and feet as the young horse grows, often completely correcting minor faults and greatly helping more serious ones. Some faults can never be fully corrected, but proper trimming and shoeing help them to such an extent that the horse can be useful and sound.

Corrective work on a horse's feet should always be done gradually and with care. The object of it all is to make the shape of the foot, the level of the foot and the angle of the foot as nearly normal as possible so that the horse will have a normal gait, but trimming should never be too drastic in an attempt to make the horse's feet normal, because you may put strain upon the bones of the leg. You cannot really change the bone structure once a horse is mature, and to attempt it will usually make the horse lame and sore or create problems that are worse than what you had to start with. In a very real sense, faulty gaits in a mature horse cannot really be corrected because you cannot make any permanent changes in the conformation of the horse's legs and feet. But corrective trimming and shoeing can modify the horse's gait and way of moving to improve it and make the horse more functionally agile in his traveling. But remember that this improvement is only temporary and the trimming or shoeing job you do must be redone with each trimming and shoeing.

Before we go into detail describing corrective trimming and shoeing, we ought to first mention some things about trimming and shoeing the normal foot. A person should always have a good knowledge of the horse's feet, legs and gait, and horseshoeing in general, before he attempts to do corrective trimming and shoeing.

The horse's foot is a unique structure with an outside horny covering that protects it and which grows continually. Ideally, in the natural state of running wild, the horse wore down his feet about as fast as they grew. Thus his feet never became too long nor were worn too short. The wild horse seems to have been almost free of serious foot problems, but with domestication, troubles began. The horse was brought from soft ground and grasslands to hard roads, from light exercise to hard work, and from healthy, sanitary pasture land to confined housing, in many cases. And under domestication, man often has the horse in situations where his feet grow too long from lack of room to exercise, or works him so regularly that the feet wear down faster than they grow. If feet grow too long they put strain on the structures of the horse's legs—bones, tendons, ligaments, and joints—and make the feet more subject to splitting and breaking. They can make the horse lame and also throw his feet and legs out of balance by uneven wear on broken feet, making strained joints or crooked legs or even a crippled horse. If feet wear down too far from excessive use, the horse loses the protective covering and he becomes lame and tenderfooted from walking on the sensitive tissues of his foot.

The horse evolved and ran wild on grassy plains where he did not need shoes, but now, due to the terrain or to the amount and type of riding, shoes are a necessary evil in most parts of the country. A horse being used by man usually does a lot more traveling than a horse on his own running wild. And unless the horse has exceptionally tough feet or is always traveling on soft ground or grassy sod, he needs some protection for his feet. Man discovered this fact quite early, almost as soon as he began to use the horse for his own purposes as a beast of burden and and as his primary means of transportation. So thousands of years ago horsemen devised ways to protect the feet of the horses they rode, using everything from socks to sandals attached to the horse's feet with straps or thongs. The practice of nailing iron shoes to the hoof was not introduced until the second century B.C. and was not really common until the end of the fifth century A.D. Shoeing as we know it today came into regular use in the Middle Ages and has not really changed much since that time. But now a relatively new concept in horseshoes is being tried—a new kind of shoe called the Nature Plate has been created. This new shoe is designed to simulate the action and effect of the perfect bare foot having sloping sides to match the natural contour of the hoof wall. The ground surface has a thin sharp edge like a bare foot. This shoe gives better traction, cuts into the ground better and therefore puts less strain on the horse's legs. The Nature Plate is the first major change in horse shoeing ideas for more than a century.

Let's take a quick look at the structures of the horse's feet. Horses are all individuals; each one is built a little differently and the feet of each horse are of slightly different shape and hardness than those of other horses, with an individual rate of hoof growth. Moisture and moist climates tend to make a hoof softer. Nonpigmented hoofs (that occur in conjunction with a white leg or foot) are softer than pigmented hoofs and are more easily broken or worn down. And because they are softer, they do not hold a shoe as well.

The outside horny shell of the hoof (hoof wall and sole) is a specialized protective covering for the sensitive living tissues inside—bones, nerves, blood vessels, tendons, etc. Nature has provided the horse with a unique foot covering that grows continually to compensate for the wear and tear of traveling. The outside covering of the foot consists of the hoof wall, the sole of the foot, the frog and the bars. The bars of the foot are the inward continuation of the hoof wall. The V-shaped frog serves as a cushion, helps absorb concussion when the foot strikes the ground (the frog is springy and expands when the hoof takes weight) and helps regulate the moisture of the hoof. The hoof wall carries most of the horse's weight and the bars serve as a brace to prevent ovexpansion and contraction of the foot. The sole is slightly concave to give grip and to allow for expansion in the foot. If any of these outer tissues—hoof wall, sole, bars and frog—are abused by injury or by excessive trimming, the normal functions and soundness of the entire hoof are in jeopardy.

The hoof wall grows downward from the coronary band and any injury to the coronary band will affect the growth of the hoof. Average rate of hoof growth in a normal horse is about one-fourth to three-eighths inch per month, so the entire hoof wall is usually replaced by new horn every eight to ten months. The horny sole of the foot continually grows downward from the sensitive inner sole. As it continues to grow, little cracks develop in the old sole and it tends to flake away, for the outer layer is dead tissue. Nature has provided that this part of the foot will trim itself, and hence the sole seldom needs trimming. This flaking away of dead tissue is called exfoliation. Most soles should never be trimmed too deeply, for they don't need it, and trimming too much will get down into the sensitive tissues. But occasionally dead material on the sole builds up, and this should be peeled away carefully with a hoof knife to the proper level, so that the layer of dead, cracked sole will not continue to build up and provide a place for the growth of bacteria and hoof diseases.

In unshod horses, the bottom of the hoof wall, the frog and the bars of the foot are all level and flat to the ground and each helps to bear the weight of the horse. On soft surfaces the sole helps to bear the weight also. The bars are very important because they act as a wedge to keep the heels from caving in and contracting. Thus the bars should never be trimmed excessively or cut away, for most horses don't have enough of this structure to begin with.

The bars and frog seldom strike the ground in a shod horse unless the ground is soft, because shoeing puts the hoof a quarter of an inch or more above the ground, by-passing nature's plan for the bars and frog to take some of the weight. The horse's ancestors roamed grassy prairie country and could run barefoot all their lives without becoming tenderfooted, but horses ridden in rocky terrain need shoes to keep from wearing down their feet too fast, and to avoid lameness from stone bruises. If their frogs and bars were left flat with the ground when shod, they would soon go lame.

Many horseshoeing experts advocate leaving enough frog when shoeing so that the frog can take part of the weight when the foot is on the ground. This is all well and good if the horse is to be used on soft ground or smooth roads where there are no rocks or gravel. Nature's plan is for the frog to take weight, for every time the frog takes weight it forces the bars apart, lifts the plantar cushion and exerts a pumping action on the network of blood vessels within the foot—aiding proper circulation. But when you ride in mountains or over rocky terrain, your horse's frog must not be so low that it strikes the ground, or your horse will go lame. This fact is often disregarded by the "professionals" and experts who learn and teach from books on horseshoeing rather than from actual experience. A frog never needs to be trimmed excessively. In most horses, trimming away any tags or dead tissue from the frog will be sufficient to insure that it does not strike the ground when the horse is shod. Occasionally horses will need more trimming.

Shoeing is an artificial condition for the horse and is not necessarily healthy for his feet, but in most parts of the country shoeing is a necessary evil. Shoes protect the hoof from excessive wear when the horse is ridden regularly. Shoes can give a horse better traction under some conditions. Corrective shoes can help correct defects of gait and make it possible for a basically unsound horse to be of satisfactory use, or make it possible for a "crooked" horse to travel straighter. Shoes can also be used to help cure diseased or defective feet, or to give relief from hoof cracks, bruised soles, or other problems.

The good horseman shoes his horse as skillfully as possible, to hinder him as little as possible, then lets the horse run barefoot at pasture part of the year to let his frog, bars, and other foot structures function normally. Most horses get this "vacation" during the winter when they are not ridden as often. A horse turned out to pasture should always have his shoes removed so his feet can function more normally; if he is not ridden, his feet will not wear excessively at pasture and the rest will be good for the health and soundness of his feet.

The first step in any shoeing job is to carefully observe the position of feet and legs when the horse is standing squarely, and then observe him in motion—at the walk and trot—to see the normal angles of his feet and to tell how he handles his feet, whether there is any paddling or winging. Notice the point at which the foot breaks over and leaves the ground (whether the break-over point is at the toe where it should be, or off to one side), or any deviations in gait. Uneven wear of the feet shows that the horse is not traveling straight but is breaking over crookedly. A horse that travels straight on sound, straight feet and legs will wear the foot or the shoe evenly, with slightly more wear at the center of the toe.

When the horse is standing squarely there should be about a 45-degree angle in the slope of the hoof and pastern. Individual horses may vary somewhat from this; some have more sloping hoofs and pasterns, others more upright. The ideal angle is about 45 degrees for the front feet and pasterns. Hind feet and pasterns tend to be a little more upright than fronts; 50 to 55 degrees is the normal angle for hinds. But trimming of the hoof should never be too drastic in an attempt to give the horse's feet a perfect angle, because each horse has his own normal angle and any radical changes can cause problems. The foot should always be trimmed so that the pastern and hoof are of the same angle, forming an unbroken line, whatever that line and angle may be for that particular horse. If you try to make a 45-degree angle on a horse that normally has less or more angle than this, you will be working against nature and the way the horse was put together in the first place; you'll just be putting strain on parts of the horse's legs. The important thing is to trim the foot so that pastern and hoof are of the same angle. If the horse's toes have grown too long, this angle line will be broken, putting strain on the legs.

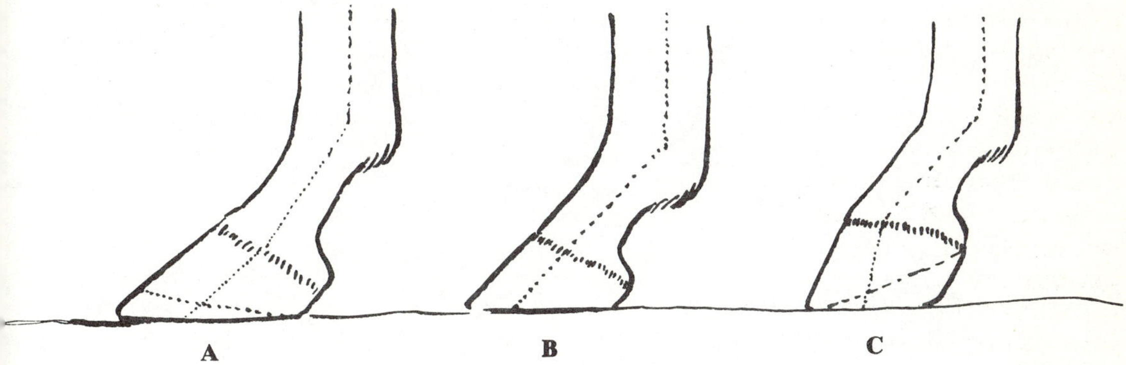

A. The toe is too long, and the foot angle is broken backwards. The toe should be trimmed as indicated by the dotted line

B. A normal foot should have about a 45-degree angle

C. The heel is too long, and the angle is broken forward

If the heel and hind parts of the hoof are too long, the line will be broken in the other direction. You will have to study the feet when the horse is standing squarely in order to judge how much hoof to take off to make the foot level at the proper angle.

When you begin to trim the foot, first thoroughly clean it. A hoof pick or the blunt edge of a hoof knife is the best instrument for cleaning all the dirt out of the hoof. Never use a sharp or pointed object for cleaning the foot; you could injure yourself or the horse if he happened to jerk his foot at the wrong time. When all dirt and debris is removed, trim any dead tissue of the sole with a hoof knife. All loose material should be scraped away, but do not trim much deeper (unless the horse has an abnormal buildup of dead material) for the horse needs this horny sole as protection for the inner tissues of his foot. After the frog is trimmed of loose tags, the outer hoof wall can be trimmed level with the sole, except at the quarters,

PARTS OF THE FOOT

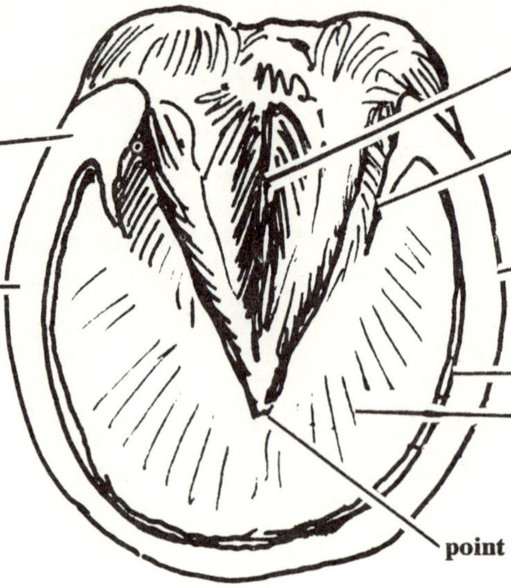

with hoof cutters. Trimming should be done from heel to toe or in a complete circle from heel to heel, for a smooth, consistent cut. The hoof cutters should be held in such a way that they make a flat, level cut, to make a level seat for the shoe. In order to make a flat surface when trimming the hoof wall, you will have to leave a little extra wall at the quarters (because the sole at the quarters is concave and a little lower than at the heel or toe); if you cut this area level with the sole you will have cut away too much and the wall will not quite meet the shoe in this place. If the bars need trimming, trim them level with the hoof wall at the heels. After the excess horn is trimmed away with hoof cutters, the hoof wall that has been trimmed should be rasped smooth to make a level seat for the shoe. The rasp should always be held flat and level so that one side of the wall is not accidentally made lower than the other.

If the horse is to go barefoot, you should leave about a quarter of an inch of hoof wall projecting below the sole when you trim the foot; don't take as much hoof off, or he will become tenderfooted quickly or sole-bruised because he will soon be walking on his sole instead of hoof wall. He needs a little extra horny tissue to keep sharp rocks and gravel from breaking into the sensitive inner structures. And the outside edges of the wall, for a barefoot horse, should be smoothed and rounded with the rasp; leaving a sharp edge will make the hoof more likely to split, crack, or chip away. Rounding the edges also prevents injury to other horses that may be in the group; wounds and cuts on lower legs can result from being stepped on or by running into the sharp snags of other horses' untrimmed hoofs, especially when several horses are in a small area (such as milling about in a corral).

If the horse is to be shod, trim the hoof wall level with the sole at the toe, and

Lynn shoeing Nikki

Leveling the foot for the shoe

Nailing the shoe on

as low as necessary at the heel to establish the proper foot angle. The shoe should be shaped to fit the foot. The foot should never be trimmed to fit the shoe, unless for the purpose of correction for some defect in gait. The shoe should fit evenly on the foot—the bottom of the hoof wall should rest flat down against the shoe. The heels of the shoe should not stick out excessively beyond the heel of the foot, especially on the front feet, or the horse may step on his front shoes when he travels. The shoe should fit the foot so that the outer edge of the shoe closely follows the outline of the trimmed hoof at the toe and around to the quarter. But at the quarters and heels the shoe should always be slightly wider than the hoof and should extend slightly beyond the heels of the horse.

The heels and quarters of the horse's foot must rest on the shoe and have a little extra width of shoe to allow for expansion of the hoof when weight is placed on the foot. If the shoe is too short, or if it does not properly fit at the quarters and heel, the shoe may cut into the foot as the hoof wall grows (the hoof may grow down around the outside of the shoe) or cause corns or other problems that can make the horse lame. To allow for hoof expansion, the branches of the shoe should extend about one-sixteenth of an inch wider than the hoof wall at the heel and quarter, and the heel of the shoe should extend this same distance behind the horse's heel. This gives the expanded hoof adequate support when the horse puts weight on the foot. If the shoe is too narrow and short, the hoof tends to expand out over the shoe when weight is placed on it, and the narrow shoe limits proper hoof expansion.

Because the foot should expand when weight is placed on it, especially at the heel, nails should never be driven into the hoof too far to the rear. The last nail hole of the shoe should not be farther back than the bend of the quarter. If the last nail is back beyond the bend of the quarter, expansion of the hoof at the heel and quarters will be quite limited.

The shoe should be properly centered on the foot. For horses with good conformation and normal feet (straight and well balanced), the shoe can be centered by using the point of the frog as a guide, for the frog will divide the bottom of the hoof into equal halves and will point toward the toe. But in horses that are pigeon-toed or splay-footed the frog usually points off center and cannot be used as a guide.

If a hoof is worn excessively on one side because of poor conformation and faulty gait, or if part of the hoof wall has been chipped or cracked and broken away, it might not be possible to make the foot level by trimming, without trimming too much away on the opposite side. If one side of the foot is broken or exceptionally low, the branch of the shoe on the worn side can be shimmed with leather so that the foot will be level when the shoe is on.

When the shoe has been properly shaped to fit the trimmed foot, the nails can be driven in. The nails should be of the proper size for the hoof and the shoe, having heads that protrude a little after they are driven in. A nail head that fits too deeply into the crease is too small and will not hold as well; the shoe may eventually work loose.

The nails should enter the hoof at the white line. In nailing the shoe on, hold the nail with thumb and index finger, resting your hand on the shoe to keep it in the proper position as you drive the nail. The location of the first nail driven does not matter greatly (whether it is on left or right, toe or heel, etc.) so long as the shoe is properly centered and stays in position during the driving of the nail. After the first two nails are driven, one on each side of the shoe, the shoe will stay in place for the driving of the rest of the nails.

Horseshoe nails are made with a straight side and a beveled side so that they will curve when driven. The beveled side of the nail point should always be toward the inside of the foot so that the nail will be directed outward as it is driven. The beveled side can be easily determined by the rough side of the nail head; the bevel and the rough side are both on the inside of the nail—thus the rule for horseshoe nails: "Rough side inside." With the rough edge on one side of the nail, you can tell which way to set the nail, just by the feel of the nail head.

The nail should be held straight, and aimed for the spot where it should come out on the hoof wall. The nail should come out about three-fourths of an inch above the shoe, or about a third of the way up the hoof wall. The acutal distance will depend, of course, on how large the horse is. A draft horse has much bigger feet than an Arabian, for instance, larger nails will be used in his shoes and the nails will come out higher in the hoof wall. If a nail comes out much too high or too low, it should be pulled and redriven (if the nail is at all bent or curved, use a new nail). A high nail may press on the sensitive tissues inside the foot and make the horse lame. A low nail may break out of the hoof and be of no use in holding the shoe. Nails driven to a uniform height give the shod hoof a pleasing appearance, but if the nails come out somewhere near the proper position it is best to leave them; a nail pulled and redriven just to make a uniform appearance may weaken the hoof wall with the second perforation and eventually cause the shoe to become loose.

The nails should be driven in the same direction as the hoof fibers, parallel with them, to minimize cutting of these fibers—which would weaken the hoof wall. The fibers of the hoof grow parallel to each other and perpendicular to the coronary band. The driven nails should go into the outer portion of the hoof wall. They should never be driven inside of the white line that separates the wall from the sole, or they will go into the sensitive part of the hoof. The white line is a very important landmark when shoeing a horse, for this line is where the hoof wall and the sole unite. Driven nails should stay outside the white line, and thus should be placed correctly (rough head and beveled tip to the inside) so they will curve outward and come out the side of the hoof wall instead of curving into the sensitive part of the foot. And the nails shouldn't be driven too straight (especially on a horse with small feet) or they may come out too high.

A high nail is more apt to prick sensitive tissues, or put pressure on these tissues if it bends a little. If a horse ever flinches during the driving of a nail, the sensitive tissues may have been pricked or punctured. If this happens, the nail should be pulled out and iodine poured into the nail hole to prevent infection.

If you use light hammer taps when driving the nail, the point of the nail will travel parallel to the horny fibers of the hoof wall and won't sever them. To drive the nail through the side of the outer hoof wall at the desired place, use light

1. Trimming and leveling the foot

2. Centering the shoe and placing the first nail

3. Driving the nails

4. Twisting off the nail tip

5. Clinching the nails

hammer taps until the nail is about two-thirds the required distance, and then strike one sharp hard blow to force the nail through the wall and finish its journey. The bevel on the nail point makes the nail curve outward, and this bevel is most effective in curving the nail when the nail is being driven sharply and rapidly through the horny fibers.

As soon as a nail is driven through and the nail head is sunk down into the shoe crease, twist the sharp tip end off with the hammer claws or cut the end off with nail cutters (and keep track of those nail tips and discard them properly so they won't lie on the ground and cause someone a flat tire). These sharp nail tips should always be taken off so that if the horse moves or tries to take his foot away from you, the sharp tip won't cut your leg.

After all the nails have been driven and the tips cut off close to the hoof wall, cut a small notch underneath each clipped nail with the rasp. This notch will keep the hoof wall from splitting when the nail end is turned over and clinched. It will also provide a place for the clinch; the nail end when clinched will be sunk down into this little notch and leave a rather smooth surface on the hoof wall. A clinch that is fairly well sunk into the hoof wall will not be so apt to get knocked loose.

To clinch the nails, pound each nail head tightly into the shoe crease while holding a piece of iron against the tip end, to bend the tip over (into the notch you cut) as it comes farther out of the hoof wall. Nails should be clinched alternately from one side to the other (the two toe nails first, or the two heel nails) rather than clinching all the nails on one side and then on the other, even though it is easier to clinch all the nails on one side and then on the other. Clinching them alternately from one side to the other gives a more uniform seating of the shoe. Nails should be clinched firmly, but not so tightly that the horse's foot is made sore and nail bound. If you clinch the nails too tightly the horse may be lame for several days afterward. Finish the clinch, bending the nails with the hammer, pounding the tips flat down against the hoof wall into the notch you cut. A short clinch holds much better than a long one, and also looks neater. A long clinch is apt to be knocked and unbent in rocks and rough terrain and will work loose faster.

With the rasp you can smooth the outside edge of the hoof where it meets the shoe, but be careful not to rasp the hoof wall too much. This horny wall is made up of tiny tubes (the horn fibers) and excessive rasping cuts them and leaves them open, allowing the hoof to dry out. If the shoe was properly fitted to the hoof to begin with and nailed on with no slipping, little or no rasping will be needed on the outside of the hoof wall to smooth it to the shoe. After the horse is shod, have someone lead him for you at both the walk and the trot so you can observe the new shoes and the horse's feet in action, to determine if the shoes are set properly, aiding and not hindering his gait, and to be sure that the horse is not lame or "ouchy" in any way from his new shoes.

Shoes should always be as light as possible, taking into consideration the wear demanded of them, to interfere as little as possible with the normal flight of the horse's foot. The ground surface of the shoe is better plain. A plain plate shoe will hinder the horse least and interfere least with his way of going. Toe and heel calks are not necessary except for ice, for draft horses that need extra traction when pulling, or for other special conditions where a certain amount and kind of traction is needed. Calks are also sometimes used on corrective shoes in certain situations.

Nails should always take a short, thick hold on the hoof wall and should not be too high. Thus they damage the least amount of hoof horn and the old nail holes can be removed with the natural growth of the hoof the next time the horse is shod. This is especially important if the horse is used steadily and will be wearing out his shoes regularly, as when he is used as a ranch horse in mountainous terrain, or in training for endurance rides, and so forth. If the horse must be shod often, you will want room in the hoof wall for the new nails for each succeeding set of shoes. If your horse is used often and you wear out a lot of shoes, a hard-surfacing material such as borium can be welded to the wearing surface of the shoe, to make it last longer. Borium is the trade name for a material that comes in a steel tube (about 15/16ths inch in diameter), filled with crystals of tungsten carbide that are very sharp and nearly as hard as diamonds. When this material is welded to an ordinary horseshoe, the steel melts and forms a bond with the shoe, while the tungsten carbide crystals retain their sharp edges. The rough surface that results will give good traction on rocks, concrete, ice, and so on, and will result in a longer wearing shoe. A shoe with a good borium weld (spots of borium added to toe and heels) may last up to ten times as long as an ordinary shoe. The borium shoe will have to be reset periodically to allow for the growth of the horse's foot, but it will not wear out before the horse needs a reset as some ordinary horseshoes will do when a horse is used hard. Borium can also be added to shoes for a corrective effect, reinforcing wear points where a horse breaks over incorrectly, wearing away the shoe on one side. Because the borium does not wear away, the shoe retains its original shape and keeps its corrective effect longer, helping the horse break over more toward the center of the toe.

A horse's feet should be trimmed, or shoes reset or replaced every four to six weeks, depending on the individual horse's rate of hoof growth, and the amount of wear on the feet if he is barefoot. A barefoot horse that gets some wear on his feet may not need to be trimmed this often. The worst thing about shoeing a horse is that all too often the shoes are left on too long. Some people actually leave a horse's shoes on so long that they eventually fall off of their own accord. A horse whose feet have grown too long because his shoes were left on may suffer such problems as leg wounds (from striking himself with his extra long or splayed hoofs), strained legs, contracted feet, corns, and so on. Since the hoof wall grows out perpendicularly to the coronary band, the horse's base of support actually grows out from under him if his shoes are left on too long. This puts great strain on his flexor tendons. Shoes that are worn too long tend to wear down and become thin and loose; sometimes bending and shifting, causing nail punctures and corns.

To remove a horseshoe, cut the clinched nails on the hoof wall with a clinch cutter, or unbend them with old hoof nippers if you have no clinch cutter. Then pick up the foot and use pulling pincers (or old hoof cutters) to pull the shoe. It is better to pull the shoe than to just cut the nails to remove the shoe. Cutting the nails will dull the tool you use, and leaves nail pieces in the horse's hoof which will have to be removed after the shoe is off. Pulling out these nail pieces is not as easy and is more likely to break and crack the horse's hoof than is pulling the shoe properly with pulling pincers. The pincers should be placed under the shoe, starting at one heel branch. Close the handle and push outward from yourself to loosen the heel branch, pushing slightly toward the middle line of the foot. After that heel is loosened, the opposite heel should be loosened. Each nail can be removed individually as it is loosened, if you wish. As you pull the shoe loose, hold the foot securely and never twist the pincers or pull crookedly, and you will not injure the fetlock joint. Continue working down both of the branches alternately until the entire shoe is loosened.

We've briefly covered the shoeing of the normal foot, so now let's take a look at corrective shoeing. Corrective shoeing is done to improve or correct faults in the horse's gait resulting from inherited conformational faults, injury, and so on, and is also done to relieve pain and to encourage healing in diseases and injuries to the legs and feet. There are many kinds of shoes used to correct foot problems in the horse. These shoes can be made by a horseshoer, or by welding additional pieces to an ordinary horseshoe. Many of the special corrective shoes were designed in the days of the draft horse, and were useful because they transformed a serious lameness into a mild lameness, allowing a horse with a bad foot problem to still be useful for work. But for today's saddle horse, even a slight lameness is too much, and a shoe that reduces lameness is not satisfactory, because the horse is only improved and not cured. Thus the great number of special shoes that have been used in the past for corrective purposes are often not fully satisfactory today, for today's horse owner is not satisfied with a horse that is merely less lame. But there are still some corrective shoes that are satisfactory for helping to straighten a horse that interferes or forges, and shoes that will help make a horse travel more normally.

A basic concept of horseshoeing is that weight (in the form of horseshoes) no matter how it is added to the foot, always reduces speed and decreases agility. Corrective shoes should always be as light as possible. The normal flight of the foot is a straight forward line. Any deviation from normal flight takes the form of an arc, either to the outside or to the inside. Adding weight to the foot in the form of horseshoes will increase the arc because of the additional centrifugal force. Ordinary shoeing thus accentuates and increases the horse's defect in gait. This is the reason that many horses overreach or interfere only when they are shod. A horse that tends to interfere (strike one front limb against the other, or one hind limb against the other) or to forge (strike his front heel with his hind toe) will do so even worse when he is shod because of the added weight of the shoes, making his strides longer and the arcs of foot flight even more pronounced. Thus he must be correctively shod when he wears shoes, to prevent these problems if possible.

When attempting to correct a faulty gait, you must try to balance the foot both at rest (when it is on the ground) and in its flight through the air, producing a straighter way of going. Thus you must fully understand the structure of the horse's feet and legs and how they move. Every horse must be considered as an individual when trying to determine how best to correct his way of traveling, for no two horses will move exactly alike. Even though several horses may have the same kind of fault that needs correcting, or the same kind of conformation that suggests a certain kind of correction, it may take several different kinds of shoeing to correct them. A horse may look as though he needs one type of shoe for correcting the problem, when actually another kind will do it better. For example, a horse that stands splay-footed (toes out) will usually land on the inside of the hoof wall

Corrective Shoes

These shoes help the foot break over more quickly:

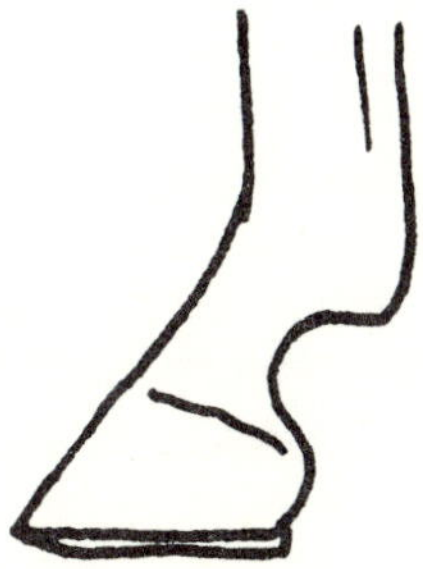

roller-toe shoe

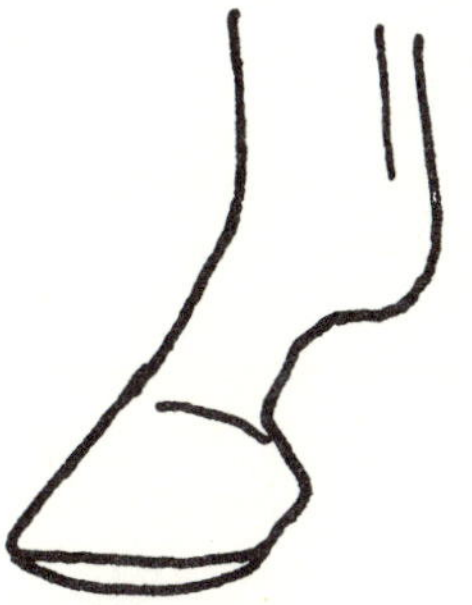

full roller motion shoe

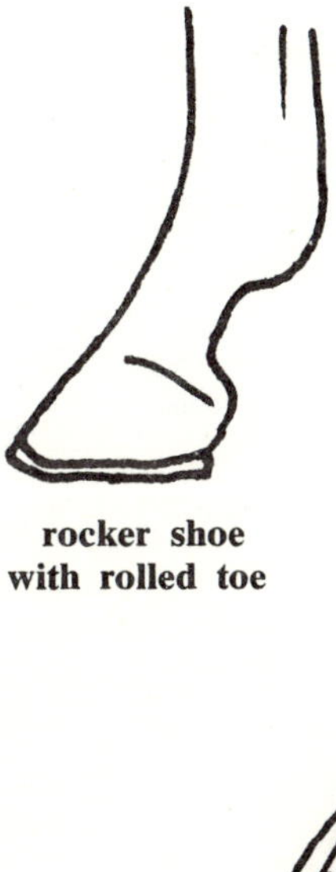

rocker shoe with rolled toe

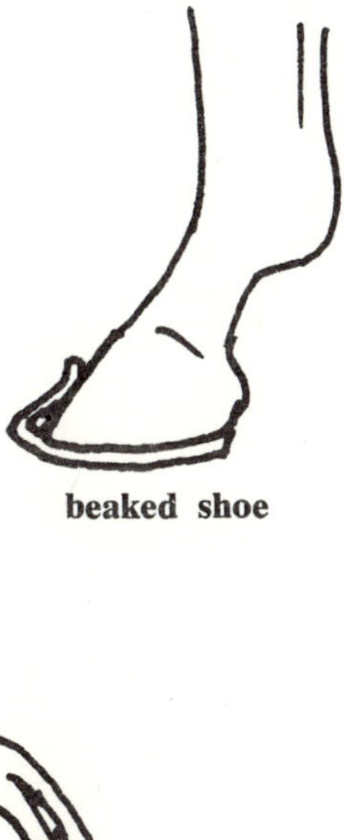

beaked shoe

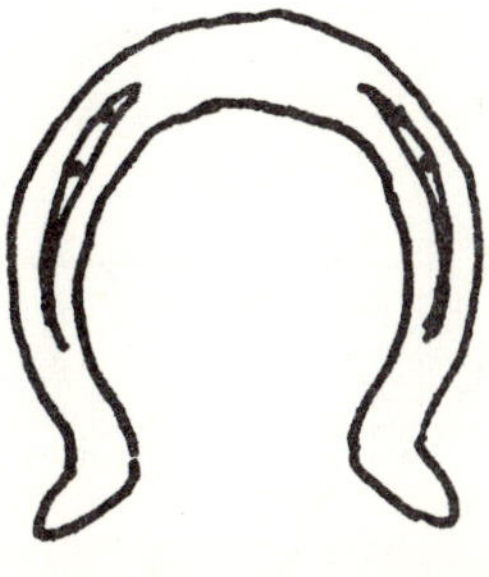

A corrective shoe with trailers on both heels makes the foot hit the ground squarely

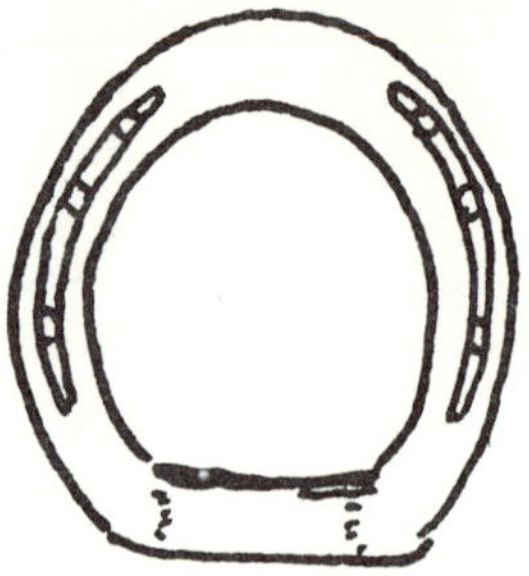

A full-bar shoe can be used to increase or decrease frog pressure

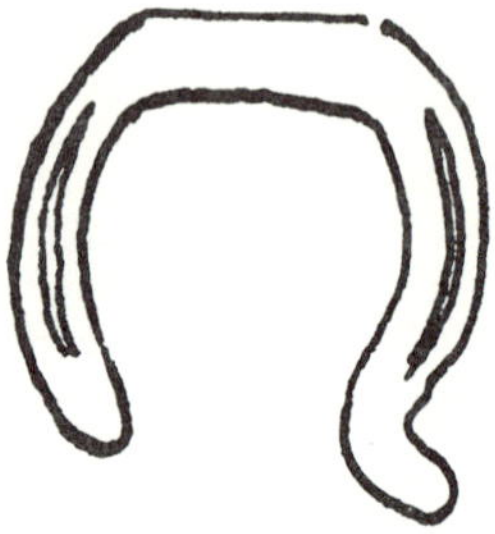

Square-toe shoe with trailer; the square toe makes the foot break over the center of the toe. A trailer is sometimes added to the inside heel of splay-footed horses, and the outside heel of pigeon-toed horses

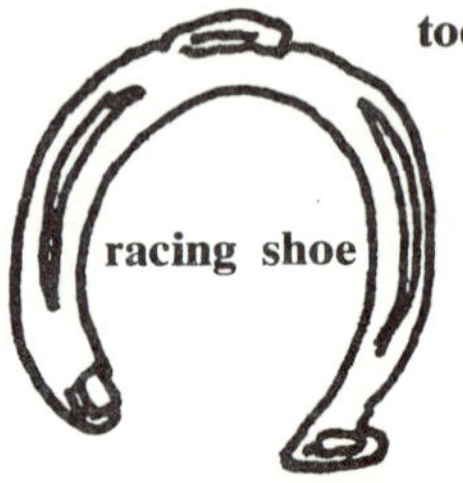

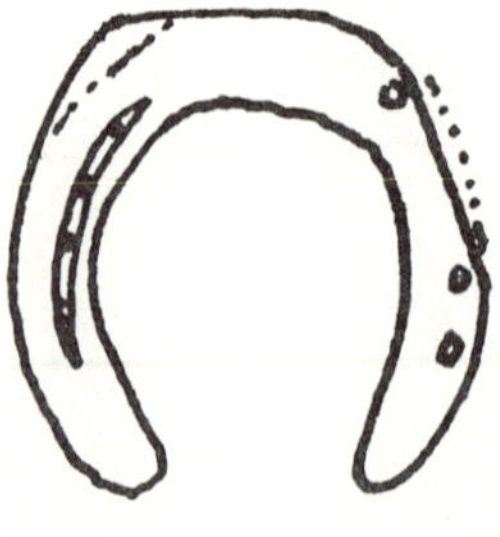

A shoe used for a horse that interferes; the lateral extension helps the foot break over the toe squarely. The spared branch and trimmed hoof wall eliminate the part of the hoof that strikes the other leg

when he travels, wearing down the inside wall more than the outside, but occasionally you'll find a horse that will be splay-footed and yet do just the opposite, wearing down the outside wall. You'll have to study the horse carefully, both while he is standing and while he is traveling, to determine the horse's way of going so that you can improve it with the proper kind of shoe. And after you try a corrective shoe, you should carefully observe the horse's movement again to see if it has been improved. Sometimes two or three methods must be tried before you'll find the best one for that particular horse. There are usually several things you can try, but as a general rule you should always try the least severe method first—a correction that will actually alter the horse's foot the least. But if this won't correct the gait enough to solve the horse's problems, more drastic methods may have to be tried. But just remember that drastic changes can cause other problems.

Here on our ranch we've only had a few horses over the years that needed corrective trimming and shoeing, but occasionally other horses are brought to us for advice. Because horses can vary so much in their individual problems, we always have to see the horse and watch him travel before we can advise.

We shoe our own horses. I've been shoeing my own horses since I was 14, and got my first experience with corrective trimming and shoeing about the same time, helping shoe our black mare Scrappy, who paddled severely, and then later my own filly, Khamette (Scrappy's half-Arabian daughter). Khamette has several problems with her front feet, as I will mention later as examples, but these problems are all greatly improved by proper trimming and shoeing.

Most horses need corrective trimming and shoeing to some degree because of faulty conformation. Because their legs and feet are not quite straight, the hoof wears unevenly. Part of the reason it wears unevenly is because the faulty conformation causes the horse to travel crookedly (feet are picked up not exactly straight, swing through the air in an arc that deviates from perfectly straight, and then are put down not exactly straight, creating more wear on one side of the foot than on the other). And since the hoof is worn unevenly, it tends to grow uneven and the problem can only get worse, the misshapen foot forcing the horse's leg out of line even more and aggravating the problem.

If a horse is never properly trimmed and shod his legs and feet may grow terribly crooked from this neglect, possibly ruining the horse for any future usefulness. But if a horse with a conformation problem is carefully trimmed and shod, the problem may be lessened considerably, or at least kept so that it never gets any worse.

When trimming a horse with uneven feet, your main object is merely to trim away the areas that have grown too long and uneven from the uneven wearing of the foot. You are trying to trim the foot so that it is more nearly level, so that the horse will tend to pick it up straighter.

If the horse toes in (pigeon toes), the outside walls of his feet are usually worn down too much. When trimming this kind of foot you should take off the excess hoof on the inside wall, from toe to heel, to help level the foot. Horses that toe in or out usually do not break over the center of the foot when the foot is picked up, but break over to one side or the other, causing more wear. If a horse travels straight the wear will be at the center of the toe. If the wear is off to one side or other you know that the horse is not picking the foot up straight. To help a horse pick up his feet straight, you can square the toe a little, making the horse break over more at the center. If a foot has a side that is flared out, you can guess that he has a conformational fault in the leg that tends to create this problem, putting uneven distribution of weight on the foot as the horse travels. The cause of this flaring out may be only minor, but as time goes on the problem only gets worse, making a weak hoof wall which breaks easily because of the way it flares out. By trimming and rasping the flared side of the hoof wall you can help the foot regain its normal shape over a period of time.

Let's take a quick look at some of the problems that are helped by corrective trimming and shoeing. First, let's consider the horse that toes out and is splay-footed. This problem can either be minor, or serious. In serious cases the horse strikes himself as he travels because the foot swings inward as it goes through the air.

When a horse toes out, his foot is not level because he wears off the inside wall. The foot breaks over the inside of the toe when it is picked up, and lands on the inside toe and wall when it is put down—after describing an inward arc as it goes through the air. These feet should be trimmed by a trimming of the outside wall to help level the foot. If the horse is to be shod, special shoes can be used to help level the foot properly. Special shoes can often be made from ordinary shoes by welding additional pieces of iron to them. Here on our ranch, Lynn makes any special shoes we need just by bending, or cutting and welding to the desired shape and thickness.

If the foot you are working on is exceptionally worn down on the side you are trying to build up, you can place a piece of leather on that side under the shoe (between shoe and foot) to help level the shoe.

There are several kinds of corrective shoes that can be used to help a horse that is splay-footed. These I will try to describe. The purpose of all of them is two-fold—to raise the inside wall to where the foot is more level, and to make the foot break over the center of the toe. A half rim shoe is just what the name implies; it is a shoe with a rim on the outside edge of the side which will go on the inside wall of the foot. When this shoe is on the horse's foot, placed upon the ground, the shoe raises the inside of the foot and tends to interfere with the foot's breaking over the inside of the toe, forcing it to breakover more nearly at the center. This kind of shoe can be made from an ordinary shoe just by welding a small rod to the outside edge of the shoe, from heel to the toe nail hole. A shoe of this kind can also be used on a pigeon-toed horse, but in the opposite manner—to raise the outside wall. In other words, this shoe is used to raise the side of the foot which is low, and to help the horse break over at the center of the toe.

Another kind of shoe that can be used for either a splay-footed horse or a pigeon-toed horse is a shoe with high inside rims and an open toe. This shoe can be made by welding a small (one-fourth inch in diameter is about right) rod to the inside of the branches of the shoe, but leaving the toe as is. The rods on each side of the shoe help make the horse break over the center of the toe. A shoe of this sort can be used if you are able to trim the foot quite level to begin with, whereas a half rim shoe can aid in leveling the foot if one side is worn too much to get the foot quite level when trimming. Shoe calks can sometimes be used on a shoe to create the same effect as the welded rod. The calks should be about one-half inch high, and are welded to the heels and to either side of the toe at the first nail position. Like the high inside rims, these calks help force breakover at the center of the toe.

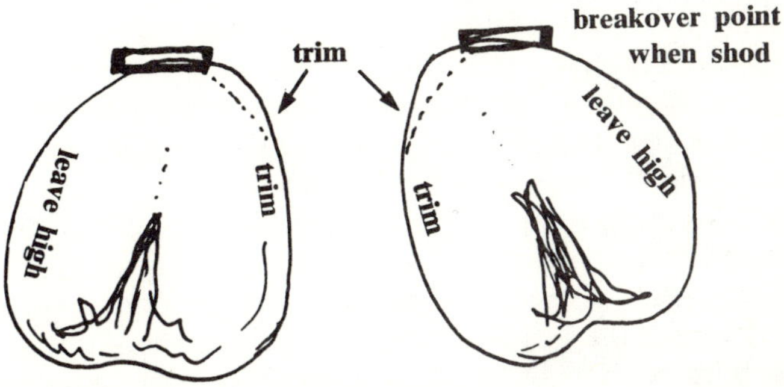

On a pigeon-toed horse like Khamette, who paddles and wears the outside hoof edges excessively, we trim the inside edge to level the foot and center the shoe so that the foot breaks over more to the center. Then the foot makes a straighter arc when it swings through the air

Our mare Khamette is pigeon-toed and paddles as she travels. She travels straighter if we shoe her correctively, trimming the inside of the hoof more and leaving the outside longer. Then we smooth off the inside toe (which grows too long) and center the shoe carefully to make sure it does not accentuate the pigeon-toed problem. Sometimes we add hard-surfacing to the toes of the shoes so that the breakover point won't wear excessively.

Sometimes a toe-in (pigeon-toed) or toe-out (splay-footed) horse can be helped just by welding a small bar across the area of breakover at the side of the toe. The bar then interferes with breakover, forcing the foot to breakover straight, at the center of the toe.

Many mild cases of toe-in or toe-out can be corrected just by using square-toed shoes. The square toe makes the foot break over straight. The foot should be trimmed level and the shoe placed on it so that the toe of the foot is even with the toe of the shoe, the square portion of the shoe sticking out on either side of the horse's toe. Do not dub off the horse's toe.

A toe extension on the shoe serves the same purpose, only more effectively. This kind of shoe is made by welding an extension to the toe of the shoe, on the side which the horse breaks over. This extension thus interferes with the crooked breakover and helps the horse break over straight. We made a shoe of this type for a horse that was brought to us because he was interfering badly with one hind foot, striking his other hind leg at the fetlock joint. The horse was a little splay-footed behind, but broke over the outside of his toe and wore the outside hoof wall. But the significant problem in this case was the fact that he struck his right hind fetlock joint as he traveled, making a terrible sore. He did not interfere at the walk or at the canter, just at the trot, and then only when excited. He had been a race horse, and didn't interfere when galloping on a race track wearing light aluminum shoes. But his owner was trying to retrain him as a trail horse and was trying to teach him to settle down and trot. When brought to me, he had a large bandage on his right hind fetlock joint to protect it.

This horse had very strange action. His legs were nearly straight behind; at first glance you wouldn't think there would be anything wrong with his way of traveling. He was just a wee bit cow-hocked and sickle-hocked, fetlock joints fairly close together when standing (base-narrow) but then his pasterns and hoofs pointed slightly out. A person would think he might break over the inside of the toe, being splay-footed, but he broke over the outside, after which the foot swung inward on its downward flight and brushed the opposite fetlock joint. We attempted to straighten the flight of the foot a little by making the foot break over at the center of the toe instead of at the side, and accomplished this purpose by means of a square-toe extension on the shoe, which Lynn made by welding a piece of metal to the shoe, as pictured. When finished, we put chalk all over the side of the shoe and hoof and trotted the horse to make sure he no longer interfered—and he left no telltale chalk on his bandaged fetlock joint.

Interference can be a serious problem. Occasionally it is rather mild, and just an annoyance if the horse merely brushes one leg against the other, or does so only when tired, or excited. But interference (the horse striking the inside of one leg with the inside of the hoof on the opposite side) becomes serious if the horse does it consistently or forcefully enough to cause injury to himself. Interference may be caused by poor conformation (a horse that is base-wide, or any horse that is splay-footed regardless of whether he is base-wide or base-narrow), by fatigue, and by poor shoeing that throws the foot off balance or makes it break over incorrectly. Interference is most common in the hind legs, and the fetlock joint is the most common site of injury. Trotters and pacers tend to interfere in front striking the opposite knee. Mild cases of interfering can be corrected by squaring the toe of the hoof or the shoe, to make the foot break over squarely at the center of the toe instead of to the inside. If the foot breaks over squarely as it leaves the ground, it will tend to make a straighter path when in flight.

When Fahleen was a 4 year old she caught her right front foot in a gate and pulled it badly. I gave her nearly a year of rest before I started riding her again. The injured pastern joint healed, but calcified and fused, limiting the motion of the joint. When I started riding her again, she travelled crookedly on that foot, especially when shod (shoeing accentuated the abnormal flight of the foot). That foot had a tendency to toe out, yet when she moved, it broke over the *outside*, was picked up with a twisting motion, and brushed the opposite fetlock joint when in

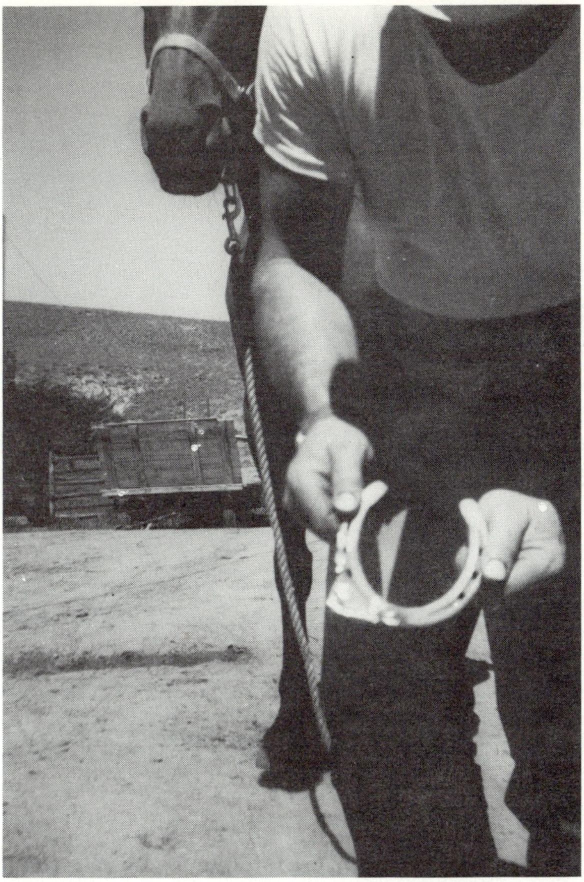

The special toe-extension shoe that Lynn made by welding an additional piece of metal to the shoe

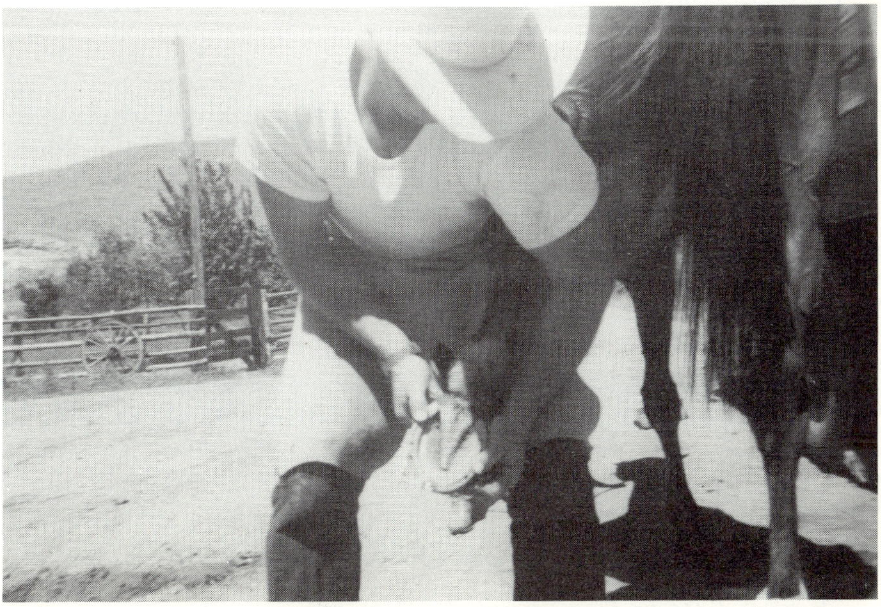

Placing the shoe on the foot

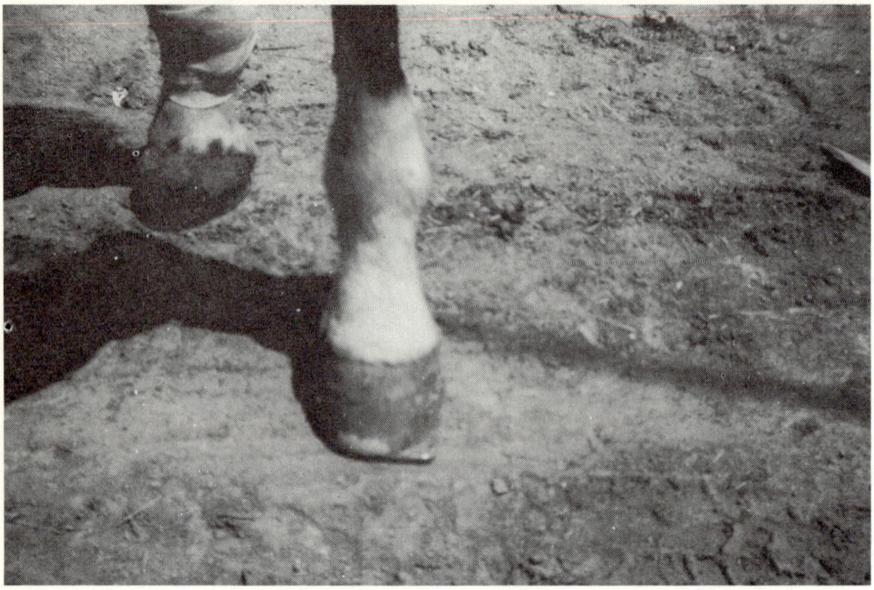

The shoe in place. This shoe prevented the foot from breaking over to the outside and made it break over to the center of the toe

flight. She did not strike the opposite leg consistently, only two or three times during a ride—but that was enough to soon make a sore spot. She struck herself mostly at the walk (taking long, gawky strides) or at the gallop in sagebrush or rough terrain. At the trot she picked her feet up smartly and they made a straight arc through the air. We were able to correct her problem (enough so that she no longer struck herself) just by making a shoe with a toe-extension on that outside part where she was breaking over. It made her foot break over at the center of the toe, and its arc through the air was straighter, no longer striking the opposite fetlock joint.

Severe cases of interfering can usually be corrected by squaring the toe (or using a toe-extension shoe which forces the foot to break over squarely at the center of the toe) and then altering the shape of the hoof and shoe so that it will not strike the opposite leg. This can be done by carefully determining which part of the foot or shoe is doing the striking (by using chalk or some other coloring substance liberally on the opposite leg in the area that the horse strikes, and then observing where the tell-tale coloring is brushed onto the hoof or shoe that does the striking) and then rasping off that portion of the hoof a little and altering the shoe to fit. Then the offending foot will not quite reach the opposite leg. Corrective shoes for a horse that interferes should always be extra light.

Other problems that can occur when the horse travels are forging and cross firing. Forging is a fairly common defect in which the toe of the hind foot or shoe strikes the heel of the front hoof or shoe (on the same side) and occurs most commonly at the trot. Forging is usually due to poor conformation—a horse that has hind legs too far underneath his body, or "sickle-hocked," or from having a short back and long legs, or long hind legs and short front legs. It can also be caused by leaving too much heel on the front shoe (the shoe extends too far back behind the front foot). In severe cases of forging, there is damage to the front foot from the continual striking. In more common, milder forms of forging, the trot is simply a noisy gait because the front and hind shoes clack together. Fatigue can sometimes cause forging. A young horse in training may sometimes forge even though he has good conformation and properly balanced feet; he forges because undeveloped leg muscles become tired.

To correct this problem you must keep in mind the fact that the front foot must be encouraged to break over and be lifted from the ground more quickly, to get out of the way of the approaching hind foot. And the hind foot must take a slightly shorter stride so that it won't quite reach the front foot. The horse should be ridden at the walk and at the trot as you observe his leg action, to determine the rate of speed at which the forging is worst. Also check the horse's conformation, and see if the feet are properly balanced. If the horse is wearing shoes, check to see if they are properly fitted and of proper weight. A heavy shoe will aggravate the problem.

When shoeing a horse that forges, the heel of the front shoe and the toe of the hind shoe should be shortened so that they will be less likely to meet. Lightweight shoes should be used, for weight increases the length of the horse's stride. A rocker or rolled toe can be used on the front feet to encourage them to break over faster and come off the ground sooner. The hind feet should be left a little longer than normal (so that the hind foot will not break over so quickly) and shod with *very* light shoes. The hind shoes can have slightly turned out heels that extend about one-half to three-fourths inch beyond the hoof (the extra heel of the shoe tends to stop the hind foot as it is being put down), or can be shod with heel calks and a rocker toe (to increase hock action—making the horse pick his foot up higher, not swinging it so far forward), the heel calks tending to stop the hind foot before it hits the front foot.

Cross-firing is about the same as forging except that it usually only occurs in pacers. The toe of the hind foot strikes the heel of the diagonal front foot. This problem is most likely to occur in a pacer that is splay-footed in front and pigeon-toed behind—the splay-footed front foot moves inward during the first part of its stride and the pigeon-toed hind foot moves inward during the last part of its stride, so diagonal feet are likely to meet. To correct this, both front and hind should be encouraged to move forward in straighter lines. The best way to accomplish this is to make the foot break over the center of the toe and thus begin the stride in the proper direction—straight ahead. If the horse is splay-footed in front he should be shod to help correct the splay-footed condition. If he is pigeon-toed behind, the hind shoes should be made to help correct that condition.

Corrective shoeing can also be beneficial in treating certain injuries and abnormalities in the foot such as flexor tendonitis, lameness from corns, chronic laminitis, contracted heels, ringbone, sidebone, cracked hoofs, wire cuts in the coronary band, and so on.

When shoeing a horse with contracted heels, remember that the frog must have pressure in order to help establish normal functions of the foot and gradually spread the heels again. If possible, the horse should be allowed to go barefoot. Shoes tend to hinder the normal expansion of the foot. If the horse can go barefoot, with his feet trimmed often, frog pressure and foot expansion will help to correct the contracted condition. If the horse has to be shod, a tip shoe (which just covers the toe area) can be used if he doesn't do hard work or travel in rocky areas. The tip shoe protects the toe from excessive wear, but leaves the heel area unshod for greater frog pressure. A full-bar shoe, half-bar shoe or T shoe can be used to induce more frog pressure, the bar or T putting pressure on the frog when the horse puts weight on the foot. A slipper shoe lets the heels of the foot slide outward when weight is put on the foot.

A contracted foot can also be helped by thinning the hoof wall at the quarters with a rasp. This thinning should start about half an inch below the coronary band in the area of the quarters and heels—gradually decreasing until the hoof wall at the ground surface is of normal thickness. This thinning helps the hoof wall to expand when the horse puts weight on the foot. If the foot is thinned, it should be shod with a pressure bar shoe (to increase frog pressure) and treated with hoof dressing daily to keep the thinned hoof wall from cracking.

Another way to promote hoof wall flexibility and expansion is to groove the hoof wall at the quarters, either with a series of vertical grooves or with one groove just below the coronary band. These grooves go into the hoof wall to the sensitive laminae, but do not go deep enough to touch the blood vessels of the foot. If a foot is treated this way, it should also be shod with a pressure bar shoe.

Corrective shoeing for a horse with ringbone can be done by shortening the toe of the foot and putting on a full roller motion shoe. This type of shoe helps transfer the action from the pastern and coffin joints to the bottom of the foot—the roller motion shoe making the foot break over quickly and easily. The same type of shoe can also help a horse with sidebones.

A horse with corns should go barefoot, if possible. But if he must be shod, the hoof wall and bar in the corn area should be cut away so that there will be no pressure on the sole. A half or full-bar shoe can be used, fitted in such a way that the frog will absorb most of the concussion.

Cracks in the hoof wall sometimes require corrective shoeing. A shoe will often help keep the slit hoof wall from splitting farther. The foot should be trimmed so that there will be no direct pressure on the split; in other words, that area of the hoof wall must not bear weight. This will allow the hoof wall to begin to grow out without further widening of the crack. The progress of the crack should be halted, if possible, by rasping or burning a groove at the farthest point of the crack. A triangle and bar design (see illustration) is satisfactory for limiting the expansion and progression of hoof cracks.

If the crack penetrates into the sensitive tissues of the hoof, the horse will be lame because of infection. If this is the case, the crack should be thoroughly

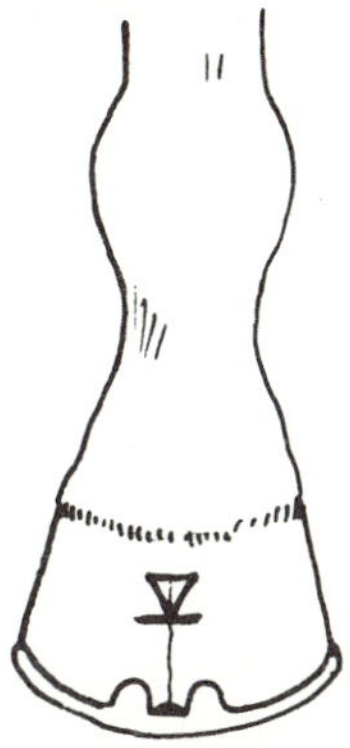

To correct a toe crack, a triangle and bar design can be cut or burned into the hoof to help keep the crack from expanding when weight is placed on the foot. The hoof wall should be trimmed at the toe so that the crack will not bear weight. Toe clips on the shoe help support the hoof wall so that the crack cannot widen

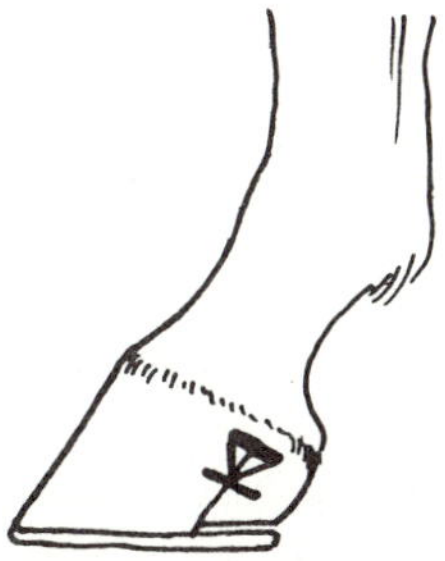

To correct a quarter crack, a triangle and bar design is cut or burned above the crack. The hoof wall is trimmed back, so that it will not bear weight and expand the crack

cleaned and disinfected with iodine, and the horse should be given a tetanus shot.

If the horse is left barefoot, the crack should be grooved and the bearing surface of the hoof wall in that area should be cut away so that it does not bear weight and expand the crack. For a toe crack, the toe should be trimmed on either side of the crack. For a quarter crack, the hoof wall at the heel should be trimmed, from the crack on back, so that the heel area will not take weight and spread the crack. Heel cracks are treated in the same manner.

When shoeing a horse with a hoof crack, the foot is trimmed so that the area around the crack does not touch the shoe and bear weight. Sometimes a clip on each side of the crack is helpful to keep the hoof wall from expanding, helping to support the hoof in the area so that the crack cannot expand. A half-bar shoe is sometimes helpful for a quarter crack, producing frog pressure—allowing the frog to bear some of the weight that would ordinarily be borne by the hoof wall that has been trimmed away from the crack.

Part of the secret of growing out a hoof crack is to trim the foot often, so that there will be very little expansion pressure on the crack. If the hoof wall grows too long, the crack will tend to split farther. Careful grooving of the top of the crack, and frequent trimming or corrective shoeing will eventually clear up a hoof crack that begins at the ground surface of the hoof wall. If the crack is kept from widening or progressing, the normal hoof growth will eventually grow it out. But a hoof crack that originates in the coronary band (from injury to the coronary band) will be a more difficult and persistent problem. Corrective shoeing may be necessary for the rest of the horse's life, for the defect in the coronary band causes distorted and defective growth of the hoof in that area. The hoof wall in that area may have to be rasped every two weeks to keep the horn growth as nearly normal as possible, and the coronary band rubbed daily with olive oil or some other suitable softening agent if it is hard and dry.

Sometimes a severe hoof crack can be repaired with a very strong glue. We tried this several years ago on a bad quarter crack in old Nell's right front foot. She had a stubborn crack originating at the ground surface of the hoof, but which was made worse by an old wire cut (scar tissue) at the coronary band above it, which weakened the hoof wall in that area. Because of the weak strip in the hoof wall, this quarter crack spread upward very readily in spite of conventional methods to halt its progress. So we used a strong, fast-drying glue to help hold the cracked area together and prevent further splitting as the hoof wall grew out. A glue used for this purpose must be very strong or it won't be able to hold the hoof wall solidly together when weight is place upon it, expanding the hoof. The hoof wall must not expand in the area of the crack.

First of all, we trimmed the ground surface of the hoof wall so that the cracked

Repairing Nell's quarter crack with glue

You can barely see the strip of fiber glass mesh

area would not bear weight, and thoroughly cleaned the hoof so that there would be no dirt or foreign material to interfere with the adhesion of the glue to the hoof wall. Then we put a strip of strong fiber glass mesh over the crack area to add strength and applied the glue, thoroughly filling in the crack. And we held her foot up for twenty minutes so that the glue would have time to set before she placed weight on the foot. She was very patient with us, for the glue we used produces some heat in its drying. After the twenty minutes, it was solid enough to hold the crack and keep the area solid and unexpanding. If done carefully and adequately, this is a procedure I would highly recommend for holding a stubborn hoof crack as the foot grows out. The foot must still be trimmed often (or the corrective shoes reset), and the glue may have to be reapplied periodically until the crack has grown out sufficiently.

When shoeing a horse with flat feet, the sole should not be trimmed much, if any, and the frog shouldn't be trimmed. The wall should be trimmed only a little, enough to smooth it for the shoe. The shoe should cover the wall and white line and just barely touch upon the sole. If the shoe is seated merely on the hoof wall, the sole will tend to drop farther as the hoof wall grows and pushes outward. If the shoe is seated in such a way that there is too much bearing upon the sole (if the web of the shoe is too wide), sole pressure may cause bruising and lameness.

If a horse has a dropped sole, a wide-webbed shoe that fully covers both the wall and the white line (touching upon the sole to prevent further dropping) should be used. But sole pressure on the shoe should be avoided.

Just a few more taps . . .

A horse with chronic laminitis (founder) can sometimes be helped a little by corrective trimming and shoeing. The foundered foot should be trimmed as nearly normal as possible, more often than for a normal horse. Wide-webbed shoes may keep the sole from dropping further. The heel can be shortened (trimmed as much as possible without making the horse tender) to help correct and counteract the downward rotation of the third phalanx (coffin bone). The toe of the hoof wall can be trimmed and rasped back to more nearly normal foot shape. The edge of the hoof wall at the toe should be shortened slightly so that it does not bear on the shoe. The toe of the shoe should be "rolled" to help the foot break over easily when in motion.

In a foundered horse, the pain is mostly in the toe area of the sole. Therefore the best type of shoe for founder is one that is designed to remove all pressure from the painful sole and transfer it to the wall and frog which are less sensitive (thus the trimming of the horse's heels and the use of a shoe that bears mainly upon the hoof wall, not the sole). A wide-webbed shoe should have a concaved upper surface so that when the horse bears weight on the foot all the pressure is upon the hoof wall and none upon the sole. But the wide web is there to help keep the sole from dropping. It may also be necessary to protect the sole by using a rubber or leather pad or metal plate.

Corrective trimming and shoeing cannot always completely correct a fault or defect in the horse's way of going, or completely cure an injured or diseased foot. But corrective trimming and shoeing, carefully done, can usually help the problem to some degree, often making the horse more useful (enabling some horses to perform light work—horses which otherwise would be useless), or alleviating his pain so that he can get around more normally. Corrective trimming and shoeing cannot always produce miracles; an unsound horse may always be an unsound horse. But if the horse's feet or gaits are helped in any degree it may lengthen the life of that particular animal as a pleasure horse or breeding animal or enable him to fill whatever other role he may be used in that will not be vitally hindered by his particular problems.

19

HORSE SENSE

HORSES ARE FASCINATING ANIMALS. PART OF THE FASCINATION comes, perhaps, from the fact there is so much potential in the horse-human relationship—the fact that man has taken a basically wild animal and domesticated him and over the centuries has worked so closely with him that there is practically no limit to what a sensitive rider upon a sensitive horse can accomplish together as a working unit. A good horse can be schooled to do many different things: leap tall obstacles with a rider on his back, race around a barrel course, perform in the show ring, outrun, outwit, and outmaneuver wild cattle on a hillside or in a sorting pen, pull a wagon or a cart, and so on. One of the horse's tremendous assets in his alliance with man has been his great athletic ability. But perhaps even more important than his speed, strength, and agility has been his receptiveness to training, his willingness to work with man. The horse has a supposedly rather limited intellectual capacity (he is not thought to be one of the most intelligent of animals); a great deal of the success in this centuries-old relationship between horse and man is due to the horse's personality. In other words, his receptiveness to man's dominance and his ability to learn are owed largely to his feelings and his willingness as well as his "IQ."

The intelligence of an animal is usually measured by scientists by the size of its brain in comparison to the size of its body, and by its ability to find food. By these standards of judging, the horse is quite low in the scale of animal intelligence. In comparison to man, whose brain weighs about three pounds (2.5% of his body weight), the horse's brain weighs only 1.3 pounds, or about .13% of the horse's body weight. Over the ages, the horse has become larger than his early ancestors, changing from the tiny prehistoric animal he once was to the 15-hand, 1000-pound creature he now is. But his brain has not increased much in size. The horse of today has approximately the same brain capacity as the little creature of the prehistoric swamps which preceded him. The early dawn horse living in swamps or prairie country had little need for a reasoning type of brain for his only needs were those of filling his stomach and staying with the group for safety. The horse's brain is quite small, compared with the

The horse is not considered to be highly intelligent—according to animal experimental psychologists—yet he is very receptive to training. He is easily taught to become useful to man. He is sensitive and willing to do great things under the hands of a sensitive, understanding horseman

size of his head, but this small brain is quite adequate for the requirements of an animal that lives by grazing and by being on constant lookout for enemies from which to flee. When testing or judging an animal's intelligence, the early backgrounds, habitat, and needs of that animal should be taken into consideration. To test a horse's intelligence by giving him tests in which he must solve a maze in order to find food is ridiculous because the horse never had any need for this kind of reasoning; his natural food, grass, grew all around him.

Thus, according to scientists, the horse has a rather low intelligence when judged by the ways in which man generally rates intelligence. Yet, when a person gets to know the horse well, he comes to realize certain unique qualities about the horse's intelligence. The horse has a very specialized way of thinking which is highly developed in its own special ways. This specialization of the horse's reasoning capacity makes him the fascinating and adaptable creature he is today—a creature that can be very highly trained, and one that can be constantly taxing the brains of his owner or trainer, in spite of his supposedly low IQ. Standards of intelligence that rate the horse in comparison with other animals (according to Columbia University experiments the horse ranks tenth in animal intelligence, being less intelligent than the chimpanzee, orangutang, gorilla, monkey, dog, cat, raccoon, elephant, and pig) are really rather meaningless, for as I mentioned earlier, there are no real tests that can properly compare different kinds of animals or compare an animal to man, even though this is the way that experimental psychologists try to compare them.

You can't properly compare a horse with another animal such as a dog, for these two animals are so completely different. When one kind of animal is compared with another that is very different, the comparison might be favorable or unfavorable to a very extreme degree, for there is really no common meeting ground, no common characteristics upon which to judge them. We can't really measure the ability of one species against the ability of another because the two species are specialized in different areas, with highly developed abilities in vastly differing roles. There are no tests of intelligence just for horses, but perhaps there should be if we want to talk about horse intelligence, or our talking has no value. Any animal, to be judged in intelligence, should be measured against others of his own kind, or else his ranking in intelligence is absolutely useless.

Each kind of animal has its own particular set of physical and intellectual capacities that suit its role in life—that enable it to survive and thrive in its particular environment, finding its food and eluding its enemies. And these physical and intellectual capacities can only be measured against individuals of the same species, for the measuring to have any meaning. Man's intelligence is measured only against those of his own kind, and more specifically against those of his own race and social group. This is only common sense. Thus we should also use our common sense in seeking to measure animal intelligence, and stop trying to compare the horse with other animals in an attempt to get an idea of his intelligence or lack of it.

The horseman who works with horses and is acquainted with their personality quirks, mental capacities and abilities to learn and to respond to human training can probably tell you a great deal more about the mental abilities of the horse than can the experimental psychologist. The horseman who has a serious interest in his horses has an inside track here, because in order to be a really good horseman, one must think like a horse and become acquainted with the ways of the horse. And the experienced horseman knows there are great differences in the mental abilities of individual horses. Just as there are differences in people, there are very intelligent horses, bright-normal ones, average, dull-normal, and also a few that are mentally deficient.

The horseman or horse trainer is not concerned with whether or not the horse ranks such and such on an intelligence scale with other animals; what he cares about is how the particular horse he is working with stacks up with other horses—whether or not that horse has the ability to grasp and accomplish the training being given, whether or not he is able to accomplish the work asked of him or able to perform in the role that the trainer has desired to put him. That, after all, is the important question when working with horses.

It is useless to argue whether the horse as a species is "smart" or "dumb." A horse with average or above average mental capacities (by horse standards) is quite able to get along on his reasoning ability. Just watch the mare out in a field with her foal. She knows very well how to take care of that foal and keep it from danger, and does a much better job of it than her human counterpart can do without the proper instructions. A group of horses in the wild are well able to take good care of themselves, communicating danger signals and eluding predators. Perhaps most convincing of all in our decision as to whether or not the horse is adequately intelligent is his capacity for learning. The horse has an ability to accumulate an understanding of many commands and signals, and can respond very readily to training.

The good horseman can feel out a horse, sensing and measuring its abilities, leading the horse on to greater feats of learning and workability without ever going beyond its ability, without ever asking too much. When horse and rider work together, the good horseman is constantly thinking with the horse and determining its ability and capacity to perform. The sensitive, understanding horseman is the best judge of a horse's intelligence, and also the most able to utilize that intelligence to get the most from the horse in terms of performance.

Let's take a look at the horse's background, and at the skills and abilities that the horse of today has inherited from his prehistoric past. The horse is not an aggressive animal, but rather a timid one. A stallion may fight to keep his leadership in a group of horses, or to protect his herd from predators, but except in the cases of stallions challenging one another, the horse is almost always on the defensive rather than the offensive, and defends himself or flees from danger rather than attacking. The early horse learned to be always alert, ready to interpret the slightest signs of danger—the rustle of a leaf, the whiff of a strange scent, movement in the bushes. He learned to remember through the months and years where the best and safest water holes and grazing areas were as he traveled from place to place with the seasons. He learned to communicate through some sort of sixth sense with the other horses in his group so that the herd could stick together when fleeing—turning and wheeling in close formation yet never colliding nor hindering one another.

The horse was essentially a gregarious animal, living as a member of a large group of horses. This fact is still true today and most horses become rather frantic when left alone, or lonely or neurotic when forced to live in an environment without other horses. Some horses adjust more readily to an isolated existance than others can. The lone horse is in a situation contrary to all his natural instincts, and is usually happier and his personality healthier if he has some sort of company—even a goat or a cow or some other animal if there are no other horses near him. But the average horse is lost if he is isolated from other horses and compelled to live as an individual in a lonely environment without leadership. Left by themselves, many horses tend to panic and gallop frantically about. Some horses you dare not leave alone unless they are in a safe corral or box stall, or they will gallop madly around the field, possibly injuring themselves in frantic attempts to get out and follow other horses.

Yet many horses can eventually adjust and adapt to this new circumstance, transferring their loyalty for the herd to man—trading and accepting man's mastery for the leadership of a herd boss. Today we find many horses in iso-

The lone horse is in a situation contrary to his natural instincts. Yet many horses can adapt and adjust to the new circumstance. This pregnant mare gets along quite well by herself

This stallion has cattle through the fence to help keep him company, and he is always in sight and sound of other horses

lated environments; the pleasure horse owner may have only one or two horses. A lone horse is completely contradictory to the horse's nature, yet many horses seem to be able to adjust, due to the fact that the horse can transfer his loyalties to man. This is one of the most remarkable aspects of the horse, and one of the most important factors involved in the horse's ability to work so well with humans and to become so highly trained in spite of a so-called low intelligence. Most horses can easily adjust to an environment dominated by humans, being capable and even quite willing to transfer the loyalty they normally exhibit for other horses to a person. It is largely because of this ability of the horse that it becomes possible to train most horses to work with man in so very many ways. It has been the horse's willingness that has made him such a valuable animal to mankind over the centuries, fulfilling all the many and varied demands that man has put upon him.

Very few animals, except perhaps the dog, have been able to adjust so thoroughly to man and his whims. Both the horse and the dog are usually quite willing to trade their freedom for food and shelter and are readily persuaded to fulfill their part of the bargain. Today's horse is often almost totally removed from his natural environment and lives a rather artificial life, depending on humans for his food and comfort.

Today's horse is rather unique among domestic animals in his ability to adapt to unusual environments and circumstances. The present role of the horse is to carry a rider, and he has adapted quite well. Weighted down with a rider, and doing everything from racing, jumping over obstacles, to herding cattle, running the barrels or doing police work, the horse is playing a role for which he was never designed by nature, and yet he does it very well. Yes, the horse of today is highly adaptable and versatile. But perhaps his greatest asset as a creature of man is his willingness to submit to human domination. Most horses, when treated with understanding and sensitivity, come along very nicely in training and actually seem to want to cooperate, eagerly progressing to new and more difficult things.

As man used the horse, he practiced selective breeding and developed several types—some that were better suited for pulling heavy loads, some that were exceptionally swift, some that were good riding animals, with agility and speed to travel quickly from place to place. As the horse was used by man and became dependent on man, he not only changed in shape and size as the result of selective breeding, he also changed somewhat in disposition and mental ability. Horses that refused to cooperate or that were unwilling to submit to human domination were eliminated, and the horses which adjusted best to the human environment and showed willingness to be handled, ridden and put into new roles were the ones most often selected to pass on their qualities to a new generation. Thus as the years and centuries went by, the average horse became a creature with a basically friendly disposition and a willingness and capacity to be used by man. Primitive responses became a little dulled and more easily controlled, but they were never entirely lost. In moments of stress or panic, most horses tend to forget some of their inhibitions and civilized veneer, falling back upon primitive impulses—running away or striking out in self-defense.

The task of the horse trainer is to overcome the natural instincts of the horse, particularly his fear of man and his suspicions about anything strange. The amazing thing about the horse is that these fears can be so easily overcome, under the handling of a skilled horseman who understands the horse's mentality and instincts. In a very short time the young horse can, with the right kind of handling, learn to trust his human captors and come to have a confidence in man, becoming willing to submit to the various strange things the man asks him to do, and ready to work with the man and do his bidding.

But "the right kind of handling" should be stressed, for the trainer who uses force rather than tact and persuasion will not have nearly the same success. This will be covered more fully in chapter 21. The horse is a willing, and to a certain extent submissive, animal yet he also has a certain amount of pride and self-respect. He will cooperate with his human handlers and trainers so long as he can retain his self-respect and is not forced into something he is afraid of, does not understand, or does not want to do. Tact, and the understanding of horses, which includes the knowledge of each horse's abilities and limitations—knowing when to punish and when to be patient—will take the trainer much farther in his attempts to teach the horse to perform under saddle in the various roles man asks of him today (doing things which are actually foreign to the horse's nature, for they are things the horse would never do on his own in the wild) than the use of force.

The young horse comes into the world inheriting all the instinctive fears of his ancestors, yet he is curious and responsive to training, receptive to new patterns of action. How well he develops, how he turns out as a saddle horse, depends on the kind of handling and training he gets

Some horses will succumb to forceful methods of training more readily than others, but they will tend to be rather sullen and resentful. The horse that is forced to submit to man, or forced to perform, will not do nearly as good a job as the horse that is led into his new role step by step by tact and respectful handling. The latter will accept his new roles much more willingly and will perform with more heart in his work, doing a much better job.

Usually one of the first things a young horse has to do after he becomes acquainted with man and has overcome his instinctive fears is to learn to accept the rider on his back as a partner, as a master with whom he must work rather than resist. Stop to think for a moment just what is involved in asking the young horse to carry a rider for the first time; this is something entirely new to him and something that is not included in his inherited assortment of stimuli and natural responses.

There is nothing in his prehistoric background to prepare the horse for this moment. Thus it is only natural that the horse should resent and resist this new experience if he is still reacting to the human out of fear. Yet if a horse is trained with understanding and sensitivity, led gradually to each new experience so that it is not alarming to him (for instance, if the young horse being mounted for the first time has already been handled a good deal by the person mounting him and has lost his fear and suspicion of that person and trusts him), he can be mounted easily for the first time and will readily accept it without alarm or resistance. This kind of acceptance of such an imposition (very few other kinds of animals will so willingly submit to this sort of thing) takes a very special kind of mentality and willingness which makes the horse unique in the animal world. Even so, the trainer is wise if he progresses with care. The horse that is forced or bullied into submissiveness may resent his new role even long after he was ill treated. Harsh training methods may leave a mental scar that may come to the surface occasionally even a long time after the new role (carrying a rider and bending to his will) has been accepted or tolerated.

To fully understand the horse and why he reacts the way he does to various things, we need to know something about his senses—his vision, hearing and touch—because in a very real way his senses, though partly physical, are closely linked with his emotions and mental reactions to physical stimuli. The horse is a very alert animal and he responds to everything around him.

A horse out in a pasture grazing can become aware of your approach in several ways. Horses have excellent vision, and since the eyes are at the sides of the head, a horse can see to both sides at the same time, as well as straight ahead. And since the eyes do protrude from the sides of his head, the horse can also see to the rear without having to turn his head. Even when his head is down, grazing, he can see quite well to all sides because of the way his eyes are situated and can even look behind himself by looking back between his legs.

If he is dozing a little or not paying much attention to his surroundings, he may hear you or feel your approach before he actually sees you. He may feel the ground vibrations of your footsteps. These ground vibrations are transmitted from his feet through his leg bones and backbone to his skull, where they are "heard" by his internal and middle ear. The horse has large air cavities inside his skull and it may be that these act as sound boxes to magnify the ground vibrations he receives.

Most horses have good vision. The way the horse's eyes work is rather unique. I've already mentioned the fact that the horse can see things on both sides of him at once, due to his eyes being wide apart at the sides of his head, giving him a different picture with each eye. He also has a remarkable way in which he focuses his eyes on near or distant objects. Most animals, including man, focus their eyes by changing the shape of the lens or by moving the lens closer or farther from the cornea, causing the rays of light to meet upon the retina at the back of the eyeball. In most animals, and in man, the retina is concave and very regular in smoothness. But in the horse, the retina is irregular, coming closer to the cornea in some places than in others. The horse makes use of this irregularity by raising or lowering his head to focus the image of what he sees; the rays of light meet upon the particular part of the eye that will produce the best and clearest image. This is why when a horse sees something of interest in the distance he raises his head and looks intently, often shifting the position of his head up or down a little as he looks.

Thus the horse's ability to focus clearly on an object depends a great deal upon his ability to move his head freely—raising it, lowering it, or tilting it a little to see the object at fullest advantage. So when you are riding the horse, you should allow him some freedom of head and neck, never using a too-tight, restricting rein. Not only does the horse make use of his head and neck for proper balance as he travels, but also for proper vision, moving his head as need be in order to focus clearly on what he is looking at.

A knowledge of how the horse sees is very important to the horseman in many ways, and especially when teaching a horse to jump. If you have ever watched a green horse being jumped over a low obstacle for the first time, you may have noticed how he came galloping toward it and then hesitated, putting his nose down as if to look at it more closely, then leaped over it with a big jump, clearing it by a large margin. This is because he probably couldn't see it clearly and to him the obstacle was of undetermined height. Only after much practice does the jumping horse learn to accurately judge the height and width of the obstacles over which he must jump.

Horses are rather farsighted. Over the millions of years that the horse has lived in the wild, this farsightedness has enabled him to see approaching enemies at some distance and have a better chance of getting away. But because of this farsightedness a horse has a little trouble seeing clearly things that are close. The human eye focuses quickly; a person can look at a distant object and then at a near one and see both very clearly. But the horse has more difficulty focusing both eyes on the same object near him because of his side vision. He can focus both eyes on the same object, but it is hard for him to do so. With his head raised he can see things far away, but a small log just in front of him may be fuzzy and out of focus until he lowers his head and examines it closely.

In many instances, the flighty reactions of the horse to apparently harmless objects can be explained by understanding how the horse sees the world around him. All of the horse's senses and nerves are geared toward flight as his best defense against danger. The horse's eye is designed to receive as much information as possible and to rush this information to the brain for immediate action. The horse has the largest eye among land animals, and his eyes are set wide apart along the sides of his head so he can see in front and to the side, and also to the rear without having to turn his head.

Because his eyes do not focus rapidly (he often must move his head a little to get the best focus) the horse may be easily startled by anything that looms up in front of him suddenly or comes up behind him. Sudden motion can startle a horse. Use caution when approaching a horse, giving him a chance to see you clearly. If you are wearing or carrying something strange, something the horse is not accustomed to he may become alarmed until he has a chance to check you over and make sure that this is nothing dangerous. And never come up suddenly behind a horse; always let him know you are coming so he will know that it is you and not something to flee from or kick.

When riding a horse, you must be alert to the things that might possibly alarm him, becoming aware of the kinds of objects and the placement of objects which might easily frighten a horse even though the average person would take no notice of them. You not only must think like a horse; you must also realize that his eyes work differently than yours do. A person's eyes are only an inch or so apart while the horse's eyes are usually six inches apart or more. As the horse is traveling along, he may see some objects with just his right eye and others with just his left.

Pretend for a moment that you are seeing things as the horse sees them. Hold your arm out in front of you, holding up one finger. Close one eye and look at the finger with the other, then change eyes, shutting the open one and opening the closed one. Your finger will seem to be several inches to one side from where you saw it with the first eye. Imagine what this same effect will do in the horse's vision. His eyes are set much farther apart than yours. If he sees a small object such as a shiny tin can lying on the ground, he might see it first with his left eye, let's say, but then as he changes position or passes it, he may see it only with his right eye. The shiny can, which the horse thought was motionless, suddenly seems to jump to the side. Thus the horse, if he is a little nervous that day, may shy suddenly at what appears to be a harmless tin can. The rider may be aware that his horse noticed the can; the horse may have pricked his ears toward it or seemed a little curious or alarmed about it. But what most riders often aren't prepared for is the fact that the horse may shy at it still—as he moves past it—even though he has already noticed its presence. For as it passes from one field of vision to the other it appears to move, and this will usually alarm the horse more than when he noticed it in the first place.

Some horses shy just because they are feeling good and have some excess energy to burn. They might like to gallop and frisk a little, but since the rider won't allow it, they take out their high spirits in shying at things. Other horses shy because they are actually afraid or timid. A good rider can usually tell when the horse is going to shy (unless it is at something totally unexpected or unpredictable, like a rabbit suddenly dashing under the horse's feet or a bird flying up from a bush or clump of grass) and prevent it or tone it down somewhat by using stronger aids and by talking quietly to the horse to reassure him.

The conscientious horseman is always on the lookout for strange objects that might startle his horse, so that he can be prepared to counteract any skittishness. Learn to predict the things your horse is most likely to shy at. Most horses are startled by moving things that they cannot understand, like clothes on a clothesline flapping in the wind, a child scooting along in a wagon, a man walking across a field carrying a canvas dam, leaves or scraps of paper blowing along the ground, moving things accompanied by strange noises, and so on.

The best way to keep your horse under control when he is skittish or trying to shy away from something will depend on the individual horse. This is why it pays to know the horse well. Some horses will be more calm and go on by an object if you allow them to take a good look at it or smell at it to make sure that it is harmless. Others will have less fear if they are turned away from it so they can't see it.

If a horse is acting skittish, try to keep him moving and get him on past the object that is frightening him. If he whirls away from it, it's often best to just keep turning him on around so that he comes to face in the proper direction again, rather than fighting him or trying to pull him back. If you fight him or whip him to make him go past something he fears, you may get him to go past, but you will also have more trouble with him next time. If you have used force, he will come to be afraid of what you are doing to him as well as afraid of the obstacle, and he will be doubly afraid the next time he meets it, fearing the punishment as well as the object and being more determined than ever not to go near it. In his way of reasoning, being near that fearful object brings on punishment, and he tries even harder to avoid it. If you are riding with someone whose horse is steadier and not so spooky, let your companion go past the frightening obstacle first. If your horse sees the other horse going calmly by it, he will lose some of his fear and follow the other horse with more confidence. Skittishness is catching; horses are very suggestible. If the horse in the lead spooks at something, the following horses will, too. Therefore it is always good psychology to have the calmest, most dependable horse go first, for if he takes the spooky object in stride, the others will be more likely to follow his calm example.

Motion tends to excite or alarm a horse. This, too, is part of his prehistoric defense pattern. In the wild, a sudden motion might mean the attack of a predator. Or, if he sees other horses running, this was usually a sign of danger, and he would want to run also. Horses are usually more excitable and nervous (ready for flight, and skittish) on windy days when there is motion all about them. The wind may also make strange sounds or distort the sounds the horse is used to hearing. Many horses that are usually even tempered or dependable and calm will be more skittish and unpredictable on a windy or stormy day. This should be taken into consideration when you are working with horses or training a young one that might be a bit nervous.

Horse's eyes are very sensitive to light. This is one reason a horse will hesitate to go out into the bright sunlight from a dark barn or stable and why he may act a bit strange when he first goes out; until his eyes adjust to the bright light he cannot see properly and even objects normally familiar to him may appear strange or alarming. A horse in a brightly lighted area will hesitate even more when asked to go into a dark or dim place. He is fearful when he cannot see, and this is an instinctive fear; in the wild the horse has been preyed upon for millions of years by nocturnal predators.

Because of the horse's unique type of vision, it is important that the rider realize how necessary it is for the horse to be able to have free use of his head and neck at times when the horse must accurately judge where to go and where to put his feet next—when jumping or in traveling through rough country, or when chasing a wild cow on hazardous terrain, for instance. To perform at his best, the horse must be able to find his own head position to coincide with his visual needs. Most riders do not pay enough attention to this and hinder the horse's head and neck when he needs to use it freely for proper balance and vision.

The horse, when traveling through terrain that is covered with rocks or obstacles such as logs or ditches, can see where to put his front feet, but he cannot see where his hind feet land. But in difficult country, the horse can usually travel so that his hind foot steps into the same track made by the front foot. When riding through rocky places or over obstacles, let the horse have free use of his head so that he can lower it if need be to judge where to put his feet.

Horses have a very keen sense of hearing. The horse can hear sounds from farther away than the human ear can hear, and can probably hear sounds outside the range of human ears. The horses that are most sensitive to sound are horses with thin skin and active ears that can move quickly in any direction. Horses with thick, heavy ears, or ears very nearly filled with hair are less sensitive to sounds and do not hear quite as well. Because the horse has a long flexible neck and can move his head in any direction, and because he also has receptive, concave ears that can also move in any direction, he has several hearing advantages over other animals. The horse not only picks up sounds, but can also pinpoint the exact direction from which the sound is coming, by moving the ears and turning them.

The horse has a good sense of smell. This sense helps him identify his food, choosing that which is palatable and good to eat, and rejecting food that may be unpleasant. Some horses are more finicky than others in their senses of smell and taste, rejecting medicated feeds, for instance. If you have ever had a horse that was fussy about eating medicated grain, you probably noticed that he didn't even have to taste it to discover that the foreign medication was mixed with it; one good sniffing of the grain was enough to tell him, and he refused to taste it.

The nasal surface of the horse is very extensive and contains a large area of smell buds and nerve endings. Horses always breathe through the nose, so all the air entering a horse's lungs passes through his nostrils. Only when a horse is extremely exhausted will he try to breathe through his mouth, and then he usually won't have much success because the large soft palate tends to drop down into the pharynx and limits his mouth breathing.

For a long time people have thought that the horse had a great fear of the smell of blood, but this does not seem to be true. But a horse may shy or seem to fear a dead animal, partly because of the unnaturalness of the creature lying there (just as he might spook away from a piece of canvas or any other strange object on the ground), and partly from the smell, if the dead animal is "overripe." Horses do not like the smell of rotting flesh. They also dislike smoke and flame and become disturbed at the smell of smoke. Fire can cause a horse to panic and become uncontrollable.

The horse is a highly emotional creature. He expresses his feelings very obviously and can also change his moods very rapidly. He can be half asleep one moment and very alert and excited the next. Or he may be nervous and jumpy, anticipating a job to be done that requires speed or precision, and very calm and exact the next moment as he goes about it, settling down into the seriousness of his work. It is remarkable that an animal that is as emotional as the horse is also so adaptable and so pliable, quite able to suit his moods to those of his handler or rider if that horseman be a capable one that can inspire confidence and obedience in the horse.

The good horseman is a person who can work *with* the horse and draw out the horse's greatest potential. He realizes the horse's emotional capabilities and is therefore careful in the way he handles the horse, always being consistent so that the horse can trust him and have confidence in him. The horse is a creature of habit, even to the way he is used to being handled. He comes to know a person and to know what to expect from that person, and horse and horseman can come to a state of affairs in which they can trust one another and work from that point.

The inconsistent horseman has less luck gaining a horse's confidence or getting the horse to perform at his best. The horse always expects his human to act in a rational, familiar fashion, and is always surprised, startled or annoyed if the person does something quite contrary to the horse's expectations. If your horse is accustomed to having you walk out into the pasture to catch him, and is used to standing quietly while you halter him, he will be startled and upset if you come dashing out on the run, and he may kick up his heels and take to the far corner as fast as he can go. Or, if he is accustomed to being spoken to softly, and in a fit of temper you shout at him, he will probably give you a startled look and forget whatever it was you were trying to get him to do.

Your horse may know you very well, but unlike a dog, he is reacting more to what you do to him, how you handle him (the outward motions and sounds) than he is actually reacting to you. The dog is also an emotional creature like the horse, but the dog seems to accept man for what he is, with all his faults and inconsistencies. But the horse expects his master to be consistent, and is usually surprised, alarmed or upset if man does something out of the ordinary that the horse is not accustomed to.

Horses are very much creatures of habit. They are true conformists. They get used to a familiar routine and it is a source of security. They like to have things done in the same way and at the same time and same order every day; presented with something out of the ordinary or a change in the routine tends to confuse and annoy a horse. He likes his own stall, his own order in the line; he likes being tied in the same place every time while being saddled. Horses become accustomed to their surroundings and like to have objects in their proper places. They don't like change, and may become suspicious or shy of something (even a familiar thing) if it is in a different place.

Horses tend to associate objects with their surroundings and situations. The horse may be familiar with a hay wagon or wheelbarrow in his barnyard, but that doesn't mean he won't shy at one or be afraid of it if he sees one unexpectedly along the road as he is traveling. A shovel being used by his master working around the barnyard won't alarm him, but if he is out along the road and sees a man carrying a shovel over his shoulder he may be startled. When riding, you must take this sort of thing into consideration, realizing that a horse tends to associate things with their surroundings rather than reason about them; you must understand what sort of things may alarm or upset a horse, so that you can anticipate problems and be prepared for the horse's reaction.

Horses are usually not so afraid of an object or a condition if they can be accustomed to it gradually, especially if they can be accustomed to it while they are young. Horses are very adaptable and can be accustomed to many strange or otherwise difficult situations with time and training, if the trainer uses understanding and patience. Constant repetition, in an atmosphere of safety, in which the horse realizes that he will not be harmed in any way by the strange object or action, can make almost anything seem ordinary and natural. Thus were military horses trained to be calm in the face of gunfire, and to take in stride situations in which most horses would become alarmed or upset.

The horse's greatest desire is to please himself and keep himself from harm. He is not as affectionate as a dog, and his loyalty is expressed in different ways. The dog will do things to please his master, but the horse does things to please himself. The trainer must realize this fact and work with it rather than against it. With tact and understanding the trainer can persuade the horse to do certain things because the horse comes to feel that what he is asked to do is not unpleasant, and then if it becomes a habit, the horse accepts it rather willingly. Most horses will readily perform any task without reward; a horse will get into the habit of performing this way, as long as what he is asked to do does not really conflict with what he likes to do.

Most horses are capable of a rather limited affection for people, especially for people who understand them and treat them intelligently. A horse you work with, if you have proper tact and patience, will come to trust you and to work willingly for you so long as you don't ask him to do things that he thinks are impossible, or things that he is afraid to do.

The horse that trusts you will not, however, come to your rescue if you are drowning or in some other dire predicament (as a dog might). The horse is not that concerned because he is not that capable of realizing you are in danger. His thoughts are more for his own safety rather than yours, but he will be your willing servant and companion under most circumstances. He will look to you as his security in difficult situations if he trusts you enough.

For instance, a horse with a good disposition and a basic trust in you will stay more calm under conditions that conflict with his own desires, and be more willing to listen to you, whether you are trying to encourage him or calm him. If he gets his feet tangled in wire and panics, his first impulse is to escape, and he may thrash about blindly trying to free himself, increasing his chances of being cut and injured by the wire. But if he trusts you and has confidence in you, your voice soothing him and your calmness may reassure him and he will calm down enough to allow you to help him and free him from his predicament.

A horse that is never abused by you will work for you very willingly, and will also be easy to catch. Most horses become difficult to catch because of two things. Either they have never completely lost their instinctive fear of man, or they have come to associate man with unpleasant things. Your horse will more readily come to you when you call him (or stand for you as you walk up to him) if he trusts, accepts and associates you with pleasant things such as a feeding of grain after being caught, pleasant exercise, a soothing massage, brushing and rubdown after a long ride. But if his memories of you are unpleasant—if he fears or dislikes you or associates you with pain, exhausting rides, confusing commands, or harsh training methods—or if you are always asking him to do things he cannot do or does not understand, he probably won't come to you and may even resent being caught.

The horse is basically a very timid animal. In the wild, his main protection was flight. Thus his reaction to most frightening or trying circumstances is to run first and ask questions later. Fortunately, to counteract this, most horses can and will develop confidence in their master. Without this ability to gain a trust in his human handler, the horse would always be very unreliable and hysterical in emergencies, or even in everyday situations which to the human are no matter of concern but which to the horse is a frightening situation.

The horse's natural willingness is perhaps his greatest asset in his alliance with man. Otherwise the horse would not permit his master to ride him for long hours when he would prefer to be grazing or standing lazily in the shade, or would refuse to do such senseless things as jumping over high fences or chasing cows or a polo ball. The relationship between man and horse is much the better because of the horse's unique mentality and willingness. If the horse had more reason and less willingness, horses would consider some of the things demanded of them to be quite insane and would probably refuse to do them. The horse has no real conception of the rider's purposes in many cases nor does he know what the rider is trying to accomplish. All he knows is that he is being required to gallop, turn, jump, or come to a standstill at the rider's commands.

If a horse respects you and associates you with pleasant things, he will be very willing to be caught

Fortunately for us, the horse goes along with all this rigamarole even though he cannot understand it, and we should appreciate the fact that most horses, after a certain amount of training, are willing to carry out patterns of behavior that are quite different from their own natural instincts, merely to please a rider whose motives are often quite beyond the horse's ability to comprehend.

The relationship of man and horse, and the success of this relationship, is partly due to man's ability to do the thinking for both and to develop with his horse a communication system which the horse can understand and to which he can respond. The man who works with horses must know how to win the horse's trust without fear, and be able to adequately communicate with the horse through use of his voice, and sense of touch. This sense of touch is used in communicating through body contact—weight and leg pressure—and through the reins and bit, and occasionally the whip.

Sense of touch is very important to both horse and rider. The rider must have a delicate sense of touch and have his reactions and reflexes under control in order to have good hands to communicate with the horse through reins and bit. The horse's skin perception and sense of touch are characteristics that have helped make it possible for horse and man to have such a close and successful relationship. Horses have sensitive skin and are capable of responding very rapidly to skin stimuli.

Most horses appreciate gentle scratching or massaging of the skin, especially over the withers, and this is a good way to calm or gentle a young or frightened horse. Young horses that are shy about having their heads handled should be handled around the back and withers first, scratching and rubbing the withers and then along the neck and mane until the head can be reached and handled.

Skin over the ribs and flanks is very sensitive. Horses are often "ticklish" in these areas. Horses respond differently to a gentle flicking of a whip over the ribs or flanks. Some horses fear the whip and will go faster when threatened with it by the rider. Other horses resent the whip and will slow down when it is used on them. Some kick out violently with the hind feet when the whip is used.

Many horses are quite sensitive about their bodies, especially the more finely bred horses, and resist any kind of abuse. The term "thin-skinned," referring to an overly-sensitive person, comes from the fact that the Arabian and Thoroughbred, having the thinnest skin of all breeds, are also the most sensitive.

The horse's withers and back, and the skin covering his ribs, are very sensitive to touch. An ill-fitting saddle that rubs or presses on the horse's withers can make him so uncomfortable he will not perform well (and can also cause sores or even cripple the horse). Because of the horse's sensitivity over his ribs and back, the rider can communicate very well with the horse by feel. The good rider keeps a delicate contact of feel of his horse with his body, and by using his thighs, knees and calves of his legs he can communicate readily with the horse and in a short time gain mutual understanding and quite a

Gentle rubbing of the skin over the withers and shoulders can help to calm a frightened horse or gentle a young one

bit of control. A horse doing intricate work or fast maneuvering, or a horse sorting cattle, can be much more easily turned and pivoted with the use of leg pressure than with reins alone. Leg pressure gets a faster, more complete response because it "moves" the horse's entire body, not just his head and neck. The horse can respond much more quickly through this body communication than he can through the use of reins alone.

To do a good job of training a horse, the horseman must understand the horse—his instincts, inclinations, and inherent reactions to various kinds of things. He must know how the horse's mind works. He must know his horse well enough to be able to sense the horse's feelings and moods and to be able to predict the horse's actions to a certain extent.

No horse is naturally mean. Mean horses are made that way by bad handling or misunderstanding. Some are more sensitive and some are more independent than others, and these, if not treated with patience and understanding, may be more likely to become mean than the horse that is naturally outgoing and friendly. But any foal that is handled with firmness, kindness and a sincere attempt at understanding him as an individual will not be mean. Young horses are often playful and will frisk and buck, nip and kick, but this is their nature when running free and is not a tendency toward being ill tempered unless it is inadvertently encouraged by improper training and poor handling.

The young horse learns to move with quickness and agility by playing these romping, teasing games. Foals in the same pasture will use up their excess energy on each other in games of tag and playful fights and will usually be much more manageable and well mannered toward the human handler than will the foal who has had no playmate with which to fight and roughhouse.

Meanness is usually the result of harsh treatment and subsequent fear and resentment, or the result of over-indulgence, or, in some cases, too much timidity on the part of the horse handler. A horse can easily sense if you are afraid of him. If he is playful and the trainer takes his playful actions as a sign of aggression and retreats, the horse will soon learn to take advantage of it. There is no place for fear in the heart of a good trainer. He must know his horses well enough to be firm but kind, for only when the trainer has confidence in the horse can the horse come to trust and have confidence in him. Firm, kind confidence commands respect.

When handling horses you must take into consideration the horse's personality and temperament as they affect his attitude and manner around other horses. Horses are gregarious. They like company. If you turn two or more horses out in a field they will generally be found together. If you are riding with a group, your horse will want to stay with the others; horses hate to be left behind. The horse wants to be with other horses, yet he is also a bit sensitive about being too close to other horses. He is gregarious, yet he wants elbow room. If another horse gets too close, he will bite or kick.

Horses have very definite likes and dislikes for other horses and also for people. Some horses seem to have preferences for certain people (you've heard of the "one man horse") or even for certain types or kinds of people. A horse trained by a woman will sometimes resent a man who rides him or be uneasy when handled by a man; men are often more heavy-handed than women. And sometimes the reverse is true; a horse used to being handled by a man may be uneasy when ridden by a woman—confused because of the difference in manner.

Some horses seem to always get along well together, while others never do, and personality quirks and individual likes and dislikes should always be taken into consideration when working with a group of horses or pasturing them together. A new horse put into the pasture will spend his first days being chased around and jockeying for position in the social rank. Horses have a very definite rank system when they are in groups, and there will be one horse that is the boss. He helps himself to the best feed, and is the one who chooses where to go. He drinks first, goes through gates first, and in general enjoys the advantages of his top position.

Horses are very suggestible and often imitate what other horses in the group are doing. If they have been eating hay and one horse decides he is thirsty and makes a trip to the water tank or creek, they may all decide they are thirsty and that it is time to get a drink. If you are riding in a group and the horse in front shies away from something along the roadway, the horses behind him will become alert to the strange object and do the same thing. If the horse in the lead jumps a ditch instead of stepping over it, the horses behind him will try to jump it, too. If one horse in a group is acting skittish or playful and decides to buck or jump a little, the others find that they would like to join in the fun, too. High spirits are very contagious. If an inconsiderate horseman gallops past you at full speed, you'll have a little difficulty keeping your own horse under calm control, for he will feel like running, too.

Horses have a strong homing instinct. If your horse gets loose while you are out riding, he will travel directly home even if you have ridden out on a roundabout way or over a new trail. Horses have a very good sense of direction; it seems that no matter where they are or how far from home, they always know in which direction home lies. But if your horse gets loose while you are with a group of other horsemen, he probably won't set right out for home but will hang around the group because he is gregarious and wants to stay with the other horses.

Each horse has a very different and individual personality. Some have a sense of humor and enjoy teasing you a little, nibbling at you or stealing your brush when being groomed. Some are playful. A horse may buck you off in a burst of high spirits and then look at you with alarm and surprise to see you on the ground. Some horses can figure out simple problems, like how to unlatch a gate or the door to the stall, or like to nibble at a knot until they have untied the halter rope. Each horse has his own capabilities, his own personality, and his own way of expressing them; each horse has his own strange habits.

A perfectly trained and perfectly mannered horse is very rare. Almost all horses have some bad habits, some of which are serious and some which are easily tolerated. Bad habits—anything from nipping to halter pulling, from not standing still for saddling and mounting, to being fussy about a certain foot being handled—are generally the result of mismanagement or bad handling, some human handler's mistake. If a horse were allowed to grow up under reasonably natural conditions, with the best of intelligent handling and training, his bad habits would be very few and very minor.

Many horses that are kept in confined places instead of large pastures where they are free to move about and eat as they desire develop bad habits out of sheer boredom. Cribbers and wood chewers, pawers and stall-walkers wouldn't exist if horses had adequate pasture room for exercise and leisurely eating. But in our modern society, unfortunately, not all horses can be kept in a seminatural environment. Many of today's horse owners are short of space, keeping horses confined in stalls and small pens and feeding them unnatural feeds such as hay or pellets, which are eaten rather quickly and the horse has little to do in his free time.

If a bad habit becomes dangerous to the horse or the handler, it is considered a vice. Biting, kicking, halter pulling, cribbing, and so on, are some of the vices that make a horse undesirable in the eyes of the horseman. Monotony, impatience and boredom are very hard on a horse's disposition, and if you must keep your horse in a stall or small pen, see to it that he has some time spent in exercise each day whether or not he is ridden. If you can't ride him, at least longe him or turn him out in a larger area for a time. A bored horse will chew fences, rub his tail, paw or kick at his stall, or chew or tear up anything left within reach. He may bang at his water tub or stick his foot through the corral fence pawing at it. Overfeeding and lack of exercise can make any horse unhappy with his situation in life. He may take out his excess energy in demolishing things around him, or he may become crankier and ill tempered. Or he may become overexuberant when you do ride him or exercise him, and be difficult to control.

The physical and mental capacities of the horse make him a very complex and fascinating creature, and one that requires our sincere, utmost efforts in attempting to understand and to work with him. To handle horses well, we must understand them well. Their individual abilities and personality quirks make them just that much more interesting and enjoyable to work with, and certainly keep us on our toes. The mind of the horse is a constant challenge, but a rewarding one as we come to know him better.

20

TO CATCH A HORSE

HORSES THAT ARE HARD TO CATCH ARE USUALLY A SOURCE OF irritation and frustration to their owners. How exasperating it is to have a horse that refuses to be caught—a horse that throws up its head and runs to the far corner of the field whenever he sees some one approaching with rope in hand. If a horse takes up as much of your time in getting caught as he does under saddle, he will be no joy to you. The ride you've planned may well be spoiled by the time it takes to catch your horse or by the mood it puts you in by the time you finally have him caught. Being hard to catch makes him very undependable. Several years ago I knew of two 4-H youngsters (nearly tearful in their desperation and frustration) who came to a horse show minus their horses and unable to participate, because they could not catch those horses. A very sad day it was for them.

There is really no excuse for a horse that is hard to catch if you have raised and trained him, for teaching him to be easily caught is one of the fundamentals of good training. A horse that is acquired from someone else can be a different matter, especially if the horse has had little of the right kind of training or a lot of the wrong kind. Being hard to catch becomes a very bad habit that is not always easily cured. Solving the problem can take a great deal of time and patience. Many otherwise useful and dependable horses are marred by the fault of being hard to catch, and many of them never get over it, partly because their frustrated owners continue to chase and corner them to catch them instead of trying to begin all over again with this aspect of the horse's training.

When I was a child, my father acquired a nine-year-old buckskin mare named Nosey. This mare was gentle and fairly well trained, but she was very timid. She feared people. She was very obedient, but apt to spook at any unusual movements, for she was afraid. And she would never let anyone catch her if she could help it. Our problem was that we had four other horses that were fairly easy to catch, a very large field (about 300 acres) on which all of our horses roamed, and now Nosey—an added flighty ingredient who always led all of our easy-to-catch horses over the nearest hilltop whenever any of us came into sight. We had several very trying episodes. Finally we got the group into the habit of coming into the corral for grain. I would take a grain bucket out and catch old Possum, then ride him back down to the corral, the other horses following. They would thunder past me into the corral where filled grain boxes awaited them and I would shut the gate on them. Once in the corral, we could usually corner and catch Nosey one way or another. She was very obedient once caught; she was apparently used to being roped, and would come to a shuddering standstill even if she just *thought* she was caught, like the time my Dad got the rope merely over her ears.

But our attempts to catch that buckskin mare out in the field were usually futile. She loved grain and would look longingly at a grain bucket, but would rarely come close enough to be caught. We never left halters on our horses, for there was too much brush and rough undergrowth to get hung up on. I once got that mare close enough to get my arms around her neck, but I didn't quite get the rope around her before she swung away. There I was hanging onto her mane, swinging through the air, and I finally just had to let go.

Then there was the time my brother and I set a trap for her. We put some grain out in a pan, laid a big lariat loop on the ground and tied the end to a stout cottonwood tree. My brother hid in the bushes at the base of the cottonwood tree and I called the horses, staying back some distance from the pan of grain so that Nosey would not hesitate to eat from it. When she got to the pan and began eating the grain, my brother pulled the rope and got her around the front pastern. That startled mare took off at a gallop and when she came to the end of the rope it threw her to the ground. Miraculously rope, tree, bones and tendons held, and we had a very surprised and shook-up, but caught, horse. We were very lucky we didn't break her neck. Needless to say we didn't ever try that stunt again; we hadn't realized she would hit the end of the rope so hard.

As the years went by, Nosey became easier to catch—perhaps because she came to know us better and came to fear us less. Time and patience helped a lot. Dad would send me out to catch the horses, perhaps because I was still a child and had all the time in the world to be patient with a neurotic old horse; I was not yet caught up in the rush of the adult world where time counts and patience runs short. And over the years Nosey responded. One rule we came to use seemed to help even with older spoiled hard-to-catch horses, and that rule was catch, then feed. If a horse knows he will always be fed a bit of grain after he is caught, he comes to accept being caught. If you take the grain out in the field with you, he will try to get a few nibbles and still keep his freedom; he'll just play games with you. The horse must be willing to let us walk right up and catch him, and then we must fulfill our part of the bargain by giving the horse his grain as we groom him, saddle him, trim his feet or doctor him. If he comes to anticipate that bit of grain, even if it is just a bite or two, he anticipates pleasant things as he sees you walking out into the field with halter in hand and he comes to meet you eagerly instead of turning tail and dashing away.

All the foals we've raised have been easy to catch, for they are handled from birth and do not fear us. Fear is one thing a horse must overcome before he can trust you and be willingly caught. With the young horse or the untrained horse, fear is usually the thing a person must first combat in order to make the horse easy to catch. You have to spend time with that horse, touching him, brushing him, talking to him, and in your training you must be gentle. If he comes to realize you will never hurt him, he will come to trust you. But I don't mean that you never punish him. There is a difference. A horse understands punishment if it is for something legitimate; this is part of his training and this is how he learns to never bite or kick. But if he is punished for things he does *not* understand, he merely thinks you are hurting him and he will come to fear you. He must be handled wisely in order to learn discipline and obedience and still have a basic trust in you.

The second thing a person has to deal with in a hard-to-catch horse is resentment. A horse usually runs from you because he fears or resents you. Your old dependable mount has no fear of you, but he may still become hard to catch if he resents the things you do to him after he is caught. If he comes to associate you with long grueling training sessions that he doesn't understand, or long exhausting rides, or painful wound treating, or whatever, he may tend to get a little harder to catch. If you are experiencing this sort of problem with a horse, take a close look at the things you are doing with that horse. If every time you catch him you take him on a long hard ride, or wear him out practicing barrel racing or some other maneuver, slow down. Catch him every now and then just to give him some grain and brush him or clean his feet and turn him loose again. Vary his routine so that some of the time you don't actually ride him after you catch him. If you merely catch him, feed him and brush him a few minutes before turning him loose again, he may look at this as a refreshing change and be more willing to be caught the next time. Vary his routine this way so that he won't come to always know what to expect—so he won't be always anticipating another hard ride whenever he sees you coming. The same can be true if you are doctoring a horse and he begins to associate you with painful wound dressing or shots every time he sees you. Some horses take this sort of thing in stride while others come to resent the doctoring and get hard to catch. Interspersing pleasant catchings with painful ones can often help.

Several summers ago my 12-year-old mare Khamette got used so hard that she began to resent being caught. She has always been a friendly, eager horse that would gallop up to you at your call. So when she began being evasive we knew she was beginning to resent the long hours under saddle and we began to vary her routine so that she could once again look forward to being caught without the certainty of another long hot range ride. A good horse usually doesn't mind a few long hard rides, but if this happens *every time* he is caught he will begin to take a second look at all this nonsense after a while. For actually, a horse is basically lazy. He works his heart out for you because he respects you and because he is trained and conditioned to do so, but if he is overworked and abused he loses some of that willingness and starts to look around to see how he can get out of some of that exhausting work. A trainer or rider who uses a horse hard (whether range riding, conditioning for endurance or whatever) must always use tact and good judgment in order to keep that horse fresh and eager and willing to put forth his best effort. And part of the secret is to vary the routine so that the horse is not constantly being asked to exert his maximum. He needs periods of rest and lighter activity in order to keep his interest and ability sharp. You don't want to wear out his body or his spirit. Sometimes it takes a little ingenuity to think of ways to keep him fresh and eager, but it is always worth it. For if he comes to resent you or what you are doing with him, you've lost the whole ball game and perhaps ruined a horse as well.

The foal can be trained to be easily caught, for foals are naturally a bit curious and in time will come up to you to look you over if you are out in the pasture. But the first few times you catch the young horse, use care and good judgment. It's best to catch the young horse in a small place like a box stall, for he can be easily cornered if need be and can't get up much speed and hurt himself. Never run the young horse to catch him. Corner him in the box stall and get him accustomed to you and the feel of your hands. Get him over his fear of you before you ever try to catch him out in an open field. Once a horse learns he can jump away from you and avoid being caught, he may try it again and again if he is still a bit afraid of you.

It helps if the foal's mother is gentle and easy to catch. If she runs to the far corner of the field with her baby when you come to catch them, her baby will come to fear you and will get into the habit of running away from you when you come into the pasture. If the mare is elusive, you have a bit of a problem unless you can coax her into a small pen or box stall for her daily grain and then go about your catching in the small area. Never run the pair around the field trying to catch them.

Weaning time is the ideal time to earn the foal's trust and confidence, for he has lost his mother and will begin to look to you for food and perhaps for some companionship as well. At this time he can be easily halter broken, taught to lead and to be easily caught, for he will associate you with pleasant things such as feed and brushing and exercise. From this point on he will be easy to catch, coming at your call, unless you abuse him.

Any horse you raise should be easy to catch if you have used tact and good judgment in his handling. Therefore most of the problem horses are those you acquire at various ages and various stages of training or lack of it. The young horse that has not been handled much is actually the easiest, for usually all you have to do is overcome his fears. Time and patience will usually win him over in a matter of days. He can come to associate his food with you and learn that you are not going to hurt him. You can feed him in the same place each day, calling him as you bring the feed, and then staying nearby as he eats it, gradually getting closer day by day until he tolerates your nearness. As he gets used to you he will let you touch him, handle him, and finally catch him. Then you gradually make the transition to catching him before he eats, so that he has actually learned to come at your call, stand quietly as you halter him, and then eat the feed that he came for.

The real problems are the horses that have been abused and have gotten into the habit of being hard to catch—horses like our old Nosey. These horses may run from fear, or from resentment, but no matter what the original reason, they now run from habit and it can take a great deal of effort and training to change their ways. A horse like this isn't about to be caught and presents a real problem because in order to persuade him that catching is pleasant (feeding grain after catching, or whatever) you first have to catch him. And if you must run the horse around to where you are both in a sweat, you've defeated your purpose. Thus, you have to outsmart the horse and catch him in one way or another that will suit rather than defeat your purposes. If you always catch him by running him into a corner or into an enclosure and wearing him down, you may always catch him (in time) but you never make him any easier to catch the next time.

With some horses it helps to keep them in a small enclosure for a few days or weeks until they get accustomed to being caught in that small enclosure. Catch the horse several times a day, and most of the times you catch him, give him a bit of grain or just brush or pet him a bit and turn him loose again. After he gets to where he will stand and let you walk up to him anywhere in the enclosure, no longer trying to avoid you, he can be tried in a

Old Nell stands quietly and patiently as she is approached

She accepts the rope around her neck

Then the halter is put on

larger pen or small pasture. If at any time a horse turns his tail to you in the small pen, threatening you with his heels, he should be slapped across the rump (with halter rope or whatever you carry if you haven't a whip). I don't believe in whip breaking, although when done correctly it can produce a horse that is easy to catch. But no horse should get away with kicking or threatening to kick. But if you slap the horse that turns his rump to you, be tactful and consistent, so that he realizes why you are hitting him. The last thing you want to do is make him fear you more. Encourage him with soft talk and kind words whenever he turns his head to you. Do not whip him if he is reacting out of fear, trying to seek an escape; whipping him on the rump at this point only increases his fear and may do so to the point that he will run right over you if you get that close again. Use the slap on the rump only when you can see he is doing it with intentions to kick or to be evasive. Your main purpose is to overcome his fear, not increase it. But if he gets to the point where he is turning tail to you to take advantage of you, he is no longer so much afraid as he is just being resentful and evasive.

Each horse is an individual and each will react differently to this sort of training. You have to feel your way along. Some horses are more timid and need lots of encouragement and patience. A loss of temper with this kind of horse will just lose you ground. Other horses are bolder and will take advantage of you if they can, and with these you must be more firm. But the main point to impress upon them all is that you are nothing to be feared (but that you must be respected) and that being caught is no crisis but is rather a prelude to pleasant things such as grain, grooming, and so on. And since for many horses one of the greatest rewards is being turned loose again, keep the first lessons short. A hard-to-catch horse may take you 30 minutes to catch calmly. Then you feed him a bite and pet him and turn him loose again and he may stand there looking at you with a puzzled expression as if to say, "What? Is this all you were going to do to me?" After a few sessions that horse gets easier to catch for he learns that avoiding you really isn't worth all the bother and that maybe being caught is not so bad after all.

If you ever have to catch a horse or group of horses that are not wanting to be caught, remember that the first step is to never run them if you can help it. A horse stays calmer if he doesn't run. Once he starts running he gets in a different frame of mind and may run just for the sake of running, circling the pasture again and again with no intentions of going into a corner or into the barn; he's enjoying himself just running. So if you are herding a horse into an enclosure or into a corner, be patient and calm and keep him at a walk if possible; he'll be more receptive to being caught or cornered.

If a horse must be cornered to be caught, often two people can be more successful than one, for he can be approached on both sides and will not be so able to whirl away. If the horse is really a bad one for whirling away or trying to dart back out of the corner again, two persons can carry a long rope between them, gradually working the horse into the corner. Most horses will hesitate to run through the rope, for it represents a sort of barrier. When the

Bambi is so accustomed to being caught that she continues to graze as Heidi approaches her

Heidi asks her to pick her head up, she halters her, and leads her out of the pasture

horse feels trapped in the corner he will usually submit to being caught. Most horses that have been handled at all will usually give in when they feel blocked on all sides. Very few horses will run over anyone or crash through barriers in attempts to get away. If a horse does this he has been badly abused in the past or else he is so stirred up and frightened that he is plunging blindly and frantically in efforts to get away. In most cases, if a horse realizes he is cornered, he will submit to being caught.

But these methods of catching a horse are usually unnecessary if you spend a bit of time and patience convincing a horse that being caught is just part of his daily routine. Some horses will be more trying than others, depending partly on their personalities and upon their past, but in most instances a habitual hard-to-catch horse can still be retrained.

21

TRAINING PSYCHOLOGY

IN THIS CHAPTER WE WILL DEAL WITH HANDLING HORSES; PERhaps another name for this chapter could be Intelligent Horse Handling. The main point I will be trying to make all the way through is that the trainer or rider must have a certain kind of attitude toward horses (especially that particular horse he is working with), a sympathetic and sensitive one. He must understand horses and be able to sense why they do the things they do and be able to correct or improve on the horse's reactions through a kind of intuitive feel for what is right for that particular horse at that particular time. He must not only be able to outthink the horse, but be able to think with the horse.

Successful riding and training is based on intelligent horse handling, using tact and persuasion rather than force. A good horseman always thinks first of his horse, making sure the horse is healthy and well cared for at all times, and when working with the horse he seeks to be sure that the horse is comfortable, that the equipment fits him properly, that his feet are well trimmed or shod and that the horse is not lame in any way, and then tries to ride or work with him in such a manner that the horse comes to understand what is wanted. A horse will be much more receptive to the training being given if he is healthy and comfortable and not irritated in any way by ill-fitting equipment, lameness or some other adverse condition. A good trainer never abuses his horse in any way if he can help it. And he is rewarded for his efforts by having a horse that responds, a horse that works for him willingly rather than grudgingly or fearfully.

"Bronc busting" or even the use of some modern gadgets and gimmicks which are designed to make horses do things have no place in the training methods of a good horseman. The person who must "break" a horse and show mastery over him is no horseman but rather a very immature person. Resorting to the use of force in training a horse or making him perform is the crudest and lowest form of horse handling, implying that the horse handler has little knowledge of how to get the best response from the horse.

Bronc busting and forceful mastery of a horse should have become obsolete with the passing of the Old West when horseflesh was cheap and horsemanship rather poor on the average. Today's horses are better bred, more expensive, and well worth the time and effort to train and handle intelligently. "Breaking" a horse implies a fight between the horse and the horse-breaker, and this kind of fighting is the last thing we want in a good horse. Horsemanship requires tact and horse sense, good hands and good judgment, rather than brute strength and daring. The ability to develop and keep a good mouth on a horse is many notches above the ability to ride a horse by sheer strength and force.

Force defeats the purpose of training. The horse has a very good memory and the effects of abusive handling will stay with him a long time. He especially remembers unpleasant experiences; things that hurt or frighten him make a vivid impression in his mind. One frightening or painful experience, such as the rider whipping him to make him do something he is afraid of doing or does not understand can undo weeks or months of gentle and patient training. It is very important for the trainer to *consistently* use tact and patience, never losing his temper and resorting to force. The trainer must constantly try to keep the horse calm and relaxed, using gentle persuasive methods, keeping from provoking resistance in the horse that might lead to a fight.

All too often a rider or trainer does not understand his horse well enough to know just how much to ask of the horse and how much is beyond the horse's ability or understanding. Resorting to force to make a horse obey is the most ignorant kind of cruelty. It is so easy to try to force a horse to do something which is beyond him (such as wanting him to keep going when he is tired, wanting him to jump an obstacle he is not ready for, or to perform a maneuver he has not yet mastered) or which he does not understand.

It is the cruelest kind of abuse to then punish the horse for disobeying when the horse has no way to know why he is being punished. The usual result of this sort of mistreatment is that the next time the rider asks the horse to do this thing for which he was once punished, the horse associates the rider's signals with the bad experience of pain or fear and automatically begins to resist, fearing punishment. If you can't control your temper, you should never attempt to train a horse, and there is also good question as to whether or not you should ever ride one.

Most horses are naturally fairly willing and docile. The main objectives then in teaching them something are not accomplished by the use of force. The horse does not need to be conquered or mastered, because he is already a naturally docile animal. The trainer's task is mainly to communicate with the horse, to make the horse understand what is wanted. Once the horse understands what is desired of him he will usually try to do it, if he is physically capable. But if a trainer is ignorant or untactful, resorting to strength and rough methods for coercing the horse to obey, he is only making it obvious that he is a poor trainer.

Very few really stubborn or mean horses are naturally that way; most are made that way by poor handling. If a high-spirited, sensitive horse is abused by the wrong kind of training he will come to resent the trainer, or even riders or mankind in general and will learn to hate the abuser. A horse treated this way may become stubborn and resist the rider in many ways, or may actually become dangerously mean. If you ever feel like hitting a horse that is acting contrary to your wishes, think twice. Force usually gets the trainer nowhere and is actually detrimental in most cases. It can backfire and be the first step in ruining a good horse.

The horse must have confidence in his rider or trainer, not fear. The horse is basically a timid animal, but also a very willing one once he has confidence in his handler. The trainer's first job is to instill confidence and do away with fear. The horse cannot understand the rider's wishes if he is afraid of his rider, and will only provoke more punishment by his fearful reactions to the rider. We cannot make the horse understand our commands until we are able to communicate with him and we cannot communicate with him until he is calm and relaxed with us. Confidence in the handler is important; without it the horse cannot surrender to the trainer's commands and wishes, cannot become relaxed enough to pay proper attention and respond willingly to signals. And, the horse must be relaxed and confident before he can be open and ready to learn. The rider and trainer must have his horse under control; the horse must not be reacting blindly and wildly out of fear.

Before a horse can become well trained, the rider must be able to control the horse's voluntary and involuntary actions. Perhaps the most difficult to control are the involuntary actions—the horse's reactions to anything that makes him nervous or afraid. Since the horse is timid and his involuntary actions are natural and somewhat uncontrollable even by the horse himself, it is the trainer's job to minimize the circumstances and situations by which these reactions and fears are provoked. A good trainer can reduce the outside stimulations that create fear or excitement in the horse. Since many of these stimulations come from the rider himself, the good horseman's primary aim is to ride as well as possible and to use tact and good judgment in his commands to the horse.

It is difficult to completely avoid occasional upsets when training a very sensitive and high-strung horse, because there will be occasional outside influences such as strange surroundings, unusual noises, and strange objects which upset or frighten a horse, but a good rider can try to make sure that no upsets or excitement are caused by his own poor riding or poor judgment. If a horse is afraid or nervous, his body tenses up and he cannot relax. And if he cannot relax, he will not do a very good job of obeying orders.

Each horse is different, with a different personality and mentality. It is up to the trainer to become thoroughly acquainted with the horse he is working with in order to understand that horse and communicate with him, constantly improving his ability to stay in sympathetic communication with that particular horse. The rider learns along with every horse he rides or trains.

By nature, the horse is willing and gentle. It is up to the rider not to spoil this. The horse is also timid, and easily subject to fear, excitement, and confusion. Confusion is one of the biggest difficulties for the trainer, for it is so easy to cause. If a horse is asked to do something which he does not understand (like a sliding stop or a fast turn on his hind legs, before he is ready for such things) confusion is the inevitable result. And if he is confused, he may resort to some other kind of action because he does not know what to do.

The horse is sensitive and a little fearful of active strength exerted by the rider. Perhaps it gives him a sense of claustrophobia. When a rider uses active strength against the horse trying to force or make the horse do something, the horse becomes doubly fearful if he does not understand what is wanted, for he does not know what to do to get away from the active force. When this happens it is not unusual for the horse to react with strength of his own, trying to get away from the rider's forceful abuse either by bucking or rearing, running away or some other devious action. The horse's strength is much greater than man's, especially when the horse is reacting out of fear. Since this kind of strength is hard to control, it is the wise trainer and rider who seeks always to avoid provoking it.

Horses have very good memories, and never forget anything. This is a quality which can work for and against the trainer, because if a horse ever learns any bad habits, they are hard to get rid of. A horse can remember unpleasant situations and bad habits as easily as he remembers the good ones. A horse often remembers the place along the road where he was frightened on an earlier ride and will be ready to shy and run away again when he comes to that place. He will remember that he was once able to avoid going over a certain log on the trail because his rider did not force the issue, and will try to get away with it again. He may remember the time a rider tied him with just the bridle reins and something frightened him into setting back and breaking them; he will remember breaking free and may try it again sometime when tied. Or, he may remember being punished excessively when seeking to perform some maneuver in a training arena, and may refuse to enter the arena again.

When working with horses, a trainer should from the very beginning try to utilize the horse's memory to the trainer's advantage, making sure that the experiences he gives the horse are those which the horse should remember. In other words, the trainer will try to ask of the horse only those things that the horse will be able to accomplish, so the horse will not have failures or setbacks to remember, or frightening situations that will cause him to become frightened again under the same circumstances. And a good trainer doesn't let bad actions become habit. This is why a horse should never be allowed to get away with biting or kicking or any other undesirable action. If he is effectively punished for these misdeeds, he will remember the punishment and won't be so apt to try a bite or a kick in the future.

The horse's memory is very good but his understanding is rather poor. He will remember things that hurt him and seek to avoid any repetitions, even though he may not know why he was hurt. Harsh treatment will be remembered for a long time and may make the horse resentful, even though he doesn't know why he was treated harshly. A horse can remember the new things you teach him, but he will also remember any bad habits picked up along the way. When training a horse, make sure that the experiences the horse encounters in being handled and ridden are pleasant ones to remember. If the horse remembers frightening experiences or situations which caused him pain, he will tend to dislike being handled or ridden and be resentful or may try to escape his duties in the future.

The horse in training can associate ideas to a certain extent, associating rewards with proper behaviour even though the reward often is merely a relaxing of the rider's aids and signals so that the horse can relax, or merely keeping the horse comfortable. Many times the reward is actually just the lack of punishment, and comes in the form of good riding and good handling which does not irritate, annoy or abuse the horse in any way.

To be effective, any punishment given must immediately follow the disobedience, or the horse will not be able to associate the two. If the punishment comes several minutes later, he will not understand why he is being punished and the punishment will do more harm than good. The trainer uses rewards continually, and these rewards can be anything from a gentle pat, or an encouraging and soothing voice, to a slacking of the aids so that the horse can relax.

Physical punishment is rarely needed, except when a horse is disobeying through stubbornness or behaving obnoxiously to a person or another horse. A rider should never tolerate any biting or kicking at another horse or person. The better the horseman and trainer, the less often will physical punishment be necessary. The trainer controls his horse more subtly, often preventing disobediences before they actually take shape. This all comes with the perfecting of the communication between horse and rider—understanding one another and predicting to a great extent what each will do next under a given situation. Harmony and unity are the key factors, horse and rider working as a team rather than as individuals at odds with one another.

The horse is a willing animal. Some are more willing than others, and most horses will be stubborn once in a while, but the good horseman never lets the horse get into a position or situation in which the horse will be tempted to disobey willfully, especially during the early training period. Disobedience is usually the result of inadequate or improper training, trying to hurry the horse too fast for what he is ready for—asking of the horse more than he is capable of performing. Disobedience can also be the result of harsh methods or cruel handling. The horse in this case is merely trying to escape the pain, or may be afraid and wants to defend himself. Bad habits and faults are generally the result of bad training, or bad experiences during training. It is actually easier to teach the horse faults and bad manners than good ones, for none of us are perfect as trainers and we all make mistakes at some time or another. To do a good job of training, constantly improving the horse instead of creating bad actions, the trainer must proceed very slowly and cautiously, without causing confusion or fear somewhere along the way.

Constructive training that constantly improves the horse requires tact, patience, experience, knowledge and much understanding of horses in general and of that horse in particular. We speak of training in the sense that the trainer takes a green horse and teaches him to be a good riding horse, but also in the sense that every rider of horses is also a trainer, because every time you ride a horse you are actually training him, improving him if you are a good horseman, or making him sloppy or bad mannered if you are a poor one. Everything a horse does when being handled or ridden is the result of training or the lack of it. So if your horse has poor manners, poor actions, or does not handle well in some way or another, he is merely reflecting your poor training.

It is not always easy to keep improving the horse, making him well mannered and tractable and highly maneuverable when under saddle. It is very easy to teach a horse faults or bad manners. This can be done merely by being unobservant, or by being lazy or negligent, letting a horse get away with things he shouldn't, such as grazing along the trail as you ride, or fussing around you instead of standing still for mounting. If a rider's actions cause the horse discomfort, the horse will usually show his disapproval by the attitude of his ears, the tenseness of his body, or the choppiness of his action, and by trying in some manner to counteract and minimize the effects of the rider's poor actions.

These actions of the horse, in an effort to counteract the rider's sloppiness or poor actions might be only slight at first, but then tend to get worse as time goes on, finally becoming a habit. Many of the aggravating habits of spoiled horses, such as grabbing the bit, traveling sideways instead of smoothly forward or instead of turning smoothly, rushing backward or backing crookedly instead of straight, etc., are generally the result of evasive action on the horse's part, in which the horse is trying to escape the poor horsemanship of the rider. It is the trainer who ultimately makes the horse (and remember that every rider is in a sense a trainer, improving his horse or making him worse).

If bad habits or poor results show up in the horse, the rider should look first to himself for the cause, rather than to the horse, and should seek to improve himself and his methods before he attempts to improve the horse. There is usually no surefire cure, just conscientious determination. And good advice from an observant horseman as to how to correct our difficulty is to no avail unless we can follow it, unless we have the ability to carry it out.

In his training every horse must start at the bottom and gradually work up the ladder of success, whether he is to be a pleasure horse, jumper, stock horse or dressage horse. And every rider who seeks to be a fine horseman must also start at the bottom and work up, gaining a good understanding of the basic elements involved in schooling horses. It is a mistake for any rider to try to skip the fundamentals, or to try to take a finished horse that someone else has trained and accomplish great things with that horse without doing some homework himself, without having within himself a great capacity for learning and improving. No matter how finished the horse when you acquire it, you must be somewhat a trainer yourself in order to keep the horse as good as when you got him. Thus every person who attempts to be a good rider must be constantly learning, constantly seeking to improve his horsemanship and his training methods.

Fahleen as a foal. The young horse is carefully handled, learning to respect, trust, and obey, as the groundwork is laid for future lessons

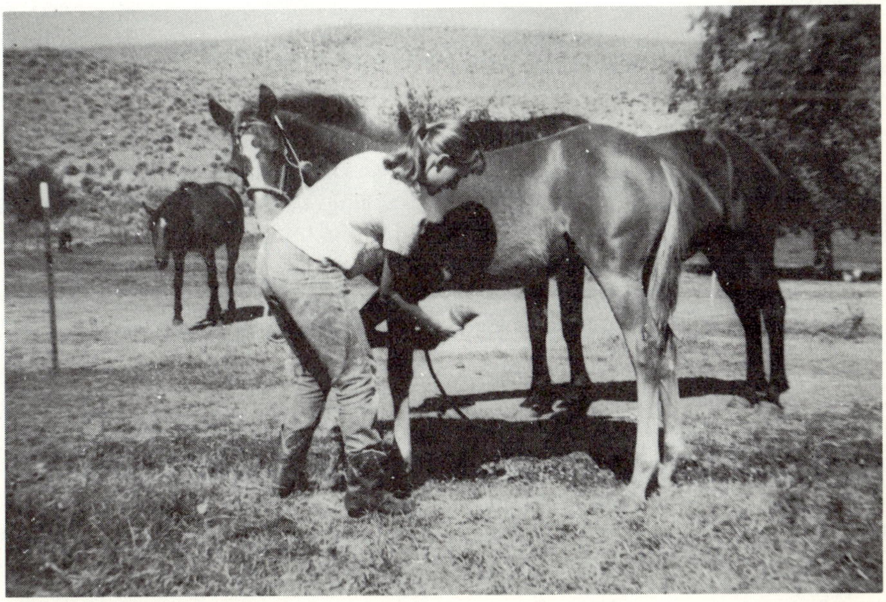

That big day comes when he is mounted for the first time. Then his schooling begins in earnest, for from that day on he will be learning, improving, and perfecting his abilities under saddle—if his rider is a conscientious one

Because in this chapter we are not actually dealing with training as such but with the attitudes and actions of the rider or handler toward the horse, we will not cover training in detail, but will merely touch on some facets of training just by way of illustration, to point out the attitudes by which a good horseman should deal with his horse. This involves the things we have been constantly speaking of—tact, patience, understanding, and a few basic principles of horsemanship.

Let's look at a few of the most elementary principles. First of all there is the aspect of movement. The horse must learn to go forward freely whenever his rider or handler desires it of him. This is so basic that the horse is obviously of no use to mankind without it. But teaching the horse to move forward freely is generally no problem at all because it is the horse's own natural inclination to do so. The difficulty comes in teaching him to move forward at our command and under our control—to move forward just when we want him to and just how we want him to (how fast and in what manner). The horse that accomplishes this, moving forward exactly when and how his rider wishes him to—at a brisk walk for traveling across country, at a highly collected trot on two tracks in dressage work, at a free canter in western pleasure classes, with a leap over an obstacle in a steeplechase course or in the hunt field, or performing a smooth maneuver to head a cow in a cutting contest or at a mad dash over the sagebrush and rocks on the range to head a wild cow—is a very well-trained horse indeed. And the well-trained horse, who accomplishes his work with ease and with a minimum of effort, appearing to be in unity with his rider at every move, is the result of fond and careful riding and training. The well-trained horse has had the kind of handling that has made him that way; he has not been abused.

Let's take a little closer look at this kind of handling. To make it simple, let's look closely at something very elemental—the problem of teaching a young horse to lead. This in itself is not difficult because most horses move freely at halter, knowing that they are being taken to water, out to pasture, into a stall for a feed of grain, and so forth. That is, most young horses that have been handled at all and have any confidence at all in the handler tend to lead fairly easily and readily even without much teaching. But in order to learn to lead well, a horse must learn a few more things. He must learn not to hang back on the rope, not to lunge ahead, or to gawk or dawdle or to head in any direction he wishes. He must learn to travel right beside his handler and to go at the pace the handler wishes. To lead a horse properly, the trainer walks beside the horse's left shoulder, the horse moving out freely.

To accomplish this, the trainer must usually instill in the horse a bit more action than the horse would normally exhibit, encouraging the horse to move forward freely, and readily. This can be done by using two aids, the voice and the whip. The horseman always makes good use of his voice in training or handling a horse, using it to soothe, encourage, calm, or stimulate. The horse that knows and understands his rider and handler can be easily encouraged or soothed by the sound of the voice. If, as he is being led, he hears that familiar voice encouraging him to move out freely and boldly, he will be more likely to do so than if he is hanging back afraid.

The whip, used as an aid in working with horses, should be thought of as an extension of the arm or hand rather than as a means of punishment. The whip, used as an extension of the hand, enables the trainer to be able to touch parts of the horse's body that he could not reach otherwise. Thus, when the trainer is standing quietly beside the horse's shoulder, he can touch the horse's hindquarters; and the horse knows at once that the touch comes from the person standing quietly beside him. This makes quite an impression on the horse as he realizes that he is controlled at both ends at the same time by the trainer, even though the trainer merely stands quietly at his head. But the value of the whip as an aid in training depends entirely on the way it is used. It should always be used quietly and sensibly, as an extension of the trainer's hand—as an aid rather than for punishment. The whip should *never* be used for punishment, for if the horse fears it, its effectiveness as an aid will be lessened and maybe completely ruined. The value of the whip in training therefore depends on the horse's confidence. If he trusts his handler and doesn't fear the whip (because he has never been abused by it) he will respect it and respond to it as an extension of the trainer, rather than fearing it and trying to escape from it.

So, when teaching the horse to lead, asking him to move forward beside you, use the whip quietly, holding it so the horse can see your hand, tapping him lightly once on the hindquarters as you ask him to move forward. If he does not understand at first, tap him again, and then as he understands and takes several steps forward beside you, praise him and let him know by your voice that he has pleased you. At the same time you are explaining to him by using the whip, urging him to move forward, you keep just a light touch on the halter rope, giving him the freedom of his head as he walks forward.

This is very important, for if the horse is held too tightly by the halter he will become nervous, almost as if he felt a bit of claustrophobia. After the horse learns to lead in the proper manner, the whip is hardly ever used, for the horse will understand what is wanted without it. It is important that the trainer stop using an aid as soon as the horse responds; the horse's reward for responding properly is the ceasing of an aid, whether it be the touch of the whip, leg pressure, the touch of a rein, pressure on the bit, or on the halter, and so on.

Your hand on the halter rope should be gentle, as should your hands on the reins when you ride. The hands are one of the key factors in determining whether a person's horsemanship is good or poor. Good hands mean a fine, light touch and a great deal of tact. Poor hands mean sloppy horsemanship, rough use of the reins and bit, and pain for the horse.

Even when teaching a horse to lead, the horseman should have good hands. Many people don't associate good hands with leading a horse, but the same rules apply to leading as to riding. It is especially important to be gentle when leading, for at this early stage in a horse's training, deep impressions are made upon him which he will never forget. Even the most elementary stages in a

The whip is used as a "third hand"—as an extension of the trainer's arm. The trainer is tapping the young filly's hindquarters to encourage her to move forward

After the horse understands what is wanted, he will walk alongside the trainer

When a horse is properly led, he walks freely beside his handler on a slack lead rope

horse's training are very important, because a horse can become very spoiled at any stage, just by the wrong kind of handling.

The hand maintains contact with the horse, by holding the lead rope or strap attached to the halter, but the hand should be passive, not active. All too often the mistake is made of pulling at the horse when teaching him to lead, either trying to pull him forward, or pulling him back if he tries to go too fast. When teaching a horse to lead, never pull. The main purpose of your hand on the halter rope is just to regulate the length of rope. As long as the horse walks along quietly beside you, there is no tension at all on the lead rope because neither horse nor you are pulling. The weight of the rope itself is enough to keep sufficient contact between the horse and your hand. The horse should not be given too much rope; 8 to 12 inches is usually sufficient to maintain good contact.

If the horse walks too slowly and drops behind, do not try to pull him forward with the rope, but use the whip as your "third hand," pointing it at the horse's hindquarters to encourage him to move forward. Trying to pull the horse forward is a very common fault; we see it in many horse show halter classes, especially when the contestants are trying to make their horses move out at a trot. But pulling at the horse to make him move forward faster is one of the worst mistakes that can be made. The horse instinctively resists being pulled; the harder he is pulled the more he will naturally pull back against the trainer. And the horse is much stronger than the man so it is foolish to begin such a useless tug-of-war. If you have ever tried to make a trailer-shy horse enter a trailer by pulling on him you have witnessed this fact—the horse resists more strongly the more he is pulled. It gives him a sense of claustrophobia, and it is his nature to resist any active force that is used against him. Even if the horse will lead as you pull him, he does so in a resisting attitude, with his head in the air and his neck somewhat braced. He is not moving freely and naturually. It is always better to encourage and persuade than to use force. The horse will respond much better and much quicker if he can respond willingly instead of under coercion.

Even though man can teach the horse a great deal, he teaches him best when working with tact and patience rather than force. Man can never change the nature and instincts of the horse, so it is the wise trainer and horseman who works with the horse, using his nature to best advantage, rather than trying forcefully to go against his nature.

So, if you try to keep your hand passive while teaching the horse to lead, you will have quicker and better success. Tension on the lead rope will only happen if the horse tries to go faster than you want him to. If the horse does bolt forward or wants to prance or trot instead of walk, he pulls on your hand. But you should resist the pull passively rather than actively. That is, do not yield as the horse pulls at you, but do not pull at the horse. Let the horse do all the pulling. Your hand acts as a solid, fixed point. The usual result is that the horse stops pulling. Most horses do not continue to pull unless they are being pulled at. The hard thing is to actually be passive, to be unyielding but not pulling. It is very easy to pull on the horse without realizing it, for to pull when pulled

This is an improper way to lead a horse. Never pull at a horse in an attempt to lead him; his reaction is to resist. Note the position of the horse's head

at is just as instinctive with humans as it is with horses. Pulling against the horse merely creates a tug-of-war, meeting force with force. Try to keep your hand steady, but not pulling.

When the horse realizes that he is meeting an unyielding object but is not being pulled at, he will cease his own pulling. The exceptions are when a young horse is exceptionally flighty and high-strung, or excited, or frightened into shying and bolting away. This kind of jerk is hard to meet by a mere fixing of your hand, so you may have to let out a few feet of rope to help take up the strain, or even travel a few feet along with the horse yourself, but still concentrating on passive resistance, letting the horse do all the pulling. Then when the horse stops his pulling, the release of strain on his halter will be immediate and automatic and he will realize that he is doing all the pulling himself—and will soon come to understand that he is supposed to travel beside you at the speed you desire.

These principles of gentle persuasion and tact rather than force will be the most successful means to develop a well-mannered, well-trained horse, and are also the best means to keep him that way. Gentleness and kindness in handling horses will pay the trainer well when it comes to getting along with and communicating with the horse. The gentle persuasion will take the horse much farther in the long run than will the use of force.

This discussion of teaching a horse to lead is just one example of how tact and patience can be used to advantage in training and handling horses. This kind of handling comes second nature to the good horseman and he need never resort to force except in rare instances or with spoiled horses (and there are all too many of them) and then only as a last resort. He will usually be rewarded for his patience with a horse that responds readily and willingly. It is always a pleasure to have a horse work for you willingly and with his whole heart, in unity with his rider instead of in opposition. This is the goal of the horseman—to have a willing horse that is pleasant to ride and open to learning. A horseman can be judged by his horses. If his horses are well mannered, quiet and willingly obedient, working with energy and spirit but working smoothly and precisely, then that horseman has used good tact and patience and has in a very real sense been kind to his horses, using them but not abusing them. If a horse we are riding or training is not performing as it should, we must look first to ourselves for the cause, and to ourselves for the correction.

To do a good job of training, you must have a certain feel about a horse; you must come into a relationship with that horse in which you understand one another and to a great extent know what to expect from one another. Whenever you start a young horse or a new horse, it takes a bit of time to just get acquainted and feel each other out. The horse is sizing you up as you size him up, and if after this brief period of getting acquainted the horse respects and trusts you, you are off to a good beginning. If the horse still fears or resents you, you still have a long ways to go. Some horses take more time than others to "come around" and gain confidence in you. Some of these need lots of careful, gentle handling, and some of them need patient, consistent discipline. But until you reach the point where the horse trusts you, you cannot really progress very far in his training.

You must know the individual horse intimately. Then you are better equipped to take him step by step and never ask too much or too little of him; you know just how far to go. When you trust a horse, he knows it, and he tends to relax and trust you a bit, too. If you know your horse well, you reach a point in your handling of him where he can be allowed to work on the "honor system"; he needs very little correction or discipline from you and in a very real sense he is working on his own, knowing what you want of him. There is a great deal of mutual understanding between you and that horse. And some of the things you do in handling that horse may look like taking chances to an outsider, when in reality you have a feel of just what you can and can't do with that horse and you can trust him. Many horses go a lot farther for you in their training if you trust them fully.

I'll cite just one small example. My chestnut mare, Fahleen, did not get much handling when she was young because at that time I was either pregnant or had a small baby to take care of. I finally began riding Fahleen when she was four years old, and didn't really use her much or get much training done under saddle until she was five. When she was a yearling she had an injured hock we had to doctor, and it made her a bit fussy and nasty about having her hind legs or feet handled. But as time went by and I trimmed her feet periodically I learned not to fight her at all. If I grabbed her hind foot and hung onto it tight so she could not take it away from me, she would dance around and kick and try to take it away. But if I trusted her (as I gradually came to do in spite of her attitude about those hind feet) and just laid that foot across my thigh to work on it, not hanging onto it at all, she would leave it there and let me clean or trim it. I had to trust *her* and then she came to trust *me*. Each individual horse has his own idiosyncrasies, and if you work with them instead of against them, you get a lot farther with a lot less problems. This mare dislikes being forced or manhandled and to have tied up her foot (as some people would have done) would have made the problem worse. Each horse is different and you have to know an individual horse well in order to handle him best. What works well for one horse won't work at all for another. This is why I stress that certain "feel" about a horse; if you know him intimately, you know what you can and can't do, what you can and can't try with him. You use all your tact, patience, judgment, past experience and knowledge, and then you'll do best if you'll throw in a bit of intuition as well.

Whenever you are training or riding a horse you must use good judgment in what you ask him to do. Don't try to push him beyond his training or abilities. The horse usually will do his best to follow your commands, so command wisely. You are responsible for the well-being of your horse; if you try to make him do something that is dangerous, he may kill himself or you trying to do it. If you try to make him do something he is not yet trained for, he may become confused and discouraged and you will have trouble if you attempt to try to get him to do it again; you'll have lost ground in your training. When riding with other horsemen, don't lose your sense of judgment. Think first of your horse, and of the everyday common rules of courtesy and safety. Don't try to make your horse do things he isn't quite ready for; this usually backfires and results in a poor showing, your loss of good humor because your horse cannot perform the job, or even worse, in an accident.

Good horsemanship is judged by your horse's manners and ability—by what he can do properly and well—and by his health and well-being. The rider who tries to be spectacular or tries to outdo everyone else, at the expense of his horse, is a very poor horseman indeed, and needs to grow up. Take your horse slowly and patiently in his training and don't ever blame the horse for your own mistakes. The good horseman has time, and much more important goals than competing in the next horse show or horse-games competition.

A well-trained, happy horse that is a joy to ride and work with is well worth the time and effort spent in making him that way. It is so easy to spoil it just by the wrong kind of riding or handling. Don't let the excitement or anger of the moment spoil that long-range goal. And never let the thrill of competition overshadow your good judgment in what is best in working toward that long-range goal. Your horse may not be able at this point to outdo the horse of your friend, but if you wait, if you refrain from attempting things with your horse that he is not ready for, some day in the future you *will* have a good horse, one that is able to compete if that is what you desire to do with him.

There is much emphasis these days on competition—in endurance riding, in the show ring or at the local "play day" featuring competitive games and contests on horseback. But the wise rider never attempts such things on a green horse, if he ever expects that horse to become a good one. Part of being a horseman is just having good judgment, knowing when your horse is ready for something new and when he isn't. Another point: if you want your horse to be merely a professional barrel racer or games horse, that's one thing. Be careful how you go about it, however, so that he will stay fresh and good at these things and won't go sour. But if you want him to be a well-rounded, well-trained horse that is good at many things (such as pleasure riding, ranch work, showing in an arena, trail riding, etc.) go easy on the games. In other words, don't spoil him for good riding. Some novices hot rod their horses so much in competitive games that they ruin their horses for anything else or even for the games afterwards. Some of these horses get so jazzed up that they are a little nutty or even uncontrollable in an arena or around other horses because they are so nervous about competing. They are overanticipating. It is sad to see an otherwise fine horse ruined in this way. There is nothing wrong with competitive games on horseback in themselves, but just in the manner in which some horsemen go about them. A good horseman can compete and do well and still come out with a well-trained, well-mannered horse. Remember that a poor performance is not the fault of the horse, but is usually just evidence of poor judgment of the rider. Good horsemanship involves a great many things, but perhaps the largest ingredient is just plain good judgment.

You need to make use of all the understanding you can get about your horse's instincts and individual personality. Horses, like people, like to be comfortable and don't like to be hurt or made uncomfortable; if you make him uncomfortable very often he will become resentful. Since the horse is naturally timid and not prone to fight or resist (unless he is made uncomfortable or is irritated or confused too many times) he will usually go along with what you ask of him as long as you make your commands very plain, and your demands accomplishable. If he understands what you want him to do and he thinks it doesn't look too difficult, he will try to do it for you.

And because he has a good memory, he will remember the things that hurt, such as a jerked mouth over a jump (because you caught your balance with the reins) or the fact that the last time you took him into an arena he got punished for not making a turn fast enough. Thus the next time he sees a jump or the next time you head him for the gate into the arena, a little warning light comes into his memory and he balks, not wanting to repeat any past misfortunes. And you as his rider will perhaps have a little difficulty reassuring him. But if he is never punished for trying to do what you ask him to do—if he is never hurt by a jump or never whipped or spurred in the reining pattern, but only encouraged by your voice and your aids so that he knows what he is doing is right—he will come to do the things you ask of him as a matter of habit.

If while he is in his uncertain stage of early training you are wise in what you ask him to do he will come a long way a lot faster than if you try to force him. Never ask him to take a jump that is too difficult for him, for no matter how well you ride him over it, it will still be a form of punishment if he fears it and jumps it awkwardly, perhaps landing awkwardly and hurting himself a bit. Never ask him to do anything beyond his ability—whether it is a fast stop, a jump, or a quick turn. If you try a fast stop from speed before he is ready for it you will punish him by hurting his mouth.

There are times that punishment is needed, if a horse is being lazy or feeling silly or skittish just to be feeling that way. In these cases you may have to be more forceful than usual and insist upon obedience. But just be sure you never punish a horse for your own mistakes or lack of judgment in what you asked of him, or through your own poor riding.

If you have a good horse, you want to keep him that way. If he is a young horse with a lot of potential, you want to see to it that his potential is developed properly and not ruined somewhere along the way by improper handling. A good horse is worth a lot, not only in dollars and cents, but in the satisfaction and pleasure he gives his rider, and perhaps in the good feeling or good reputation he gives that rider if he does well in shows or other competitive events. The really good horseman takes good care of his horse not only in the stable but also in the saddle and in everything he does with the horse. Good training psychology pays off in many ways, resulting in a horse that will work well for you. Intelligent horse handling can thus bring satisfaction—pleasure in a horse that can do the job, and pleasure in a job well done. Think first of your horse and the horse will respond by being a well-trained, well-mannered mount—a source of joy and an asset to his rider.

22

GAITS OF THE HORSE

KNOWING SOME OF THE FACTS ABOUT HOW A HORSE MOVES CAN be a great help in becoming a better horseman, trainer, or judge of horses. A working knowledge of the anatomy, gaits and athletic ability of the horse shows us the "why" of good horsemanship; if we know something about the horse's gaits and how he moves, it is easier to understand how and when to use certain aids and signals to persuade the horse to do what we want him to do. This knowledge also helps us use good judgment in what we ask of the horse, not asking him to do anything that is physically beyond his ability in a given situation.

Let's take a brief look at some general facts about how the horse moves. The horse has strong muscles in the upper part of each leg, but below the knees and hocks there are no muscles. The muscles from the upper leg extend downward as long tendons. The muscles in the front of the limb pull it forward. The muscles at the back of the forearm bend the knee, and so forth.

The horse's center of gravity lies right behind his withers; his front end, with his head and neck, is heavier than his hindquarters. About two-thirds of his weight is carried by his front legs. Thus a horse that is standing squarely on all four feet cannot lift a front foot until he gets some weight off his front end, and he does this by raising his head a little (shifting his weight back) or crouching a little on his hind legs or moving a hind leg forward to take more of the weight. You've probably noticed this if you have ever started to lead a horse that has been standing squarely on all four feet. If you didn't have any slack in the rope as you moved forward, the horse probably pulled you back a couple inches when he raised his head and shifted his weight to the rear in order to pick up a front foot. To move in any direction, the horse has to overcome the fact that he is front-heavy. This is why he must be collected in order to move well balanced, with his head up, nose down, and his weight carried back farther on his hind legs so that he can be light on his front end.

The horse is capable of four or five gaits, but most horses do only three gaits naturally. These are the walk, trot, and gallop. The canter is a slow, collected gallop and is not really a separate gait. Some horses, through selective breeding and proper training, are capable of performing other gaits as well, such as the running walk, rack (also called the singlefoot), the slow gait or stepping pace (a highly collected rack), the amble, and the fox trot. Then there are also the distinctive gaits and leg actions of the various Paso breeds (Paso Fino, Colombian Paso, Peruvian Paso, etc.).

Some horses, partly due to heredity and conformation, naturally do a pace instead of a trot. Some horses are capable of both a trot and a pace, but these are rare. Horses that do merely the natural walk, trot and gallop are sometimes called "straight-gaited" as opposed to "gaited" horses. The gaited horses are the American Saddlebred, Missouri Fox Trotters and Tennessee Walking Horses. These horses are bred for smooth, easy, ground-covering gaits that are comfortable to ride as well as beautiful to look at. Gaited horses are mainly used in shows and as pleasure horses. They are not usually used for fast maneuvers where quick changes of direction are required (such as when working cattle or performing in timed contest events). This is because the gaited horse is usually longer geared than the ungaited horse and built for straight traveling. The ungaited horse usually has more natural ability to be collected for fast maneuvering or quick changes of direction.

Standardbred horses are either trotters or pacers, and are horses whose fastest gaits are the pace or the trot. These horses can pace or trot as fast as other horses gallop. Most horses are either natural trotters or natural pacers; very few do both gaits. The trot is a diagonal gait with diagonal legs moving forward at the same time, either the left front and right hind or the right front and left hind. The pace is a lateral gait, with the legs on the same side coming forward together, right front and right hind, then left front and left hind. The pace can be a very swift gait. Because the pace is a lateral gait, with the horse supported first by legs on one side of his body and then on the other it produces a rolling motion of the body from side to side. This gait is not good in mud or snow or other slippery surfaces because of the danger of sliding or falling. The pacer doesn't have the balance and support that the trotter has. For this reason pacers are not often used for anything but traveling on flat country; they are not as good as straight-gaited horses for rough terrain or precarious footing or quick maneuvering and fast changes of direction.

The amble is classified as one of the slow gaits and is usually thought to be a slow, mincing pace at walking speed. It is a comfortable gait. Some authorities state that it is not truly a lateral gait (as is the pace) because the horse's hind foot may come to the ground slightly before the front foot does.

A five-gaited horse must be able to do the three natural gaits (walk, trot and canter) and in addition to these must be able to do a rack and one of the slow gaits. The slow gaits are the running walk, fox trot, amble and stepping pace. The stepping pace is a highly collected slow rack in which the horse lifts his knees high, making a flashy, showy gait. In some of the show gaits of five-gaited horses there is danger of overreaching, and many

The trot—diagonal legs work in unison

The pace—lateral legs work in unison

gaited horses are shown in quarter boots to protect the heels of the front feet from being struck with the hind feet. The fox trot is a slow, broken trot in which it looks as though the horse is walking with his front feet and trotting with his hind feet. It is faster than a walk. The horse's head nods as in the walk, and he brings each hind foot to the ground just an instant before the diagonal forefoot strikes the ground.

The flat foot walk is the natural gait of the Tennessee Walking Horse. In this gait, the feet land in the same sequence of a normal walk, but because of the loose action of the horse, he can move quite rapidly at this gait. The running walk is the fast walk of the Tennessee Walking Horse and is a four-beat gait somewhat between a normal walk and a rack. In the running walk the hind feet overreach and land in advance of the foot prints made by the front feet, overstepping them from just a few inches to 18 inches or more. This long striding gives the horse a gliding motion. In the Tennessee Walker, this gait is characterized by a nodding of the horse's head, a flopping of the ears and a snapping of the teeth in rhythm with the walk. The running walk is a smooth gait that is easy on both the rider and the horse. When observing the horse doing the running walk, the gait resembles a pace, but the hind leg comes to the ground before the front leg on the same side.

The rack or singlefoot is also sometimes called a broken amble. It is one of the show gaits of a five-gaited horse. The rack is usually an artificial gait (a gait that the horse is trained to do) but in some cases it is a natural gait since a few horses singlefoot instead of trotting. We once had a black mare (Scrappy, dam of Khamette) that racked instead of trotted. She was not capable of doing a natural trot, but her singlefoot was the smoothest gait we ever rode, as well as being ground-covering. Scrappy's mother could not trot either; she was a natural pacer.

The rack is an exaggerated walk that has four distinct beats. It is a fast, flashy gait in which each foot comes to the ground separately at equal intervals. At first observation, this gait looks similar to a pace, but it is a four-beat gait rather than two-beat; the hind leg lands just before the front leg on the same side. In this way it is similar to the running walk, but in the rack the legs are lifted higher; there is more up-and-down motion. At the rack, the hind feet do not overstep the track of the front feet as much as in a running walk. In some horses the hind feet do not overstep at all at the rack, landing behind the track of the front feet. The rack is smooth and easy on the rider, but rather difficult and tiring for the horse, unless the horse does this gait naturally.

Over the centuries, horses have been working animals, useful for various purposes. Their value has been determined by their ability to move—to carry a rider, to pull a carriage, to catch a calf. By selective breeding and training over generations and generations, horses of different breeds have been somewhat specialized. They have developed a type of action or gait that was best suited for a particular job, such as the trot for pulling carriages smoothly and swiftly, the fast gallop of the race horse, the running walk of the Tennessee Walker for the plantation owner who wanted a smooth riding horse that could still cover a lot of ground in a day for overseeing large estates, the slow, steady walk of the draft horse for pulling heavy loads, and so on. This is one reason that certain types of horses today have specialized gaits. When riding a horse, you can tell by listening to the beat of his hoofs on the ground what gait he is in and how his legs are moving in sequence.

For the rest of this chapter we will look in more detail at the three natural gaits of the horse—the walk, trot, and gallop (or canter). These are the gaits that most horses do naturally, and are the gaits you should become very familiar with when riding or training a horse.

When the horse moves, he shifts his weight from one leg to another, raising or lowering his head to lengthen and shorten his neck to help him keep in proper balance for moving freely and easily. Whenever he moves, he has to overcome the fact that he is front-heavy. When a horse walks, he brings each leg forward at the exact moment to keep him balanced. He crouches slightly as he takes extra weight on his hind feet in order to lift a front foot and move it forward. He uses his head and neck for balance, just as we swing our arms when we walk or run. At the walk and at the canter and gallop, the horse's head bobs at each step. His head drops each time a front foot comes to the ground, and his hip drops when a hind foot comes to the ground.

The horse makes balancing movements with his head and neck at the walk and at the gallop, but not at the trot because his balance is steadier at the trot. The trot is a two-beat gait, with diagonal legs moving together, striking the ground at the same time and pushing off together. Thus at each stride of the trot the horse is supported by diagonal legs (a leg on each side and on each end of his body) instead of by one leg or two or three as at certain phases of his stride at the walk and canter.

Contrary to the popular belief that all the power of progression comes from the thrust of the hind feet, the front feet help in pulling the horse along at all gaits. When a front foot is extended and strikes the ground, the toe grips the ground and actually pulls the body forward, cutting into the ground somewhat if the horse is barefoot. Until the sole of the foot reaches a position where it is vertical with the ground (just before the foot breaks over and is picked up again) the effect of the front foot is that of pulling the horse forward. Any unnatural interference with the front feet, such as shortening the toe or lowering the heels can make a difference in how fast the horse can travel; it lessens the traction and cutting action of his front feet and can make the difference of several feet or even yards in a mile race.

Let's take a closer look at the walk. The walk is a four-beat gait in which the horse's legs move separately (no two legs moving in unison) in a definite sequence. The speed of the walk makes some difference in how the feet are picked up and set down. Horses are all individuals; they are all built just a little differently and each one has a slightly different way of going. Some are naturally fast walkers, some are naturally slow. Some horses take long strides while others take short choppy ones. Some horses pick up their feet smartly, while others are "daisy cutters," making a very low arc over the ground. Some horses' hind feet overstep the tracks made by their front feet anywhere from a few inches to a foot or more, while other horses step into the same track with their hind feet that they made with the front feet. How a horse is built (see the chapter on conformation) determines to a large extent how he picks up his feet and moves them. But no matter how a horse walks, he has the potential of being taught to walk fast. Any horse can be trained to walk more briskly; it just takes more time and patience for teaching some horses than it does for the ones that are naturally fast and eager walkers.

There are many advantages to having your horse learn a fast walk; a brisk, free-flowing walk is a desirable thing in any horse, whether you ride him for pleasure, or in show competition, or use him for endurance riding or ranch work. It's no fun to ride a horse that merely plods along, nor is it very enjoyable to have to pump him all day to make him move out; you'll feel like you've gone more miles than the horse.

To teach your horse a faster walk, it helps if you understand how the horse moves at the walk. In teaching him to walk faster, you'll be using your hands and your legs. Your hands check him from breaking into a trot, and your legs urge him to move more energetically. So let's look at the action of the horse's head and neck (which you'll be controlling with your hands) and the action of his legs (which you'll be controlling with your legs) as he walks.

As mentioned earlier, the horse's head bobs each time he takes a step at the walk. His head drops each time a front foot comes to the ground and then raises in order to shift his balance each time he prepares to pick up a front foot.

During the slow walk, the horse usually has three feet on the ground at once. When walking, he moves each leg in a definite order, each moving separately (only one foot coming to the ground at any one time), thus creating a definite four-beat gait. When the right front leg moves forward, the next leg to move forward is the left hind, then the left front, then the right hind. This sequence continually repeats itself: right front, left hind, left front, right hind, right front, left hind, and so on. As a hind leg comes forward, the front foot on that same side prepares to take off, and leaves the ground a split second before the hind foot lands, getting out of the way of the approaching hind foot. When the horse begins to walk from a standstill, he moves a hind foot first. If he starts with his left hind, the next foot to move will be the left front, and so on.

At the slow walk, the hind foot lands several inches in back of the front foot that is leaving the ground, but at a faster walk the hind usually lands in the same track or a little in advance of where the front foot was. As the speed of the walk increases, each foot is raised earlier. The fronts are picked up more swiftly, to get out of the way of the approaching hinds. At the fast walk each foot is raised before the diagonally opposed foot comes to rest (in other words, the right hind is lifted before the left front foot strikes the ground, and the left hind is lifted before the right front strikes the ground). Also, the front foot of one side may be lifted before the hind foot has reached the ground on that same side, causing the horse to roll a little toward that unsupported side. Thus at the fast walk the horse has only two feet on the ground at once, and changes his weight alternately onto a lateral pair (two feet on the same side) and then to a pair of diagonals. At a good fast

Slow walk—the horse has three feet on the ground at once

Right front has grounded. Left hind is in the air and will come to the ground next—to be followed by the left front, then the right hind

walk, the hind foot of most horses will land well ahead of the track made by the front foot that has just been picked up. At the fast walk and at faster gaits, the horse's heel comes to the ground first, but at the slow walk the foot is usually put down squarely.

When you are teaching a horse to walk faster, or pushing any horse into a fast walk, you can make use of the knowledge of how his legs move in sequence. Since much of the horse's propelling power comes from his hind legs, you can encourage him to walk faster by stimulating each hind leg at the proper time to make him push off more vigorously.

The horse already knows that he is suposed to move more energetically in response to your leg pressure, but how you use your legs can make a great deal of difference in the ultimate response of the horse. You want him to take longer strides, not just faster ones. Short fast choppy strides tend to tire a horse sooner than long strides. When the horse pushes off with a longer swing of his leg, he will cover the ground faster, yet still seem relaxed and unhurried. So if you stimulate the horse at the proper time, at the proper phase of his stride, you'll get much better results than just squeezing or kicking your legs at random.

Remember that when the horse moves at the walk he moves one leg at a time; each foot is landing in a definite sequence and no legs work exactly together in pairs as they do at the trot. Using your legs more strongly will speed up his walk; using your legs encourages him to move faster, and if you check him to keep him from breaking into a trot, he will have to respond by walking faster. The best results will come from using your own legs alternately, stimulating the proper hind leg of the horse at the right time.

Squeezing with your right leg stimulates his right hind, and squeezing with your left leg stimulates his left hind. By squeezing your legs alternately at the proper moment, you can encourage him to move each leg more energetically as it prepares to leave the ground, and get better results than by intermittently squeezing both legs together. The only time you can really encourage the horse to move a certain leg more vigorously is at the instant that leg prepares to push off from the ground. Once the leg is in the air already, you are too late with your urging for there is not much the horse can do in response. You must use your own legs at just the proper time.

The horse's left front leg comes forward just an instant before the right hind comes forward, so if you are going to stimulate the right hind, catching him before he pushes off with the right hind, you must use your own right leg as his left front foot comes to the ground or you will be too late. The simplest way to do this at first, until you get into the rhythm of it, is to watch the horse's shoulders, timing yourself to their movement. As the left shoulder comes forward, the horse's left front leg is in the air. As that left front foot grounds, squeeze with your right leg, and as the right front foot grounds, squeeze with your left. It all happens so fast when a horse is walking swiftly, that you will have the proper timing if you squeeze each side just as that opposite shoulder moves forward.

The fast walk. At the fast walk the horse only has two feet on the ground at once. He alternates his weight from a pair of diagonals (A, left front and right hind) to a pair of laterals (C, right front and right hind). Sequence: left front, right hind, right front, left hind, left front, and so on.

Teaching the horse to walk faster. A and B: Rider squeezing with right leg (as left front leg is on the ground; cue to squeeze is as left shoulder comes forward) to stimulate the horse's right hind foot which carries the horse's weight and is preparing to push off

C: Horse's right front foot has just grounded (right shoulder is forward) which means that the left hind is coming to the ground. Thus the rider begins to squeeze with the left leg to stimulate the left hind leg as it takes weight and prepares to push off again

When squeezing the horse with your legs, use your legs where they hang in the stirrups, not back toward the flanks. Don't kick. The leg pressure should be exerted through the upper calves of your legs. After you get the feel of the horse's rhythm at the walk, you'll find that you can squeeze your own legs alternately at the proper time without having to watch his shoulders, for this comes naturally with the rhythm and roll of the horse's walk; you'll find that it doesn't take any concentration or thought (after you learn the feel of it) and not much exertion on your part. Once a horse learns to do a fast walk, all you have to do is swing with him, flexing at the hips and waist and rolling your weight in the saddle a little as you alternately squeeze with your legs. In a sense you are pushing the horse with your whole body as you merely move in rhythm with him. It doesn't take long to get the feel of it, especially after you discover how naturally it comes with the rhythm of the horse's walk.

This kind of urging to a fast walk is much easier on you than constant kicking and is more effective. Constantly kicking a horse is tiring to the rider. Also it is just a form of nagging; after a while the horse gets immune to it and doesn't respond. Proper use of your legs will get much better results, especially when coupled with light hands and a good feel of the horse's mouth.

Good hands are as important in teaching a fast walk as knowing how to use your legs. Your hands can make the difference between a calm or a nervous horse, between a responsive horse or a frustrated one that is always fighting the bit. You need to keep the right amount of feel on his mouth for proper control and adequate communication, without hurting his mouth or making him resent the bit. You don't want him to break into a trot when you urge him with your legs, but neither do you want him prancing or throwing his head because of too much restraint through the bit.

Lightly finger the reins if the horse starts to break into a trot. You have to be able to feel when the horse is going to break gait and check him slightly before he does or you will be too late. Each time you check him with the reins, immediately continue squeezing with your legs so that he will know you want him to walk faster and not slower.

This feel is accomplished only when you ride "on contact," your hands holding the reins so that you have continual communication with the horse through the bit. Your hands must go with the horse, giving and taking as his head and neck move back and forth. If you make the mistake of riding with a fixed hand (always in the same position), you will bump the horse's mouth with the bit at every stride, or else have so much slack in the reins that you never are in direct communication with the horse's mouth.

You don't need a tight rein to keep a feel on the horse's mouth. If you have a heavy set of reins, or a small length of chain between the bit and the reins, this feel will still be there even though there is some slack in the reins. A heavy rein is good for teaching a horse to walk fast; you can keep a feel on his mouth without actually tightening the reins (still on a slack rein, so to speak), and the rhythm and swing of the heavy reins as he walks helps to maintain the feel on his mouth and helps keep him from breaking into a trot. A heavy rein with a little bit of slack and swing to it keeps a good constant feel on his mouth if you use your hands properly, without ever bumping his mouth. Never have your reins too slack, for you need constant control of your horse to keep him at exactly the gait and rate of speed you want him to go. Once you get that rhythm going with your body and legs moving with the horse's walk, and a little swing to your reins to keep the horse from breaking gait, the horse will perk up, stretch out his legs, and really walk.

Many horses do not learn to walk fast with a rider on their back until after they have learned collection. A collected walk is not necessarily a fast walk since you can make the horse do a slow collected walk if you wish, but teaching a horse collection is a great aid in teaching him a fast walk, too. Once a horse learns collection, you can collect or extend his walk at will. A collected horse will respond readily to both bit and legs, enabling you to urge him properly and yet at the same time maintain a delicate feel on his mouth to keep him from breaking gait into a trot. The collected horse is alert. Many horses are not alert at the slow walk; they are often extended, often just daydreaming along, and are much more apt to stumble than if they were somewhat collected. By collecting the horse you get much more alertness and animation, both of which are necessary before the horse can do a good fast walk.

Now let's look at the trot. The trot is a two-beat diagonal gait, with diagonal pairs of legs moving forward together and striking the ground at the same instant. During the fast trot there is a moment of suspension in which there are no feet on the ground. During the slow trot this moment of suspension is not so obvious; one diagonal is coming to the ground at nearly the same instant that the other diagonal is pushing off. This gait has constant, equal rhythm, and the horse has good balance. In other words, he does not make balancing movements with his head as he does at the walk and at the canter or gallop. At the trot his head comes up and stays steady rather than bobbing up and down.

The speed and animation of the trot can vary a great deal. The jog is a very slow, unanimated trot and seems to be a natural gait of the horse. Many horses prefer to jog rather than walk fast, and must be taught to walk fast. The jig is also a slow trot, but it has more up and down movement than the lazy jog; a horse usually jigs when he is nervous and doesn't want to settle down and walk. An excited horse may jig and refuse to walk. This is usually a sign of bad manners or lack of training on the horse's part, or inability of the rider to use his legs and reins properly and gently to make the horse walk.

A trot can be slow, medium or fast, and can be collected or extended, or merely uncollected. A collected trot is not necessarily a slow trot; it can be fast or slow, depending on the wishes of the rider. In the collected trot, the horse is more animated than at an uncollected trot; he is traveling with his head up and his chin down, and he moves with his hindquarters more underneath his body. In other words, he has his weight balanced farther back and is not traveling so heavily on his front end. A collected horse picks up his feet more smartly and springily than the uncollected horse.

At the slow trot, the horse makes short strides. He makes longer strides at a medium trot, and very long strides at the extended trot. He covers the ground faster, not so much by speeding up his leg movements as by lengthening his stride. You will discover this if you trot the horse a certain distance first at a slow trot, then a medium trot, and finally an extended trot, counting the number of strides each time.

In the extended trot the horse makes extremely long strides and had a tremendous amount of propulsion. He almost seems to float above the ground. This is the racing trot of the Standardbred horse. Most horses must be taught to do an extended trot, but some will learn more quickly and do it more naturally than others. The extended trot feels quite different from an ordinary trot. It is almost as if the horse has shifted into high gear. It is not as easy for a horse to break into a gallop from the extended trot as it is for him to depart into a gallop from a medium trot.

When riding at the trot, you should sit the slower trots and post the faster ones. A slow trot is sometimes called a sitting trot and a fast trot a posting or rising trot. When you sit the trot, you keep your body in a balanced position (upper body slightly forward) but you can keep from "bouncing" by rolling your weight on the fleshy areas of your buttocks. If you try to keep a rigid seat you will have more difficulty. The good rider should be able to sit the slow trot gracefully without bouncing or losing proper balance and forward feeling of movement.

The trot is the best gait for covering ground swiftly without tiring the horse. A good medium-fast trot is easy on horse and rider (once the rider learns how to ride it properly). It is much swifter than a walk, and less tiring than a gallop or canter. If you ever use your horse in trail rides or endurance rides, you will discover the value of the trot and of learning to ride it properly. In most competitive rides, the distance cannot be covered in the time allowed just by walking; it is best covered by wise and judicious use of the trot. In endurance rides it has been found that the horses which are walked on the very steep and rough places and trotted wherever traveling is good finish in best condition, much better than the horses that are galloped. The cavalry used the trot a great deal in traveling; on long marches they used the walk and trot alternating a walk after each long trot, and rarely used the gallop. When riding range here with our ranch horses we have found the medium fast trot to be our horse's most useful gait when we have many miles to cover in a day.

When riding anything faster than a slow trot, you will want to post. It is easier on the horse as well as being easier on the rider. Posting is rising in your stirrups to one beat of the trot and sitting on the next one. In other words, the rider is moving up, down, up, down, etc., in rhythm with the horse's trot. At the fast trot it is much easier to stay in proper balance with your horse by posting than by trying to sit and keep from bouncing. You can keep your upper body more forward over the withers and center of gravity of the horse, where he can carry the load with the least effort.

When you post, there are two ways of moving with the horse as he trots, and you should know both for you may want to use one or the other for different purposes. One way of posting is called posting behind the trot. When you post this way, you let yourself be thrown upward from the saddle rather naturally by the thrust of the horse, at the same time pushing yourself (your upper body) forward to keep in balance with the horse, and then letting yourself down again on the next stride. The other way of posting is harder to learn but it is actually easier on the horse and rider and is a smoother kind of action. You do it by leaning your upper body forward, rolling your weight forward on your thighs, and by pushing yourself forward and then up. When you get good at it you can post very close to the saddle this way. Posting in the first manner (going up and forward, posting behind the trot) is the best way to post if your horse is overambitious and you are trying to keep him at an even speed. The second method of posting can be done if your horse is tending to go too slow at the trot; pushing your body forward in this way has the effect of making the horse go faster, for it is a form of urging. You are a little bit in front of his center of gravity for an instant and his tendency is to move forward to keep in proper balance. Posting in this way, and using your legs for encouragement, you can push the horse right along when trying to increase the speed of his trot.

One other aspect should be mentioned about riding at the trot: It is important that you learn to post correctly on the diagonals. The main reason for learning to distinguish between the diagonals is so that you can change occasionally when riding at the trot; it will be less tiring to the horse and will aid him in equal muscle development. If you unknowingly post always on the same diagonal, it will be tiring to the horse and will tend to make him one-sided in his muscle development.

When the horse trots he alternates from one diagonal pair of legs to the other diagonal pair. We say that the horse is on the left diagonal when his left front foot is on the ground, and on the right diagonal when his right front is on the ground. When you are posting, you are posting on the left diagonal when you are sitting in the saddle as the left front foot is on the ground (up in your stirrups when the right front is on the ground) and posting on the right diagonal when sitting while the right front is on the ground (up in your stirrups when the left front is on the ground). This can be a little confusing at first to try to figure out, because you cannot really see the horse's legs as you trot, and he is moving them so swiftly that it can be hard to figure which diagonal you are posting on.

It helps if you watch the horse's shoulders, and practice at first at a very slow trot until you catch onto the way his legs are moving as you are rising and sitting to post. When you are trotting in circles, you should always post on the outside diagonal, for this is most natural to the horse (unless he has already been made one-sided by a rider who unconsciously posts always on the

Posting on the left diagonal. The rider is down when the horse's left front foot is on the ground

The rider is up out of the saddle when the left front foot is in the air. (Heidi Smith and Surrabu)

same diagonal) and is less tiring. Thus, if you are circling to the left (with the center of the circle to your left), you should be posting on the right diagonal. If you are watching your horse's shoulders as you trot, you should be rising as his right shoulder moves forward, and sitting as that right shoulder moves back. You are moving with his right front leg, up when it is up (in the air) and down when it is down. The reason you should post on the outside diagonal when the horse is trotting in a circle is because the horse actually has a lead at the trot; his outside legs are having to travel farther than his inside legs when he is circling (his body is curving somewhat to fit the circle). Thus you tire him less when you post with his leading foot. When you are traveling in a circle to the right, the outside diagonal will be the left one.

You should change posting diagonals whenever you change direction. If you are making a circle to the left, posting on the outside diagonal, you should change diagonals when you start a circle to the right, so that you will again be correctly posting on the outside diagonal. To practice and get the feel of it, you can trot your horse in figure eights. When you come to the center of the eight after having made a circle to the left, you can change diagonals by either sitting an extra stride, or standing for an extra stride, whichever is easiest for you. You will automatically be posting on the opposite diagonal then as you change direction and start the circle to the right. If you are practicing figure eights with your horse in learning to post the proper diagonals, make your circles well rounded and of equal size, and have the horse go straight for a stride or two as you change and begin the circle to the opposite direction. Don't make circles that are too small when your horse is trotting; the horse needs room too move his legs properly without striking himself.

When traveling in straight lines or across country at the trot, you should make a conscious effort to change diagonals periodically so that the horse will not be ridden always on the same diagonal. If a rider posts more on one diagonal than on the other, the horse may come to prefer the often-used diagonal and become one sided in his development. Many horses that have been ridden this way will feel unbalanced and unnatural on the other diagonal and will shift their strides and alter their steps so that the rider is again posting on the preferred diagonal. On long rides the horse will not get as tired if his muscles have been equally developed by a rider who posts on both diagonals.

Now let's look at the canter and gallop. The gallop is the fastest natural gait of most horses. The gallop is a fully extended gait; the horse stretches his head and neck forward and takes long strides with his legs. The gallop is a four-beat gait, with a brief moment of suspension in which all four feet are in the air.

Whenever a horse gallops or canters, he is in the left or the right lead, which means that during the time he is putting his feet down onto the ground, he is reaching farther forward with his left or his right feet. For instance, let's say that the horse is on the right lead. His four feet will land in sequence and then be all off the ground in an instant of suspension. The first foot to land again will be the left hind. Then the right hind will reach farther forward and lead. Next to strike the ground will be the left front foot. Then the right front will reach forward and come to the ground, leading. When the right front is picked up again there will be another brief instant of suspension and the process repeats itself. At the fast gallop the horse never has more than two feet on the ground at once. The two feet on the ground are first both hinds, then the leading hind and nonleading front foot, then both front feet, then the period of suspension. But each foot strikes the ground individually to produce the definite four-beat gait; no two feet come to the ground at the same time.

If a horse is galloping in a straight line, he will normally change leads every once in a while to keep from tiring. If he is galloping in circles or making a turn, he normally leads with his inside legs. In a sense this helps to brace him and he is less likely to fall. If a horse is circling to the left he should be on the left lead; if he is circling to the right he should be on the right lead.

The canter is a slow and collected gallop. The horse is not so extended; his head is up, his neck is more arched, and he carries himself with weight balanced farther back on his hind legs. He is more ready for a quick change of direction at the collected canter than he is at the extended gallop.

When the horse canters, he has one or three feet on the ground at once, except for the phase of his stride in which all four feet are off the ground. The canter is usually a three-beat gait, unlike the gallop, which has four distinct beats. If a horse is cantering on the right lead, his left hind leg lands first, then his right hind comes to the ground farther forward and leads, landing simultaneously with the left front leg for the second beat of the cadence. Then the right front leg reaches forward and comes to the ground for the third beat. When the right foreleg is lifted again there is a brief period of suspension, as in the gallop, in which all four feet are off the ground. At a medium-fast canter, the left hind leg is lifted from the ground before the right foreleg comes to the ground, making a brief instant in which the horse has only two feet on the ground instead of one or three as in the slow canter. If the canter becomes quite animated it may result in a four-beat cadence like the gallop; in this instance the paired front leg and opposite hind leg do not land together in the same instant. The hind foot lands slightly before the front foot.

When a horse is cantering or galloping, the diagonal set of legs that are on the ground together get the most stress and are more subject to tiring. In the right lead, the right hind and left front take the stress; in the left lead the left hind and right front take the most stress. For this reason it is good to change leads often when you are cantering or galloping a horse; then the stress will fall for a time on the opposite diagonal pair of legs and give the others a rest.

When a horse changes leads, he does it during the instant of suspension when all four feet are in the air. If he has been cantering on the right lead (with his left hind leg landing first and his right hind leg reaching farther and leading) he changes during this instant of suspension and the foot that comes to the ground first is the right hind, followed by the left hind, which reaches farther, leading. Thus he is now on the left lead and he also follows suit with his front legs.

When a horse changes leads properly he does it smoothly, always changing with his hind feet first, for they are the feet that come to the ground first after the period of suspension. If a horse changes leads in front first (as he will do if cued improperly by the rider), or changes in front and not behind, he will have a rough, unbalanced gait. When riding the horse at the canter, you must give him the proper cues so that he will be able to change leads smoothly and correctly.

Your horse will be fairly well along in his training by the time you and he have mastered collection at the canter and have gained the ability to take either lead from a standstill and to change leads smoothly when changing direction at the canter or cantering on the straightaway. As rider, you should be familiar with all the details of the canter and gallop, for these gaits bring out your skill or lack of it, and also demand good training and obedience on the part of the horse. If you are familiar with the way the horse moves at the canter and gallop, you can more readily understand how and when to use certain signals to make the horse perform as you want him to; you will also be more able to use good judgment in what you ask of the horse, not asking him to do something that would be physically impossible for him to accomplish in a given situation.

When you wish to put the horse into a canter on a certain lead, or make him change leads while cantering, you must consider the effects of your legs and

Cantering on the left lead: The horse's right hind leg will come to the ground first, then his left hind leg will come to the ground and lead. His right front will then strike the ground and his left front will come forward and lead

Cantering on the right lead: The horse's leading hind foot is still on the ground and his right front foot has just come to the ground farther forward of his left front

reins. He will learn to change leads most easily, or to take a certain lead easily, if you first put him in such a position that leaves no room for hesitation or confusion. You can put him in the proper position or take advantage of him being in the right position by proper use of your legs, weight and reins. It is easier to teach a horse to change leads or to pick up the canter on either specified lead after he learns collection. He should be taught collection at the canter before you try to teach him the change of leads.

As mentioned earlier, a canter properly executed is a slow and collected gallop. There is a difference between a collected gallop and a slow gallop. Some horses that have not been taught collection can indeed gallop slowly, but they are still extended, looking as though they are about to fall apart at every stride. They lack animation and collection, are strung out and travel heavily on their front feet. The uncollected horse is not in a position to change direction suddenly or gracefully; to do a good job of changing leads while cantering, the horse must be collected.

Whenever a horse is cantering or galloping, he will either be on the left or right lead, or disunited, which means that he is on one lead in front and the opposite lead behind. When a horse is cantering or galloping disunited, it produces a rolling motion of his body that can be somewhat uncomfortable for the rider. It is also dangerous, for the horse is unbalanced, and when his feet are moving in improper sequence, he is likely to strike a front foot with a hind and throw himself. If your horse ever starts the canter or gallop disunited in his leads, stop him and start him over again.

When mounted, there are several ways in which you can tell what lead the horse is on, without leaning over to try to look at his feet. It is not a good idea to lean over to look at the horse's feet, for this throws the horse off balance. It is better to learn by feel and by watching the shoulders and listening to the cadence of the hoof beats what lead or gait or diagonal (and so forth) your horse is on.

If the horse is in the left lead, a glance down at his shoulders will show you that his left shoulder is always moving forward about four inches in advance of his right shoulder; the left shoulder, like the left leg, is moving farther forward at each stride than the right one. Also, your inside leg (the left one in this case) will be in motion whereas your outside leg will be comparatively still.

When you start teaching a horse to canter, with most horses it is usually best to pay little attention to leads at first, letting the horse lead with right or left as he chooses (just make sure he is never disunited). When first teaching the horse to canter you will not be making small circles where the correct lead is essential. During the first cantering lessons it is best to canter on the straightaway or out in an open field, teaching the horse to move out freely and easily. If you try to concentrate on leads from the first, some horses become confused; there are too many things for them to think about all at once. The horse first needs to learn new balance and how to handle himself at the various gaits with a rider on his back.

When teaching a horse to canter or to take proper leads, you make use of your legs and weight. A horse responds to your shifting weight because this changes his center of gravity. When you lean forward, the horse tends to go faster in order to restore his balance, and when you lean back he will slow down for the same reason. To teach him a slow and collected canter, urge him with your legs sufficiently to keep him cantering, and finger the reins to keep him from going faster. You use your weight to advantage here, sitting more upright than you would at the gallop; if you were to lean forward too much the horse would take it as a signal to go faster. Using your weight farther back at the canter keeps the horse collected and traveling slowly; his head will be raised instead of extended and his hind legs will push farther underneath his body. He will have more spring to his movements. His propelling power will be directed up and down as well as forward. Checking him lightly with the reins while urging him gently with your legs tends to bottle up his energy, encouraging him to move more springily, his weight carried more on his hindquarters, traveling lightly on his front end.

With some practice, a horse should soon be able to do a slow and collected canter. You should be able to put him into a canter from a walk, trot or standstill by leaning forward a little and squeezing with your legs. Then, as the horse takes his first cantering stride, you lean back slightly so that your upper body is inclined just slightly forward, and continue using your legs and hands to keep him from slowing or speeding.

After the horse canters well, you can begin to teach him to take the proper leads at your signals. Most horses take either lead naturally when running free without a rider on their backs. The added weight of a rider, plus the fact that the mounted horse isn't sure what to expect next from his rider, makes most young horses rather awkward at first, especially with their leads.

Most horses prefer to canter in the left lead when ridden, and some will not take the right lead at all until taught to do so. This is because most horses, like people, are "right handed." When a horse goes into a canter or gallop he raises his front quarters and gives a push with his hind legs, most of the pushing power coming from the hind leg on the side opposite the leg with which he intends to lead. Therefore, because he is "right handed," his main driving power comes from his right hind leg and he naturally goes into the left lead. At the same time that he gives this push with his right hind leg, his muzzle will be slightly tilted upward to the left.

There are several ways to put a horse into a certain lead. One method may work well for one trainer and not for another. Or one horse may respond better to one method than another horse would. One method may work well for you, but occasionally a problem horse might come along on which you'll want to try another method.

Regardless of which method you use with your weight and reins, you should (as you give the signal to canter) squeeze hardest with your outside leg, the leg opposite from the side on which you wish the horse to lead. In other words, if you are making a circle to the right (center of the circle on your right) and wish the horse to pick up the canter on the right lead, you squeeze hardest with your left leg. This stimulates the horse to give his greatest push with his left (outside) hind leg, and thus he will most naturally lead with his right (inside) leg. Also, since most horses are trained to move away from leg pressure, squeezing with your outside leg helps to signal the horse to make the turn to the right, the direction he will be going as he picks up the canter on the right lead.

Let's say you are asking the horse to pick up the canter on the left lead. The best way to accomplish this on a green horse is to ask for the canter while he is making a circle to the left, and can be most easily accomplished the first few times if he is trotting.

One method of asking for the canter on the left lead is to turn the horse's head very slightly to the outside of the circle (in this case turning it to the right since he is circling to the left). Turning his head to the right and leaning back just a little at the same time as you push hardest with your outside leg will usually result in the horse picking up the canter on the left lead; turning his head to the right hinders his right shoulder action and frees his left one to move forward and lead. Also, pressing with your right (outside) leg tends to push the horse's haunches slightly toward the left (toward the center of the circle) and this helps him naturally lead with his left hind foot as well as stimulating his right hind to push off hardest. As you do this, you should also lean slightly to the left. If you use this method, turning the horse's head to the right when asking for a left lead (tilting his muzzle slightly to the right) be careful not to turn his head too far to the right or you will put him off balance and he will take the right lead instead of the left. This is because he will push his left shoulder forward to catch his balance and the result will be a right lead because he will be at the wrong phase of his stride to take the left lead at that moment.

When giving the horse the signal to pick up the canter from a trot, you should give the horse the signal as his outside front leg is on the ground. This means that the inside hind leg is on the ground also and the outside hind is moving forward, ready to strike the ground. If you give your cue, pressing with your outside leg at this proper phase of his stride, he will be in the correct position to obey you by beginning the canter on the proper lead. He will be able to push off more strongly with his outside hind leg (right in this case, where you are asking for a left lead) that is coming right now to the ground, and thus go into the left lead. If you are posting on the outside diagonal (the right one in this case), you should give the cue to canter, pressing with your outside leg, as you settle down into the saddle (right front leg striking the ground). If you are not posting, but sitting the trot, you should use your outside leg strongly as the horse's inside shoulder moves forward.

Another method of asking the horse to pick up the canter on a specific lead is to lean forward and to the inside, in this case to the left. Leaning to the left and forward and using the reins and legs to indicate left direction at the same time help urge the horse to turn to the left and he will lead more readily with his inside legs (the left ones in this case) as this is his natural tendency when making a circle. If you use this method, don't lean excessively.

Putting the horse into the canter on the right lead: Rider leans to the inside (right) and turns the horse's head to the inside. The rider squeezes with his outside leg to put the horse into the canter on the right lead from a walk. The horse will push off hardest with his outside (left) hind leg

Putting the horse into the canter from a trot, on a turn: The rider leans slightly into the turn and squeezes with his outside leg

Many riders make the mistake of leaning too much, throwing their weight over the horse's shoulder and putting him off balance. Again, if you are taking the canter from a trot, use your cues on the outside diagonal. In some ways it is better to try the canter departure from a sitting trot since you are more in contact with your horse at the sitting trot and able to use your hands and legs more harmoniously. If it is hard for you to tell what phase of his stride the horse is in while you are sitting the trot, it is better to use a posting trot and give your cues on the outside diagonal. Which method you use to put the horse into the canter on a certain lead (whether you turn his head to the outside or the inside) will probably depend on the results you get from the horse. Some respond better to the first method, some more readily to the second. Whichever method you use it is better to do it only slightly and to lean only slightly and concentrate most on using your outside leg strongly, a little behind the girth to push the haunches over toward the center of the circle and to stimulate the outside hind leg. As the horse becomes more proficient at taking the correct lead at your signal, your cues can become more and more refined. As you and the horse work together on leads and communicate better and better, you may need merely to shift your weight a bit, finger a rein, or move your leg slightly and your horse will know what is wanted. Whatever the final cue will depend on you and that particular horse as you work together on the canter and leads.

Putting the horse into the canter on the left lead from the trot.

A

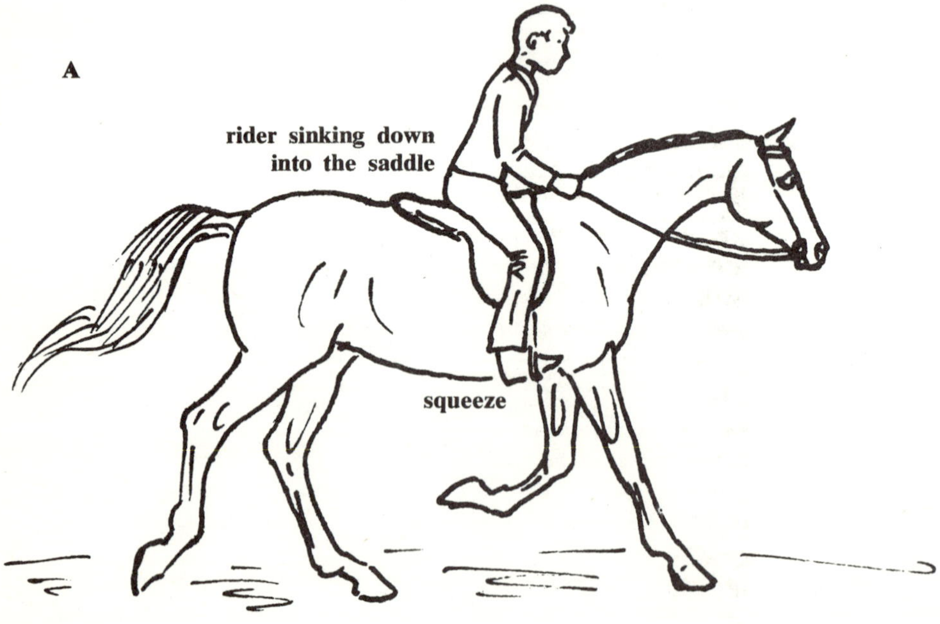

A. The rider can be sitting the trot or posting on the outside diagonal (right in this case). He gives the signal to canter on the left lead, squeezing with his outside (right) leg as the horse's right front foot is on the ground

B

B. The horse begins to change from trot to canter. His head comes up and his left leg reaches forward to lead

C

C. The first real cantering stride. The rider raises his body to cantering position and shortens the reins because the horse's head has come up

For the first few lessons, the horse should be trotting when you put him into the canter, and the best way to make him take a certain lead is to give him the signal to canter when going around a turn or making a circle. Here you can take advantage of his natural tendency to lead with his inside legs. Be sure to make him take the right lead as often as the left lead and vice-versa, so he won't become a one-lead horse. Give him the signal to take the canter on a specified lead at various places in the area where you are working him, so that he won't anticipate your signals always at the same place or at the same turn.

As the horse becomes better trained and more accustomed to taking the correct lead at your signal, your cues can become more and more refined (as mentioned earlier) to where they are almost imperceptible. When giving the cues to canter, it helps to push with your inside pelvic bone as you shift your weight to the inside. Pushing with your pelvic bone increases the horse's impulsion on that side and is better than excessive leaning, for too much leaning will throw him off balance, impeding his leading shoulder. You don't want to hinder the inside shoulder at all, but free it so that it can move forward freely and easily and let that front leg become the leading leg.

When the horse is taking his leads well on cue from the trot in a circle, you can gradually work him into taking either specified lead on the straightaway as well as on a turn, and from a walk or standstill as well as from the trot. At first you may have to push his hindquarters a little to one side or to the other (to the left when taking a left lead, right when taking a right lead), but with practice he should get to where he will depart into the canter when his body is in a straight line. If at any time he becomes confused or won't take a certain lead, go back to putting him into it on a turn. After a while he will associate your signals with the correct lead, especially if you give the signals at the proper phase of his stride and use his natural inclinations to advantage.

None of the lessons you give should last so long that the horse resents them, or you will sour the horse on these maneuvers and he will refuse to respond or may start fighting your cues. If he responds properly and takes a lead successfully, go on and do something else with him. Don't keep drilling him needlessly or he'll soon become tired of it.

When the horse takes the proper leads readily at your cue, you can go one step farther and teach him the "flying change of leads"—the change of leads at the canter, in which the horse must change his lead while he is in that phase of his stride in which all four feet are in the air. Most horses, cantering freely in a pasture, change leads this way naturally, but they usually must be taught to do it with a rider on their backs. Some horses will change leads properly when working cattle, following the turns and dodges of a cow, without training or without cues, but usually must be taught the change of leads for arena work or any other riding in which the horse does not always know when to anticipate the rider's desire to change direction.

The figure eight is one of the easiest movements in which to teach a horse the flying change of leads. It is best to have each loop of the eight be a circle of at least thirty feet or more in diameter. Canter the horse in a large circle on the proper lead, then give him the signal for the other lead as you change directions and begin making the other loop of the eight. It is best to drop down to the trot for a few strides before you make the change, bringing the horse to the trot in the center of the eight and straightening him (you don't want his body awkward and crooked, or even bent a bit as it is while he's circling) then giving him the signal to canter on the opposite lead after you have put him in position to begin the next circle. This way the horse will pick up the proper lead readily, because you have already taught him to take his leads from the trot, and a few trotting strides will give him time to get his feet ready to go into the new lead.

The more you practice, the fewer trotting strides you will need, until you can sense that the horse is ready to try the flying change. When you try the flying change, canter one circle of the eight and just as you begin the next circle, check the horse for an instant with the reins and then immediately give him the cues for the new lead. He should make the change of leads in mid-air. Don't prolong the lessons. Flying changes are tiring to a horse, if overdone.

In teaching the flying change you may find it helpful to keep the horse's head bent just slightly in the direction of the old circle for an instant before you make the shift to the new direction. Use your legs strongly as the horse makes a stride, using your outside leg most strongly behind the girth, signaling for the next lead. Use your legs as you are farthest down in the saddle (at the phase of the stride in which your weight is pressed down hardest in the stirrups and your body is lowest), and be careful not to lean too much in indicating the shift of direction; you don't want to put too much weight over the side that is to become the leading shoulder. In time the horse will learn to respond readily when you ask for a flying change, with only a minimum of cues.

When the horse is trained to this point he should be able to change leads at your signal whether he is making a circle or cantering in a straight line. After he is this far along in his training he will be more useful for any kind of fast work, whether as a stock horse, or in fast reining or any other type of exhibition or contest work. One of the main purposes in training a horse is to have him always in control, working willingly as you indicate, so that the two of you are working together instead of against one another. Your horse can't do a good job until he is always where you want him, balanced and with his feet under him properly, until you and he work together as a unit, and this can be accomplished only after you, as his rider, have control of his cantering and his leads.

A correct stop is almost as important as correct movement. We'll touch briefly on the stop in this chapter even though it is not a gait. When the horse is asked to stop while traveling, he uses his front heels as brakes while he straightens his front legs and plants one or both hind feet on the ground more forward of his normal standing position. He flexes his hocks and stifles to change the center of gravity of his body, bringing some of the weight back to his hind feet. When the horse stops from a fast gait, such as the gallop, or when he is forced to stop unexpectedly, his weight must be transferred back to the hind legs before he can regain control of his front legs. If he didn't shift his weight back, he would somersault over his locked front legs because he is front-heavy and because the momentum of moving forward is still propelling his body in a forward direction. When he stops from fast speeds, the horse must first crouch down on his hind legs with his stifles, hocks and fetlock joints flexed. Thus his front legs are extended in front of him and his hocks give and cushion the stop, acting as brakes until the momentum of his speed has been reduced and his weight can settle back down to his balanced standing position. When you see a horse make a smooth, well-balanced stop, you'll notice that he raises his head (but his chin is down; if his nose is in the air, his rider was too harsh with the reins), flexes his hocks and crouches on his hindquarters a little to shift his weight to his hind feet and thus come to a stop. If he doesn't place his hind legs well underneath himself to take some of the weight and shift it back, he bounces to a stiff-legged stop on his locked front legs.

Most horses are fairly good athletes; they can handle themselves with agility, dexterity, and good balance when they are running free. There are exceptions of course, for some horses are clumsier than others, but many horses that are good athletes when running loose do not handle themselves nearly as well with a rider on their backs, especially young horses in training. This is partly because the added weight of the rider makes the balance different, and also because the horse does not always know exactly what is going to happen next and may not be prepared for a sudden stop or change of direction or some other new maneuver that the rider asks of him. The horse may be at the wrong phase of his stride to execute the new maneuver immediately or properly. This is why a rider, and especially a trainer, needs to understand the way a horse moves so that he can give his horse the proper signals at the proper time for best results.

For this same reason, the horse needs a lot of work and training at all gaits in order to develop dexterity and balance with a rider, and to gain the ability to adjust his stride to the situation. The horse and rider, with experience and practice, should both greatly improve their abilities. The horse learns to travel well with a rider on his back, and the rider learns to collect the horse and to give the proper signals at the right times for the best response from the horse.

23

STAKING OUT

TEACHING A HORSE TO STAKE OUT FOR GRAZING IS NOT DIFFICULT, and staking can be a very handy thing for a horse to know about. Here on our ranch over the years we have taught several of our horses to stake out, usually at some early stage in his training, for now and then a situation comes up in which we want to be able to stake him out—on an overnight trip into the mountains, a hunting trip, or just to be able to use him as a super "lawn mower" occasionally to clean up the tall grass in our lane or around the barn yard.

Staking can be handy if you ever travel with your horse, even to rodeos or horse shows, for sometimes there just isn't an available stall or pen. If your horse has been trained to stake out, there is never much of a problem concerning space to keep him. And staking is especially convenient if you take your horse on overnight trail rides or camping trips.

Let's look at a few of the advantages of staking out. If you do camp out or travel with your horse in the mountains, it is inconvenient to leave the horse tied up all night because he needs feed—and it can be rather unhandy to pack all that feed with you. You'll prefer to keep the horse confined in some way during the night, because if you just turn him loose to graze you can't really be sure whether or not you'll still have a horse in the morning.

Hobbles are handy when you are in the back country, but hobbles have a few disadvantages. A horse can become accustomed to hobbles and learn to travel quite well in them—and be a long way off in the morning. A group of hobbled horses can get to running and travel rather swiftly after they get the hang of moving in hobbles. And, it is rather easy for a horse to travel downhill in hobbles, but not so easy for him to go back up the hill. Come morning you might have to look for him down at the bottom of a mountain slope. If hobbles are used often on a horse they can make his pasterns sore and stiff. About the best way to camp out with a horse and make sure he'll still be there in the morning—and also have a belly full of grass from the night's grazing—is to stake him out.

But if you ever plan to stake your horse, it's a good idea to give him several lessons first. And during those first staking lessons, you should stay nearby—to be on hand to get him out of a bad tangle if he happens to get fouled up in the rope and down on the ground. Most horses learn pretty fast. After they've gotten their feet in the rope a time or two, they'll watch where they put their feet and begin to graze systematically in a circle to avoid getting mixed up in the rope.

Staking can be good preliminary training for a roping horse, for the horse soon gets used to having a rope touch various parts of his body and learns that it isn't going to hurt him even when he gets fouled up in it. Staking can also be good basic training for practically any horse, if you want to use it that way. Even the nervous and excitable horse can usually grow a little wiser and calmer after a few tangles in the rope (if stake training is done carefully and tactfully) and learn where to put his feet. Most horses get pretty good about staying calm and getting themselves out of a potential tangle if they do start to get a foot over the rope.

When Lynn and I were first married, we spent our first summer on a dairy farm in southern Idaho before we started buying our present ranch. We had two horses with us at the time—Lynn's Morgan-Saddlebred mare and my three-year-old Arabian-Thoroughbred filly, but we had no satisfactory place to keep them. The rest of my horses were still at my father's ranch, but I wanted my filly with me because I was starting her training. The dairy farm we were renting had very little pasture area; the one small pasture we did manage to fix up that summer was for our dairy calves and heifers.

So we kept the horses in a small corral at night and decided to stake them along the ditch banks in the fields during the daytime, where there would be grass for them to eat. We didn't want to keep them continually confined in the small corral, for then they would have to be fed hay. Hay was rather scarce on our farm that summer, and so was the money to buy any. Out of necessity, our horses learned to stake.

I had never staked a horse before in my life and neither had Lynn. Somehow the job fell to me, and the two horses and I learned about staking together. In the years since that time, there have been several occasions on which we were glad we had taught those two horses to stake out, so we've trained a few of our other horses to stake, too.

Some horse trainers and packers halter break a green horse by staking him, tying him with a large soft rope to a big log or some other kind of heavy weight such as tractor tires. When the green horse gets through fighting the rope and being tangled up a few times he is pretty well halter broken, and broke to stake out as well. But this method can be pretty rough, especially on a high-strung horse. It's usually better to teach a horse to stake out after he's already thoroughly halter broken and handled a little. He won't be so apt to hurt himself.

Once a horse has learned to stake and knows what it's all about, it's best to stake him by a front foot. He won't get tangled up so often that way, and

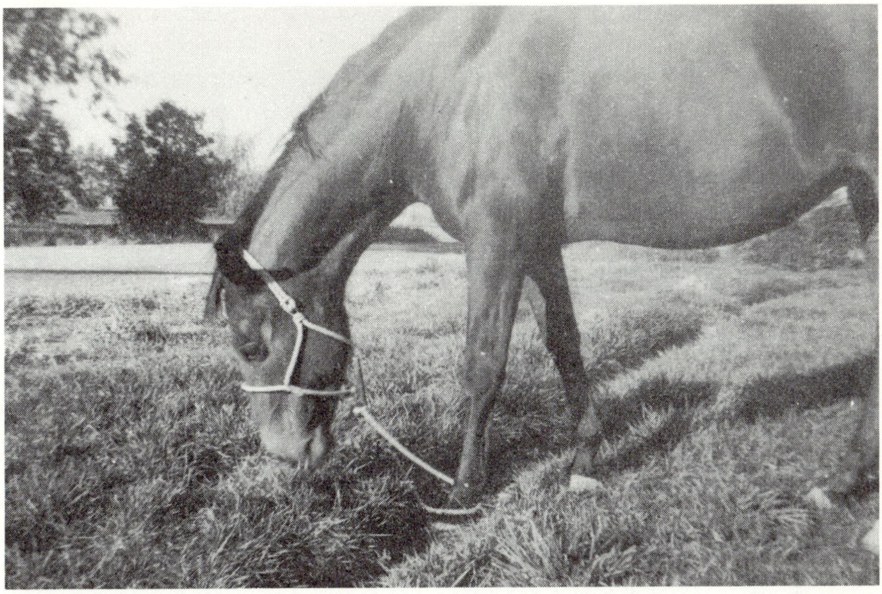

Lynn's Saddlebred-Morgan mare on our dairy farm at Gooding, Idaho, learning to stake out

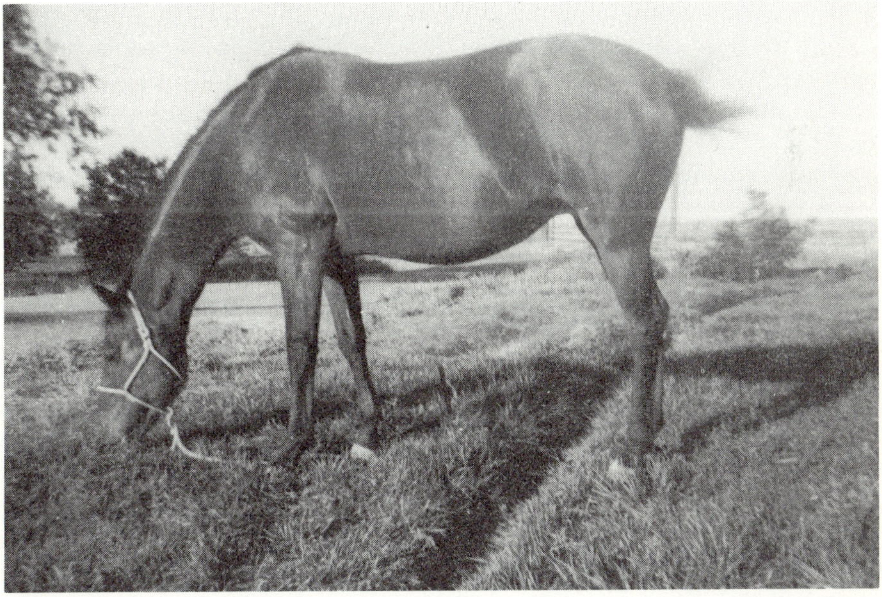

She stays calm when she gets her foot caught in the rope, holding her head down, and discovers that by moving the proper foot her head is soon free again to continue grazing

he'll have a lot less trouble. But start him tied by the halter; it's much easier for him to learn what it's all about.

Some horses are quite spooky about ropes and you might hesitate to stake them, but there are sometimes ways to get around their fear before you actually try staking them. I had serious doubts about Nikki, my Anglo-Arab filly, for she was quite high-strung and nervous about strange things and I had visions of an upside-down, thoroughly entangled, rope-burned, hysterical filly. I probably would never have considered teaching her to stake out if it hadn't been for the fact that staking was about the only way to feed her that summer. Necessity can teach us a great deal.

With a little thought I discovered that there might be ways to get around a rope-shy horse's fears. A few sacking out lessons with a rope can help, rubbing parts of the horse's body with the rope, letting the rope move around the horse's legs, and so on (very gently and tactfully at first) until the horse realizes that the rope isn't going to hurt him.

And with Nikki, I found that it helped to let her graze a few times in the yard just dragging a rope attached to her halter, so that she could get used to having a rope underfoot and around her legs and learn not to step on the rope. After dragging a rope for several sessions, the horse learns that when he steps on the rope and it holds his head down, it's really no cause for great alarm or excitement (some horses seem to have a sort of claustrophobia at having their heads caught down low like that) and that by picking up the proper foot his head will be free again.

It's not impossible to start a horse staked just by a front foot instead of giving the first lessons with a halter, but if you decide to do it this way you'll want to prepare the horse for it gradually by letting him get accustomed to having his foot tied to something. A few short lesson with hobbles can help him get used to having his front feet restrained. After he's used to the hobbles he can be put in a corral or small enclosure (or preferably, a confined area with grass where he can graze) with a short length of rope fastened to one front pastern, preferably by a foot strap. As he moves about, he learns not to resist or fight when he steps on the rope and inhibits his front foot; he soon learns to maneuver his feet accordingly.

When you actually stake the horse, you'll need a large soft rope at least an inch in diameter. It won't burn him as badly as a small rope or hard-twist rope. The rope should have two swivels in it to keep it from twisting up as the horse moves around and around the stake. If you are staking the horse by the foot, be sure to use something around his pastern that won't cut into it or rope burn. A leather strap with some kind of padding on the inside works quite well and you can put a ring around the strap or through the buckle of it to fasten the stake rope to it. Sheepskin lining or foam rubber inside the strap makes good padding.

The stake can be anything that you can tie the horse to and drive in the ground deep enough to hold him—a sharpened green stick, a sharpened iron bar, etc. It's best to have a stake with a flattened top that will not injure the horse

Lynn and Khamette—getting the mare acquainted with the rope, letting her feel it around her legs, so she will discover that it will not hurt her and that it is nothing to fear

Foot strap, chain and swivel, and stake rope

if he bumps it or steps on it. The stake should be driven down fairly close to the ground so that the horse will be less likely to injure himself upon it.

If the stake is driven in well, the horse won't be able to pull it out because he won't be pulling directly upward; the stake rope is long enough that even if the horse is staked by the halter and pulls on his rope, his pulls on the stake are generally sideways and not straight up.

The first few times you stake him, do it in a flat grassy place that has no rocks or sharp objects to hurt him if he should happen to get tangled in the rope and throw himself down on the ground. It might also be a good idea to keep him hungry for a few hours before his first real staking; then he will probable be too busy grazing to try and get loose.

If you have started the horse staked out by the halter, change to staking him by a front foot after he knows what it is all about. It is much safer to stake a horse by the foot if he is to be staked out for long periods of time. A horse staked by the halter can get into trouble. If he tries to scratch an ear with a hind foot he can get his foot caught in the halter, especially if he is shod, and he can get into a really bad situation. I know of one horse that had to be destroyed because he got a hind shoe caught in his halter and struggled and injured himself very badly.

Staking a horse by the halter is generally inconvenient for the horse because he is almost sure to step on the rope with a front foot, inhibiting his head, or get the rope around a hind foot, and sometimes he can get into a problem that is hard to get out of because his head is caught and held down low by the entangled rope. When staked by a front foot, the horse's stake rope doesn't come up off the ground each time he raises his head, and he is less apt to get his hind legs entangled as he turns. Staking by a front foot saves a lot of trouble.

After your horse has learned to stake, you'll find camping or traveling with him less complicated and quite a bit easier. You won't have the problem of having to pack along a lot of feed for him, the problem of wondering whether or not you'll be able to find a place to keep him, nor the worry of whether or not you'll be able to find him in the morning. All you'll need is an adequate rope and foot strap, a stake and a grassy place, and your horse will appreciate the chance to fill his belly instead of being tied up all night.

24

TRAILER TRAINING

ALMOST EVERY HORSE OWNER AT SOME TIME OR ANOTHER has cause to transport his horse—to a horse show, to a trail ride, to a fair, to be bred or to a new home. It helps if the horse is easy to load into a trailer. All too often the frustrated horseman resorts to force, attempting to pull the horse into the trailer, either by hand or with a system of pulleys and rump ropes, or attempts to load the horse by making it so uncomfortable for the horse outside the trailer by whipping him (or backing the trailer up to the barn door and proceeding to annoy the horse with the whip until he attempts to "escape" into the trailer because it is the only place to go to get away from the whip), that he will enter the trailer.

These methods usually get the job done eventually, and sometimes succeed in making a horse easy to trailer load, but all too often the use of force backfires, making the horse more fearful of the whole procedure. It is better for the horse's training to use gentler methods, and the end result is more desirable; the horse trailer loads willingly and without fear.

Let me clarify this somewhat. There is a difference between training the young horse or the inexperienced horse to trailer load and retraining the spoiled horse—the one that has already been made stubborn or fearful by improper loading procedures. The young or inexperienced horse is usually no problem to teach—if the trainer has lots of patience and time—and is often much easier than the older, spoiled horse. Let's talk about training the young horse to trailer load, producing a well-mannered horse that loads easily and willingly, and then take a quick look at loading (and retraining) the problem horse.

When the horse is a foal, he can be trailered a time or two with his mother, gaining confidence from her. If the early trailering is always a good experience (no frightening incidents, no fighting or struggling) the young horse will be off to a good start on his trailer loading lessons.

But often the horse is a weanling or older before any trailer training is attempted. This is fine, for when the horse is this age you can easily lure him into the trailer with good things to eat. Park the trailer in his corral or pasture and block up the wheels and tongue securely so that it will be steady and unmoving as the horse goes in and out of it. If it is a single-axle trailer you can block the wheels and tie the tongue securely to a fence, next to a sturdy post. Now you can feed the horse in the trailer, letting him take several days to overcome his fear of it if need be.

Some horses, after watching you put the grain or hay in the trailer manger, will go right in without much hesitation. Others will start in, back out, start in again, and finally stay in after they discover that there is really nothing in there that is going to hurt them. Others will not go in at all the first few times and you'll have to place the feed on the floor of the trailer where the horse can reach it without actually getting into the trailer. Let the horse become accustomed to the trailer in his own good time. If he is timid or suspicious and you have to feed him on the floor of the trailer, you can then gradually put the feed farther back so that he will have to really reach for it, then have to put his front feet in, and finally, with the feed in the manger, have to get clear in. The main thing is that the horse is learning that the trailer will not hurt him, that he is not being forced in any way, and that he can go in and out with ease.

Feeding a horse in a trailer for several days or even weeks is about the best way to get him accustomed to getting in and out of a trailer with no fuss or fight. But sometimes the young horse's first experience with a trailer is when he has to be loaded and taken somewhere. Perhaps you have no trailer of your own, or no trailer not in use that can be left in his corral. If you have to combine his first experiences in a trailer with actually taking him somewhere, just make sure you have a little extra time in which to try to load him, and be sure that you have a lot of patience and understanding as you go about it, for you want to make sure that he will have no bad experiences that might make future loadings difficult.

Every horse is somewhat fearful of a trailer the first time he is loaded, because of his natural fear of strange things. If there is a ramp to walk up, he is instinctively suspicious, for the artificial footing is not compatible at all with his idea of safety. If he has never been in a trailer before, or seen one at close quarters, the enclosed trailer is also viewed with suspicion. It may be rather dark inside the trailer, and instinctively the horse hesitates to go into dim places. When loading a horse, always try to put the trailer in a place so that the sun shines into it, giving the horse a better view inside it. And if the trailer is placed in this way, the sun will not be shining into the horse's eyes. Another problem the horse encounters when approaching a trailer for the first time is the low roof. This takes a little getting used to, for most horses are afraid of low roofs, afraid of striking their heads. Do not use a low-roofed Quarter Horse trailer for a tall Thoroughbred. Some horses get a bad case of claustrophobia being hemmed in by a low roof and literally climb the walls. So, the average horse, when faced with the problem of entering a trailer for the first time, is somewhat suspicious and fearful, and has no confidence at all in this new contraption.

Obviously then, the trainer's first task is to gain the horse's confidence, to disperse his fears. All too often the opposite is done, by loud words, whips and ropes, making the horse all the more certain that his suspicions and fears were well founded. Some people may argue that the patient and gentle approach takes too long, but what are 10 or 20 minutes extra during that first crucial loading when compared with the value of producing a well trained horse that will load easily in the future. I have seen several nervous, sensitive horses that took more than an hour to be put into a trailer by forceful methods. If successful loading and a calm horse are the goals (to say nothing of an uninjured horse and/or trainer), it is well worth the while to be patient.

In the first place, the trainer must persuade the horse that he is patient, so that the horse will come to have confidence in him. It is usually better to work alone with the horse, with no extra people to distract the horse, and especially with no one standing behind the horse to make him nervous or afraid. If he feels "boxed in" by someone standing behind him, he may want to get clear away from the trailer.

Lead the horse up to the trailer, leading him on a loose rein, the way you have taught him to lead—walking freely beside you. As the horse gets near the trailer he may stop, so you stop too, giving him time to look things over. The horse may decide that he doesn't like the looks of the situation, and may back up a step or two. He knows very well (or at least has a suspicion) that he is meant to go into that contraption by the way you led him up to it, and since he is wary and somewhat fearful of this development, he has decided he will *not* do such a thing; he won't go stepping into the unknown.

Here is where your patience and persuasion will pay off. If the horse decides to step back a few steps, let him, always being careful not to pull on the horse, putting no tension on the halter rope. Pulling on the horse or trying to hold him in place when he steps back only makes the horse more afraid. If you have ever tried to pull a horse to make him approach or enter a trailer, or have ever watched anyone do it, you will know what I mean. Pulling on the horse only makes him pull back all the harder. He is afraid to be forced into something he does not trust. Tension on the halter rope as he approaches the trailer only makes him think he is being forced all too close to something dangerous. He may become nervous or afraid. If he is on a loose rein he won't be so fearful, for potential danger that he can jump away from if need be is not nearly so bad as potential danger when he is being held tightly.

So if you understand how the horse thinks, you will be patient and let the horse step back if he wishes, moving back with him, talking to him pleasantly and letting him know that there is no cause for fear. Pat the horse and praise him; help him relax. Soon he will stand quietly in front of the trailer without fear, having time to look around and think things over. Let him smell the trailer if he wishes; let him satisfy his curiosity and overcome his apprehensions. When you feel that the horse is relaxed and ready, encourage him to take another step or two toward the trailer, or up the ramp if there is one. One step at a time is sufficient, and it does not matter if the horse decides to back up again a time or two.

Place the trailer so that the sun shines into it

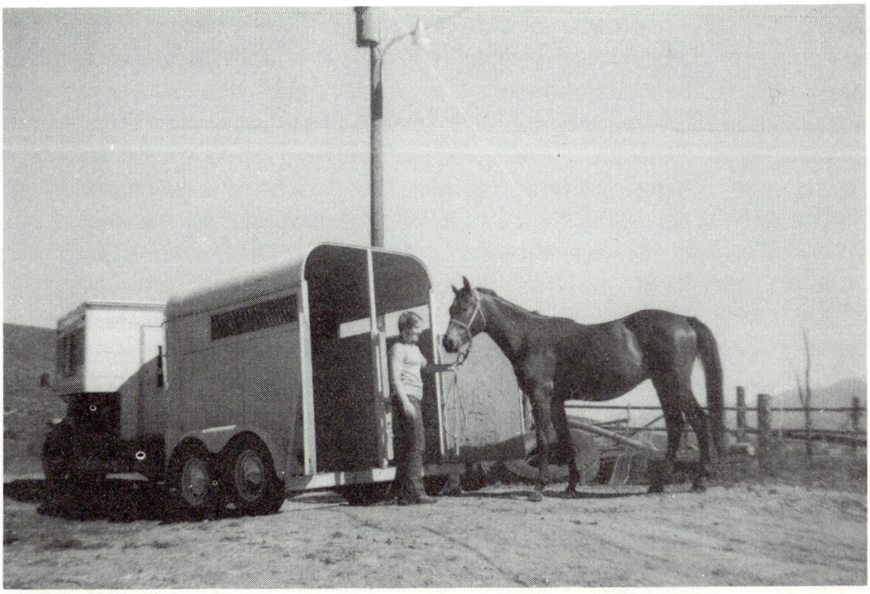

Lead the horse up to the trailer and give him time to look things over and think about the situation

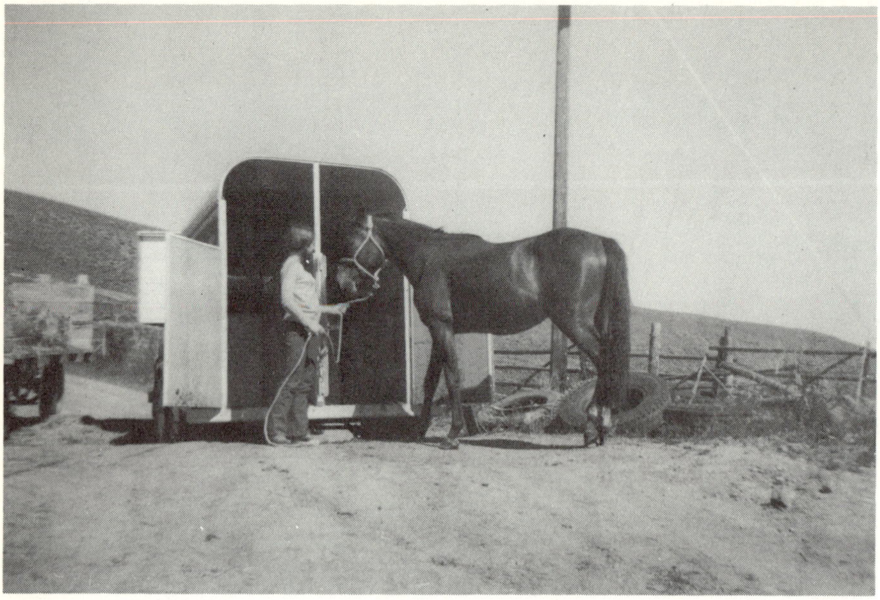

After he has had a moment to think things over, with the encouragement of his handler, he begins to feel that entering this contraption may not be so bad after all

With patient tact, you are changing the horse's mind, convincing him that the trailer is not a thing to be afraid of and convincing him that he may actually decide to enter it after all. A good trainer can sense the horse's change of mood and will know when the horse begins to reconsider. This indication of the horse's changing attitude may be no more than a mere twitch of a muscle or a slight stretching forward of his neck to take a new look inside the trailer—just the beginning of new interest that shows you that the horse is beginning to reconsider.

The idea of moving forward into that strange contraption is now coming into the realm of possibility in the horse's mind. From that moment on, the battle is won so long as you continue to be patient and tactful. Give the horse a little more time, a lot of encouraging words, and try to relax him and set his mind at ease. Then begin again, step by step, to encourage the horse into the trailer. You may use the whip, just as you used it in teaching the horse to lead. As long as the horse does not fear the whip, but accepts it as sort of a third hand of yours, it can be used. If the horse decides to back out of the trailer a time or two, let him. He is just trying out this new situation and making sure that he really can accept it with safety. If you keep your patience, and if there is a little grain in the trailer manger for the horse to eat, he will begin to relax enough to stay inside the trailer without fear, and his rump chain can be snapped in place and his halter secured in front. To insure good loadings in the future, the horse can be led in and out of the trailer several

In we go!

Let him back out a time or two if he decides to. Give him all the time he needs to decide that the trailer is a safe place to be

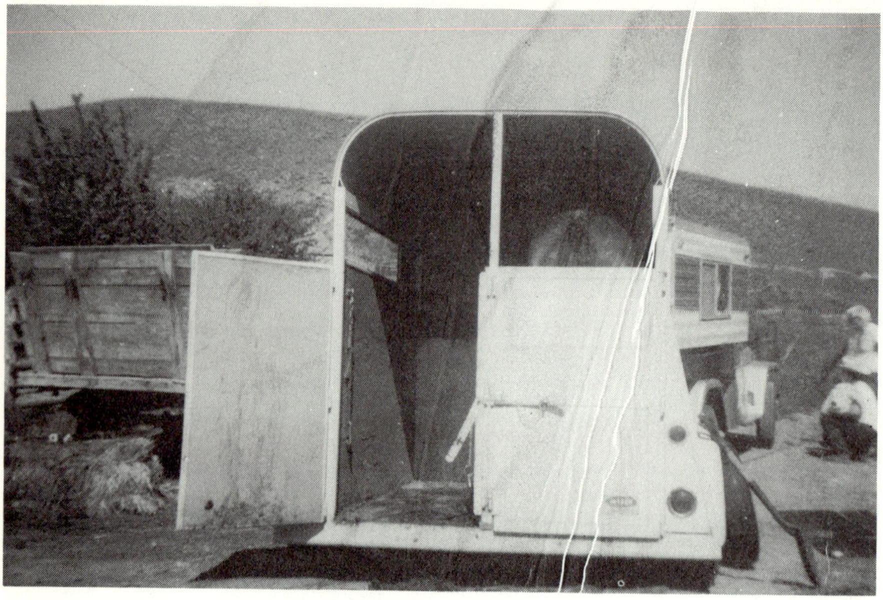

All loaded and ready to go

times, and the lesson repeated daily for several days, the horse always getting a reward of grain when he is in the trailer. Just a few lessons of this kind will make him very willing to load into a trailer for the rest of his life, so long as he never has any frightening or hurtful experiences while in a trailer.

The same procedure can be used on a horse that has been spoiled and made afraid of trailers by poor handling. Only this kind of horse may take longer, and may need a few preliminary lessons with the whip as if teaching him to lead, to acquaint him with it properly and to let him know that it is nothing to fear. It's usually best if you work alone. The horse can understand and have confidence in the one person handling him; he will only become confused and fearful if he is confronted by several persons at once.

After you get the horse loaded and you are ready to take him somewhere, remember that you want to make his first experiences in a trailer pleasant ones. If at all possible, take him on a few short "practice" drives before you have to make a real trip with him. Give him some hay and/or grain to pick at and preoccupy him as he travels, and give him a brief ride. These first few trips are very important, for they will help build up his confidence if he is trailered slowly and carefully, or they can shatter his confidence (if he is jostled about) and make him more trailer shy than ever. Plan your first few trailer trips over good, smooth roads, and limit your top speed to about forty miles per hour. Start as slowly and smoothly as possible and build up your speed gradually. Slow down on the turns (right-angle corners should be taken at about five miles per hour) and be sure that the trailer is completely around the corner before you begin to accelerate. All too often a driver begins to accelerate

This filly is hesitant to enter the trailer, even though her stablemate is already loaded

She is accustomed to a "rump rope" from her earliest leading lessons as a foal, so the rope is placed around her buttocks

With a little encouragement, in she goes

after the car or pickup has made the turn, not realizing or stopping to think that the trailer following is still making the corner. Take curves very slowly. The horse has to learn how to balance and brace himself around the turns and you don't want him to become upset and start scrambling because you took a corner too fast.

Horses that climb or scramble are usually horses that like to spread their legs wide to brace themselves on corners or curves. In many trailers, especially in a two-horse (side-by-side) trailer with a dividing partition, there is not quite enough room to spread the legs as wide apart as some horses want to. Thus when a corner is taken too fast, the horse tries to brace himself and can't, so ends up leaning against the wall or partition and scrambling with his feet on the other side, literally climbing the wall. This can get to be a very bad habit, and for some horses that really feel confined and unable to brace themselves properly on a turn, it can get to be a very serious problem. Some horses get to where they throw violent fits around the corners. This kind of horse may ride beautifully in a truck or in a stock trailer where there is plenty of room to move about and to brace himself as he wishes, but becomes a problem when trailered in a two-horse trailer with partition, where he is more confined. This sort of horse may kick and struggle, tremble and sweat, just because he feels too hemmed in. Some horses travel quietly on the straightaway and scramble only on the corners.

One solution is to remove the divider in the two-horse trailer. If you haul two horses and cannot remove the divider, you can remove the lower part of it, or replace it with a partition that does not come clear to the floor, giving the horse extra foot room to spread his legs and brace himself. Some horses like their extra foot room to the right, others always to the left. Whatever the case may be, load them on the side of the trailer that gives them their extra foot room in the middle, under the partial partition. This will often solve the problem and enable the horse to travel comfortably in a two-horse trailer.

For a young or inexperienced horse, it is sometimes a good idea to load an older, experienced stablemate first when first introducing him to the trailer. If he sees the stablemate go right into the trailer without hesitation and begin eating grain, he will begin to think that this contraption is not so bad after all, and he may begin to think about hopping right in there wanting some of that grain for himself. And the presence of a calm experienced horse alongside him on his first few rides can be very reassuring. He may travel much better those first times if you take along his older, experienced stablemate.

The older, spoiled horse that refuses to trailer load is much more of a problem than the young and inexperienced horse. The spoiled horse may have been in an accident, or may have had poor handling when being trailer trained. At any rate, he does *not* want to enter a trailer, for it means frightening things to him. Once a horse gets it into his head that the trailer is a bad, bad thing, it can be very hard to change his mind and he will fight unreasonably sometimes to keep from going into a trailer.

There are many devices, gimmicks and theories as to how to get the problem horse into a trailer. Some will work on some horses and not on others. The success of a trailer loading method depends in part on the individual horse and what his problem is, and also will depend in part on how he reacts to the method used. Some horses will accept forceful handling more readily than others. Some will put up more of a fight if handled forcefully, knocking themselves about and perhaps injuring themselves or their handlers. You've got to know the horse quite well and be able to out-think him—able to work with his individual personality and idiosyncrasies instead of against them. Some horses are partly bluff. They refuse to enter a trailer, but start to get serious with them (show them a whip or put a rump rope around the hindquarters) and they go right in. Others will just get more frantic at the sight of a whip or the use of ropes.

It's not worth fighting a horse. You may get him loaded, but loaded under protest; the struggle that got him in there is just one more bad experience that might make him hate the sight of a trailer the next time. Sometimes a horse can be forcibly loaded (provided he is not hurt in any way) and become easier to load as time goes on—but only if his experience while in the trailer is good. In other words, if when he is in the trailer he gets good things to eat, and he is always trailered carefully, never frightened by poor driving or rough roads taken too fast, bad corners, etc., he may actually gain confidence. But this will depend in part upon his personality.

It is usually best to start over from the very beginning with a spoiled horse and feed him in the trailer for several days or weeks until he begins to gain confidence and will hop right in very readily on his own. Because he is fed in the trailer, he begins to associate it with good things instead of bad things, and because he is not being hauled anywhere he gradually loses his fear. Then after he has accepted the trailer and will get into it readily, you can begin taking him on very short rides, gradually increasing them if all goes well. A little bit of time spent in retraining the horse to be a good passenger is worth taking. What are a few days or weeks of a few minutes feeding daily in the trailer, when compared with the entire future loading ability of that horse? Indeed it is worth taking a few moments' time each day for however long you think he needs it, if it will save the struggles of loading an unwilling horse (which often take up much more time) or perhaps the injury of that horse or yourself in the process of that struggle.

Just a few more comments on horse trailers and trailering in general: Make sure your trailer is safe. I had a friend whose horse refused to load on her usual side of the trailer; the horse was finally loaded on the other side and my friend began the trip, only to soon discover that one of the tires was flat on the trailer. The horse had sensed this as she first put a foot in on that side, and refused to get in. Before you load the horse, make sure everything on the trailer is working properly. Check the tires. Check the springs, the door latches, and so on. A broken spring can give the horse a bad ride. A wooden floor may be cracked or rotted to the point where a horse could put a foot through it. Make sure your floor is sturdy and in good shape, double thickness preferably. And then use a nonslip rubber mat of some kind that gives the horse good footing. A wooden or metal floor can become quite slippery. A thick rubber mat gives better traction and is also easy to clean.

Tie the horse short, so he can't turn his head back around or get his head down between his front legs. And always use a butt strap. A butt strap as well as a door is a good safety precaution; if the tailgate or door comes unlatched, the horse cannot back or fall out. Make sure the horse is comfortable. If you are traveling with him in cold weather, blanket him and use an enclosed trailer or put on canvas curtains to cut out the draft and snow. In warm weather you might want to use insect repellent on the horse. You don't want him bothered by flies that he cannot cope with in his confined condition. Give him feed to chew on, but don't stuff his manger so full of hay that it gets in his way. A scoop of grain is all right to encourage him in, but a flake of hay will keep him busier longer and happier on a long trip.

Use a trailer that is large enough (and tall enough) for your horse. Give him plenty of head room. If his ears hit the ceiling or if he has to keep his neck bent, he won't be very comfortable and may feel too hemmed in. If a horse does suffer from claustrophobia, use an open stock trailer or a big trailer with lots of head room and window space so that he can look out.

If you are on a long trip, stop periodically to give the horse feed and water, and take him out to stretch his legs every six hours. Some horses will not urinate in a trailer and will become quite uncomfortable if left in for too long a time. On a long trip you'll be glad to have a horse that is easy to trailer load, for you'll want to unload and load him several times. You don't want a horse that you are afraid to let out of the trailer for fear you'll never get him back in. Introduce each horse slowly and gently to trailer traveling and he'll be a willing traveler ever after.

25

PRACTICAL DRESSAGE

WHAT IS DRESSAGE? THIS IS A WORD THAT HAS OFTEN BEEN misunderstood. And how useful can dressage be to the average rider, the pleasure rider, or the western rider? In this chapter we shall attempt to answer both questions.

To many western riders dressage is just a lot of foolishness that is only good for show ring exhibitions. In one sense this is partly true, because some kinds of riding and training, such as the type that produces the circus horse and his various acts and maneuvers, have been mistakenly classified as dressage by the novice. Because of this false notion of what constitutes dressage, the whole idea of dressage is often discarded by many horsemen as being impractical and ridiculous for their own purposes. This, in most cases, is unfortunate, for almost any horse, in whatever role he performs, can be made better by applying some of the elementary precepts of dressage.

The word dressage, literally translated, means "training," but it means a kind of training that goes beyond the simple stages of breaking a horse and making him willing to carry a person around. Dressage means specialized training—the kind of consistent and understanding horsemanship that is necessary to develop perfect obedience in the horse and perfect lightness, balance and agility, so that the horse can respond best to the rider's cues. It means the developing of the horse's understanding and ability, so that he becomes more maneuverable with the rider on his back, able to perform as the rider wishes him to.

The use of the word dressage in the English language is actually quite recent. It is a French word that has no exact English equivalent in meaning. This is perhaps one reason why there has been so much misunderstanding concerning the meaning of the word. The French word dressage is derived from the verb "dresser," which is used when speaking of animals that are trainable. The word means teaching or schooling an animal, but it implies teaching an animal certain types of actions (in other words, training means teaching the physical body—teaching the animal to perform certain maneuvers) while "dressage" puts more emphasis on the mental capabilities of the animal, or working with the animal through its understanding.

Therefore the term dressage is not used when speaking of the process of gentling or breaking a horse or teaching him the essentials of carrying a rider on his back (starting, stopping, turning, moving at various gaits, and so on), in which the horse becomes useful to man. Dressage is therefore the art of improving the horse beyond this stage, making him more agile, more willing, easier to control and to communicate with, more pleasant to ride, and more graceful and balanced in his actions.

To be able to school a horse to this degree, the rider must have a certain amount of talent and natural ability, but he also needs determination and the willingness to put in a great deal of hard work. And of course he needs a thorough knowledge of dressage. But this knowledge, no matter how much a person tries to learn through reading, is gained primarily through experience, through working with horses and discovering for oneself what it is actually all about. Reading about dressage is very helpful, but a person must then attempt to apply this as he rides and schools the horse, coming to realize in reality what dressage is all about. A good horse, in the hands of a good trainer, can keep improving, coming a shade closer to that unattainable ideal of perfection, developing a greater harmony between horse and rider.

The finely schooled horse is light and handles himself with excellent balance, gracefulness and good action. Achieving this kind of lightness and agility in a horse goes on beyond the realm of practical, ordinary training of a horse. The ordinary horseman can do well to profit from this highest art of riding, for it can provide the guidance that is needed by every horseman who wants to improve his horse and his horsemanship. Even though dressage itself begins where ordinary types of training and riding leave off, it is dressage that can provide the most valuable principles in improving our methods of handling and training horses, even for the person who never goes on into dressage seriously.

Every good horse, whether he is a pleasure horse, a police horse, a show jumper or a cow horse, to be effective in his role of work and to be able to perform above the low standard of just plain usefulness needs a little dressage. Dressage is a progressive series of gymnastic exercises for the horse which will gradually develop impulsion, suppleness and balance, and develop his natural gaits to their fullest extent. Any trainer who can develop the fullest potential of an athletic horse such as a cutting horse, reined stock horse, or performing horse in any field requiring collection, good balance and precision, has used a bit of dressage, even though he may not call it that. That good trainer keeps his horse working and improving because he develops the horse slowly and gradually, and searches out and corrects any faults or problems as they develop. He takes the time to develop the horse as fully as possible, not souring him or burning him out. If more trainers used a little dressage, along with the time and patience required, they would turn out better horses—better athletes that could do a variety of things well and willingly.

Dressage training is progressive; one must start from the bottom and work up. Each new exercise can only be done correctly when the foundation has been properly laid by the previous one. You can't collect a horse until he has impulsion. Gymnastic exercises in dressage are designed to develop muscles, poise, dexterity and balance. Some dressage exercises are very beautiful to watch. Good dressage training is always very gradual, for nothing good or lasting can be done in a hurry. Some trainers who turn out "thirty-day wonders" use gimmicks and other methods to rush a horse through his training period to produce a finished cutting horse or reining horse or barrel horse or whatever, but more often than not the horse does not stay trained and develops bad habits or goes sour. One cannot really skip the basics in horse training (developing a good solid foundation of training in which the horse learns gradually, building on what he has learned earlier) and get away with it for long. Sooner or later the rushed horse regresses or goes sour.

The horseman's first goal in training a green pupil is to strengthen the horse's muscles (especially of the back and neck) so that the horse can perform easily while carrying the additional weight of the rider. This is done during the period when the horse is being rudimentarily trained to a stage of plain usefulness, learning to obey signals and to stop, start and turn as the rider desires. The horse is developed and strengthened gradually, first at the walk—doing special exercises and traveling out across country. The horse is encouraged to extend his head and neck and push his hind quarters well underneath himself, reaching strongly with the hind feet. Then this is also done at the trot—a strong trot rather than a weak jog. The rider tries to develop in the horse a strong trot at rather slow cadence in which the horse takes long strides. In other words, the trot is not fast, but the horse is moving vigorously, with good impulsion and long strides. His neck is extended but his head is nearly vertical.

Great impulsion is one characteristic of a good dressage horse. The horse gives one the impression of being a coiled spring, with much repressed energy that can be directed in any direction at any moment by the rider. Impulsion is energy, and a special kind of quality that is present when the horse moves with a good degree of animation. This quality in the horse is only naturally present when he is excited (as when prancing vigorously out in his pasture) and therefore it must be developed gradually in the horse under saddle to be present in his way of going when he is working calmly.

The second purpose of dressage (after strengthening the horse to enable him to perform easily under saddle) is to make the horse agile and flexible. This requires exercises which make him bend his spine. Some of these exercises are the shoulder-in, haunches-in, and so on. Suppleness and good balance can be demonstrated by a horse that can canter a very small circle smoothly and easily. In order to canter the small circle smoothly, the horse must be able to bend from poll to croup, the bend in his spine matching the bend of the circle. When the horse is bending properly, his hind feet exactly follow the path of the forefeet. Horses do not travel this way naturally, and must be taught these exercises to develop greater agility and flexibility. He must be taught to follow his head and bend his body around the curves. When running at liberty in the pasture the horse seldom follows his head, but actually inclines his whole body slightly away from the direction he is traveling. If he is cantering in a circle to the left, his body is canted slightly to the right. He keeps his balance fairly well this way, until he gets into an unusual situation, such as slick ground or an unexpected change of direction (as may happen when he has a rider on his back that demands a change of direction before the horse expects it or makes him canter in a very small circle). Good training can actually make the horse *better* than he was naturally, even though this seems impossible when you first consider it. A bit of dressage, in which the horse learns to collect and handle himself with balance and dexterity, can actually improve upon his natural abilities and inclinations when running at liberty.

You've probably seen horses fall down when playing and bucking in a pasture, especially if the ground was a bit slick. This is because the horse does not naturally handle himself with perfect balance, or control. I have a good little half-Arabian mare, Nikki, (the other half being mostly Thoroughbred) who has developed into the best rough country cow horse I have ever ridden. She has never fallen with me even though we've done some hair-raising things chasing cattle over some pretty steep, rough and slick hillsides. Yet when she was a young horse I saw her fall down more than once when playing at top speed in a slick pasture. But now, even when playing at liberty, she has better balance and control; when she comes to a slick corner even at top speed one can see her balance herself and adjust her stride to meet the problem. She actually moves with a greater degree of collection and dexterity even at liberty now than she ever did before she was trained as a cow horse. The things she learned about stopping and turning, heading a cow over the rocks or sagebrush or mud, with a rider on her back, have actually improved her athletic ability and made her more agile even when romping at liberty. I've noticed this in several well-trained horses and it makes me realize how terribly valuable good training can be to the athletic ability of the horse. I do not claim that my little cow horse is well trained in all areas, but I do know that she is the most agile and surefooted horse who has ever chased a cow for me. Good natural ability, combined with proper development and training, can make a truly outstanding performer. And you need both. A horse with the most athletic potential in the world may still become merely a mediocre performer if he is not developed properly.

The horse that can bend his body to canter in a smooth small circle shows a good example of lateral suppleness or lateral flexibility. Longitudinal flexibility involves working in straight lines, such as the horse being able to halt from a trot or a canter in just one stride, change gaits smoothly whether slowing to a lower gait or changing to a faster one. These speeding up, slowing and stopping exercises (and stopping means coming to a square halt with no resisting or head-tossing) require perfect balance; the horse must have the ability to shift his center of gravity forward or backward as the rider requires of him. This is also part of the third purpose of dressage which is to make the horse willing and completely obedient—teaching him to move in any direction, at any gait, at any speed, either laterally (sideways) or on a diagonal, or with his forehand active and hindquarters passive or vice versa (rotation on the quarters and on the forehand). The horse is taught the pivots and the half pass (also called the two-track) and also to be able to change easily from an extended gait to a collected one and back again.

Besides trying to improve upon the horse's natural way of moving in some areas, dressage schooling also strives to simulate under saddle what an athletic and agile horse can do naturally when free and unburdened with a rider. If you have ever watched good horses playing in a field or dashing excitedly about, you may have noticed some beautiful and fantastic maneuvers. The free horse prancing excitedly with head and tail in the air will sometimes collect himself even to the point of doing the passage, and at other times may extend himself as fully as any advanced dressage horse under saddle. The free horse is capable of trotting and cantering with great lightness, good balance and cadence (rhythm). Thus dressage training is primarily an attempt to teach the horse the best of what he is already capable of doing, but striving to teach him these movements under saddle, wanting him to do them calmly and at the rider's bidding. The horse must learn to balance himself with the extra weight of the rider, requiring the development of muscle and agility.

A dressage horse is in complete harmony with his rider, so much so that the aids the rider uses are practically invisible to the observing eye. The horse is not forced to do his maneuvers and he does not resist the rider in any way.

The art of fine riding is vastly complex because it does involve this problem of complete harmony, both physical and mental, between horse and man. The action and interplay between rider and horse is constant and changing, made up of many actions and reactions. The art of good horsemanship can never really be taught, it must be felt. Dressage, the highest art of riding and training, is based upon common ideals. Dressage of the best kind has its principles laid down in the rules of the F.E.I.—Federation Equestre Internationale—the international group governed by the most competent horsemen in the world. Yet the interpretation of these theories and ideals, and the putting them into action must be ultimately left up to the rider and his individual horse because each horse and rider are themselves individuals, with individual capabilities and talents.

Dressage training is usually categorized into three stages. Light, or basic dressage does not include any highly collected work and mainly consists of teaching the horse to be obedient at all gaits on straight lines and in circles. Intermediate dressage takes the horse farther, through various schooling exercises, shoulders-in, pivots, and two track, including the flying change of leads and the beginning of the passage. Heavy dressage, the highest level, includes the highly collected movements such as piaffe, and the airs (off the ground movements).

Heavy dressage is called high school or *haute école,* and includes the special movements called the piaffe (trot in place), the passage (a beautiful slow, rhythmic trot with high leg action, the horse carrying much of his weight on his hindquarters), the flying change of leads at every stride, and the pirouette (a turn on the haunches at the canter, the horse cantering in place as he makes the turn). When these movements are done properly, they represent the highest possible degree of collection, flexibility and suppleness. The classical school of dressage is a very old tradition, and it includes the "airs above the ground." It began as a military school and the horses were highly trained to perform in battle. Some of the famous and ancient "off the ground" movements such as the levade, courbette and capriole were used to great advantage when the mounted soldiers were fighting foot soldiers. But today these specialized movements are no longer taught in dressage, except in a very few places in the world.

Modern dressage has some important differences from the classical tradition. The horses of the old school were small-type horses with fairly high knee action. All of the movements in the old school were done in the highest degree of collection, and the highlights of these displays were the movements on two tracks, in which the horse was always presented with his head and neck bent, looking at the audience. As an art, and in the perfection of seat, harmony and precision of action, the classical school is still an excellent model and one that is very difficult to equal. But this is no longer the school of today. Horses used in dressage today are bigger, faster, more powerful, and move with a lower, long-striding action. Modern dressage is mostly performed outside and is not so concerned with indoor presentations as was the old school. Modern dressage puts more emphasis on extended gaits; in fact it puts as much emphasis on extended gaits as it does on the highly collected movements, and uses continuous transitions from one to the other.

The dressage horse is truly an athlete, with suppleness and strength in every muscle. In the classical sense of the word, dressage means the suppling and gymnastic training of the horse's body. The result of this training will be a supple horse whose gaits and movements will be just right and whose performance will be in perfect harmony and balance. He can show the highest degree of collection and can also achieve the greatest degree of extension, extending or collecting promptly at the rider's command. This is the kind of perfection that is required for the Olympic dressage tests.

A good dressage horse must be handled carefuly and intelligently from the very beginning of his early training, for he must have confidence in his handler. In preparation for dressage training, the horse must learn a good walk. If he can learn to do a free, unrestrained walk, with freedom of his head and neck, the idea of jogging will never even occur to him. A good walk is the basis of all later development of the horse, and it is almost impossible to gain if the horse has been started out wrong, developing bad habits in his walk (such as a lazy walk, or prancing or jogging instead of walking out freely).

The trot should be calm, confident and energetic, and never a lazy gait. The trot is perhaps the most important gait in developing a dressage horse.

The trot can have many variations, and these are not just variations in speed, but also in collection, animation, and quality. The horse at liberty, trotting naturally, without a rider, can vary his speed, animation, collection and extension as he wishes, but most horses lose this natural way of going when ridden and thus must be encouraged and guided to regain this same sort of suppleness when under saddle. For a horse to trot well under saddle, he must be supple and at ease with his rider, flexing well and accepting the rider's leg pressure without excitement or resentment. A good trot contains a certain degree of impulsion and animation; the horse's body is propelled upwards as well as forward. Calmness is not the same as laziness. The trot is always calm, but not lazy. The rider strives to cultivate calm energy in the trot as well as in the walk, using his legs softly whenever needed, making the difference between a rhythmic gait and a sloppy one. Energy of gait means rhythm and animation, not speed. The horse needs to gain the ability to increase his energy without actually increasing his speed. To achieve this, his rider must use a good deal of tact and patience to develop animation and impulsion in the horse's gaits without actually restraining the horse with the bit.

Teaching the horse to go forward freely at the use of the rider's legs is perhaps the first principle of control and must be taught from the very beginning of the horse's training. And, of course, equally important is the second basic principle—acceptance of the bit. Control of and communication with the horse depends upon the rider's proper use of bit and legs, and it takes a lot of tact to determine just how much (and when) each is needed. Proper care should be taken in the horse's early training to develop a good mouth, a mouth that is responsive to the bit, so that the rider with light hands can instantly communicate through the bit to the horse, and without any resentment or resistance on the part of the horse.

In developing a well-schooled horse, the rider's seat and balance are quite important. The horse can carry weight best at a point just behind his withers. The rider must try to develop a good seat, putting his weight where the horse can carry it most effortlessly—well forward. The loins are the weakest part of the horse's back. So, the best seat on a horse is right behind the withers, in the lowest part of his back—a deep seat in the saddle (not back toward the cantle).

The rider's body should be supple, relaxed but erect. The balanced seat is used by dressage riders, a seat that is easiest on the horse. In the balanced seat, when the horse is standing still, the rider's upper body is vertical. At walk, slow trot and canter, the upper body is slightly forward. At the ordinary trot, extended trot and gallop, the body is farther forward.

The rider must be relaxed, flexible, and well balanced. Any stiffness in the rider will relay stiffness and tenseness to the horse. The best training cannot be accomplished without the best riding. Therefore the person who wishes to train a horse in dressage work must himself concentrate on becoming an excellent rider. The rider who is in perfect unity with the horse does not hinder the horse's action nor the rhythms of his gaits. And with a good seat, the rider's hands and legs can act delicately and independently, with true lightness of control.

The rider communicates with his horse through a special language of actions, called the aids, but these—hands, legs, voice, and body weight—are used in many ways and in varying degrees. The horse should be calm and confident, having trust in his rider, and should be able to move freely when asked. He should also be able to slow or stop, collect or extend, when asked.

Collection is a word that is understood by any good horseman. But there are several degrees of collection. The highest degree of collection, used with the highly schooled dressage horse is called *rassembler*. This is a French word, which in its widest meaning describes the entire attitude and bearing of the horse in all his actions.

COLLECTED HALT

The first and most elementary degree of collection is that in which the horse has a proper head carriage, accepts the bit with willingness and travels forward freely, his head and neck flexed at the poll and giving to the bit. The horse is ridden into the bit and he becomes lighter on his forehand through an increase in action of his hindquarters. Head carriage alone is of no value unless it is the result of increased energy and impulsion in the hindquarters, producing free and improved forward movement. Collection, even in this first and most elementary degree, cannot be accomplished until the horse has a good mouth, is confident and accepting his bit.

The second degree of collection comes when the horse learns to travel forward freely with a perfectly balanced and energetic stride, using his hocks well underneath his body. In other words the horse is shortened a bit; he brings his hindquarters a little closer to his head. This makes him lower his hindquarters a little and slightly raise his front end, lightening his forehand. The rider can accomplish this degree of collection in the horse by proper use of his legs and hands, working at shortening the base upon which the horse moves. This must be done tactfully, gradually and properly or the horse will tend to react by going slightly sideways instead of shortening himself.

This collecting and shortening of the horse can only be done if the horse is pliable and supple in all his joints and muscles, and is best accomplished through a careful course of gymnastic exercises through which the horse can gradually progress. The horse's ability to collect himself at his rider's wishes can only develop gradually, and if the rider tries to force the horse into something beyond its physical ability, all the horse can do is some type of evasive action (which is usually "transversing himself" or traveling slightly sideways).

If a horse has been carefully schooled and has a good mouth, relaxed and yielding to the rider's light hands, he will be yielding and supple in the rest of his body. But if his mouth is set or stiff, not accepting and giving to the bit, he will also be stiff in his body. It takes very good hands to keep a happy relaxed mouth. Working to increase the horse's impulsion and collection and working to create a relaxed, yielding mouth are two aims that are inseparable. The rider must use his legs and hands together skillfully to create impulsion and collection but must use his hands in such a manner so as to keep the mouth relaxed.

The third degree of collection, or "*rassembler*" describes the complete harmony and bearing of the horse in all that he does. This can only be developed from collection that is already very good. The rider must have a sensitive feel for the harmony of perfect rhythm, proper action, correct forward impulse, and he must have a very sensitive and delicate rein control. This final degree of collection is not something which the rider can teach the horse, but is a quality which the horse may gradually develop himself through increasing suppleness, dexterity and muscle control. The rider generally achieves a glimpse of this fineness of collection only after many years of work and the training of many horses and I must admit that I am, myself, not that far along yet.

When working on collecting a horse, the rider aims at creating proper flexion in the horse's head and neck, by increasing the action of the hindquarters and riding the horse gently into the bit. The rider therefore requires of the horse increased energy of gait without an increase in speed. If the rider tries to create flexion and collection by pulling on the bit, the horse will react adversely and try to avoid this pulling, usually by sticking out his nose, opening his mouth, or dropping behind his bridle (tucking his chin a bit farther) and avoiding contact with it. Using an ordinary snaffle bit, light hands and sufficient leg pressure at the proper times are usually sufficient to teach the horse the first and second degrees of collection. The double bridle has some advantages over the snaffle when trying to progress any farther than this, for finger signals are possible with the use of two bits working in different ways. With the double bridle there are four reins that can each be used singly or in combination with one or more of the others, or in conjunction with the legs. The best development of the horse depends upon the rider's ability to properly use the combination of reins and legs.

The problem many riders have in this respect is in realizing and adequately putting into effect the part that the hands and bit must play in achieving collection. The horse's hindquarters must become more active in order to shorten his base of movement and this requires an increase in energy and impulsion, but this extra energy must not be directed into more speed. It is the job of the rider's hands to keep the horse from going faster, yet the hands must not restrain. The rider must not pull, for the horse must move into the bit; the rider merely indicates the speed by the length of his reins, therefore determining where the horse will meet the bit, and thus regulating the amount of tension that the horse will exert. Only by gradually developing a good mouth on the horse and at the same time developing increased energy in the hindquarters will the rider be able to properly collect the horse and create a longer, more animated and balanced stride.

We shall now look briefly at some of the movements that are used in schooling a dressage horse. Keep in mind that this chapter on dressage is too brief to give a full and adequate description of the step by step school of the dressage horse; it may merely serve as an introduction to the subject. The serious student of dressage will want to look farther, for there are several good books on the subject of dressage which can be very beneficial to the person who wants to go deeper into this subject. All I am attempting in this chapter is to brush over dressage as best I can in this short a space, hoping to define and clarify the meaning and purpose of dressage, and hoping to awaken a deeper interest in this finest art of schooling a horse in those horsemen who try to take their horses and their training methods seriously. For, as mentioned earlier a few elementary principles of dressage can help to improve any horse and rider, in whatever role they do perform. And the serious horseman may wish to pursue the subject even farther

One of the first movements that a dressage horse is taught is the turn on the forehand. Ideally, this movement is a small circle on two tracks; the hind legs make a large circle around the front legs, which make a very small circle, with the inside front leg almost marking time in place. The forehand is in the shoulder-in position. In its most elementary form, the turn on the forehand is a very simple lesson in which the horse learns to obey the rider's leg and move his hindquarters around to the side. The horse learns this movement easily, yet it is not a natural movement because whenever a horse turns on his own, naturally, he makes a turn on his center, moving his forehand one direction and swinging his hindquarters the other way. The turn on the forehand teaches the horse lateral control (moving to the side with the hindquarters) but also serves as a valuable exercise for developing the muscles of the loins; the hind legs must step sideways in large steps. Developing the elasticity of the loins is very important in the process of developing an athletic horse. And, apart from dressage as such, the turn on the forehand is a valuable movement for any horse, because it enables the rider to control and position the horse for movements such as opening gates from horseback, passing through them, and shutting them again without ever dismounting.

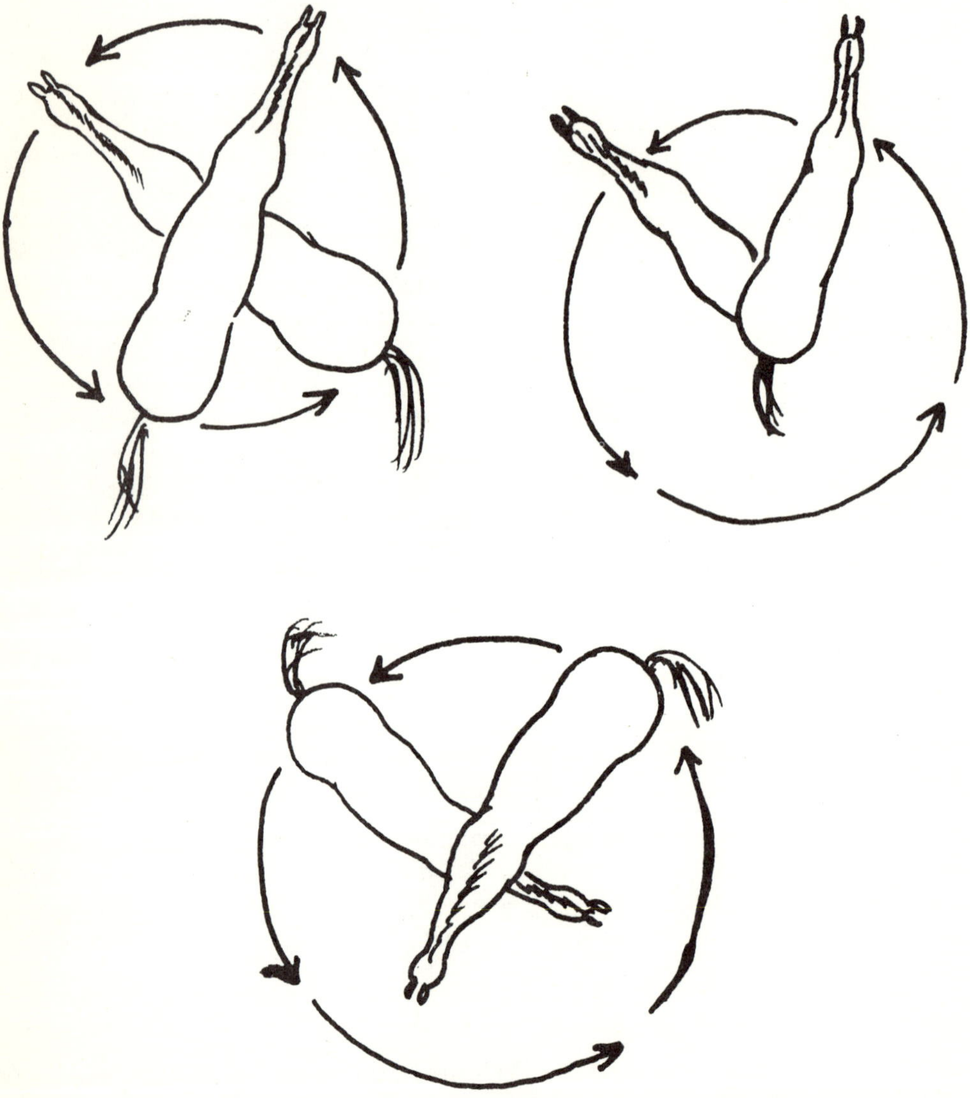

Above left, **turn on the center—this is the way the horse turns naturally;** *above right,* **turn on the haunches;** *below,* **turn on the forehand**

To execute a turn on the forehand, the rider stops the horse squarely, and to move the hindquarters to the right, uses pressure with the left leg, and uses the left rein back toward the withers, pulling in the direction of the right hip. The horse's head and neck should be bent slightly to the left, and when the movement is done properly it produces good flexion of the neck and jaw. The rider's hands hold, while the leg pushes the hindquarters over, the horse retaining a feeling of forward movement.

Riding in circles is very important in the training exercises of a dressage horse, teaching the horse to bend his body to the shape of the circle, which is not exactly easy; the horse's spine is fairly rigid and has little flexibility laterally, and his loin muscles are not accustomed to this kind of movement. Riding in circles is one exercise that improves upon the horse's natural ability (when done correctly) teaching the horse to flex and move in a more balanced way than he does at liberty. Most dressage training merely seeks to simulate under saddle what the horse is capable of doing naturally, but traveling in small circles is something that no horse does very well naturally, especially at the faster gaits. This may be because the horse in the wild had no reason for traveling in small circles; he had lots of room for his ramblings. But the horse cavorting in confinement (such as a small pen or corral) or the horse under saddle that is required to perform various maneuvers, either for reasons of showing his ability in the arena or for some other purpose where agility and good balance are required, may have to maneuver in circles or small corners, at various speeds. Thus he will be more able to accomplish these feats if he has had some good exercises in lateral flexibility, increasing his ability to bend his body to fit the circle and to balance himself properly as he circles.

So, let's take a closer look at the exercises performed in circles. If the horse makes a circle, the outside loin muscles (if he is bending his body properly) must extend, while the inside ones contract. The trot is a very good gait for doing exercises in circles, for it is a two-beat gait in which diagonal legs work together in unison. On a normal trot in a straight line, each diagonal pair would travel the same distance in a stride, but when the horse trots in circles this is no longer true for the inside legs must then travel a shorter distance than the outside legs; the outside legs must take longer strides. And the fact that the legs are working in diagonal pairs requires quite an adjustment in the horse's movement to keep good balance and rhythm, especially in small circles. At each stride the inside leg of the diagonal pair must travel a shorter distance than its outside counterpart. The problem is not so great at the walk and at the canter because the legs can move more independently of one another; there are none working in diagonal pairs.

Trotting in a circle is an excellent exercise for developing the horse physically, for it demands proper bending, adjustment of the horse's stride and rhythm, increased impulsion on the inside leg which carries more weight than the outside hind leg. It isn't easy for the horse and therefore should be worked into gradually, trotting in large circles at first and gradually attempting smaller ones. The horse must have good head carriage and bend his neck to correspond with the bend of the circle. In circling, the horse should be collected and on the bit, with even contact with both reins. Since his neck must bend to conform with the circle, the rider should slightly lengthen the outside rein. If the horse is well collected, he will seek to re-establish equal contact with both reins, on both sides of his mouth, as the outside rein is slightly lengthened; thus he will tend to bend his neck just a bit to maintain equal contact, bending to conform to the circle.

While talking about small circles, we should mention the volte. A circle may be any size, but the volte is a small circle of a specific size; it is the smallest circle that the horse can make on one or two tracks. Thus the radius of the volte is the same as the length of the horse. In other words, the radius of this small circle is about ten feet; it's diameter is about twenty. Riding a circle this small requires a very highly schooled horse. Once a horse is balanced and agile, performing circles as small as ten or twelve yards in diameter, he can more readily begin working on various figures such as loops and serpentines and figure eights. These figures which require changes of direction are rather difficult to perform smoothly and should not be attempted until the horse is ready for them.

Let's look now at movements in which the horse works on two tracks, his hind feet following a different track than his front feet. Working on two tracks (in effect, going sideways) is another gymnastic exercise which can aid the horse's flexibility and physical development.

When you begin teaching a horse to sidestep and move on two tracks, he will do it rather poorly at first because the action of moving his legs forward and sideways at the same time is not natural to him. The shoulder-in is a good exercise in which to start working on two tracks, and can often be best accomplished by first riding the horse in a circle, bending his neck and body to conform to the circle. Then, at some point in the circle, begin pushing the horse sideways, still keeping him bent in the attitude he was traveling on the circle. Thus he is now traveling on two tracks. You should keep his body bent at about a 30-degree angle with the line of his progression.

The inner hind leg carries much of the load, moving sideways under the body, and this requires a lot of hock flexion. If the horse is clumsy and his hock action is poor so that he loses his impulsion and rhythm, continue forward again in a circle and when his hock action and impulsion are good once again, try taking the two track again. To start with it is best to use a large circle and to try the shoulder-in on just a slight angle; it will be easier for the horse.

After the horse can accomplish the shoulder-in on straight lines, walking on two tracks, movements in a circle can be tried. When two-tracking in a circle with the shoulder-in position, the hind legs must take longer strides than the front legs—ever more so as the circle becomes smaller, and reaches the ultimate on the smallest possible circle which is the turn on the forehand.

The quarters-in movement is similar to the shoulder-in, the horse walking on two tracks with his body flexed, but is very different in that the horse is flexed in the direction he is going; in the shoulder-in the horse is flexed away from

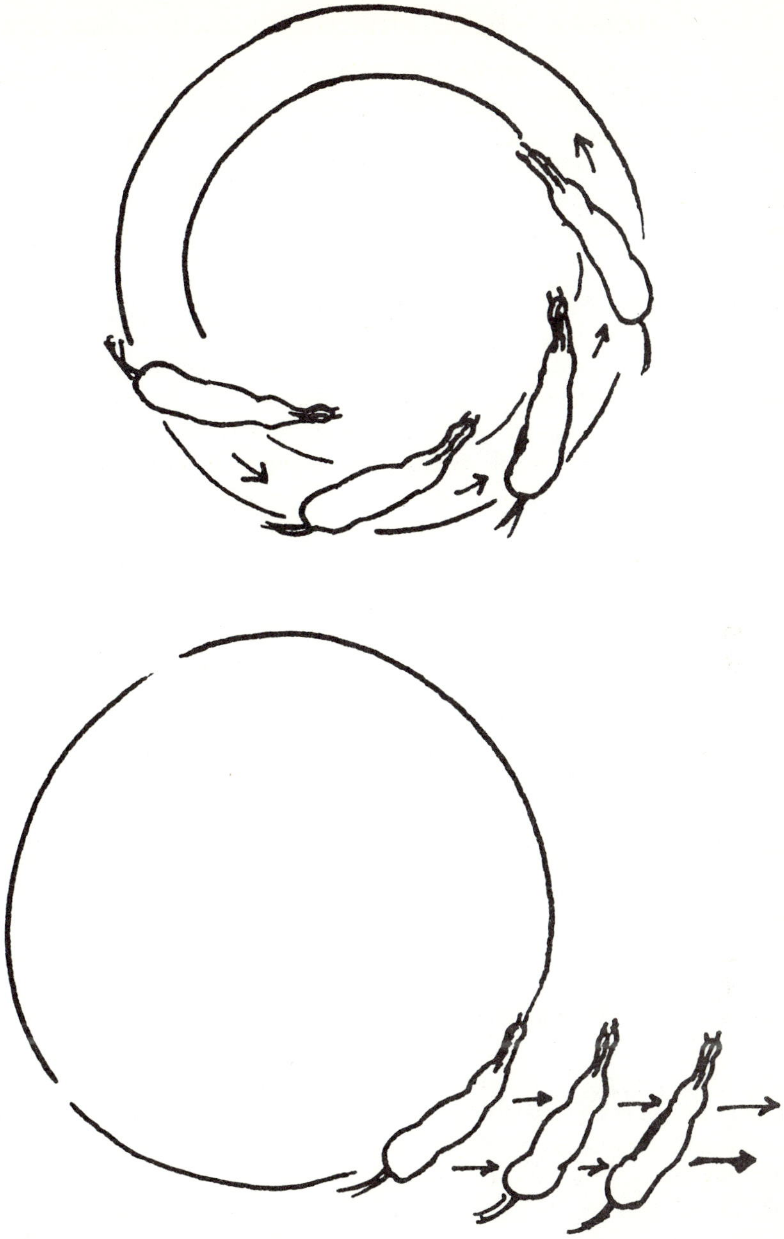

Teaching the horse to work on two tracks; *top,* **shoulder-in on a circle;** *bottom,* **shoulder-in taken from a circle**

the direction he travels. The quarters-in position for two-tracking is less natural to the horse, but is a great aid in exercising the same muscles used in performing the shoulder-in, using them in a different combination. So it is the rider's advantage to use both types of exercises in order to best develop the muscles and flexibility of the horse.

To attempt the quarters-in, ride the horse in a circle again as was done in preparation for the shoulder-in, but leave the circle when the horse's head reaches the desired point, pushing him sideways. If circling to the left, you will at the desired point use your right leg to push the horse to the left (continuing to keep his body bent and flexed as it has been in the circle), and your reins to move him sideways to the left also. But be careful not to impede the horse's forward impulsion by the use of the reins.

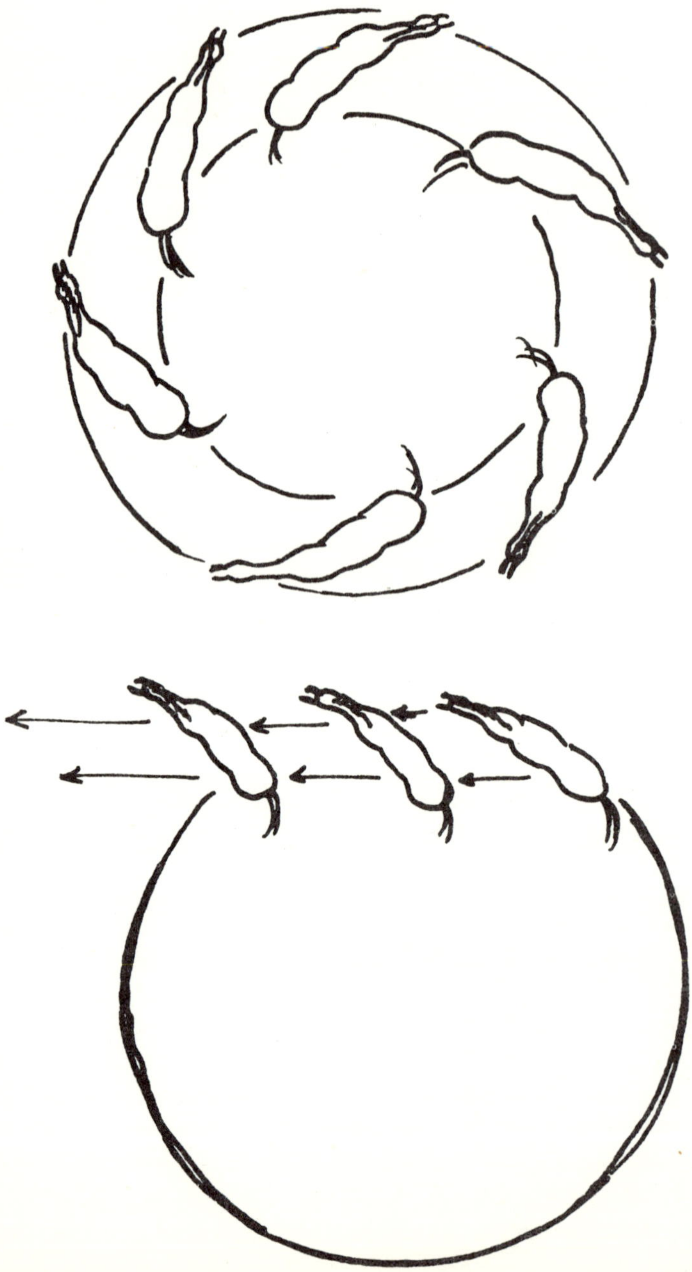

Top, **quarters-in done on a circle;** *bottom,* **quarters-in (head to the wall) taken from a circle**

Quarters-out is the same movement exactly as quarters-in except that the horse is moving with his hindquarters toward the outside of the arena or riding hall (tail to the wall) whereas in quarters-in he moves with his head to the outside (head to the wall). The horse walks on two tracks with his head flexed in the direction he is going. If he is flexed, his head will be carried properly, forehead perpendicular to the ground. If his head is bent—his nose coming up sideways—you are using the wrong rein effects.

When the horse can perform the two track quite well along straight lines in the quarters-in or quarters-out position, you can attempt it in a circle, the front legs making longer strides than the hind legs in the quarters-in position, the hind legs making longer strides in the quarters-out position. However, the quarters-in position is the one most often used. In the quarters-in position the front legs take longer strides, shoulders and front legs extending while the hind legs flex. The horse is quite light on his front end. The ultimate in this movement, on the smallest circle possible is the pirouette, or turn on the haunches. Before trying this, you may wish to teach the horse to take corners in the head to the wall position, for this will aid him in learning to pirouette.

The half pass is a movement in which the horse's whole body is *not* flexed. There is just a slight flexion of head and neck, and this flexion is always in the direction of movement. In the half pass the horse moves obliquely forward, his body parallel with the sides of the arena or riding hall. The horse, even though he is moving on two tracks and traveling obliquely, must have much forward movement and animation. This is very difficult to do until the horse is schooled to such a fine degree that he needs very little help in the form of aid from legs and reins. Too much rein in an effort to make the horse move across the arena in the correct position will usually tend to destroy good forward impulsion, especially in a horse that is not highly practiced in this movement. As the horse progresses, the legs can be used less and the reins more, as the horse begins to associate the aids with what is wished by the rider.

Crossing from one side of the arena to the other, on a diagonal, is called a change of hand. If, when you are part of the way across, you reverse direction and return to the original side, you are executing a counter change of hand. A counter change of hand performed on two tracks means reversing the horse's bend and flexion and sideways movement and takes considerable skill on the part of horse and rider to do it correctly and smoothly, especially when done at repeated intervals.

Let's look now at the gaits of the horse as we school him for advanced work. We are aiming at a certain degree of perfection in the walk, trot, and canter, and must work at their gradual development. All three should be developed equally as the horse progresses.

The walk is discussed in some detail in the chapter on gaits; it is a four-beat gait in which each leg moves separately. It is a quiet gait and useful for giving the horse periods of rest and relaxation between more difficult movements. The walk is naturally a relaxed gait. In a good free walk, the rider should not hinder the horse's head and neck movements. The good trainer cultivates

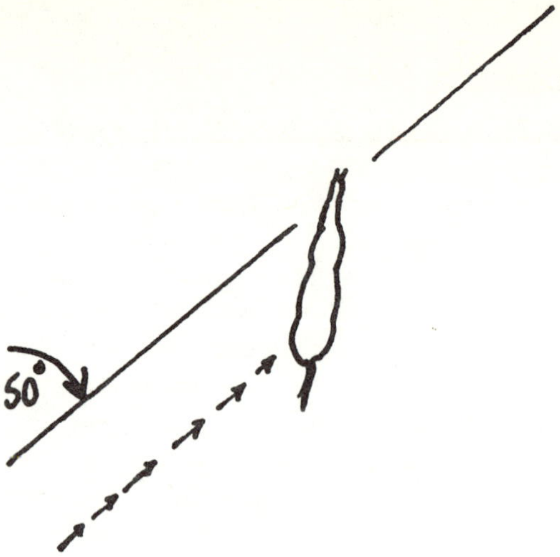

Half pass—crossing from one side of the arena to the other on a diagonal, the horse moving on two tracks

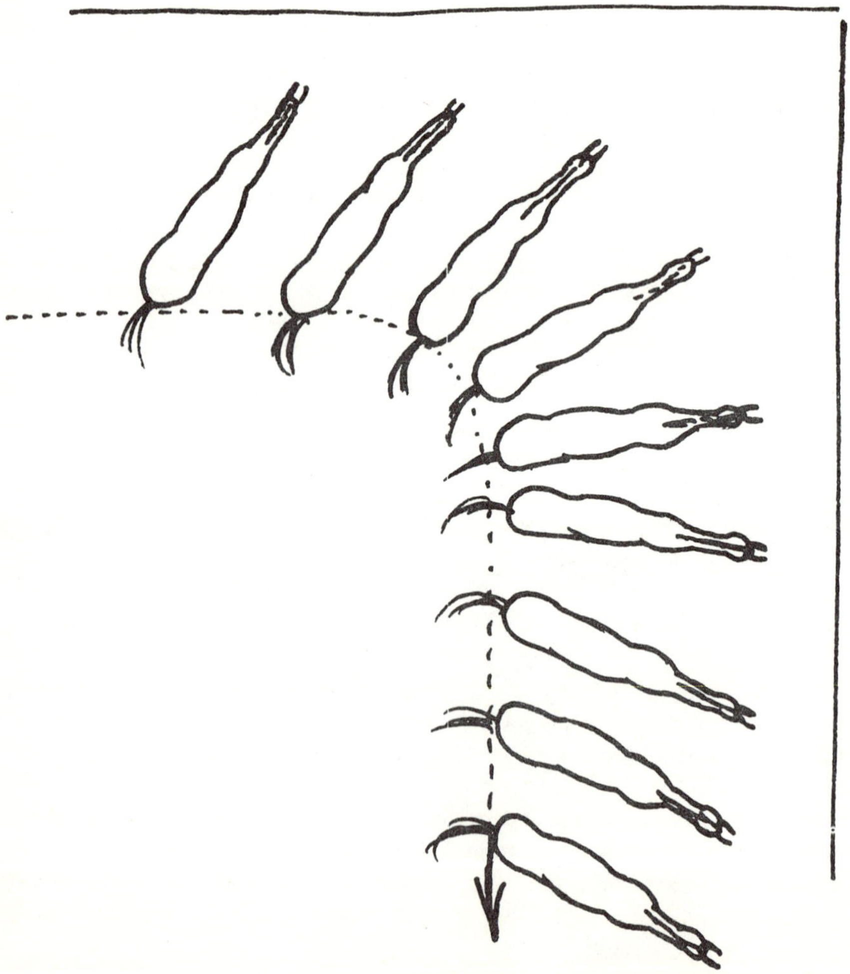

The quarters-in can be a good exercise before attempting the pirouette

and improves the natural abilities of the horse; he never supresses them.

In dressage work there are several types of walk. In the free walk the horse goes with a long rein; there is just the slightest contact and no attempt at collection. The walk on a loose rein is nearly the same except that there is virtually no contact except through the weight of the reins, which are dropped down on the horse's neck. The horse has freedom of head and neck and can stretch or carry his head and neck in any position he wishes; this can be a form of rest and reward after highly collected movements or difficult maneuvers, giving the horse a chance to relax.

The free walk, and the walk on a loose rein, should be free and easy with no great speed or extension. If the rider tries to make the horse go faster in a free walk, the horse will take faster strides, but not longer ones. There is really not much a rider can do to improve the horse's free walk, but he can keep from hindering it.

The ordinary walk is one in which the horse has a more positive contact with the bit, a higher head carriage and the beginning of collection. The horse walks along at a natural gait, not greatly collected, but ready to become collected instantly if the rider asks it. It is difficult to achieve real collection at the walk until the horse first learns collection at the trot, because the walk is a gait in which the horse's weight and balance constantly shift.

Once the horse has learned collection at the faster gaits, which more readily lend themselves to this sort of work, the rider can carefully use these same principles at the walk and develop in the horse a true collected walk. The extended walk is another variation of the walk that can only be developed after much work at the trot. In the extended walk the horse greatly lengthens his stride without hurrying his rhythm of step. If the rider tries to create an extended walk before the horse is ready for it, the horse will usually only hurry his stride without extending it. Speed should not be confused with extension.

The trot is the most important gait in the development of a dressage horse. Besides its value for developing the horse gymnastically, the trot by its very form makes some kinds of movement more easy for the horse. When trotting naturally, as the horse does when he prances in excitement in his pasture, he moves with spectacular grace, cadence and lightness. This is what the trainer tries to duplicate under saddle.

The trot is a two-beat gait, very regular and balanced, and the horse uses his hocks energetically, keeping his head high and his forehand light. Because his head does not make balancing movements at the trot, the rider's job is simplified and he can keep his hands still, not interfering with the horse's mouth.

At the trot the horse's muscular development—of shoulders and hocks and all his joints—progresses as a whole; no limbs, no one part of his body is developed more greatly than the rest. The head and neck are still and there is very little natural swing to the back when the horse trots. This stiffness of back makes the horse unpleasant to ride, and limits his lightness. Until a horse is quite supple he won't have the cadence, balance, and lightness that makes for a nice trot under saddle. Suppleness can be developed better at the walk and at the canter than at the trot. When the horse is supple, his trot will then be comfortable to ride. In fact, on a supple horse, the rider will find it not uncomfortable to sit in the saddle at even the most extended trot. But for practical purposes during schooling of the horse, the rider should post the trot whenever it is uncomfortable or difficult to sit it, for if he insists on a seat that is uncomfortable to both horse and rider, much of the harmony and balance of that horse and rider will be lost.

No horse is ready for dressage work until his ordinary walk, trot and canter under saddle have been well establshed and developed along good, sound, lines. Before starting to develop the trot in a dressage horse, the horse must be able to perform a good ordinary trot. A good ordinary trot has fairly long strides of even length, with regular tempo and rhythm.

In dressage work, the rider's purpose in working with the trot is to make its range variable, from great extension to great collection, with perfect rhythm, action, and flexibility. To begin this work, the rider starts with an ordinary trot, with soft contact in the horse's mouth with the bit. The horse is fairly animated and light. Now the rider closes his legs gently on the horse, making him speed up a little, using his hind legs more energetically and lengthening his strides momentarily. Then the rider's hands gently contact the horse with the bit and the horse is quietly brought back to the ordinary trot. In time and practice this gradually leads to collection and lightness.

Gradually the rider works on lengthening the horse's stride for a short time, then coming back to the ordinary trot, slowing and collecting. In time the rider can produce a longer stride in the horse while still at the speed of the ordinary trot, adding a measure of brilliance and animation. The rider concentrates on maintaining the same cadence and rhythm even during slight variations of speed. The secret of progress in this type of work is to go gradually, never attempting more than the horse can do and never trying to do more than the rider knows how to do.

The rider can begin to teach the horse to associate the posting trot with extension and the sitting trot with collection and slow work. The horse will tend to extend on the rider's rising alone without the need of leg aids. And in slowing the trot, rhythm and cadence must be maintained, in fact accentuated; the horse marking time with great energy. The schooled horse uses much more energy at the slower gaits then the untrained horse uses for these speeds. The rider tries to obtain cadence (the horse's legs working with definite rhythm and energy) even at the slowest speeds, so that the horse has action with a bit of snap and zest to it. The slower the horse is traveling, the more the rider must concentrate on maintaining cadence. Exercises on circular movements can help the rider develop the moving ability of the horse, for when circling, the horse's inner hind leg must push farther underneath the body and take on more of the weight.

At this stage the rider must also work on smooth transitions from trot to walk and walk to trot, and then gradually work on transitions from halt to trot and trot to halt. In the earliest stages of dressage the rider lets the horse slow gradually, the trot subsiding into an ordinary walk. As his training progresses, the horse will be able to make the transition in just a stride or two.

The collected walk is energetic; the horse moves vigorously even though his stride covers less ground than in the ordinary walk. This is because his legs move higher and his joints flex more. He maintains the regular four-beat gait but his head is high, almost in the position it is carried in a good collected trot; his attitude is one of mobility and energy. When the horse can easily make the transition from trot to walk, it is not too hard for the rider to maintain the horse's mobility and energy that was being expressed in the trot.

As the rider works on developing the horse's walk, he can begin to vary its speed and length of stride at will, just as he did with the trot. The length of stride can be varied from pronounced movement to practically no forward movement at all, to the rein-back and then forward again. This is all done by the use of the rider's hands and legs. Smooth starts and stops come naturally with this work also. For instance, if the horse is standing still and the rider gently presses with his legs as he moves his hands forward and lengthens the reins, the horse moves forward. Or, if the rider presses gently with legs while the horse is moving, without moving his hands, the horse moves forward more, comes into contact with the bit and finds that it does not yield, then discovers that the rider's leg pressure stops as soon as he contacts the bit and realizes that he has done all that the rider wished him to do and stands still in a collected halt. Then if the rider presses his legs, moves his hands forward a little to suggest forward movement, the horse will begin to move forward again.

When backing, the horse should have his head up, and bent at the poll, his face nearly vertical. He should move backward in two-beat time, his steps deliberate and his legs working in diagonal pairs. If he backs too hurriedly he won't have time to flex his hocks enough and may back crookedly.

The rein-back is valuable as an exercise for strengthening and suppling the loins, haunches, and hocks; the horse is shortened from front to back, putting more weight on the hocks. The quarters drop and the loins rise. The rider should be able to halt the horse smoothly or make a smooth transition from backward to forward with no pausing.

Advanced work at the trot includes the passage and the piaffe, two movements which embody the highest degree of cadence, balance and control. The passage is a trot with very high action, in which the horse seems to travel in slow motion. There is a definite instant in which one diagonal pair of legs hangs suspended in the air, hesitating, as the other pair is on the ground. The horse prances from one diagonal to the other, moving forward slowly with perfectly cadenced steps, about a foot at each step. The piaffe is this same brilliantly majestic "trot" but the horse lifts his legs in place without moving forward. A well-schooled horse can make a transition from passage to piaffe and back to passage again with smoothness and ease. The goal of these movements is to achieve absolute regularity of rhythm, with great animation and good elevation of the legs as they take their strides.

The passage and the piaffe are both expressions of the suspended trot, the

PASSAGE

PIAFFE

horse moving in slow motion. The biggest problem in obtaining these movements is the fact that great energy and animation must be combined with slow motion. The slower the forward speed of the horse, the greater the difficulty, for the horse tends to lose movement as he slows down. Thus the piaffe is more difficult than the passage to do well. The forward urge must be maintained even though the horse is not moving forward.

The passage should be taught first, gradually developing it through the perfection of the slow cadenced trot, slowing it while yet retaining its energy. The horse's stride in passage or piaffe must be absolutely regular, and this requires keeping the horse calm. Force will accomplish nothing and only confuse and irritate the horse. The horse must come to realize what is wanted and then perform it on his own. And this takes much time and tact on the part of the rider.

So the rider gradually works at slowing the trot, his legs maintaining the energetic movement of the horse. Yet the use of the reins in slowing must not oppose the leg pressure and create confusion and opposition in the horse. This can be done by using a slow animated trot and alternating it every few minutes with a free walk to help keep the horse calm and relaxed. The horse will begin to understand that speed is not wanted. His transitions from slow trot to walk and back again will become lighter and easier and he will begin to slow the speed of his trot as he begins to understand what is wanted. As he progresses, the walking steps in between periods of trotting can be gradually decreased, the rider keeping the horse collected now during his walking strides as well. While making the transitions from trot to walk, the rider tries to maintain the rhythm and cadence as long as possible, and to pick it up again as soon as the trot is resumed, creating a cadence so slow that there begins to be a hint of hesitation. When the horse is able to do several steps with this marked cadence and moment of hesitation as each diagonal is in the air, he has attained what the French call the soft passage. Even though he doesn't yet lift his feet very high, he is well on his way to doing a passage if his stride is regular and calm. If he is hurried or forced, his steps will not be regular and he won't learn a correct passage.

The passage puts great strain on the horse's hocks. Hocks and fetlocks develop strength and elasticity only gradually, so this kind of work should not be overdone. If overdone, the horse will try to avoid the strain and will take short little jumps with his hind legs instead of using them diagonally. Thus the passage and piaffe, if done correctly and worked into gradually, are excellent exercises for strengthening the horse's hind legs, especially the hocks.

As the horse becomes good at doing the soft passage, he will begin to do a regular passage naturally on his own with a little encouragement from the rider. The horse will begin to use his legs more energetically and with more elevation. Once he does this, he is ready for the piaffe. The rider cannot merely try to slow down the passage into the piaffe, for the horse will lose his animation. So the rider must try to create variations of forward speed at the passage, and transitions from longer to shorter strides and vice versa. These variations are only slight, but the horse learns that he can vary his speed and can be always ready to move forward again, rather than just learning to go slower and slower. Gradually the passage is slowed, each slowing followed always by moving forward again. Finally the horse will do a step or two nearly in place. When he will finally do several steps in place, he is doing the piaffe. And each time he does, the rider then moves him forward again before the difficulty of this movement makes the horse lose proper cadence. The forward urge must be maintained throughout this movement or the hind legs will not be lifted properly; the horse must feel this forward urge, the desire to move forward again at any moment the rider bids him to.

Now let's look at the canter. The horse's natural canter, out in a pasture, is slow and collected with his head high and weight balanced back toward his hindquarters. The action is upward as well as forward and he is light on his forehand. The horse does not naturally carry himself straight at the canter, but rather somewhat sideways, his haunches flexed a little to the side.

When schooling a horse, the most usual way to ask for a canter departure on a certain lead is to use diagonal aids, conforming with the horse's natural tendency to flex his body a certain way when he goes into a certain lead. In other words, when asking for a strike-off into the canter on the right lead, the rider flexes the horse's head in that direction and uses his left leg to stimulate the left hind leg (which does the pushing off when the horse goes into the right lead) and which will also tend to push the quarters a little to the right. This puts the horse in the position that he naturally takes when striking off at the canter. But using the aids this way makes it impossible for the horse to start off straight, which he must learn to do when he is farther along in his schooling, to be able to change leads at frequent intervals. In order to change leads often, the horse must canter straight. So; when the horse is far enough along in his schooling that the rider begins to think about encouraging him to canter straight, the leg and rein signals must be somewhat modified.

The rein is used as the first signal, slightly before the leg signal is used; a slight touch on the left rein means a canter on the left lead and a slight touch on the right rein means a canter on the right lead. When the horse understands this sort of signal, the leg is not needed strongly behind the girth, but can be used close to the girth. Then as the horse progresses in his understanding, both legs can be used, and the horse will know which lead to take from the rein signal without risk of confusion from leg pressure. Now the rider can use one leg or the other more strongly to encourage the horse to canter straight. As the horse becomes more and more familiar with his rider's aids, he will come to associate the canter on a certain lead just by his rider's position in the saddle.

The horse's natural canter is a fairly slow and collected gait, but the collected canter used in dressage requires a very high degree of collection and lightness. The problem is to develop in the horse a very slow and collected canter, without the loss of characteristics of a natural canter—still exhibiting a great amount of animation and energy.

Many riders try to slow the horse's canter by pulling on the reins, but this interferes with a good mouth and hinders the freedom and action of the horse's stride, making his strides short and choppy, his action cramped and stiff. It isn't easy to develop a slow and collected canter that still has nice graceful strides and total freedom of movement; it takes much time and patience. Here again, the horse cannot be forced into this movement but must come to it gradually, with perfect understanding of what is wanted so he can do it himself with ease and grace.

To accomplish this the rider should work on the canter out in the open, not in an arena, and should never concentrate on too much work at one time. He should canter the horse in a natural canter even though this is too fast at first, alternating the canter with a free and quiet walk. Above all, the horse must be kept calm and happy. Never fight him or get him excited. Don't restrain him with the reins or he will become upset and frustrated. If he is cantered and walked, cantered and walked, always kept on a loose rein and relaxed, he will soon get the idea about slowing down.

The first transitions from canter to walk will be gradual; the rider is never forcing the horse. But as the work progresses over a period of several days the horse will begin to understand that the rider wishes only a short distance at the canter, at a slow canter, and begins to comply, slowing down without any real restraining on the part of the rider. And because he is unrestrained and unhindered, his canter will still have a free and graceful stride and he will carry himself naturally.

During this work, the rider will use both leads so that the horse won't get into the habit of just cantering on one. Work at the canter can be done in the early stages of dressage training, in conjunction with the work at the trot. Cantering helps to supple the horse's back and loins and can help improve the trot. In dressage work it is best to make canter starts from the walk instead of from the trot because if the horse gets used to going into the canter from the trot it will be more difficult to develop his extended trot; he will be inclined to break into the canter. If the rider starts his horse into the canter from the walk he will be better able to place the horse correctly and use the aids correctly for a good canter departure. These canter departures at the walk, taking either desired lead, are good practice for later work at changing leads. The horse should be very good at taking either lead from the walk before the rider tries changes of leads at the canter.

When slowing the canter and changing from the canter to a walk, the horse as he progresses will need less and less trotting strides and will eventually be able to go directly from a canter into a walk. If the rider is careful and always uses tact and patience, the horse's transitions will become smooth and easy and he will become automatically more light and collected.

There is just one phase of the horse's cantering stride in which he can change directly from canter to walk, and this is at the moment in which he has three feet on the ground—both fronts and one hind. This phase of his stride corresponds roughly with the phase of his walk in which he also has both fronts and one hind on the ground. At this moment of his canter, the horse can "shift gears" and go directly into the walk. The rider must give his signal to the horse with enough time for the horse to then make the transition at the correct phase of his stride; the horse must be given enough time to obey the signal. And, at the moment the horse is ready to make the transition (when his leading leg comes to the ground in front of the nonleading leg and one hind leg is still on the ground), the rider should cease his checking signal and use the aids for continuing at the walk, giving a slight yielding of the reins and a squeeze with the legs as the horse lengthens and begins his first walking stride.

Making the transition from the canter into a halt is not too hard when the horse already can do a smooth change from canter into walk. But to make the transition into a correct halt with the horse standing squarely instead of crooked, is a bit difficult. The horse must be checked from the canter, but allowed to round off his stride so that he stands square; the rider uses a half halt to put the horse into the first walking stride, then checks him fully to bring him to a smooth, square stop.

When first teaching the horse to change leads, the rider brings the horse to a walk (smooth transition from canter to walk), walks a few strides, then starts again at the canter on the opposite lead. When the horse is ready to change leads at the canter, he makes a flying change, during the phase of his stride in which all four feet are off the ground. In changing leads, the horse lengthens himself for an instant, his head and neck stretch out a little, so the rider must be careful not to interfere with his movement by using too much rein pressure. The rider gives his signal for the change, giving the horse an instant in which to obey the signal the next time he has all four feet off the ground.

The amount of time needed for the horse to respond will depend on the ability of the horse and the quickness of his reactions. To do a good job with the flying change, the horse must be collected and light, but cantering naturally, not pulled together and hindered by the reins. If his stride is too short, due to the rider's inability to get a good canter without pulling on the reins, the horse will have trouble making the change smoothly. It also helps in the beginning of this work to make the change when the horse is making a change in direction. Later, as he becomes more capable, he will change equally well when traveling in a straight line, which he must learn to do in order to change leads every few strides.

When the horse changes leads readily and smoothly, the rider can begin working on the change of leads at short intervals, asking for two or perhaps three changes during one canter. The secret of this work is gradual progress; any hurry or overdoing of this difficult work will cause problems. Once the horse is changing leads well, the strides in between can be gradually reduced. But the rider should not try too many changes, never more than just a few during one canter, followed by some relaxed walking. To keep the horse from anticipating the rider's signals, never ask for the same amount of strides

Last Cantering Stride

First Walking Stride

between changes, but ask for three, then five, or four, or whatever.

Changing leads at every two strides is hard and requires very finely timed signals and much understanding between horse and rider. The Spanish Riding School (Vienna) uses this movement, changing leads at every other stride, and does not go any farther than this. But in Olympic dressage, one requirement is that the horse change leads at every stride, a very spectacular feat and very difficult to master. The rider must be constantly changing his aids and yet keep them very light and very precise, never interfering with the horse through the mouth, nor hindering his impulsion.

To begin this work, the rider signals for the change while cantering (on a straight line of course) and as soon as he feels the horse begin to change he reverses the aids and asks for the next change. Thus the horse makes just one stride and changes again. Then the rider walks him awhile and tries it again, then goes gradually on from there, asking for a few more changes, over a period of days and weeks. The Olympic test calls for 15 changes at every stride, no more and no less, so the rider must have very accurate and precise control. It takes much tact and patience in this work to keep the horse calm and working well, avoiding confusion and excitement.

The counter lead or false canter is when the horse canters in a circle on the wrong lead (outside feet leading instead of the inside). This movement is used in dressage for two purposes—to make the horse understand that he is only supposed to take the lead desired by the rider (and not change leads on his own, but just at the rider's signal) and also to make the horse more supple. Cantering false makes the horse find his balance while traveling in an unnatural way and uses his muscles differently. Thus it develops his muscles more fully and helps him to better balance himself under any sort of situation, making him more agile.

Teaching the horse to canter false requires some tact, making the horse understand what is wanted rather than trying to force him into it. To begin teaching the horse this movement it is best to use very large circles at first, with only a slight curve, and then by riding a serpentine, where the horse stays on one lead, being on the correct lead half the time and the counterlead the other half. It is important to use the aids very clearly so that the horse will understand that he is supposed to keep cantering on a specific lead even though his inclination is to change. When doing the counter lead correctly, the horse's body is bent in conformity with the circle he is traveling, but he is still flexed in the direction of his lead. In other words, his jaw is relaxed to the action of the rein; he is looking slightly in the direction he is leading (to the outside of the circle in this case). But this relaxation and flexion is only in his head and jaw, not his neck. The horse should be able to bend his body to the circle and still have his head and jaw flexed to the outside rein.

Cantering on two tracks is similar to trotting on two tracks and training the horse in this movement can be approached in the same way. When doing the counter-change, the horse must change leads, so he must be very good at changing leads as well as cantering on two tracks. Repeated changes, or zigzags, become more difficult, and the work should be done gradually so that the horse can progress as he becomes more capable of changing his lead and his flexion while changing direction at two-tracking.

The pirouette at the canter is a full turn on the haunches (360 degrees) while the horse keeps cantering. To do this he must not lose movement or his hind legs will cease to be cantering and he will be merely pivoting on stationary hind legs. While pirouetting, the hind legs canter in a very small circle, retaining forward impulsion. To teach the horse this movement, the rider practices a few strides at a very slow canter, always moving the horse on faster afterward. If the horse is always ready to move freely forward he will keep his momentum and animation and will not tend to stop cantering when he is slowed.

The horse must also be good at cantering on two tracks. Teaching the pirouette comes gradually, and only after the horse is very supple and strong in his loins and hocks. The rider can begin making turns, only a step or two at a time, immediately moving the horse out again at a faster canter so that the horse will realize he is supposed to keep cantering. Soon the horse can make a 90-degree turn at a slow canter and from there it is gradual progress to a full circle.

The extended canter is natural to the horse and very easy to teach in dressage work, if the rider doesn't neglect it in his concentration on more collected work. The horse should be able to make smooth transitions from collected to extended canter and back again, with no abruptness. The change from one to the other should be easy and effortless, fluent, the horse retaining good balance.

There are many kinds of dressage tests, varying in difficulty. When entering a test, the rider should thoroughly memorize the test and the points in the arena at which certain movements are to be done, and in what order. But in practicing the test with his horse, he should not go over the exact order of the test or the horse will memorize it too and begin to anticipate, becoming less careful and less free and easy in his movements. He will go through it rather routinely and get sloppy. So when practicing for a dressage competition, vary the order of movements for the horse so that he will always stay on his toes. And get him accustomed to the arena and markers that are used beforehand so that they will not distract his attention when he sees them and travels over them.

But even if you never intend to seriously compete in dressage tests, dressage can be of much value to you and your horse. Any schooling you can give your horse that makes him more agile, more able to respond to your cues, more maneuverable under saddle, is beneficial in whatever role you use your horse. The pleasure horse will be more pleasant to ride, more able to perform in any capacity that his rider wishes to use him. The show horse will have more "presence," more animation and vitality under saddle, and his gaits will be smoother, better balanced and more flexible. He will have better response to his rider and put forth a better showing whether he is doing a fast walk, a figure eight, a roll-back, a good transition from one gait to another, or a smooth stop. The jumper will have more control and agility and be better able to maneuver himself over and between jumps. The trail horse will have better gaits, better muscle development for endurance and for agility over difficult terrain. And he will have better maneuverability, whether traveling through difficult terrain (down timber, rocks, and so on) or through man-made obstacles (gates to open from horseback, etc.). The stock horse will have better balance and control when performing fast maneuvers in the show ring or chasing cattle under natural conditions; he'll have his hocks underneath him and will be able to move in any direction at any speed he needs to—gracefully and smoothly.

A little bit of dressage is helpful for any horse; it makes him more rounded in his education and he is less apt to become one-sided or spoiled. And usually, the deeper you get into dressage work, the farther you want to go, for the results you get with the horse make it seem increasingly worthwhile. A little bit of dressage makes a better-trained horse, and a more full experience of dressage makes him more than trained; it develops the horse physically and mentally so that he is truly at "one" with his rider—able to understand whatever the rider asks of him, and physically able to perform it.

Horsemanship is the fine art of communicating with your horse so that you can work together in perfect unity. And dressage, through the use of tact and patience, with the goal of communicating through the horse's understanding rather than through force, is the highest form of that art.

26

SHOWING YOUR HORSE

COMPETITION IS A MAJOR INTEREST IN THE HORSE INDUSTRY today, both professionally and just for fun. If a person has a good horse, he usually competes on him, at a horse show, in the trail rides, at the local county fair or Sunday afternoon rodeo, or he may become serious in his showing and take his horse on the show circuits to major horse shows across the country. But whether a person shows his horse seriously or just for fun, there are some good pointers and facts about showing that he ought to know. A good horse is a good horse, but sometimes it is the little things that make or break a showing—the little extra things of showmanship that determine whether or not the horse will be in the ribbons.

There are many kinds of classes in which a horse can be shown, everything from halter classes in which the horse itself is judged, to all kinds of performance classes. In some of these, the rider's ability is judged (equitation classes, seat and hands classes, etc.), and in others the horse's performance alone is judged (western pleasure, stock horse classes, cutting horse classes, open jumping classes, trail horse classes, etc.).

Horse showing is not only popular and fun; it is big business. Many horse breeders use the showing of their horses to make the public aware of their horses' quality and ability. A horse may be a good horse, but if he also has a good show record behind him, he will be worth more. Many people are more interested in buying or breeding to a horse that has some national fame. The best in the breed get that kind of reputation because of their winnings in many shows, or because of the good showing of their offspring.

Let's take a look first at halter classes. In a halter class the horse is being judged, primarily on his conformation but also on a few external things that add or detract from his appearance, such as the amount of flesh he is carrying or the way he is groomed. He is judged on conformation and on how well he exhibits breed type. If he is a Quarter Horse, he'd better look like a Quarter Horse in a halter class, though it will depend somewhat on the judge's preference whether he should look like a racing Quarter Horse, a bulldog Quarter

Horse or a middle-of-the-road type. Appearance is the main thing in a halter class, but there is also quite a bit of training involved in making a top halter horse.

A halter prospect should have excellent conformation. Really, that comes first, and breed type comes second, but you've got to have both to do much winning. If a horse has good feet and legs and travels straight, as well as having a well-balanced overall appearance, you've got the first problem licked. If you can detect a weakness in your horse's conformation or way of traveling, you can be sure that the judge will, too. Halter classes are becoming more and more competitive as more really good horses are being bred and shown.

But a halter horse must also be developed as well as bred. The competition is so keen that you may not win even if you have a very good horse unless you also take the time to develop him (feeding, training, grooming) so that he will really look good every moment whatever you are doing with him. If there are several horses in the class that are equally outstanding conformation-wise, the horse with the best appearance and manners is going to get the ribbon.

When selecting a halter prospect, things like breed character, size for age, conformation and way of going are things to keep foremost in mind, but once your selection has been made as to the horse you'll show, your goal is to develop that horse to his highest potential in appearance and manners, so that you can convey that horse's excellence to the judge.

Ideally, the time to start training your halter horse is right after birth. You should handle the foal, get him accustomed to you, easy to catch, halter and lead, for training a foal is much easier than training an older horse. The foal learns quickly and never forgets his early lessons. And it is very easy to teach a horse to lead correctly, walking beside you, when he is a foal.

But if you haven't been preparing the horse for halter showing since the time he was young, you should allow at least 60 days for getting him ready. It takes time to get him in top physical condition and to get him perfectly trained. Getting a horse in top physical condition requires proper feeding and plenty of exercise. Feed, exercise and grooming are all necessary ingredients to put the bloom on a horse. The best breeding and conformation in the world still won't win in the halter class unless the horse is also properly fed, exercised and groomed to look his best.

A feed ration must be one that will meet all the nutritional requirements of the horse. Good feed will help get a horse looking well, and good exercise will help keep his muscle tone—changing the feed to muscle rather than fat. Too many show horses are too fat. Fat may cover up a few conformational faults, but a judge of horses should be judging halter classes with riding animals in mind, not beef.

A good horse will look much better carrying the proper amount of flesh, with good muscle tone rather than flabby fat and a big belly. If your halter horse gets some good exercise he will have better muscle and be able to move more freely and easily. If the horse is of riding age, he can be ridden to exercise him; if he is a young horse he can be led or longed daily.

Exercise, as with anything you do with the halter horse prospect, should be

When selecting a halter prospect, keep in mind things like breed character, conformation, and way of going. This foal is of good size and muscle for his age, and is a Quarter Horse type

Exercise your halter prospect. If you can't ride him, longe or lead him. This filly is working on a longe line

built up gradually, starting with just 15 or 20 minutes a day. As his exercise increases, his feed may have to be increased. But if he is too fat, he should be exercised sufficiently to trim down a little and get in better shape, musclewise. If the horse gets sweaty and hot, don't keep exercising him; cool him by walking and brushing, then daily increase his exercise gradually as he builds better muscle and wind. When a horse begins to get in good shape, 30 or 40 minutes of exercise a day will help keep him that way.

Grooming is an essential part of the horse's bloom and should be a part of each day's routine. Lots of brushing removes dirt and dandruff and tends to put a shine on the coat by stimulating and spreading the horse's natural body oils. A pound of elbow grease is worth a pound of grain when it comes to putting a good bloom on a horse's coat, but the horse must first be in good physical condition. The bloom must come from both the inside and the outside.

When grooming and cleaning up a horse, brushing is better than washing. Water should never be used all over the horse's body. Horses have a natural body oil in the skin and hair that can be destroyed by washing, leaving the coat dull and rough. Washing can also lower the horse's body temperature too much and chill him, especially if there is any breeze, making him susceptible to catching cold. With light-colored horses or horses with white legs it is sometimes necessary to remove dirt and grass stains with water, but it isn't necessary to wet or wash the whole horse.

A smooth, sleek hair coat comes from several things. First of all the horse must be healthy. It is always a good idea to worm your horse when you start to condition him for showing; he'll utilize his feed better and slick up faster. Good feed, good exercise and good grooming are also important in getting a good hair coat. If it is winter and you live in a cold climate, your horse will naturally have a long coat of winter hair, but if you try to get him shed out and sleek earlier in the spring than normal—in preparation for showing—be careful. Stabling a horse in a warm barn, blanketing him and grooming him will help him shed out and get a summer hair coat even if it is still winter, but don't overdo it. Don't get him sweating under his blankets and then chilling (from the moisture) and catching cold. And don't expect him to spend much time outdoors in cold or stormy weather if he just has a short coat of hair.

If you are planning to show a foal, and wanting to get the foal's hair coat looking its best, don't keep the mare and foal confined in a barn all the time. The foal needs sunshine and exercise. This may actually be more important to his growth and well-being than a carefully pampered coat of hair. If the weather is cold you might keep the mare and foal in the barn at night, but turn them out during the day whenever the weather permits.

Let's look more closely at the grooming that goes into preparing a horse to look his best in the show ring. Grooming—a regular daily program of it—is important for any horse you plan to show. Good grooming removes the surface dirt and dandruff from the horse's coat, has a massaging effect on his muscles, stimulates the oil glands of the skin that keep the hair looking healthy and shiny, and helps open the pores of the skin.

When grooming the horse, work from head to tail in a systematic manner so that you are not just spreading the dirt around. Clean your brush often as you go along. A fairly stiff dandy brush is a good one to start with, removing the most obvious dirt or mud. A curry comb can be used on dried mud or sweat to remove it easily, but a brush is much better than a curry comb to use on the horse's body; many horses are rather sensitive and a brush is not as harsh as a curry comb. A good brushing will remove the dirt, especially if you use a flip of your wrist to actually get out the dirt, flipping it out and away from the horse's body instead of merely spreading it around. After you have gone over the horse with the dandy brush and have removed most of the dirt, you can go over him again with a fine body brush to get any dust or dirt that might have been missed by the dandy brush.

Use a brush on the horse's legs rather than a curry comb, for legs are sensitive. A soft cloth can be used around the horse's head and eyes and other sensitive places such as the loins, flank, and sheath. And when you are done grooming the horse and he is clean, use a soft cloth as a finishing cloth to take off the last little bit of dust on his body and give his coat a shine.

A brush should be used on the horse's mane rather than using a comb, for most horses have rather thin manes to begin with and a brush will not pull out so many hairs as a comb. If a horse's mane does need thinning, this can be done with razor or scissors, just so you be very careful to thin it uniformly and leave the ends uneven so that the mane will not have a chopped off appearance.

The mane and tail should be styled according to the horse's body type and breed. But think twice about roaching the mane clear off and shortening the tail to just a wisp; the horse's mane and tail are his primary weapons against biting flies and mosquitoes. If a horse's tail is too long or thick it can be trimmed and thinned. Some people pluck a horse's tail, taking a few hairs at a time and jerking them out. This accomplishes the thinning, but it is painful to the horse. Some horses will be rather patient about the ordeal but others won't, and it's better to trim the tail and thin it with scissors or razor rather than pulling out the hairs. But when trimming the tail, care must be taken to trim and thin it artistically, never trimming very many hairs at the same length; you don't want the tail to look chopped off. It should taper gracefully and look natural.

The long hair at the fetlock joints should be trimmed. This is called booting the horse and gives his legs and feet a much neater appearance. The long hair in the horse's ears can be trimmed. The best way to do this is to trim the hair flush with the ear while the ear is held closed. The hair inside the ear should never be trimmed out, for this destroys its purpose; it is there to keep out flies and dirt. Long hair under the horse's jaw can be trimmed to give his head a neater appearance, but the whiskers on his chin and lips should never be removed for these are the horse's feelers. He can't see the end of his nose and these whiskers help keep him from bumping the ground or the side of his stall or the bottom of his oat box.

The horse's hoofs can be cleaned with an old stiff brush. When they are completely clean, a hoof dressing can be applied. Clean the bottom of his feet regularly and keep his feet properly trimmed. If the horse's hoofs are dry, a good hoof conditioner should be used. When you are showing the horse, don't use an oil or hoof dressing that will collect dust in the show ring. A good clean hoof is best, but if you want to use a hoof dressing be sure to wipe it off well before you take the horse into the ring.

The trimming you do on a horse (clipping fetlocks, long hair in ears, etc.) should be done just a couple days before the show so that the horse will have a fresh, neat appearance and the hairs you clip won't begin to grow out again. If the horse is being shown in full mane, clip a small section just behind the ears. This is called "bridle path" and removes the hair that would be under the headstall or halter. A well-trimmed bridle path gives the horse's head and neck a neater and more graceful appearance.

When preparing the horse to go into the show ring, it might be wise to spray the horse's body lightly or wipe on lightly a good fly repellent. This will keep the flies from pestering and irritating him when you want him to attend to business, and will also give a shine to the horse's coat if applied in the proper amount. Too much fly spray will give the coat a plastered appearance. If you are using a spray, make sure the horse is accustomed to it before the show so you won't upset him with the hiss of the spray just before you enter the ring.

In the halter class the horse will have to stand squarely so the judge can look at his conformation, and will have to be able to lead freely and smoothly at both the walk and trot so the judge can look at his action and way of going. Lessons in standing squarely and in leading properly should be given well in advance of the actual showing. Manners in the show ring are learned at home. The horse should also become accustomed to the halter which will be used in the show ring. It should be a good looking halter—one that adds rather than detracts from the horse's appearance—and should fit him well.

When the horse is to stand still in the show ring for the judge to evaluate, he should stand square and be alert, with his head up and his ears attentive. Most trainers teach their horses to come to attention and stand squarely when the trainer stands in front of them and snaps his fingers, giving a slight lifting motion to the halter rope at the same time. In a short time the horse will learn to associate standing squarely and assuming an alert pose with these actions, especially if a training whip is used correctly to encourage the horse to come to attention and arrange his feet squarely.

If the horse is standing crookedly or resting a hind foot, the whip can be used lightly to tap or tickle the offending foot until the horse picks it up and puts it down in the right place. If the horse's front foot is too far back, you can tap him gently with the whip on the back of the heels or cannon until he moves the foot forward. Care must be taken not to overdo the use of the whip; you don't want to make the horse fearful or resentful. Stop irritating the foot

If the horse is not standing squarely, the trainer stimulates the desired foot with the whip

immediately when the horse moves it and accomplishes the desired position. If the stimulation with the whip is overdone the horse may become excited because you have confused or annoyed him, and then he will do nothing right. The first few lessons don't have to be perfect. He needs to first understand what is wanted, and this may take several lessons.

Feel satisfied that the horse is progressing if he gets used to the training whip in a couple of lessons and begin to move his feet when you use it. Even if he isn't perfectly square, but has improved his position, call it good for that day. He will get better as he comes to understand what is wanted. In time you won't need to use the whip on his legs at all; by merely pointing to a leg with the whip he will understand that you want it moved and will put it in a better position. But this all takes time and patience. Many short lessons are better than a few long ones. The horse will stay fresher and learn faster and won't come to resent what you are trying to do.

If a horse gets bored or a little tired of what you are doing with him and won't look alert and perky, talk to him and maybe tap him on the nose with the lead strap—not hard, just enough to get his attention. Or snap the whip in front of his front legs or chest (don't make him think you want him to move his front legs however). You don't want to startle him or make him move, but you do want him to look alive and get his ears up.

Depending on the class, you may not be able to use a whip when you are in the show ring, so the horse must be trained to the point that you don't need

Heidi and her first purebred Arabian filly, Shezadahl 67565

Many trainers teach the horse to assume a square stance and look alert when the trainer gives a lifting motion to the halter rope and snaps his fingers

In this photo the filly is relaxing too much, resting a hind foot and losing her alertness

one. He should be thoroughly accustomed to standing squarely. If he slouches, just the sound of your voice should make him square up. If a front foot is too far forward, you can touch it with the toe of your boot and tell him to move it back.

The time spent at home getting a horse in shape for a halter class—the grooming, feeding, exercise and training—will all pay off in the show ring because there your horse has to really look good and be good, for the judge has only a few minutes to look at him. A few lax moments in the show ring and the ribbons will be going to someone else.

When you lead the horse for the judge, remember that you are trying to show the horse to him. The judge will want to see his action so he will stand directly in front of you or directly behind you as you move the horse out. Therefore you never try to lead by walking out in front of the horse. You'll obstruct the judge's view.

The proper way to lead a horse is to walk beside the left shoulder, holding the lead strap about eight inches from the halter with your right hand (but not pulling at the horse—you are letting the horse walk freely beside you on a loose strap) and carrying the balance of the lead strap neatly coiled or looped in your left hand. Your right foot should come forward as the horse's left front foot comes forward. In other words, the inside legs of horse and handler come forward together. This way there is no chance of the horse stepping on your foot or tripping you with his leg. Your closest feet are always moving together, moving forward at the same time.

Lead the horse, walking beside him, letting him move out freely on a loose lead strap

Do not pull him along. The horse tends to resist, and this puts his head and neck in an unnatural position

The horse should learn to lead properly at the trot as well as at the walk, with you trotting along beside him. He should move out freely and smoothly at the trot. It is very aggravating to a judge to watch a horse that hangs back at the trot, with the handler out in front trying to pull him along. If you are in front of the horse the judge cannot see the horse's action clearly —you are in the way, and if you have to pull the horse along he probably won't be traveling straight anyway. If the judge can't see his action, he won't guess, he'll place another horse instead. So if your horse doesn't lead well at the trot, some time should be spent in teaching him.

You can signal the horse to trot by moving your own feet faster (trotting in place, so to speak). If your horse balks, you may have to carry a switch or training whip and tap his hindquarters with it as you ask him to trot. Soon he'll get the idea.

Every time you stop the horse, whether from a walk or trot, he should stop squarely and should not roam from side to side. The more training you can give him on standing squarely, moving out properly at walk and trot, and stopping squarely, the more likely he will be to put on a good performance in the show ring. If it is the horse's first attempt at showing, he may be a little nervous or fidgety just because it is strange to him. You may want to do a little practicing at home with other horses nearby, or take him to a show arena and get him accustomed to strange sights and circumstances before you show him. But if he has had good training, trusts and respects you, he should put on a creditable performance even if it is his first show. However, the more you show him, the more accustomed he will become to the show ring and the better he will be. The only problem you'll have then is if he takes it so much in stride that he is dull and bored. Then you'll have to work on keeping him alert.

When you are actually at the show and in the ring with your horse, it is your responsibility to try to keep your horse alert and looking his best, trying to show the horse to the judge advantageously at every opportunity. Pay attention to your horse as well as the judge. Don't get so busy watching the judge that you forget about your horse. You may think the horse is standing correctly and then turn around to find that he has moved a little or is slouching or resting a hind foot.

You should have the horse so well trained that he will be able to stand squarely for 15 or 20 minutes if need be without getting fidgety or tired and unattentive. But don't wear him out. When the judge is close by, even if he may not be looking at your horse at that particular moment, keep the horse at attention and looking his very best. But if it is a large class and the judge is elsewhere, you can let the horse relax a little so long as he is still standing squarely and looking good.

Keep aware of where the judge is at all times, and when he approaches don't stand in his way. Let him see the horse. Move away from the area which the judge will wish to see; never block the judge's view. Have your horse ready whenever the judge looks at him. The horse will look better if you turn his head slightly in the direction of the judge, and if he is standing

Don't concentrate so much on what the judge is doing that you forget about your horse—he may move and start slouching

Lead the horse straight away from the judge so he can see the horse's action

If he is led crookedly, the judge can only guess

squarely at all times. Never let him move away from the judge. When the judge asks you to walk and trot your horse, lead the horse from the left side and move straight away from the judge at a good brisk walk, push him around the turn at the other end of the ring so that the judge can see him

Push the horse around the turn so that the judge can see the horse

But don't restrain the horse as much as in this photo—the horse is not moving as freely and smoothly around the turn as he should

as he turns; the judge wants a good view of the horse, not of you. Trot him straight back to the judge. Keep the horse alert, but with his head in a natural position. A good showman stays out of the judge's line of vision himself, and moves decisively, not looking back. When you have finished trotting the horse back, go to the end of the line and stand your horse squarely, where he won't be in the way of the remaining contestants.

Good manners on the part of the showman are almost as important as having a good horse. A good showman does not overshow his horse. He is alert and attentive to the judge's wishes, but he doesn't put on a vaudeville act at the end of the leadstrap. And always remember that good sportsmanship is as important as good horsemanship.

When you stop to think about showmanship in a halter class, there are a few other points that could be mentioned. If you enter the ring first, you will be looked at first and your horse had better be good and do everything right. If you go in last, you'll also get looked at first when the judge asks you to line up.

If you are a novice, the safest place to be in your first show or two is in the middle, for you can give yourself a little time to get oriented and ready; there are horses on either side of you and you can take cues from them, and position your horse correctly in the lineup when the time comes.

When the horses are asked to line up, avoid congested spots and try to get in a place where your horse can be seen well, but don't be too obvious about it. In other words, try to show your horse to best advantage, but don't be conspicuous. Overshowing a horse can be as bad as undershowing. Don't be aggressive and obnoxious; that kind of behaviour will be disliked by the judge and contestants as well. You are competing, but you must be a good sport.

When preparing a horse for showing, teach him to trailer load, well in advance of the show date. You want to be able to get him there without a hassle. When you get to the show, try to get good facilities for your horse. If there is

You don't need to put on a vaudeville act at the end of the lead strap

straw bedding in his stall make sure the straw is fresh and clean, not dusty or moldy. And don't let him eat straw. If you feed him some hay and let him fill up on that, he might not be so interested in eating his bedding. Check his stall for hazards—splintered boards, nails, wire, etc.—and make sure there is nothing in the stall that might hurt or scrape him.

Some horses become temperamental about eating and drinking when they are in a strange place. It is wise to take along all your own feed so that the horse can eat what he is accustomed to. If he doesn't like strange water you can put a little salt in his feed to make him thirstier and more likely to try the water. Another thing you can do is accustom him to a trace of vinegar in his drinking water before you start going to shows. If he gets used to the strange taste, he will usually drink any strange water if you put a few drops of vinegar in it. Take along your own buckets for feed and water. Water your horse in your own bucket, out of a faucet. It's not always safe to water your horse in public watering troughs for he may pick up disease. Vaccinate him against all the common communicable diseases before you take him to shows, but even then, take no chances when you are among strange horses.

If you are at a show for several days, get your horse out and exercise him every day just like you would at home. If it is winter, take a few horse blankets along, for you can't depend on the weather. If it is very cold, blanket him in his stall. But if he is in a large barn and there are many horses, the barn will be warm just from their body heat and you won't need to blanket him.

Let's look at some of the other kinds of classes you might wish to show a horse in. Western equitation and western pleasure classes are somewhat similar in the things you are required to do, but there is a major difference; in a western pleasure class the horse is being judged on his performance, and in a western equitation class the rider is being judged on how well he rides and handles his horse.

The western pleasure class is one of the most popular classes in western horse shows, having large numbers of entries. Thus the competition is keen. And since there are no great feats of skill involved, no reining, roping, or jumping, the horse being judged on his abilities as a pleasure horse, a horse has really got to be good to win. A lot of horses are good pleasure horses. So a horse that wins the western pleasure class must be exceptional. In an open pleasure class it doesn't matter if the horse is purebred or grade, young or old; in fact his conformation need not be exactly perfect. What counts is his ability to give a pleasant ride, and this is what the judge should be looking for.

But there are really no clear-cut rules for a western pleasure class and not all judges judge all western pleasure classes alike. There are many variations and interpretations of the existing rules. The American Horse Shows Association created the existing rules, but these are not specific in all cases, are not always widely known, and are sometimes misinterpreted.

For the western pleasure class, various rule books will state: "Horses are to be shown both ways of the ring at the walk, jog and lope on a reasonably loose rein without undue restraint." The horse is judged on disposition, manners, gaits and the ease with which he changes gaits, his conformation, condition, grooming, and the way in which he responds to the rider and what the rider demands of him. But just how much he is judged on one thing as against another varies with the rule book and the judge. For instance, one rule book states that the horse should be judged 70% on performance and suitability and 30% on conformation and manners. Another rule book says: "Ease of gaits, ease of handling, and change of gaits, 75%, conformation 25%". Still another rule book lists several rules for western pleasure, but no percentages:

"1. Western equipment must be worn by horse and rider. 2. Horses to be shown at walk, trot and canter both ways of the ring. 3. Horses to be ridden on a reasonably loose rein. 4. Use of stallions as pleasure horses shall be at the discretion of the show committee. 5. Tie-downs are optional with the rider." Still another rule book gives this guideline: "Ten points each for appointments (rider's dress and horse's tack) and conformation. Twenty points for manners. Ten points for each gait each way of the ring. Perfect score 100. Riders may be requested to back their horses to break a tie."

There are still more rule books with still more variations. This gives you some idea what is required in a western pleasure class, but you can also see that it leaves room for some confusion and argument. And it means that very likely no two judges will judge a western pleasure class exactly the same way. This can be very confusing and sometimes disappointing. One judge may advocate horses really extending (doing a good trot and gallop) in a western pleasure class, while another judge may prefer the horses that stay more collected, never moving out very fast. Some judges prefer a looser rein than others; some like to see the horse on the bit. The western pleasure rider must keep this fact in mind, and try if possible to determine what type of riding the judge wants to see.

Let's look at the gaits of the pleasure horse. Most judges will want to see a good, free-flowing walk. A horse is more of a pleasure to ride across country if he can really get out and travel. Having to constantly urge a horse to make him keep up with the group is no pleasure. A good walk is one of the necessary characteristics of a good pleasure horse. The horse is traveling swiftly, stepping out with long strides, but still relaxed. He doesn't look as though he is being pushed, nor does he look like he is about to break into a trot at any second. The walk is a gait in itself to be developed to its fullest potential; it is not merely a pause between jogs and canters.

The trot in most western pleasure classes, in the preference of most judges is merely a jog. That is, the horse is supposed to do a slow and easy trot and the rider sits the trot. This jog is an important gait to the western pleasure horse because it is a medium-fast gait and is less tiring than a canter. A jogging horse is usually calm and sedate; he's not as likely to get excited or out of control as he might at the extended trot or at the gallop. The jog is easy, relaxed, unhurried. It is not a jig. A jig is a short trot but is prancing, choppy, and nervous.

In the show ring the jogging horse should maintain a steady even gait,

The western pleasure horse must travel freely and easily and have comfortable gaits. He should change gaits smoothly and be a pleasure to ride

The western pleasure rider may be asked to back his horse

not speeding up or slowing down. The rider sits this trot to show how easy and smooth it is. Even if the judge asks the riders to extend the trot, in a western pleasure class you should sit this faster trot and never post, for you are trying to show the judge how easy and smooth your horse's trot is. Any roughness of gait is undesirable in a pleasure horse. So, even though good horsemanship advocates posting the fast trot when you are out on the trail, sit that trot when you are in a western pleasure class.

The canter, or lope, whatever the judge calls it, should be free and easy. The horse should be relaxed and smooth, traveling with ease and gracefulness. He should not fight the bit trying to go faster, nor should he be lazy and reluctant, making the rider urge him to keep him cantering. He should be able to respond readily and smoothly to a change of gaits—able to speed up into a gallop and back down to a canter, or drop to a trot easily, or even to a walk when the rider wishes. The western pleasure horse should be able to go from walk to canter and from canter to walk smoothly. And in the show ring your horse should always be on the correct lead at the canter.

Sometimes a judge will merely ask you to walk, trot, and canter both ways of the ring and pick his winners from that alone. But most western pleasure classes today are so large and the competition so tough that often a judge will ask for several other things as well, in order to sort out the best horses. He may ask the class to extend the trot, for instance. Or he may ask for individual work, such as mounting and dismounting, backing the horse, or individual work along the side of the arena.

If you are asked to back your horse, the horse should back smoothly and in a straight line. His head and neck and jaw should be relaxed; in other words he should not be rigid or braced against the bit. He should not open his mouth or throw his head up, nor should he drop his head and rush backward.

If the judge asks you to dismount and mount again, your horse should be trained to stand perfectly still. He should stand quietly for you to mount from either side, and continue standing still until you ask him to move forward. A horse that fidgets or tries to move forward while you are mounting will be penalized.

Individual work at the side of the arena may be called for if the judge has singled out several of the top horses and wants to see them perform individually. To be prepared for this your horse should have some preliminary training so that he won't be herd bound (reluctant to leave the lineup of horses and move out away from them). Your horse will not score well if he must be urged out of the lineup or if he continually keeps angling back toward the other horses while he is working alone. If the judge asks you to leave the lineup and do some individual work, he might ask you to travel partway around the arena at a walk or jog and then change gaits at a designated point, or walk or jog down one side, reverse and then canter back on a certain lead. If a judge doesn't specify which lead he wants you to take, always take the proper one in conjunction with which side of the arena you are nearest. He might also ask you to canter down one side, stop your horse, and then walk back. Your horse should canter freely even though he is leaving the other horses, and walk calmly back even though he might want to hurry back.

Sometimes a judge will call for a canter departure from a standstill. In other words, he may ask you to stand your horse for a few seconds, then move him out into a canter on the correct lead. Or he may ask you to halt your horse abruptly from a gallop. If this is the case, your horse should stop quickly and smoothly and stand still. Or, the judge may ask a few riders to perform a flying change of leads or figure eight. If the class is a good one and it is hard to pick the winner, the judge will probably have to resort to some individual work to pick the best horse, so you and your horse should be prepared.

Your appearance will add or detract from the overall picture seen by the judge. If you are sloppy and your horse is not well groomed, you won't win the class even if your horse performs well. Your horse should be in good flesh, have good conformation, and be neat and well groomed. You should be neat and clean, your tack appropriate to the class and should be clean and in good shape. Before you enter the class, find out what is considered correct equipment for that class.

In a western pleasure class you will have to ride with a western saddle and curb bit. Some rule books do not allow curb chains. Some require the wearing of chaps or the carrying of a slicker and lariat. So check with the rules to make sure you and your horse are correct.

Now let's look at western equitation. This type of class is a horsemanship class, in which the rider is being judged. The rider will be judged on his hands, seat and balance, how well the horse responds to his cues, the appearance of the horse and rider and the suitability of the horse to the rider. The horse and rider should be a well-balanced pair and should look well together, complementing one another.

In western classes there are often two kinds of equitation classes, the stock saddle seat equitation classes and the seat and hand classes, but they are very similar. The rider should be able to perform the regular ring routine as well as any additional maneuvers that the judge may ask for. A western equitation class is often thought to be a modified version of an open stock horse class.

The horsemanship is in a sense more important than the performance of the horse. The methods used by the rider to obtain the performance are what is being judged. Thus, a good rider on a poor horse should actually place higher than a poor rider on a good horse. Credit should be given to the rider who is able to get a creditable performance out of a flighty or somewhat green horse rather than given to the good horse who does a creditable performance in spite of a poor rider. The rider is being judged, not the horse, so conformation and soundness of the horse should not be taken into consideration unless it is so serious that it will impair or interfere with the horse's ability to perform the required movements.

In western equitation classes, the AHSA rule book states that:

> The saddle must fit the rider. It may be slick or swelled fork, have a high or low cantle, but must definitely be sized to the rider. A half-breed, spade, or curb bit may be used, but no wire, chain or other metal or rawhide device is permissible in conjunction with or as part of the leather chin strap. A hackamore is prohibited. Tie-downs, running martingales and draw-reins are prohibited. Spurs are optional in all Stock Saddle Seat classes. . . .tapaderos . . . are optional. Silver equipment may be used but shall not be given preference over good working equipment. Riders to wear western hat, chaps, shotgun chaps or chinks, cowboy boots, and carry rope or reata. If closed reins are used, hobbles must be carried; if split reins are used and the horse is broken to ground tie, no hobbles are necessary. Bosals or caveson-type nosebands are prohibited. The use of shoes other than standard horse shoes are to be discouraged and may be penalized by the judge.

This rule book goes on to say that:

> The rider should sit the saddle with legs hanging straight and slightly forward to the stirrups. These should be just short enough to allow heels to be lower than toes. Body always should appear comfortable, relaxed and flexible. In repose, arms are in a straight line with body, the one holding reins bent at elbow. Only one hand is to be used on reins and hands shall not be changed. Hand to be around reins. Fingers between reins not permitted. Reins to be above pommel and as near to it as possible. Use of both hands on reins or one hand braced against saddle horn or coiled reata will be penalized. Rider should sit to jog and not post. At the lope he should be close to the saddle. All movements of horse should be governed by reining and shifting rider's weight in saddle. Feet should be placed in stirrups with weight on ball of the foot.

The horse is to be shown both ways of the ring at the walk, jog and lope and should always be on the correct lead at the lope. The horse and rider should also be prepared and able to do anything else the judge asks of them, such as mounting and dismounting and some of the other maneuvers which have already been covered briefly when discussing western pleasure classes.

There is some argument over some of the rules governing the western equitation class, such as the rule prohibiting any fingers between the reins. Most good horsemen use a finger between the reins; it gives better flexibility and control, and it seems ridiculous to prohibit this in a horsemanship class. Some of the rules may change if enough horsemen protest them. Be sure to check on the rules that will be used in the class you wish to enter, so that you can be sure what you will be judged on.

Another type of equitation class is the bareback class. This again, is primarily a judge of the rider and not the horse, but a good horse can help a lot. The rider will be asked to walk, trot and canter both ways of the ring, and perhaps to back his horse, halt from a canter, start cantering from a standstill and so on. The main thing here is the rider's seat and balance—how well and how gracefully he can sit the horse bareback at various gaits and changes of gaits and still be in proper balance with his horse at all times. It takes good balance and seat, and of course, good hands.

The horse should work smoothly and easily and always be in perfect control. If it comes to a tie, or if the judge finds it hard to decide among the top riders, he may ask them to dismount and mount. The proper way to mount a horse bareback is to spring lightly up, springing high enough that you can balance your upper body across the horse's back just behind his withers. Then you turn and twist your upper body and bring your leg up so that you can assume the mounted position. The horse should stand still, of course, while you are accomplishing this.

Another class which is becoming popular in western shows is the trail horse class. In this class the horse must stay calm and steady even when confronted with strange new obstacles, so your potential trail class horse should first of all have a good disposition, and then should be given some good and careful training in preparation for the show ring.

A good trail horse should first of all be a pleasure to ride—alert and active, moving well and responsively at walk, trot and canter, changing gaits smoothly at the rider's cue. He should have good manners and be under perfect control. Then, he must be able to go through an obstacle course set up in the show ring to simulate possible obstacles and problems that he might meet along the trail. Here are a few of the things that might be included in a trail class: crossing

Bareback riding. Horse and rider should be in perfect unison at all gaits. Here is the walk

The trot

Here at the canter

Mounting bareback

a bridge, trailer loading, going through a gate (the rider is to open and close the gate from horseback), crossing water, riding through a ditch or gully without lunging, carrying two people, dragging firewood or a bale of hay with a rope dallied around the saddle horn, crossing logs, leading another horse, going through a narrow lane of brush, backing between two logs, standing still for the rider to dismount and mount again from either side, walking through unnatural obstacles that might be along a road or trail (such as old car tires), standing still and letting the rider put on a jacket or slicker, approaching and passing a strange animal or a fresh hide, etc.

Once in a while a trail class may call for the horses to be briefly hobbled or ground tied. There are any number of obstacles that might be used, or actions called for, so if you plan to show your horse in a trail class you should prepare him for as many things as you think might be encountered.

If a trailer is used, your horse should load readily and with very little hesitation; a good horse should load into any strange trailer. More points will be given if your horse will go right into the trailer without you having to lead him in. Loading will not earn a lot of points; it is how the horse loads that will help or hinder your score. Obviously, a horse that balks and needs some coercion and encouragement to get into the trailer will not score as highly as a horse that goes right in with no fuss or bother.

Many trail classes will have a bridge for the horse to cross. It will usually consist of a board platform for the horse to walk over, and occasionally it is rigged so that it will tip a little when the horse walks over it. The tipping effect is rather ridiculous, but you may want to have your horse prepared, just in case. A horse shouldn't be penalized if he stops and looks at the bridge before crossing it, though some judges will knock off points if he does. Under real trail conditions you will want your horse to stop and look at a bridge before he crosses it, so you can get a good look at it, too. You don't want to cross a bridge that is too old and rotten or falling apart. When crossing a trail class bridge obstacle, it should not count off if your horse pauses to look at the bridge if he then moves forward across it boldly at your signal.

Most trail classes will have a gate that the rider must open, move through, and close again, all without getting off his horse. The horse must be trained beforehand so that he will know exactly what to do and will accomplish it calmly and smoothly. He must be able to work up close to the gate so his rider can reach down and unlatch it, push or pull it open, side-stepping smoothly, walk through, and then close it again and latch it. The horse must therefore be able to move his hindquarters to one side or the other to get up close to the gate, and must be able to sidestep when needed as the rider pushes or pulls the gate. When preparing to open the gate, the rider should note whether the gate opens to left or right and be ready to maneuver accordingly.

Carrying two people is another test that might be required of your horse. Under real trail conditions this might occasionally be necessary if someone's horse goes lame or gets sick. Carrying double, traveling through obstacles, dragging firewood, etc., can all be practiced at home, as can putting a big slicker on

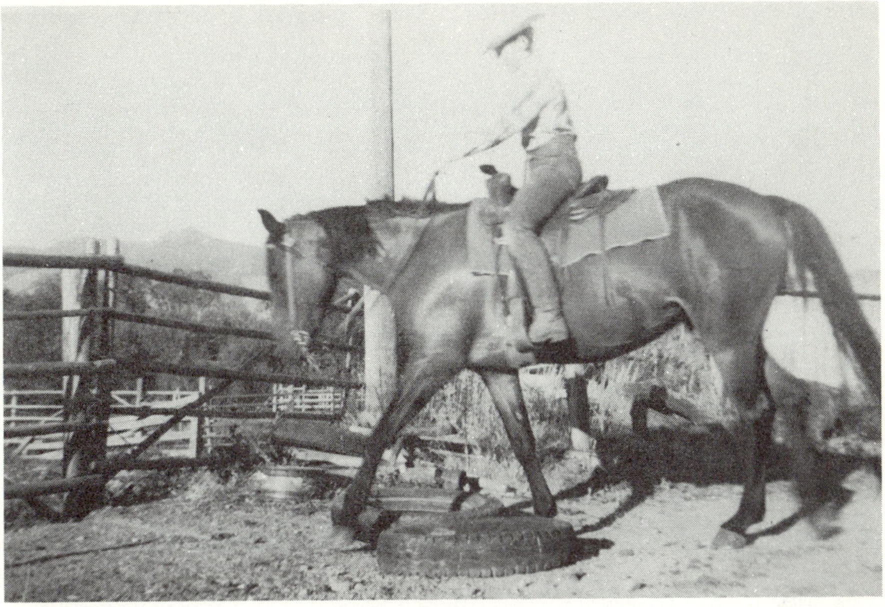

Some trail classes may have you walk through an obstacle course of old tires

You may want to practice riding your horse over poles or logs so that he will travel over them confidently and with agility in a trail class

Rider unlatches the gate and pushes it open

She rides the horse through it

Then she pushes it closed again and latches it

and off. If you don't have a slicker, use a piece of canvas or plastic—anything big and flapping and spooky so that your horse can get accustomed to you moving around in the saddle with a flapping object.

Many trail classes include a live animal or a fresh hide as one of the obstacles. The horse must go past without refusing, balking, or shying. The obstacle may be a caged animal, or a hide draped across the trail or across a log beside it. This is supposed to represent any of the unexpected spooky situations your horse might encounter along the trail, a wild animal starting up in front of your horse, or a dead animal along the trail. It is ridiculous when a trail class demands that the horse walk right across a dead hide, for you won't do this out on the trail anyway.

It is logical to ask your horse to go around a dead animal or to go past something spooky. When you come to this kind of obstacle in a trail class (often the obstacle will be situated near a fence or a log or piece of brush and you are supposed to ride your horse between them) it is all right if your horse hesitates and looks the situation over, but then he must step right on by the spooky situation at your signal and encouragement. He will be counted down if he refuses or shies or creates a commotion. A good trail horse is well trained, sensible, alert, yet responsive to his rider's wishes.

Most western horse shows will also have classes in reining, stock horse classes and cutting classes. These are rather specialized and a contestant must thoroughly acquaint himself with the rules of the class and know exactly what will be required of him and his horse. The reining classes will have set patterns and movements in which you will show your horse, and include such things as a good sliding stop from a gallop, a roll-back in either direction with the horse coming out of it at the canter on the correct lead, figure eight and change of leads and backing correctly.

Now let's look at some of the other types of classes you will find at major shows—English classes, gaited classes, and jumping classes. The English pleasure class is in a sense similar to a western pleasure class in that the horse is being judged as to his suitability as a pleasure mount; he is judged on his manners and smoothness, ease of gaits and change of gaits. But of course he is shown in English tack and the rider is suitably attired.

In an English pleasure class the horse is shown at a flat-footed walk, a slow trot, and a comfortable canter, both ways of the ring. He should always be on the correct lead at the canter, and should also be somewhat collected at all gaits. English pleasure horses should always be a little on the bit; the rider should not use as loose a rein as is permissible in western pleasure classes.

Proper riding attire is very important in English and hunter classes, as is the use of appropriate tack. The rider's clothes should not only be appropriate to the class but should also be neat and well fitting. Specifications for tack in various classes sometimes change from year to year, so be sure to check on all the details and requirements for the class you wish to enter.

In an appointments class in which horse and rider are shown in hunt attire, such things as breast plates and braided reins are preferred, a sandwich case is carried on the saddle, rider carries hunting crop and spurs, uses pigskin or other suitable leather gloves, and carries string gloves. The idea of this class is to require correct and neat equipment, such as the rider would use and wear in actual hunting when correctly turned out.

There are several types of hunter classes, for there are several ways to classify hunters. They can be classified according to weight and according to experience. When classified according to weight, the lightweight hunters are those horses carrying up to 165 pounds. Middleweight hunters carry from 165 to 185 pounds and heavyweights carry over 185 pounds.

When classified according to experience, the hunter classes may be divided like this. The maiden class would consist of those horses that have not won a first place at a recognized show in this particular division. The novice class would consist of horses that have not won more than three first places. Limit classes are for horses which haven't won more than six firsts. A "suitable to become" class is for horses four years old or younger who have not won a first in the hunter division, except in classes that require no jumping or classes where the jumps were less than 3½ feet.

The green hunter class is for horses that haven't been hunting with a recognized hunt for more than one season and have not won a first in the hunter division prior to the January first of the current year (except in classes requiring no jumping, or classes whose jumps are less than 3½ feet high). The qualified hunter class is for horses that are regularly hunted with a registered pack. The model hunter class is a class in which the horses are judged on conformation

Proper riding attire is very important in English and hunter classes. This is Heidi on Surrabu, dressed in hunting attire

English Pleasure Class. Horse is Las Trad 31156, Arabian stallion ridden by Trudy Slifka, Boise, Idaho

Waiting for their class

Hunter Class

and soundness only; the horses are judged in hand—shown at halter without saddles. In the model classes the horse or pony is stripped and led into the ring to be judged on conformation, quality, fineness, and the look of good breeding, substance (sufficiently heavy bone, good depth of chest, enough sturdiness to guarantee that the fineness and quality are balanced with strength and stamina) and soundness.

Then there are working and handicap classes. The working hunter class is open to all hunters regardless of their classification and is judged mainly on performance, way of going, the manners of the horse and his style of jumping, and hunting soundness. If the horse is sound and performs well he has a chance of winning; conformation and "honorable blemishes" (blemishes gotten in the course of his hunting career and which do not in any way interfere with his soundness) do not count against the horse in this class.

The handy hunter class is open to all hunters, too, and the tests in this class are to determine the horse's agility and manners. The rider will have to open and close gates, mount and dismount, lead the horse over a jump, etc.—meeting with obstacles that he might have to deal with out in the hunt field. To show a horse in this type of class the horse must be calm and well trained and have good confidence in his rider.

The handicap hunter or jumper class is open to all hunters and jumpers and the jumps are set according to the age or experience of the horse. The hunter hack class is open to all hunters and the jumps are no higher than 3½ feet. The horse is also judged on his manners as well as his jumping ability, and his gaits should be comfortable and easy. He is shown at the walk, trot, and canter as well as over jumps.

On the hunt course the jumps are made to represent the various kinds of obstacles met in the hunt field. Jumps are 4 feet high for regular horses and 3½ feet high for green horses in their first two years of showing. Large ponies (over 13 hands) compete over 3 foot jumps and small ponies compete over 2½ foot jumps. Hunters ridden by juniors and small hunters (under 15½ hands) use the green hunter course.

The conformation hunter class is a class in which the horse is judged on his conformation and manners as well as his performance. He should have a good way of going and a certain amount of brilliance. He should show good quality and stamina and be perfectly sound. He must take his jumps cleanly and properly, smoothly and in the proper stride. Jumping too fast or too high will be penalized.

Jumping classes differ from hunter classes in that a jumper is judged mainly on his ability to clear a jump whereas a hunter is judged on many other things as well—manners, conformation, style, etc. Jumpers may be any breed or height. Blemishes and conformation don't count. The only thing that counts is whether or not the horse clears the jumps without committing a disobedience, and clears them within the time allowed. In a jumping class the horse is scored on actual jumps—scored on whether or not the horse clears them. In case of ties, the jumps are raised or widened by the judges. This class judges performance only.

Judged on performance, way of going, manners, style of jumping and soundness

In some jumping classes the horse is penalized if he touches a jump; in other classes he is only penalized if he actually knocks down a jump or doesn't clear it. There are many kinds of jumper classes—maiden, novice, limit, open, handy and so on.

A class for scurry jumpers is a class with a time limit, with the fastest horse winning. One second is added to the horse's time for each fault he commits. In a touch and out class the winner is the horse that clears the most obstacles without a touch. In a knock down and out class the winner is the horse that clears the most obstacles without a knockdown.

In several types of jumping classes, faults are scored this way: A horse touching obstacle with any part of his body behind the stifle—½ fault; touching obstacle in front of the stifle or with any part of the rider or his equipment—fault; a knockdown—4 faults; a foot in a water jump—4 faults.

Then there is the saddle horse division. This includes the three and five-gaited classes. The three-gaited saddle horse is shown at the walk, trot and canter both ways of the ring and is judged on his quality, performance and manners. Attire and tack and style of riding for saddle seat and saddle horse classes are entirely different from that of other classes. If you plan to show a horse in saddle seat or saddle horse competition you must become familiar with this type of showing in order to become correct.

The three-gaited park saddle horse is judged on style, finish, quality, manners and all around brilliance. His gaits must be very pronounced and flashy. The road hack class is quite different. In this class the horse is shown on a loose rein at the flatfooted walk, easy trot, extended trot, easy canter and gallop. In

Jumping Class—judged mainly on the horse's ability to clear the jump

this class the horse must also stand quietly for mounting.

The five-gaited saddle horse is shown at the walk, trot, and canter and two additional gaits. His walk should be graceful and animated. His trot should be smooth and slow. His slow gait, which can be an amble, running walk, or fox trot, should be slow and animated with very high action. His fifth gait, the rack (a four-beat gait sometimes called a singlefoot) should show speed and good form.

Special dress for the rider is required in five-gaited classes. For morning and afternoon classes the dress is informal—jodhpurs, coat and hat in light blue, gray or tan. For evening classes the rider's dress is formal—dark blue or black jodhpurs and tuxedo-type coat of the same color, with a black derby. In evening classes for three-gaited horses the riders wear a silk hat.

There are several types of horsemanship classes for English riders. In these equitation classes the performance of the rider is being judged, not the horse though a good horse that performs well can certainly aid the rider's performance and show his skill to advantage. There are hunter seat classes, saddle seat classes, and so on. There are lead line classes for children under six years old. The hunt seat is a forward seat. The saddle seat is more upright and to the rear.

Horsemanship over jumps is another type of class that judges the rider rather than the horse. The rider is judged on his position on the horse over a jump, his control, balance, and use of his aids over a jump course. The rider must not interfere with the horse's performance. He must be good enough that he doesn't hang on the reins and restrict the horse's head and neck in ba-

Hunt Seat Class

Horsemanship over jumps. The rider is judged on form, balance, control, and use of aids

lancing. He must not lose contact with the bit nor pull too hard. His weight must be forward and balanced and he must not look down at the ground or at his horse; he should keep his head up and his eyes on the next jump. The rider's body must follow the movements of the horse; he and the horse must be in perfect harmony and unity. He should keep his knees well bent, his heels down, and his arms should be extended, his hands following the movements of the horse's head. In this class the rider is judged on horsemanship alone, regardless of how his horse performs.

In regular jumping classes, in which the horse's performance is being judged, fault penalties are counted against the horse for every blunder he makes. Most of the worst faults are knockdowns, which technically are any lowering of an obstacle's height, such as a pole knocked off. A knockdown by any part of the horse's body behind the stifle counts as two faults. A knockdown by any part of the horse's body in front of the horse's stifle is four faults. A knockdown of the jump standard or wing of the jump by any part of the horse's body or rider's body counts as four faults.

Showing a horse can be educational and rewarding. Much preparation goes into a good showing. There is still the sad fact that politics sometimes enters the picture, and the fact that no two judges will judge exactly alike—the fact that personal preference can be intermingled with horse show rules and breed standards, but as long as there are horsemen dedicated to producing and competing in good shows, horse showing will continue to be a worthwhile sport. It provides entertainment for spectators, but much more important than that, it provides good and honest (we hope) competition among horsemen. Horse showing at its best helps encourage better horses and better horsemanship. It encourages the improvement of breeding stock. When horse show standards are high and competition is keen, it encourages developing better horses—better conformation and breed type, better ability and training.

The American Horse Shows Association regulates the rules of all the major shows. This is a good situation, because then there can be some standards all over the country for horses to compete on equal terms in similar classes. The ASHA is a good thing, but horsemen need to go a step farther and work toward more unity on standards.

Even though the rules are stated and are to apply at all ASHA shows, there still needs to be more clarification, and less room for individual interpretation and prejudice. Horsemen must work together to keep the rules up to date and clear so that horsemen entering various classes can know what will be expected of them and won't be confused or disappointed when the class is judged.

If you plan to show a horse, study the rules of the classes you wish to enter, observe some classes if possible, ask the advice of a professional showman if need be. Get your horse ready for the class; never try to show a horse that is not conditioned or trained for the event. When you think your horse is ready, get your entry in early, get your horse to the show on time, and remember that good sportsmanship is what makes showing enjoyable. Happy horse showing, and good luck!

27

COMPETITIVE TRAIL RIDING AND ENDURANCE RIDING

AN AGE-OLD SPORT (AND ONE WHICH WAS, AT TIMES, A NECESsity rather than a sport) has come back into popularity during the last decade and a half. This sport is the long-distance ride. It can take the form of competive trail riding or endurance riding, and it requires a well-conditioned horse and rider. In this day and age of super technology and jet-age wonders, it is refreshing to see a sport like this gaining popularity, for the long-distance ride is an event that takes much time and training on the part of horse and rider, and is a type of riding that strives to work with and improve upon the horse's natural ability—to carry a rider and to travel, over all kinds of terrain. An endurance horse is not made overnight. It takes hard work, a lean, fit horse. a durable rider who is a good enough horseman to know how to condition the horse and to get the most from that horse while hindering or stressing him least, and it takes time, lots of time.

This type of sport is not for the weekend horseman nor for the horse owner who likes a fat, sleek animal that can stand around in a show ring for twenty minutes. This sport is for the horseman who enjoys working with his horse daily, enjoys strenuous work and prides himself on a horse that is sound and healthy, able to stay sound and healthy even under rigorous conditions—a horse that has the go power and staying power to undertake strenuous work with ease.

Down through history horses and horsemen have left amazing records of endurance and there are many famous instances of long-distance riders and spectacular horses. Some of these were planned rides and some weren't; some were during war or were valiant attempts to get help, send news, and so on. The U.S. Cavalry conditioned good horses for their long marches (sometimes a cavalry unit had to be able to travel 250 to 300 miles in a week) and sometimes held their own endurance rides as contests in which their best horses and riders could compete.

One of the first of the modern day endurance rides is an annual 100-miles-in-one-day ride often called the Tevis Cup Ride. It was originally called the Western States 100 Mile-One Day Ride, and was begun in 1955 to prove a

Salmon River Competitive Trail Ride, 1971

By the dawn's early light—riders get their horses out of their stalls and saddle up for the ride

Rider no. 6 sets her watch and waits for the signal to go

point. Many people were of the opinion that horses today in our modern age are not as tough as their predecessors. Wendell Robie, a breeder of Arabian horses, disagreed, and got into a good argument with a friend. At last they decided that talking wouldn't settle the argument, and Wendell Robie volunteered to try to ride 100 miles in 24 hours or less. The route he picked was from Lake Tahoe across the old Sierra Nevada Pony Express route to Auburn, California. This would be a ride covering all kinds of terrain, and quite a variation in elevation as well (Tahoe City is over 6000 feet, Emigrant Pass is 7930 feet, Auburn is only 1000 feet). The route included some of the most rugged hills in the West, and several deep and difficult canyons. Robie was sure he could make the ride, but in order to make the test even more conclusive, he talked four more men into riding with him. They decided to make the ride at the time of the full moon nearest the end of July, since it would take them longer than the daylight hours to go that distance; after the sun went down they would be guided by the light of the "riding moon" (Plains Indians' name for that time of year).

Four of the five men rode the 100 miles in less than 24 hours. Enthusiastic friends felt that these men had done the impossible, and wished that they had gone along with them. The usual question was, "Will you do it again next year?" And so the 100 mile ride was born and became an annual event, all riders completing it ("One Hundred Mile-One Day-One Horse-One Rider") receiving a silver and gold belt buckle in honor of their accomplishment. Then in 1959 Will Tevis (the "Iron Man", who held several endurance records himself—257.5 miles in 11½ hours, 200 miles indoors, etc.) and his brother donated the sterling silver Lloyd Tevis Cup to be presented to the rider making the fastest time (with his horse in good condition). And in 1964 Louis Haggin donated the James B. Haggin Cup, in honor of his great grandfather, a noted race horse breeder, this cup to be presented each year to the horse among the first ten finishers that came through in the best condition.

The Tevis Cup ride has helped to pave the way for many more. Today there are quite a number of good endurance rides, of varying distances. There are also quite a few competitive trail rides. Perhaps we'd best take a quick look at the difference. Both are competitive events, both are long-distance rides, and both require horses to be in good condition for traveling long distances in a short time. But there the similarity ends. The endurance ride is a race; the rider who gets his horse from start to finish in the shortest time, while still maintaining his horse's physical condition to pass the veterinary inspections is the winner. The competitive trail ride is not a race, although the riders must usually complete the ride within a specified time limit. The judging in a competitive trail ride often places some emphasis on good horsemanship, well-trained and well-mannered horses, as well as good physical condition. There is often a point system in which the horse's condition, time, and so on are all taken into consideration when judging the winner.

The North American Trail Ride Conference (NATRC) includes about 30 different rides in different parts of the country that are all carried out and judged

Salmon River Competitive Trail Ride, 1971

Mary Lou Forbis from Riverton, Wyoming, a woman crippled from birth who had only been able to walk (with the help of canes) for about a year before this ride

Here she mounts her horse with a bit of assistance—but once aboard, she does fine

under a strict set of rules. Participants can accumulate points from ride to ride in their various divisions (heavyweight, lightweight and junior, and some rides also have an amateur division) and at the end of the season these points are totaled and awards are given. Endurance rides are an individual thing. Each one is individually governed. Until 1971 there was no overall body of rules nor any organization concerning these rides, but in 1971 the American Endurance Ride Conference came into existence, which is an organization which helps to standardize rules, set up dates for the various rides so that they won't conflict with other rides already in existence, and helps spread information about endurance rides. Let's take a quick look at some of the differences between competitive trail rides and endurance rides.

NATRC rides have a minimum and maximum time allowance and a rider who comes in too early or too late is penalized. Endurance rides have no minimum time limit and the rider is disqualified if he takes longer than the maximum time.

NATRC rides are usually judged on condition of the horse (40%), soundness (40%), manners (10% or 15%) and way of going (10% or 5%). The winner is usually (all other things being equal) the horse that comes in closest to the minimum required time in the best condition. There are usually awards for horsemanship in the various divisions as well.

Endurance rides are judged on speed, soundness and condition, and ability to proceed. Manners or way of going are not counted, and a horse is not disqualified on these points unless he bothers other horses along the trail or is

Eddie Forbis trotting down Withington Creek toward the noon stop.

Kelly Robbins, age 78, from Hamilton, Montana, riding his horse Copper as he approaches the noon stop. Kelly was the oldest rider on this ride. Our dog Pepper is in the foreground

Noon stop on Withington Creek

too nasty to be examined by the vets. In both endurance rides and NATRC rides there are check points where the horse's condition is checked by veterinarians (taking temperatures, pulse and respiration rates) and he is checked for lameness. The rules of both rides are to protect the horses, pulling any horse from the ride that is not in fit condition to continue.

An endurance ride is usually longer than a competitive trail ride. It is an athletic event in which the same horse and rider cover a set course, usually 50 to 100 miles, with the first horse to finish in acceptable condition being the winner. There is no minimum time limit. There is an award for the best conditioned horse as well as for the horse that finishes first. There are no weight divisions. At the end of each year a national champion endurance horse and rider are selected according to a point system, the points being earned on any AERC sanctioned ride. Points are earned for every ride finished and also for top placings. An endurance ride is controlled by licensed veterinarians. This rule is to protect the horses, for only a trained person is capable of judging the horses' fitness before and during the ride. Points toward winning the Endurance Horse and Rider of the Year trophy are earned by the horse and rider as a team. If a rider uses a different horse, he cannot add those points to any tally earned with another horse.

If the first horse to cross the finish line on an endurance ride is in no condition to continue on, he will not be the winner. This rule is to protect the horse from an over-eager rider who would use up the horse in order to be first across the finish line. Once in a while this happens; the first horse across the finish line is not declared the winner because the vets feel he was used excessively or came through in unsound condition.

Conditioning a horse for a competitive trail ride or an endurance ride is a challenge. It takes time; you can't begin a few weeks before the ride and hope to have the horse ready. Conditioning takes good management, good feed, and lots of work, built up gradually.

If you intend to compete seriously in this sort of riding, you should first of all choose a good horse. Almost any kind of horse (as long as he is sound) will do all right in some of the shorter rides with good training and conditioning. But if you are going to tackle some of the longer, tougher rides, take special care when selecting your mount. Choose a horse that is at *least* five or six years old. If a horse is mature you won't be so apt to injure him with hard work and he will hold up better under stress and strain. Choose a horse of good conformation—good sloping shoulders and pasterns, correct feet and leg angles, etc. Conformation is discussed more thoroughly in an earlier chapter. The horse with good conformation will hold up better than the poorly conformed horse; he has better "shock absorbers" and is not so likely to go lame or unsound under strenuous and continuous work. Also, the horse with good slope to shoulders and pasterns will have a freer, smoother gait, more able to "eat up" the miles and more able to give you a comfortable ride. Pay special attention to how your horse travels, because any tendency to interfere will become more pronounced when he is tired. A horse may strike himself toward the end of a long ride even though he doesn't interfere when he is fresh.

Horses being checked (temperature, pulse, and respiration) upon arrival at the noon stop

Patricia Murphy on her Arabian gelding Roco Snip (from Boise, Idaho) leaving the noon stop

If you are a serious endurance rider, look hard at the Arabians and part-Arabians. These tough little horses seem to have a corner on guts and stamina and weight-carrying ability for long strenuous rides. This is a general statement, of course, for you can find outstanding individuals in every breed who will make endurance horses too; Arabians are just more numerous in the winner's circle at the end of the tough rides. The big, stout horse has advantages over the little horse in heavy roping or other events where size, weight and strength count, but out on the long trail, the little horse has the advantage over the larger animal. There is an old saying that "A good little horse can beat a good big horse" and in endurance riding it seems to hold true.

A big horse has a lot to pack around. All that muscle may be fine for a quick sprint, but on a 50-or 100-mile hike it can get to be a bit burdensome. Exertion produces energy, and also waste products. The conversion and elimination of these wastes, especially in the muscles, seems to be more efficient in the lighter-muscled horse. The by-products of the production of energy accumulate rapidly in the muscles during exertion and these must be carried away by the blood stream. In the heavily muscled horse this clearing away process is slower, and the heavy muscles may create too much sarcolactic acid to begin with. Azoturia and "tying up" occur more often in the heavier horse; the small, lighter-muscled horse is not nearly so prone to this type of metabolic problem.

The smaller horse, especially the Arabian-type horse with his short, strong back, has more weight-carrying ability for his size than does the larger horse.

Leaving the noon stop: Heidi Smith on Khamette and Peg Platt from Hamilton, Montana, on Popsy, who was Grand Champion horse

Kelly Robbins and Copper coming into checkpoint 3 on Mulkey Creek

Trail riders climbing a hill where the trail goes through our cattle range

Some of our cattle watch as the riders wind their way through our hills

Look at the various horses that have won the tough endurance rides over the years and you'll see a lot of small Arabians in the 750-900 pound bracket, and most of these were carrying good-sized men. So do not underestimate the small horse when selecting an endurance prospect.

In my own opinion, guts count for as much in determining the stamina of a horse as will conformation, soundness etc. You want a horse that is well built, a good traveler and absolutely sound, but then you also have to have drive and willingness to make a top endurance horse. If you have to choose between the "perfect" horse in bodily form and a maybe not quite so perfect horse with guts and willpower and energetic drive, choose the latter if you can condition him carefully and keep him sound. To compete successfully on a long ride, a horse has to *want* to go. If he hasn't got the inclination or ambition—if you have to urge him every step of the way—he won't get as far and you'll be beat before you ever get near that finish line. You want a horse that travels willingly and easily, not one that wears you out trying to make him go.

This is what made our Khamette (half-Arabian, age 12 at the time) such a natural in the 1971 Salmon River Competitive Trail Ride and helped her and Heidi to place second in the junior division and win the trophy for best Idaho horse. She was not conditioned very long (one month) and hasn't the best conformation (she is bench-kneed, pigeon-toed, has splints and windpuffs), but she makes up for all that in will to go and just plain toughness. This mare has a jaunty, fast, ground-covering walk, and a powerful, long-striding trot that she can keep up for miles. This mare might not hold up on a 100-mile ride with her type of conformation, but her temperament and go-power made her ideal for a shorter one, and she did admirably for a first attempt.

In the summer of 1972 Heidi conditioned her six-year-old Arabian stallion Surrabu for the ride, and I conditioned Nikki. Nikki is a tough, 900-pound little mare that had only been used for ranch work—range riding and cattle work—and that was the only kind of conditioning she got, but with a bit of emphasis on trying to lengthen out her trot when we covered distance riding range. She's the most willing cow horse I've ever ridden, but she's just an "average" traveler, and we had to transfer some of that cow-chasing enthusiasm into her traveling for her to make the grade in a 6½ hour 35-mile ride. Like I say, enthusiasm in a horse makes all the difference, and in that aspect, Khamette is the better traveler. But Nik has straighter action and more agility and surefootedness, so I tried to spark her enthusiasm for this additional role in life. She's put in many a 10- and 12-hour day in the hills and working cattle at high speeds over difficult terrain, so I knew she had the stamina; it just took some work on her traveling gaits, to extend her walk and trot. Nikki and Surrabu both made it through the ride in good shape. Surrabu placed third in his division (junior), but Nik didn't place because she pulled a muscle sliding down a hill and went temporarily lame (for about 8 miles) on her left hind leg. Even though she didn't bring home any ribbons, we felt a sense of accomplishment just in finishing the ride and discovering that a 10-year old cow horse can change roles and make a competitive trail ride.

Nikki got all her conditioning riding range and moving cattle

How does a person go about conditioning a horse for an endurance ride or competitive trail ride? First of all, you start early, preferably several months before the event, or even longer than that for one of the longer, tougher rides. If the horse is quite soft when you begin his conditioning, you will want to start out rather slowly and gradually; a soft horse is more likely to become overworked or injured than a horse whose tendons and muscles are in good shape. Start out with short rides at the walk, gradually legging up the horse and strengthening his muscles. Then gradually add longer rides, with more trotting, and finally some galloping to help build up his wind. As the horse becomes fit, you will pay special attention to his trot, for this is the most important gait for the long-distance horse. The trot is a well-balanced gait, and is also the most effective gait the horse has for maintaining good speed over long periods of time without becoming over-tired. You'll want to build up your horse's trot so that he moves out freely and easily and willingly, at good speed.

Early in your conditioning program you should have your horse wormed, and his teeth checked. He will do a better job of utilizing his feed and will get in good condition faster (and be able to stay in good condition better) if he is not harboring a heavy infestation of internal parasites or if he has no difficulty eating. Internal parasites can greatly sap a horse's reserve. An experiment was conducted in 1968 at the third annual Bitterroot Competitive Trail Ride (at Hamilton, Montana). A blood count and fecal sample was taken from every horse. For the fecal samples, an evaluation from 0-4 was given to show the degree of infestation by internal parasites, specifically strongyles (bloodworms). There are more than thirty species of bloodworms, and these parasites deprive the horse of nutrition and can affect his blood count. They can also cause colic, as well as being a detriment to his ability to perform well. No horse on this particular trail ride that had a plus-4 reading (meaning a very heavy degree of infestation), and there were several, performed very well at this ride. If you are going to demand a great deal from your horse, give him half a chance by initiating a good worming program early in your conditioning schedule, to knock out those worms and keep him wormed. Nikki and Surrabu were both tubed early in the spring, with a follow-up program of low-level wormer given daily in the grain for 3 weeks to completely break the worm cycles, then fed a daily low-level wormer for the first 21 days of each month.

Feed is also very important as you try to build up the horse's energy and endurance. He should have an adequate, well-balanced diet, the amounts you feed gradually increasing as his work increases. If he is on good pasture, he may only need a daily grain ration, especially if you are still in the early stages of your conditioning program. As you get farther along in your conditioning and are working him harder and longer, add some hay to his diet. Most irrigated pasture grasses make a relatively soft feed and the horse will not do as well as he will if fed some hay. Clover pastures are especially poor if you are working the horse hard and may make him prone to colic. If you are feeding him hay, select the best quality hay you can get—good grass hay or

Heidi and Khamette at the end of the ride—back at the fairgrounds having TPRs taken

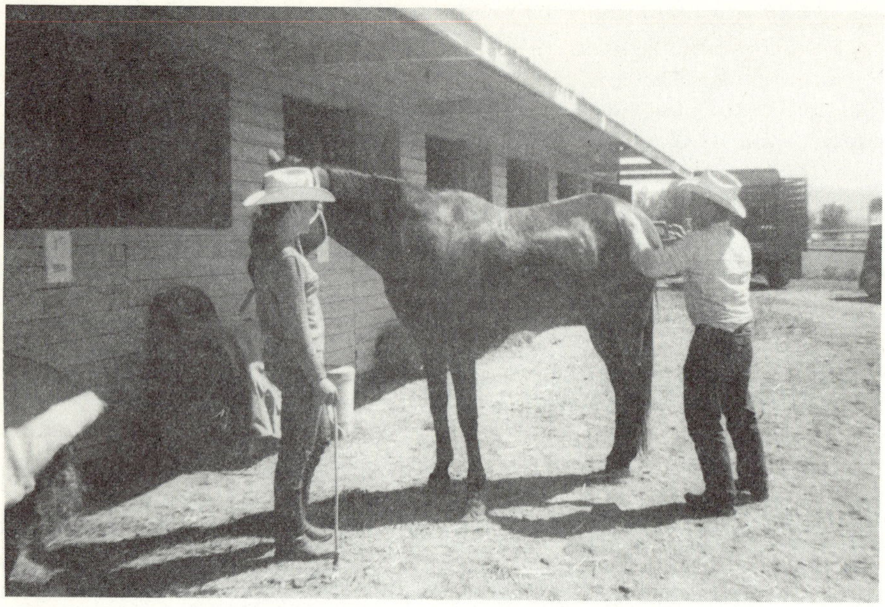

Lynn helps give Khamette a good brushing after the ride

Idelle Moore from Ledger, Montana, crossing the finish line on Skookum Parchico. They placed fourth in the lightweight division and won the Appaloosa trophy

a mixture of grass and legume hay. You may wish to give him some type of vitamin and mineral supplementation as well. And make sure that he always has salt available. The amount and kind of grain you feed him will depend somewhat on your individual preferences and upon the individual horse. Some will need more grain than others. When feeding grain, remember to cut down the feed on days he doesn't work. And on days you work, feed him his grain ration well before you start to ride. It isn't wise to give a horse a big grain feed and then work him hard immediately after. Give him an hour or two to start digesting it.

As you begin your conditioning program, you should have your vet check the horse to make sure he is healthy and sound. A blood count is a good idea if you are seriously wanting to get your horse in shape. A complete blood count will show what condition your horse is in, whether there is any indication of anemia. If his blood count is low, there are vitamins and injections that can be given to help correct the problem. The red blood cells are important in transporting fresh supplies to the muscles and in removing waste materials from those muscles. If the horse's blood count is low, he won't be able to get in top shape. A blood count at the start of your conditioning program and another later on as the horse is getting in shape should help tell the story and let you know how the horse is doing in that area, giving you time to work on the problem if he is low.

Hot blooded horses—Arabians and Thoroughbreds—have higher blood counts than cold-blooded horses. In Arabians and Thoroughbreds a normal blood count is often above ten million, while colder-blooded horses may be lower than seven million. This is another reason why hot-blooded horses are better suited for long rides—they are better able to keep muscles supplied with energy and the wastes adequately carried away, for they have more red blood cells with which to do it.

Notice the color of your horse's gums and the inside of his nostrils. If they are a healthy pink, he probably has good blood. If they are pearly white, you have problems. Lack of color in gums and nostrils usually means a low blood count or overwork. Low blood count can be due to lack of vitamins or minerals, or too many parasites.

While you are working a horse in preparation for long rides, the horse should have a good appetite. If he loses his appetite, there's something wrong. He may be sick. Or he may be overworked. Loss of appetite during the conditioning period is often the first sign of overwork and leg-weariness; leg weary is a term that means the horse is psychologically as well as physically tired—he has been worked so much that he doesn't have any enthusiasm or drive. He doesn't care anymore. This can be a real problem and it may take many weeks of rest to cure it. If a horse is droopy and lethargic after a long ride, you'd better find out why. If he is still perky and alert and is looking around for something to nibble on, that's good. Good appetite is always a good sign.

The real test of your abilities in conditioning a horse comes in your judgment of how much stress you give your horse. You must build him up gradually, but you must stress him so that he becomes able to do more and more work. You don't want to overdo him, yet you want to challenge him, to stress him, so that he continually develops his muscles and wind. In other words, you travel a fine line, with an intimate feel of how much your horse can take. You have to know his abilities and capabilities and to be able to tell how he is coming—how much more he should be able to do this week than he did last week, without hurting him or discouraging him. Keep him interested, don't overdose him. Don't do the same old thing on the same old trail day after day. Vary your routine, travel different country, tackle new and different obstacles.

You can partly condition a horse by giving him workouts at the gallop, building up his wind, but your main aid to conditioning that horse will be the 10 to 30-mile hike—a lot of it at the trot as your horse gets in shape. And climb a few hills. A two-mile gallop on days you don't take a long ride can help keep him in shape, but it is really a poor substitute for miles on the trail.

Be careful not to overcondition your horse. If you are conditioning him for a 50-mile ride, don't ride him that far every day. Take him on rides of 25 to 35 miles, say two or three times a week, with shorter rides, gallop workouts or days of rest spaced in between. You are shooting for a maximum effort the day of the ride. Don't use up that horse before the ride; save something for the ride. You want to peak your horse that day; you want to have him coming along great and built up to where he can take it, to where he can give you his maximum effort the day of the ride. You don't want him leg weary and discouraged when he gets there. You are trying to bring him to his peak physically and psychologically so that he will perform at his best on the day of the ride.

You will want to learn how to take your horse's pulse, respiration and temperature. He must be accustomed to this sort of thing so that he will not object to the handling the vets and TPR girls give him the day of the ride. Also, a knowledge of his pulse, respiration and temperature can help you in your conditioning program to learn how he is coming along. Respiration rate can be easily seen by watching the horse's nostrils or flanks. Pulse can be taken with a stethoscope at the horse's girth, or felt with the fingers at the jaw, root of the tail, the pastern area just above the heel bulbs. The temperature is taken rectally (you will want to tie a string to the end of your thermometer). Find out what your individual horse's rates are when he is at rest. Then you can take readings when he is stressed, to see how high they go, and how quickly they return to normal. Heartbeat in a healthy horse can vary from about 36 to 50 beats per minute, 40 is about average. Respiration at rest is usually about 8 to 24, with 16 breaths per minute being average. Normal temperature in a horse can vary from 99 degrees to 100.5 (Nikki's normal temperature is 99.6 and Surrabu's is 100.3).

When the horse is working hard his TPR rates go up (respiration may go up to 90, temperature may climb to 102 or even 103 if it is a hot day) and pulse rate may go up to 120), but if he is in good condition, these rates should begin to drop as soon as he stops to rest and should be down to nearly normal within ten minutes. As you condition your horse you will notice that it takes him less and less time to cool out after a workout, as he begins to get in good shape. If your horse's pulse or respiration does not drop markedly, or increases after a ten-minute rest period, you are working your horse too hard. In any horse, if the respiration rate becomes faster than the pulse rate—when the horse is panting and breathing very rapidly and shallowly—he has reached a point of exhaustion in which his overall condition is endangered. If he is pushed beyond this point, his temperature (which up to then has been above normal because of exertion) will drop suddenly into subnormal ranges and he will be in a state of shock. A horse should never be ridden to this point; not only his health, but his life is in danger.

When a horse is in good condition his TPRs will return to normal very quickly after exertion. This recovery rate is what the TPR crew and vets are checking during a competitive ride. It isn't so important what the horse's reading is when you go into a check point; the important thing is how much the TPR rates drop and how close they come to normal within ten minutes.

When you are making a long ride, rest the horse whenever you feel he really needs it. And if you are getting off, loosen the cinch for a few minutes, lift the saddle a bit so that air can circulate beneath the pad to cool his hot back, and let him catch his breath. On a long ride it is all right to let the horse drink when he is thirsty, as long as he doesn't drink too much, and as long as he keeps moving afterward. We like to let our horses drink at every stream or water trough. It refreshes them and then they don't try to tank up all at once when they get hot and dry. It's not good to let a hot horse drink a great deal of cold water all at once, especially is he is going to be standing around afterward.

Help your horse all you can, by riding correctly and staying well balanced. You have to condition yourself, too, so that you are able to take a long ride without getting tired. You can't have someone else condition your horse for you and then expect to be able to ride that horse well on a long ride. You've got to get in shape along with the horse. Don't sag or slouch in the saddle. A tired rider makes a tired horse, creating a drag on him that can hurt the horse's kidneys or give him a sore back. Keep your weight well forward going up and down steep hills; he can carry it with less effort that way. If you hang back, you not only are a drag, but you hinder his loins and hindquarters. Keep the horse collected in rough country, especially going downhill, with his hind feet well under him; he'll be more sure-footed.

Make sure your saddle fits him properly. You might get by with ill-fitting equipment if you don't ride the horse very often, but the daily work you'll do conditioning him for endurance can make a lot of sore places if your gear doesn't fit or isn't kept clean. Make sure your gear fits him, use a saddle pad that is thick and soft, keep it very clean, and your horse always clean and well groomed, and you'll have a minimum of problems. You may want to use equipment that is very light, especially if it makes a difference on your weight division. It's better to be in the lighter end of a division than the heavy end. Also, your tack is dead weight and you want to make your horse's load as easy as possible. This is why many endurance riders use flat saddles, McClellans, and so on. On a long ride, every pound counts. Choose equipment that is strong, but light.

You'll also have to pay special attention to your horse's feet as you condition him. Sound feet and legs are a must or he will never hold up. The horse should have good feet to begin with, with wide heels and concave soles. Keep him properly trimmed and shod. You don't want hoof cracks, brittle feet, thrush or soft feet. If he is kept in a corral or soft pasture, take him out in the rocks and break him in gradually. He needs his feet toughened up before he tackles a trail ride, for there is usually some rocky going on most competitive trails. If you live in a dry climate you may need to use a hoof dressing or conditioner to keep your horse's feet from becoming too dry and brittle.

Your shoeing program should be well planned. Your horse's shoes may be the reason for success or failure on a long endurance ride. If your horse's feet are not in good shape, or are poorly shod, he may go lame on you. You will want to shoe him about every six to eight weeks, depending on his rate of hoof growth; and if he wears shoes out more often than this, you will need to apply some sort of hard-surfacing to the ground surface of the shoe to make it last longer. If shoes are put on too often, your horse will run out of hoof for new nail holes. The last set of shoes you put on the horse before the ride should definitely be hard-surfaced, and should be put on about three weeks before the ride (at least two weeks). You want your horse to have plenty of hoof and sole for the ride because he may have lots of rocky ground to cover. It's also a good idea to take along a set of shoes (shaped to fit your horse's feet) in case your horse does lose a shoe on the ride. When your horse is shod, leave as much frog and sole as possible. And fit the shoe closely to the foot, especially on the inside hoof wall; you don't want the horse to strike himself with a shoe when he is tired near the end of a long ride. Shoe him as carefully as possible, to hinder his way of going the least, or to correct any minor defects in gait that he might have.

As you condition the horse, you may have to work on certain aspects of his training as well as improving the fitness of his muscles. You have to accustom him to new and different things, in preparation for what he will have to go through at a competitive ride. If you'll have to haul him some distance to get to the ride, make sure he is well trailer trained and accustomed to traveling. You don't want him so upset or worn out from the trip that he won't be able to perform well on the ride. If you are going to take up competitive trail riding, remember that your horse will be judged on manners as well as condition, and spend some time with him getting him used to thorough handling. You don't want him to kick a TPR girl or bite the judge!

Teach him to stand still, and to tolerate people walking around him, touching him, picking up his feet. At checkpoints the TPR crews will put a stethoscope at his girth and put a thermometer in his rectum. The horse should stand quietly for this with no fuss. The vet will feel his back (especially over the loins) and other places (such as the girth) after the ride to check for soreness or swelling. Some vets may want to check his teeth, look at the membrane around his eyes, feel for sores around his ears, and so on. If any of this bothers your horse, you'd better work on it before the ride and get him used to it so he'll take it like a gentleman.

Get him used to being around other horses, and also get him used to working alone. He'll have to be able to travel alone on the trail, able to settle right down to work. And he should pay little attention to other horses that pass him or that he passes. If he wastes a lot of energy fretting and worrying over the other horses, he'll get tired sooner. This was the problem I faced with Nikki on her first trail ride; most of the riders were trying to make fast time on the first part of the ride, and Nikki got pretty excited with all the horses around her and going past her. It took a few miles for her to settle down and realize that she should travel at her own speed regardless of what the other horses were doing.

You'll also want your horse to be accustomed to being in a stall or tied up overnight. At some rides the horses are stabled in stalls the night before and after the ride; at other rides they are just tied up. Get your horse ready for this. You want him to be able to rest the night before the ride; you don't want him fretting because of strange conditions. If he's never been in a barn before in his life he may get claustrophobia in a stall and go completely buggy. Nikki was this way and I was glad I tried putting her in a barn a couple weeks before the ride. I found out how nervous it made her, and still had time to work on the problem, putting her in the barn for only a little while at a time until she got used to it.

Tie your horse up all night a few times before the ride. On some rides, the horses will be tied up at night and during the lunch hour. Tie him to a trailer or a tree and give him his feed and water there. Give him about four feet of rope so he can reach the ground to eat, drink or even lie down if he wants to. On the ride, you can bring along some straw and bed him down well where you tie him. If he is accustomed to being tied, he'll lie down at the end of his rope and rest almost as well as if he were loose in a corral. Most horses will urinate sooner if they are standing in grass or soft bedding than they will on hard ground; they don't like the urine splashing on the hard ground. If you tie your horse overnight, you'll probably be sleeping right there close by to take care of him and able to make sure he doesn't get himself in a predicament.

Another thing to get him accustomed to is traffic. If the trail ride leaves from a fairgrounds, the first part of it may have to be along a highway, and maybe the last part as well. This was another problem for Nikki. She had never been around a busy road, never been off a ranch in her life. It took a bit of adjusting to get used to big trucks roaring by, but she took it quite well.

Get him used to walking through water, over downed timber, and any other natural obstacles you can think of—perhaps a bridge, steep gullies, big rocks, and so on. Take a hoof pick along with you; the kind that folds up and fits in your pocket is very handy.

As you condition him, speed up his trot and learn to post if you haven't already. Posting is easier on you and the horse both and keeps you better balanced over the horse's back where he can carry your weight the easiest. Your horse should be developing a fast, long-striding trot that really covers the ground, and you shouldn't have to urge him much. He should be getting into the spirit of things and learning that when you get out and travel, you really get out and go. You can alternate gaits to help keep him fresh. Walk up the steeper slopes, walk after a particularly long fast trot, canter occasionally where the going is good. A change of gaits is restful to a horse as long as the canter is not overdone. Basically, you are working on developing that horse's trot to its very best, and alternating that trot with a good walk that is restful, yet ground covering. Find out how many miles per hour your horse can walk and trot, so you'll have a good idea about gauging his progress on a ride—so you'll make your time limit with ease.

On a long ride, the horse should urinate at least once. If he doesn't, find out why. Some horses need to stop and relax a bit before they will urinate. If your horse is one of these, plan a strategic rest stop somewhere along the

End of the ride. Horse on the right is Popsy, ridden by Peg Platt of Hamilton, Montana. Popsy placed first in the lightweight division, won the awards for the Best Out-of-State Horse and Best Unregistered Horse, and Grand Champion of the ride. Peg also won the Sportsmanship Trophy

way, even take off your addle if need be and give the horse ten or fifteen minutes.

After a lot of time spent in the saddle, be careful how you take the saddle off. After riding several hours there has been quite a bit of pressure on the horse's back and the blood has been somewhat pressed out of that area. If you get off suddenly and take your saddle off, this pressure is released all at once and a lot of blood rushes into the area. This can sometimes cause permanent lumps and bumps on the horse's back. If you've been on a long ride, get off and lead the horse that last bit of the way before you take off his saddle, relieving some of the pressure on his back. Or, loosen the cinch a little and lift the saddle a few times to let blood begin to return to the area. Then take the saddle off a few minutes later. Massaging the horse's back after the saddle is removed can prevent heat bumps and can also help to remove them if they do form.

You may be aiming at some competitive trail rides (these are anywhere from 25 to 60 miles long, the 60-mile rides usually being two-day events) or at the longer endurance rides. The conditioning is the same; the difference is only that of degree. You may be getting your horse ready for one particular ride only, or you may be preparing him for several over the course of the season. If you are shooting for a national championship or a high-point regional award, you may have to take in rides from early spring to late fall and this means keeping your horse fit and up to his maximum potential from March til November. But this is hard to do. If you get your horse in top condition and do well on his first few rides, he may become over-conditioned and leg weary and start to slip. If you are going to participate in just one ride, try to peak your horse for that one ride, developing him so that he'll reach his top potential the day of the ride. If you are going to try to make rides all season, bring him almost up to his full potential and try to maintain that level of fitness throughout the season. Keep him fit without overconditioning, so you can maintain him at a level of condition that will allow him to keep campaigning without burning him out. Even if you don't win any grand championship trophies, if you place second or third in the majority of rides you enter, you will probably have enough points to win a national or regional trophy at the end of the season.

When is a horse ready for a 100-mile ride? This question can only be answered for each individual horse in terms of his conditioning—when he has reached the point that he can travel for long periods of time at good trotting speed without becoming excessively tired. In other words, after a short rest his recovery rate is good and he has the ability and willingness to continue on.

Next, if you think that your horse is ready, ask yourself the question: "Am I ready? Am I horseman enough to be able to measure my horse's capabilities without hurting him?" Endurance riding tests horsemanship every bit as much as it tests the horse. The rider who uses good judgment will do a better job. He has to be able to sense his horse's fatigue and give him rest or

Frank Platt on Liddle Fiddle. They won the heavyweight division

Nikki and I upon arrival at the noon stop—1972 Salmon River Competitive Trail Ride

TPR girls checking Nikki after her arrival at the noon stop

slower work when he needs it. This kind of rider will always bring in a fresher horse than the rider who cannot sense or who ignores the horse's fatigue. The rider who is tired himself often fails to judge his horse accurately and often "runs out of horse." Good seat and hands, balance and weight are all very important. The better you are able to ride, the better you can help your horse. On a long ride, every advantage counts.

Just how much can you ask of your horse? On a long ride, you try to feel constantly for your horse's capabilities as you go along. On a long ride, the horses that fail to complete the ride are horses that go lame or are used up too soon, either due to lack of conditioning or lack of judgment on the part of the rider, overusing his horse on certain parts of the trail and not saving enough for later. Judgment in pace is very important, and experience is the best aid to learning this kind of judgment—experience and knowing your individual horse very well. On a long endurance ride you will learn more about your horse in that one day than you might learn in a whole lifetime of ordinary riding. Endurance riding takes a special kind of rider with a special way of thinking when it comes to using horses, for the endurance horse is used and conditioned to a much greater extent than the average horse.

It takes time, and a lot of hard work, to get a horse ready for a competitive event like this. And it's not unusual to have a few setbacks or major problems along the way. But this is part of what makes the challenge. There's always the element of uncertainty, and of risk, especially if your horse does have a specific problem that you are trying to overcome. For instance, there was the spring and summer of 1971 when my sister was trying to condition Khamette in spite of a nasty saddle sore. She conditioned the mare bareback almost entirely (which greatly aided her seat and balance, for she had to do a lot of trotting and a lot of cow chasing over difficult terrain in the mountains) and gambled on the saddle sore—hoping it would heal up in time to use a saddle the day of the ride, hoping the use of the saddle on the day of the ride would not reinjure the sore, and hoping that Khamette's back would not be bothered by the use of a saddle it had not been conditioned to. The gamble paid off and that ride had a happy ending. Heidi and Khamette finished the ride in good shape and even brought home trophies and ribbons (Best Lemhi County Horse, Best Idaho Horse, 2nd place in junior division, and 5th place in junior horsemanship).

And then there was the summer of 1972. Heidi and Surrabu had no major problems, but Nikki and I had our share. First, in early spring, Nikki cut her heel bulb (right front foot). Nothing deep or serious, just enough to temporarily make her lame and enough to make part of the tissue at the heel bulb slough away. Then later in the spring we struggled through her vaccinations. This mare is somewhat allergic and sensitive to several kinds of vaccines so I gave her shots at separate intervals, not all at once. But she reacted very adversely to every shot that year except her lepto shot; it just produced local swelling that went down in a few days. But her encephaloid shot (for sleeping sickness) made her choke up and cough with any exertion; her lungs were congested for three days, and I couldn't ride her during that time. Her strangles shot made all her legs swell up, and her influenza shot made the muscle where she received it swell up as big as a grapefruit—a swelling that took more than a month to go down. I was very glad that she had her vaccinations early—she had time to recover from them before the trail ride.

A month before the ride I was bringing home a cantankerous black Angus yearling heifer from the range, and had to gallop down a rocky hillside to head her off as she made a dash for freedom. Nikki hit a bad rock with her right front foot and went instantly lame; she bruised her heel and cracked it (same heel she cut earlier in the spring). We managed to bring the heifer home, but that nearly ended our trail riding attempt for that year. I took Nikki's shoes off and turned her out to pasture, rested her, and applied a hoof dressing daily to her heel crack. After ten days she was limping less and I put a shoe on that foot, with no pressure in the heel area so that movement in the crack would be limited, and in a couple more days she wasn't favoring it. So I began carefully to ride her again, going slow and easy at first and praying through every rock pile. Occasionally she would favor it, but only for a few steps. We had just a bit more than two weeks left before the trail ride.

She was still on pasture as I began riding her again and I soon found that she was allergic to clover. She was dull and stumbly on a couple of rides (perhaps she had come close to foundering) and was mildly colicy after another. So I took her off pasture and put her back on hay (well-watered-down hay, since she's a bit allergic to dust) but let her graze ten or 15 minutes a

Heidi and Khamette did all their conditioning bareback in 1971

day (no clover) so she'd have a bit of green feed. We had no more problems in that department.

Then we had some cold rainy weather for about a week. After the first night of cold rain, Nikki had a nasal discharge, and a cough when she exercised. Her temperature was not abnormal so I suspected allergy rather than disease. That was when I decided to put her in the barn, and perhaps it was a blessing in disguise for I found that she was very frightened in the barn; it gave her claustrophobia. I accustomed her to it gradually and put her in the barn whenever it rained. She no longer got wet and chilled and this seemed to solve her respiratory problem. Nikki had more allergy problems that spring than she had at any other time in her life, perhaps because she was pregnant. We bred her a couple months before the trail ride (for her first foal), thinking a pregnancy in such early stages would not make much difference, but apparently it changed her body chemistry a bit and in her case made her more sensitive to things for awhile.

Then four days before the ride I was trotting home from checking cows on the range, coming down a ridge covered with sagebrush, when Nikki suddenly went very lame on her right hind leg for about three steps and then was fine again. I examined her leg and foot and found nothing wrong, so continued home. She did it a couple more times on the way home, always when coming down hill and always when trotting. Nothing seemed wrong; all I could find was a small crack in an old scab on her fetlock joint. Perhaps she struck it with the other foot and caused the severe but temporary pain. Why, I don't know, unless it was because she was going through sagebrush. She never did it again. But you can bet it nearly gave me ulcers, that close to the trail ride and not being sure why she'd done it. The next day I rested her. The following day I took her for a five mile fast trot just to see if she would be OK and she passed the test. The next day was the day before the trail ride and Heidi and I rode our horses the 12 miles to town to the fairgrounds and the preride check, Nikki and I feeling exultant that we were actually going to make it to the ride in spite of all we'd been through.

And we made it through the ride, in spite of our freak accident along the way. It was our first attempt, but it made us aware of our potential. Nikki's TPR ratings were better than the average in her division, and would have been even better if she hadn't had the extra stress of being lame for 8 miles (her finish-line TPR, especially her temperature, reflected that stress, though her rates were good again at the check an hour later). The day of the ride you begin to realize just what your horse can do, and what kind of reserves he has. I had conditioned Nikki carefully, watching her progress chasing cattle on the range, and knew she was a tough little horse. I knew she could make it through the ride. But I didn't fully realize just how well she could do until we were actually out on the trail competing. I was afraid she was overdoing herself (she got excited on the first part of the trail and wanted to keep up with some of the horses who were traveling faster than I wanted to) until we'd passed the first two check points and found out how good our TPRs were;

she was in very good shape and standing it very well. And then we had our bad luck (slipped sideways down a steep hill) and went lame, but Nikki gamely pushed on for ten miles to finish the ride anyway and actually went the last two miles and through the finish-line vet check without limping. But I discussed her lameness with the vet and had him thoroughly examine her to discover what she had done to that leg. His conclusion was a pulled muscle. He commended her ability as a trail horse but we both agreed it would not be fair to place a horse that had been lame. So Nikki and I chalked that first ride up to experience, for we both learned a great deal. She put her heart into that ride better than I thought she would. Yes, it made us aware of our potential, and somehow got into our blood at the same time. I'm afraid those of us who have tried trail riding or endurance riding once are then forever hooked, addicts for life. It's a great challenge, a great sport, where you meet great people. And the nice thing about it for us is that we can condition a horse just doing our everyday work riding range and working cattle.

Those of you who have good horses and enjoy long hours in the saddle really ought to try long-distance competition. Maybe you already are. Really all it takes (besides addiction on the part of the rider) is a good horse that can go on and on and on, and stay sound and happy doing it. Best of luck to you when you try the Tevis Cup!

INDEX